Structural Mechanics

STRUCTURAL MECHANICS

Fifth edition

A revision of
Structural Mechanics
by W. Morgan and D. T. Williams

Revised by
Frank Durka

Chartered Civil Engineer
formerly
Senior Lecturer, School of the Built Environment,
Liverpool Polytechnic

LONGMAN

Addison Wesley Longman Limited
Edinburgh Gate, Harlow
Essex CM20 2JE, England
and Associated Companies throughout the world

First published in Great Britain by Pitman Publishing Limited 1980
Fourth edition published by Longman Group UK Limited 1989
Fifth edition published by Addison Wesley Longman Limited 1996

British Library Cataloguing in Publication data
A catalogue entry for this title is available from the British Library

ISBN 0-582-25199-0

Set by 21 in 10/11 pt Ehrhardt
Produced by Longman Singapore Publishers (Pte) Ltd
Printed in Singapore

Contents

Important tables

Preface

In the preface to the first edition of this book the late W. Morgan and D. T. Williams expressed the opinion that structural mechanics, the elementary stage of the theory of structures, 'is often rendered unnecessarily difficult by a highly mathematical approach and, therefore, the mathematical treatment here has been kept as simple as is consistent with accuracy. In all cases a full understanding of the basic principles has been considered more important than mere mathematical agility.'

Whilst the original authors' simplicity of the mathematical treatment has been maintained throughout subsequent editions, more emphasis has been placed on the relevance of structural mechanics to the process of structural design as outlined in Chapter 1. This approach makes the text suitable for students taking degree and diploma courses in architecture, building and surveying and those in the early stages of civil and structural engineering courses. The text also covers the syllabus of the BTEC Higher National unit 8162E Structural Mechanics.

The initial chapters of this book deal with the concept of forces in their various aspects – concurrence, non-concurrence, moments, etc. – and their effect on structural materials and elements in terms of stress and strain. The significance of the shape of the cross-section of structural elements is considered in Chapter 8. The remaining chapters are devoted to the design of simple structural elements (beams, columns, gravity retaining walls) based on fully updated relevant data as given in the current editions of British Standards and codes of practice.

The preparation of this and the earlier editions of this book has not been confined to a single person, as the title page may suggest. In this work I have been helped by many, either directly or indirectly, and my thanks are due and gratefully given to them all.

None of this, however, would have been possible without the generous co-operation of my wife. I am, therefore, extremely grateful to her for putting up with the inconvenience caused by the whole exercise, and not least for all those innumerable mugs of coffee she so uncannily provided at the right moments when, in spite of the willingness of the spirit, the flesh grew weak.

I hope that many students will find this current edition as helpful in their studies as the first edition was to me and the successive ones to my students. Then all the effort put into this work will have been well worthwhile.

Frank Durka
Liverpool 1996

Acknowledgements

We are grateful to the following for permission to reproduce copyright material:

British Steel: Sections, Plates and Commercial Steels for our Tables 9.3 and 12.4 from *Structural sections to BS 4: Part 1 and BS 4848: Part 2*, available from British Steel, Steel House, Redcar, Cleveland TS10 5QW.

Extracts from British Standards are reproduced with the permission of BSI. Complete copies can be obtained by post from BSI Customer Services, 389 Chiswick High Road, London W4 4AL; Telephone: 0181 996 7000.

1 Introduction

The fundamental purpose of a structure is to transmit loads from the point of application to the point of support and, ultimately, through the foundations to the ground. That purpose must be fulfilled within the constraints imposed by the client's brief, which will inevitably insist on low initial costs and low maintenance costs and may vaguely stipulate the functional needs of the project. Consequently the process of structural design begins with the appraisal of the client's requirements through collaboration with other members of the project design team – the architect, services engineer, quantity surveyor and, in some cases, the builder may be consulted at this stage as well.

The structural design itself is a combination of art and science in that it consists in the creation of a structural form which will accommodate the often conflicting aspects of cost, function, services, not forgetting aesthetics, and be capable of being quantified to produce dimensioned details for the purpose of erection. There are, therefore, two distinct stages in the process. In the first, the structural engineer draws on his experience, intuition and knowledge to make an imaginative choice of a preliminary scheme in terms of layout, materials and erection methods. In the second stage, the chosen scheme is subjected to detailed analysis based on the principles of structural mechanics. The resulting scheme must be consistent with the engineer's basic aims to provide a structure which satisfies the criteria of **safety** and **serviceability** at reasonable cost.

A safe structure may not be easy to define but it must, at least, not collapse under an applied load. The standards of safety have changed considerably since the Code of Hammurabi (c.2000 B.C.) declared the life of the builder forfeit should the house he built collapse and kill the owner. It was not until the 19th century, however, that the concept of a **factor of safety** was first formulated in terms of the ratio of ultimate or, in certain cases, yield stress to the working stress. This is explained in Chapter 6. Some decades later an alternative approach was introduced in terms of the **load factor** based on the ratio of loads rather than stresses, i.e. load factor = ultimate load/working load. Using factors of safety ensures a satisfactory performance under working loads, but only assumes a reasonable margin against failure, while load factors ensure a definite margin against failure and assume a satisfactory performance under working loads.

Both factors operate on the implicit assumption that the determined values both for the loading which the structure is expected to carry, and for the strength of the materials of which the structure is made, remain constant, thus inferring a guarantee of absolute safety. Whilst this may be satisfactory in certain cases, it is generally recognized that a more realistic measure of the safety of a structure can be achieved through an assessment of the **probability** of its failure. In this method, called the **limit state design**, partial safety factors are used, separating the probability of failure due to overloading from that due to variability of strength of the materials.

Whereas safety deals with the structure's ability to carry its loads without unwarranted risks to human life and limb (i.e. with the ultimate limit states, ULS), the requirements of serviceability of the structure (i.e. the serviceability limit states, SLS) insist on its fitness for use without excessive deflections, disturbing vibrations, noticeable cracking or other local failures which may demand costly remedial work.

The limit state design method, recently introduced into the codes of practice, is briefly outlined in Chapters 9 and 10.

Mathematical models   To ensure the safety of the structure it is necessary to provide it with sufficient **strength** and, to safeguard its serviceability, it must have adequate **stability**. In order to assess the sufficiency of its strength and the adequacy of its stability, the structure needs to be presented in quantitative terms before it can be analysed and, since it does not yet exist, its behaviour has to be simulated. A scaled-down version of the proposed structure may seem the obvious choice here, but now even fairly complex structures can be analysed by means of mathematical models.

Fig. 1.1   Loading a spring

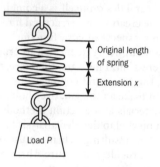

Consider, for example, the behaviour of the steel spring shown in Fig. 1.1. As the load P is increased, the extension x in the length of the spring will also increase. From several pairs of measurements of P and x it will be found that x is proportional to P. Therefore, the relationship between the two quantities can be formulated by means of the equation $P = k \times x$, where k is a constant representing, in this case, the stiffness of the spring. This equation serves as the mathematical model for the behaviour of the loaded spring since it allows the extension in the length of the spring to be determined for any given load.

The behaviour of a structure is clearly more complex than that of a steel spring. It calls, therefore, for more complex models. In fact, even a simple structure may contain too many unknown factors to be analysed completely. In such cases it becomes necessary to replace the real structure by a simplified, idealized version of it.

An example of this simplification can be shown in the case of a steel roof truss. Figure 1.2(a) represents the real structure. Note that the roof loads are applied to the truss at the joints; the members of the truss are made up of standard structural steel sections and are welded together. Figure 1.2(b) shows an idealized truss. Here the roof loads are indicated by arrows directed vertically downwards, and the truss members are shown as simple bars with pinned connections.

These two structures differ in several aspects, the most significant being the substitution of pinned connections for the welded ones. A welded con-

Fig. 1.2 Steel roof truss:
(a) illustrative representation;
(b) idealized representation

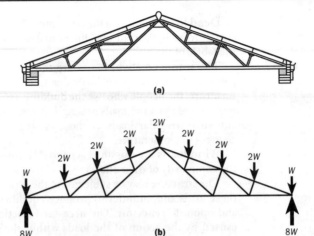

(a)

(b)

nection gives some degree of rigidity which complicates the behaviour of the truss members. A pinned connection, on the other hand, is assumed to be frictionless allowing the members free movement relative to each other, thus simplifying their analysis. This idealization is made on the assumption that the truss would function satisfactorily if the connections were in fact pinned, and that the difference in the behaviour of the members of the real and the idealized trusses is sufficiently small. This has been shown by experiment and experience to be the case for the ordinary range of trusses.

It is, of course, of vital importance that the structural engineer keeps these assumptions within strict limits in the knowledge that each assumption tends to reduce the accuracy of the subsequent analysis. The engineer must be aware, for example, that the application of the model for the behaviour of the steel spring, discussed earlier, is restricted by the fact that the stiffness of the spring remains constant only within a certain range of values for P. The validity of the model, after all, depends on how closely it represents the real behaviour. The engineer is required, therefore, to make grave and far-reaching choices which must be founded on sound experience, supported by an intuitive understanding and fostered by a broad knowledge of structural behaviour. In the search for the appropriate solution, the engineer may find assistance in the codes of practice, which are based on the collective opinion and corporate judgement of experienced designers. They are intended to provide a set of guidelines within which design decisions can be formulated and the analyses carried out in accordance with the principles of structural mechanics.

Forces

The term mechanics, according to a dictionary definition, refers to 'that branch of applied science which deals with the action of forces in producing motion or equilibrium'. In the context of structural mechanics, forces represent the loads that the structure is expected to carry. These are generally classified into three groups: the dead loads, the imposed (live) loads, and wind loads.

Dead loads comprise the permanent loads due to the static weight of the structure itself, the cladding, floor finishes, and any other fixtures which form the fabric of the building.

The **imposed (live) loads** are those produced by the intended occupancy of the building, i.e. loads due to the weight of plant and equipment, furniture, the people who use the building, etc. They also include snow loads, impact and dynamic loads arising from machinery, cranes and other plant, and such irregular loads as those caused by earthquakes, explosions and changes in temperature.

Wind loads are classified separately owing to their transitory nature and the complexity of their effects.

The structure may be subjected to the action of almost any combination of these loads and, in order to produce equilibrium, it must provide an equal and opposite **reaction**. The necessary reaction is generated by the **stress** caused by the action of the loads within the material, and by the ensuing **strain** in the elements of the structure.

The two concepts (stress and strain) are discussed in detail in Chapter 6. They are the direct outcome of the action of **forces** and the **deformations** they produce.

The definition of a force is generally derived from Newton's first law of motion as that influence which causes change in an object's uniform motion in a straight line or, as is more appropriate to structural mechanics, its state of rest.

Most of the loads supported by a structure exert forces on it by virtue of their mass which is subjected to the gravitational pull of the earth. This pull, as was demonstrated in the case of the steel spring in Fig. 1.1, acts downwards, thus indicating that the force has a direction as well as a magnitude and is, therefore, a vector quantity.

As such, a force can be represented by a straight line of a given length indicating the magnitude of the force. The direction of this straight line is drawn parallel to the line of action of the force and denoted by an arrow. Such a line is called a **vector**.

The SI unit of force is the newton, denoted by the capital letter N, and is that force which, when applied to an object having a mass of one kilogramme, gives it an acceleration of one metre per second per second, i.e. $N = kg \times m/s^2$.

Attention must be drawn here to the ambiguity that exists in the use of the term *weight*. In this textbook, weight is considered as the force exerted on a body by the gravitational pull of the earth and is, therefore, stated in force units (newtons). However, in accordance with the Weights and Measures Act 1963, the weights of structural materials are specified in mass units (kilogrammes). Hence, in order to calculate the forces due to dead loads, the kilogrammes have to be converted to newtons by multiplying them by the gravitational acceleration. The value of this acceleration is 9.81 m/s^2, but it is generally accepted that, to simplify the conversion for the purpose of structural calculations, the figure may be taken to be 10 m/s^2. For example the gravitational force due to a 100 kg weight is

$$100 \text{ kg} \times 10 \text{ m/s}^2 = 1000 \text{ N} \quad \text{or} \quad 1 \text{ kN}$$

To counteract the action of the gravitational (and other) forces, the structure must provide an equal and opposite reaction in order to remain in a state of rest, i.e. in equilibrium. The action of the forces may manifest itself in any of the following ways:

- *Tension*   The forces acting on the object act away from each other; the object (string in Fig. 1.3(a)) is being stretched.
- *Compression*   The forces acting on the object act towards each other; the object (brick pillar in Fig. 1.3(b)) is being squeezed.
- *Shear*   The forces acting on the object cause parts of the object to slide in relation to each other (Fig. 1.3(c)).
- *Bending*   The forces acting on the object produce a bending effect as shown in Fig. 1.3(d). Note that this action also causes sliding (shear).
- *Torsion*   The forces acting on the object cause it to be twisted (Fig. 1.3(e)).

Force, however, cannot be observed directly. It is recognized and, therefore, measured by the deformations it causes. The deformation which is of particular interest to the structural engineer is that resulting from bending, i.e. deflection.

The control of deflection is of great importance in safeguarding the stability of the structure. The deflection must, therefore, be quantified. To this end the study of the deflected forms of structural elements or frameworks provides a useful basis for the appreciation of structural behaviour and it is advisable to acquire this helpful ability.

A number of simple examples are illustrated in Fig. 1.4(a)–(g).

Fig. 1.3   Action of forces:
(a) tension; (b) compression;
(c) shear; (d) bending;
(e) torsion

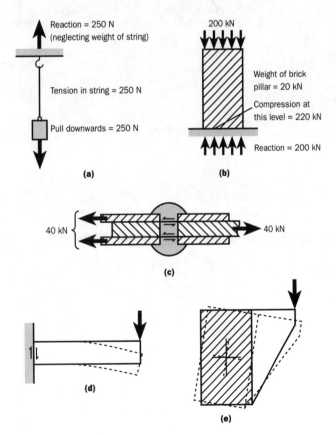

Fig. 1.4 Deflection: (a) straight cantilever; (b) propped cantilever; (c) encastré beam; (d) simply supported beam; (e) continuous beam; (f) three–hinged arch; (g) portal frame

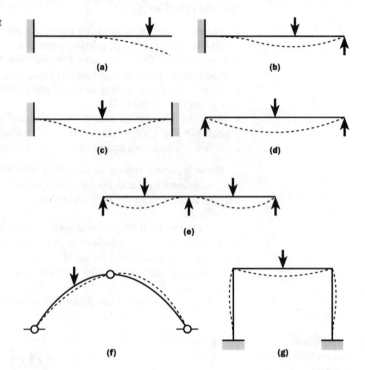

2 Concurrent coplanar forces

All the problems in this chapter are concerned with forces which:

- meet at a point, i.e. they are concurrent
- act in one plane, i.e. they are coplanar

The chapter deals with the graphical methods of resolving such forces into convenient components, and then determining their resultant effect, for the purpose of establishing the equilibrium of the system of forces.

Triangle of forces AB in Fig. 2.1(a) represents a horizontal wooden beam containing a number of hooks. By means of string, a small ring, and two spring balances a weight of 35 N is suspended as shown. It is assumed in this example that the readings on the balances are 20 N and 25 N respectively. The ring is in equilibrium (at rest) under the action of three forces, i.e. the vertical pull downwards of the weight and the pulls exerted by the two strings. The condition can be represented on drawing paper by Fig. 2.1(b) which is called the **free-body diagram** with respect to point O.

Fig. 2.1 Triangle of forces: (a) load diagram; (b) free-body diagram at point O; (c) force diagram

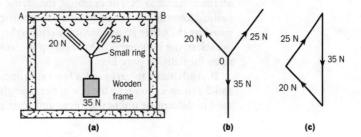

The three lines are called the **lines of action** of the forces and the arrows represent the **direction** in which the forces are acting on the ring. In order to plot on paper the lines of action of the forces, the angles between the strings can be measured by means of a circular protractor. An alternative and simpler method of transferring to paper the lines of action is to hold a sheet of paper or cardboard behind the strings and, by means of a pin, to prick two or three points along each string. Pencil lines can then be drawn to connect these points and thus to fix on paper the free-body diagram.

The magnitudes of the forces can be represented on paper by vectors, i.e. lines drawn to scale. For example, if it is decided to let 1 mm represent 1 N then the force of 35 N will be represented by a line 35 mm long, the force of 25 N by a line 25 mm long, etc.

Now, if the free-body diagram is drawn on paper to represent the lines of action and directions of the forces, and three lines are drawn, as shown in Fig. 2.1(c), parallel to these forces to represent to scale their magnitudes, it will be found that a **force diagram** will be formed in the shape of a triangle. In an actual experiment the result may not be quite as accurate as indicated in Fig.

2.1(c) because of possible lack of sensitivity of the spring balances and the difficulty of transferring accurately to paper the free-body diagram. The fact is, however:

> If three forces meeting at a point are in equilibrium, they may be represented in magnitude and direction by the three sides of a triangle drawn to scale. (See note on p. 14.)

This is the **principle** or **law of the triangle of forces**.

By placing the spring balances on different hooks, e.g. as in Fig. 2.2, further illustrations of the principle of the triangle of forces can be obtained.

Fig. 2.2 Second example of a triangle of forces: (a) load diagram; (b) free-body diagram; (c) force diagram

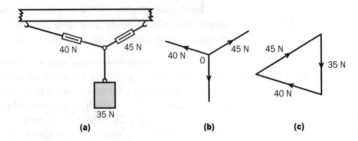

Action and reaction are equal and opposite, yet in Fig. 2.2 it appears that 85 N are required to support a downward force of 35 N. This is because the strings are inclined in opposite directions and react on each other as well as on the force of 35 N. For example, the string which has a tension of 40 N is pulling towards the left in addition to pulling up. This pull to the left is resisted by the other string pulling to the right, so that only a part of the force in each string is effective in holding up the weight. This will be demonstrated more formally on page 13.

It is advisable to carry out a few experiments similar to those of Figs. 2.1 and 2.2 in order to verify the law of the triangle of forces. The law can then be used to determine unknown forces and reactions.

Example 2.1   Determine the tensions in the two strings L and R of Fig. 2.3(a).

Fig. 2.3 Example 2.1: (a) load diagram; (b) free-body diagram; (c) force diagram

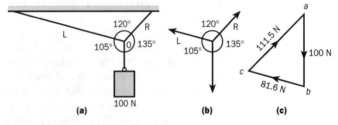

Solution   Point O is in equilibrium under the action of three forces as shown in the free-body diagram of Fig. 2.3(b), therefore the triangle of forces law must apply.

Draw a line *ab* parallel to the force of 100 N to a scale of so many newtons to the millimetre. From *b* draw a line parallel to string L. From *a* draw a line parallel to string R. The two lines intersect at point *c*. The length of line *bc* gives the tension in string L (i.e. 81.6 N) and the length *ca* gives the tension in string

R (i.e. 111.5 N). (A student with a knowledge of trigonometry can obtain the magnitudes of the forces by calculation.)

Example 2.2

Two ropes are attached to a hook in the ceiling and a man pulls on each rope as indicated in Fig. 2.4(a). Determine the direction and magnitude of the reaction at the ceiling (neglect the weights of the ropes).

Solution

Since the hook is in equilibrium under the action of the forces in the ropes and the reaction at the ceiling, the triangle of forces principle applies. Draw *ab* parallel to the force of 250 N to a suitable scale (Fig. 2.4(c)). From *b* draw *bc*

Fig. 2.4 Example 2.2: (a) load diagram; (b) free-body diagram; (c) force diagram

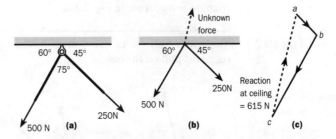

parallel to the other rope to represent 500 N. The closing line *ca* of the force triangle represents to scale the reaction at the ceiling and is nearly 615 N, acting at approximately 7° to the vertical. This force can be called the **equilibrant** of the two forces in the ropes, since the ceiling must supply a force of 615 N in order to maintain equilibrium.

In the last problem, the reaction at the ceiling due to the pulls in the two ropes equalled 615 N acting as shown in Fig. 2.5. If we substitute one rope for the two ropes and pull with a force of 615 N in the direction indicated, the effect on the ceiling will be exactly as before, i.e. the result of the two forces is equivalent to one force of 615 N. This force is, therefore, called the **resultant**. Note also that this resultant is equal in value to the equilibrant and acts in the same straight line but in the opposite direction. The resultant of a given number of forces is therefore the single force which has the same effect on the equilibrium of the body as the combined effects of the given forces.

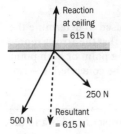

Fig. 2.5 Resultant force

Parallelogram of forces

The law of the parallelogram of forces is in essentials the same as the law of the triangle of forces. Any problem which can be solved by the parallelogram of forces law can be solved by the triangle of forces law, although it may sometimes be slightly more convenient to use the former method.

Example 2.3

Referring to Example 2.2 determine the resultant of the pulls in the two ropes by using the parallelogram of forces.

Solution

Measure a distance OA along line *a* in Fig. 2.6 to represent 500 N and measure a distance OB along line *b* to represent 250 N. From A draw a line parallel to OB and from B draw a line parallel to OA thus forming a parallelogram. The length of line OC, which is the diagonal of the parallelogram (from the point

Fig. 2.6 Example 2.3:
parallelogram of forces

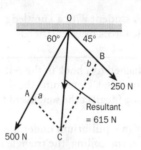

where the two forces meet), gives the value of the resultant of the two forces
(615 N) and also its direction. Note that the triangle OBC is identical with the
triangle of forces *abc* in Fig. 2.4(c).

Example 2.4   Two ropes pull on an eye-bolt as indicated in Fig. 2.7(a). Determine the
resultant pull on the bolt.

Fig. 2.7 Example 2.4: (a) load
diagram; (b) resultant force
diagram

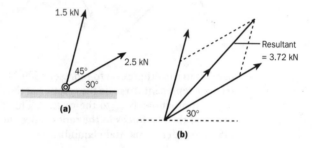

Solution   The solution is indicated in Fig. 2.7(b).
     Note that when the parallelogram of forces law is used to determine the
resultant of two forces, the two forces must be drawn so that the arrows representing their directions both point towards the meeting point or both point
away from it. In other words, a thrust and a pull must be converted into two
thrusts or into two pulls.

Example 2.5   Determine the resultant force on the peg due to the thrust of 2.5 kN and the
pull of 1.5 kN (Fig. 2.8(a)).

Fig. 2.8 Example 2.5: (a) load
diagram; (b) and (c) resultant
force diagrams

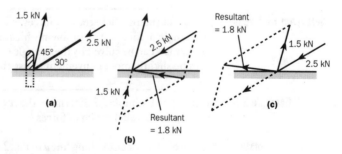

Solution   The resultant is found either as shown in Fig. 2.8(b) or as shown in Fig. 2.8(c).
Note that the effect on the peg is the same whether it is pushed or pulled with
a force of 1.8 kN.

Example 2.6   A rigid rod is hinged to a vertical support and held at 60° to the horizontal by means of a string when a weight of 250 N is suspended as shown in Fig. 2.9(a). Determine the tension in the string and the compression in the rod, ignoring the weight of the rod.

Fig. 2.9 Example 2.6: (a) load diagram; (b) free-body diagram; (c) force diagram

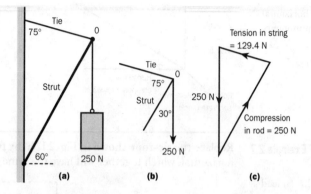

Solution   Point O is in equilibrium under the action of three forces which meet at the point, i.e. the weight of 250 N, the tension in the string and the compression in the rod. The solution is given in Fig. 2.9(c). Note that a tension member is called a **tie** and a compression member is called a **strut**.

Rectangular components   Figure 2.10(a) shows two rods of wood connected by a hinge at the top and supported by two concrete blocks. If a gradually increasing pull is applied to

Fig. 2.10 Rectangular components: (a) load diagram; (b) force diagram of forces meeting at hinges; (c) force at each compression member

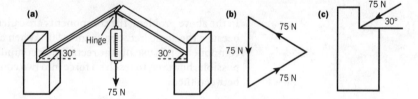

the spring balance, the vertical force necessary to cause lifting of the blocks can be recorded. Assume this downward force is 75 N when the angles are as given and when the blocks just begin to lift. The compression acting along each member is 75 N as obtained by the force diagram, Fig. 2.10(b). Now consider what happens at the bottom end of one compression member. The member is pushing on the block with a force of 75 N (Fig. 2.10(c)). Since this force is inclined it has a twofold effect. It is tending to cause crushing of the concrete as well as to cause overturning. If the force were vertical (and axial), there would be crushing effect only and no overturning effect. If the force were horizontal, there would be overturning effect and no crushing effect. In order to evaluate the crushing and the overturning effects of the inclined force it is necessary to **resolve** such a force into its **horizontal component** and its **vertical component**.

It was demonstrated on page 10 that the resultant of two forces meeting at a point is given by the diagonal of a parallelogram. By a reverse process, one force can be replaced by (or resolved into) two forces. By constructing a rectangle (parallelogram) (Fig. 2.11) with the force in the rod (75 N) as the diagonal, the horizontal and vertical components are given by the two sides respectively of the rectangle. The components can be found either by drawing to scale or by calculation.

Fig. 2.11 Resolving horizontal and vertical force components

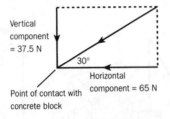

Example 2.7 Replace the tie-rope shown in Fig. 2.12(a) by two ropes, one vertical and one horizontal, which together will have the same effect on the eye-bolt.

Fig. 2.12 Example 2.7: (a) load diagram; (b) horizontal and vertical force components

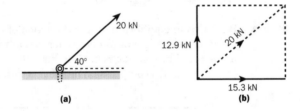

Solution By constructing a rectangle with the force of 20 kN, drawn to scale, as the diagonal (Fig. 2.12(b)) the vertical component (and therefore the force in the vertical rope) is found to be 12.9 kN. The horizontal component is 15.3 kN.

In the above problem, the components are called **rectangular components** because the angle between them is 90°. When a force has to be split into two components it is usually the rectangular components which are required. It is possible, however, to resolve a force into two components with any given angle between them.

Example 2.8 Replace the rope X which is pulled with a force of 500 N by two ropes L and R as indicated in Fig. 2.13.

Fig. 2.13 Example 2.8: resolving a single force into its components

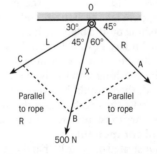

Solution Measure a length along line OB to represent 500 N and assume this line to be the diagonal of a parallelogram by drawing lines from B parallel to ropes L

and R respectively. The forces in the two ropes which together have the same effect on the bolt as the single rope can now be scaled off, i.e.

Force in rope L = 450 N (length OC)

Force in rope R = 365 N (length OA)

These two forces can be said to be components of the original force of 500 N.

Example 2.9 Determine the tensions in the two chains X and Y of Fig. 2.14(a). Then determine the horizontal and vertical components of these tensions. In addition, determine the horizontal and vertical components of the reactions at A and B. Neglect the weight of the chains.

Fig. 2.14 Example 2.9: (a) load diagram; (b) free-body diagram; (c) force diagram; (d) horizontal and vertical components within the chains; (e) reactions at points A and B

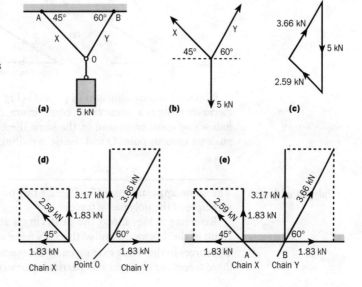

Solution The free-body and force diagrams for the point are given in Fig. 2.14(b) and (c). The tensions in the chains are respectively 2.59 kN and 3.66 kN. The horizontal and vertical components of the tensions in the chains are given in Fig. 2.14(d).

Note that the sum of the two vertical components is equal to the value of the suspended load of 5 kN (action and reaction are equal and opposite). Note also that the horizontal component of the force in chain X is equal to the horizontal component of the force in chain Y and acts in the opposite direction (action and reaction are equal and opposite).

Now consider the reactions at points A and B, which equal the tensions in the chains (Fig. 2.14(e)). Again, neglecting the weights of the chains, the sum of the vertical components of the reaction is equal to 5 kN and the horizontal components are equal and opposite.

Referring back to Fig. 2.10(a) on page 11, it was stated that the concrete blocks were on the point of overturning when the compression acting along each member was 75 N. The horizontal component of this force is 65 N, and this component can be prevented from having an overturning effect by connecting the two inclined members by a horizontal tie member (Fig. 2.15(a)). It

14 STRUCTURAL MECHANICS

will now be impossible to cause overturning of the concrete blocks by pulling downwards at the apex. The tension in the tie member will be equal to the horizontal component (65 N) of the thrust in the inclined member.

Another way of finding the tension in the tie is by applying the triangle of forces law to the joint between the inclined and horizontal members (Fig. 2.15(b)). Neglecting the weights of the members, the vertical downward force of 75 N is balanced by the reactions of the blocks (37.5 N each block).

Fig. 2.15 Tied truss: (a) and (b) load diagrams; (c) free-body diagram; (d) force diagram

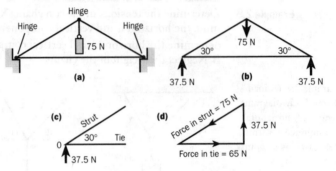

Consider the equilibrium of point O (Fig. 2.15(c)). Note that, when three forces meeting at a point are in equilibrium, the arrows in the force diagram follow one another around in the same direction (Fig. 2.15(d)). The strut is pushing towards point O and the tie is pulling away from the same point.

Polygon of forces

The same apparatus used in Fig. 2.1 can be employed to demonstrate the principle of the polygon of forces.

Referring to Fig. 2.16(a), the small ring is at rest (i.e. in equilibrium) as the result of the downward pull of the 35 N force and the pulls in the four strings. The forces in the strings are given by the readings on the spring balances and these forces are given for one particular arrangement of the strings.

Fig. 2.16 Polygon of forces: (a) load diagram; (b) free-body diagram; (c) force polygon

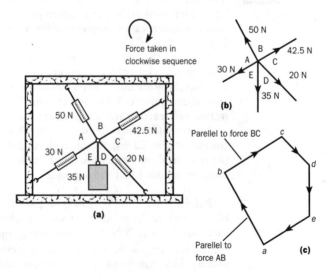

The lines of action of the forces can be transferred to a piece of paper or cardboard (in the manner described on page 7) to give the free-body diagram (Fig. 2.16(b)).

It will be noted that the *spaces* between the forces in the free-body diagram have been lettered A, B, C, etc., and each force is therefore denoted by two letters, e.g. force AB (which separates space A from space B), force BC, force CD, etc. (If the forces are taken in anticlockwise sequence they will be called force DC, force CB, force BA, etc., but it is more usual to take the forces in clockwise sequence.) This method of notation was devised by an engineer (R. H. Bow) about 1870 and is known as **Bow's notation**. It is not essential to use a notation for simple problems but some form of notation is indispensable when a large number of forces are involved, such as in roof trusses (see Chapter 5).

Having plotted the lines of action of the forces in the free-body diagram, the next step is to choose a scale of so many newtons to the millimetre in order to plot the force diagram. Starting with the force AB, a line *ab* is drawn parallel to it, the length *ab* representing to scale the magnitude of the force (50 N). From *b* a line *bc* is drawn in the direction indicated by the arrow on force BC (42.5 N).

The other forces are drawn in the same manner and, if the experiment has been performed accurately, it will be found that the last line drawn parallel to force EA (30 N) will finish at *a*, where line *ea* represents to scale the force EA.

It is not necessary to start the force diagram with force AB. Any one of the forces can be chosen as the starting force.

Further experiments can be performed by connecting the strings to the other hooks in the framework, and it will be found in every case (apart from small experimental errors) that, when a number of forces are in equilibrium, the force diagram will form a polygon, the sides of which represent to scale the magnitudes and directions of the forces.

Study carefully the force diagram or polygon in Fig. 2.16(c) and note that the arrows 'chase each other' in the same sense around the diagram, i.e. the tip of each arrow is pointing to the tail of the arrow in front of it. This will always be true for force diagrams when the forces are in equilibrium, and use can be made of this fact to discover the direction of unknown forces, as in the following example.

Example 2.10	A rod (the weight of which is negligible) is hinged to a support at S and is supported by a tie (Fig. 2.17(a)). From the point O three ropes are pulling with the forces indicated. Determine the forces in the rod and tie.
Solution	The point O is in equilibrium as the result of the action of five forces, i.e. the three given forces and the forces in the rod and tie.

The free-body diagram at point O is given in Fig. 2.17(b).

It is an advantage to start Bow's notation in such a manner that the two unknown forces are last. First, draw a line *ab* parallel to force AB to represent to scale 2.5 kN, then draw lines *bc* and *cd* to represent the other known forces. From *d* draw a line parallel to force DE. Since point O is in equilibrium the force diagram must form a closed polygon. Therefore from *a* draw a line parallel to force EA to intersect the line drawn parallel to force DE. The intersection point of these two lines is point *e*.

16   STRUCTURAL MECHANICS

Fig. 2.17 Example 2.10: (a) load
diagram; (b) free-body diagram;
(c) force polygon

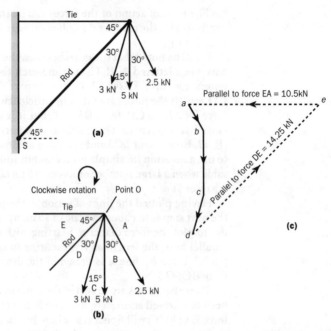

The force in the rod is given by the length of line *de* and is 14.3 kN approximately. Similarly, the force in the tie is given by the length of line *ea* and is 10.5 kN approximately. The arrows on lines *de* and *ea* must follow the general direction; therefore the tie is pulling away from point O and the rod (which is a strut) is pushing towards point O.

Note that the forces in the rod and tie have been stated to be 14.3 kN and 10.5 kN approximately. The forces can be obtained by calculation and may be slightly different from the values given, since calculation is more accurate than drawing. Graphical methods are, however, sufficiently accurate for most structural work and should be used when they are quicker and more easily applied than calculation methods.

The resultant of two forces meeting at a point can be found by the triangle or parallelogram of forces as described earlier. The resultant of any number of forces can be determined by the polygon of forces.

Example 2.11 Three ropes pull on a bolt as indicated in Fig. 2.18(a). Determine the resultant pull.

Fig. 2.18 Example 2.11: (a) load
diagram; (b) free-body diagram;
(c) force polygon

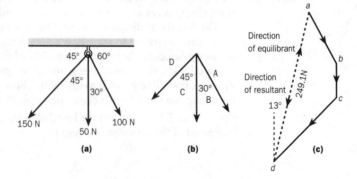

Solution   The force polygon is constructed from *a* to *d* by drawing lines to scale parallel to the three given forces (Fig. 2.18(c)). The closing line *da* of the polygon gives the magnitude and direction of the reaction (equilibrant) supplied by the support. The resultant of the pulls in the three ropes is, of course, equal to the equilibrant and is 249.1 N pulling downwards at an angle of approximately 13° to the vertical.

Summary   *Triangle of forces*   If three forces, the lines of action of which meet at one point, are in equilibrium, they can be represented in magnitude and direction by the sides of a triangle if these sides are drawn parallel to the forces.

Parallelogram of forces   If two forces meeting at a point are represented in magnitude and direction by the two sides of a parallelogram, their resultant is represented in magnitude and direction by the diagonal of the parallelogram which passes through the point where the two forces meet.

Polygon of forces   If any number of forces acting at a point are in equilibrium, they can be represented in magnitude and direction by the sides of a closed polygon taken in order.

Components of forces   A given force can be replaced by any two forces (components) which meet at the point of application of the given force. When the angle between the two forces is a right angle the components are called rectangular components.

Equilibrant   The equilibrant of a given number of forces which are acting on a body is the single force which keeps the other forces in equilibrium. Any one of the forces can be considered as being the equilibrant of the remainder of the forces.

Resultant   The resultant of a given number of forces is the single force which, when substituted for the given forces, has the same effect on the state of equilibrium of the body.

Exercises

Note: All pulleys in these exercises are **smooth** (i.e. frictionless).

1 Determine the tensions in the two chains L and R in the case of Fig. 2.Q1(a), (b) and (c).

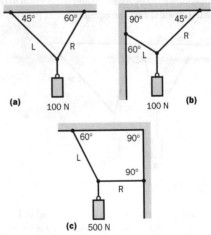

Fig. 2.Q1

2 The angles marked A in Fig. 2.Q2 are equal. Determine the minimum value of this angle if the tension in each rope must not exceed 1 kN.

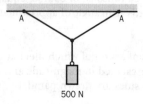

500 N

Fig. 2.Q2

3 The string marked L in Fig. 2.Q3 passes over a pulley. Determine the tensions in the two strings L and R by considering the equilibrium of point 1 then determine the reaction at the pulley by considering the equilibrium of point 2.

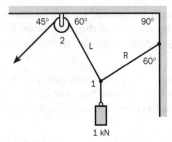

1 kN

Fig. 2.Q3

4 The string marked L in Fig. 2.Q4 passes over a pulley and the system is in equilibrium. Determine the tensions in the strings L and R, then determine the magnitude of the weight W and the tension in the string M. (Consider first the equilibrium of point 1.)

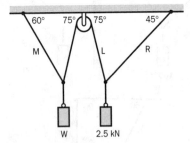

W   2.5 kN

Fig. 2.Q4

5 Two struts thrust on a wall as indicated in Fig. 2.Q5. Determine the magnitude and direction of the resultant thrust.

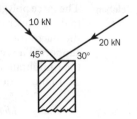

Fig. 2.Q5

6 The top of a pole resists the pulls from two wires as indicated in Fig. 2.Q6. Determine the magnitude and direction of the resultant pull.

Fig. 2.Q6

7 A bolt resists the pull from three wires as indicated in Fig. 2.Q7. Determine the resultant of the two pulls of 5 kN and 2.5 kN then combine this resultant with the remaining pull of 4 kN to determine the resultant pull on the bolt.

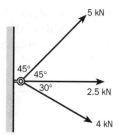

Fig. 2.Q7

8 Referring to Fig. 2.Q8, determine the value of the angle A so that the resultant force on the wall is vertical. What is the value of the resultant?

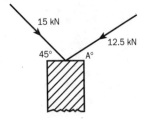

Fig. 2.Q8

9 The top of a pole sustains a pull and a thrust as shown in Fig. 2.Q9. Determine the resultant force.

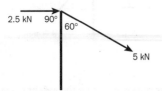

Fig. 2.Q9

10 Two men pull on ropes attached to a peg as shown in Fig. 2.Q10. Determine the resultant pull on the peg.

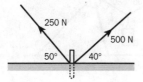

Fig. 2.Q10

11 In trying to move a block of stone resting on the ground, one man pushes and another man pulls as shown in plan in Fig. 2.Q11. Determine the resultant of the two forces.

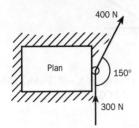

Fig. 2.Q11

12 Determine the resultant pull on the bolt in Fig. 2.Q12 and the reactions at the pulleys.

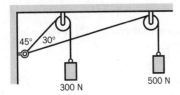

Fig. 2.Q12

13 Determine the forces in the tie and strut in the case of Fig. 2.Q13(a), (b) and (c) neglecting the weights of the members. All joints are hinged.

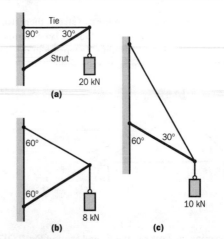

Fig. 2.Q13

14 Referring to Fig. 2.Q14 a weight of 5 kN is suspended from A by a flexible cable. The rope C is parallel to the member AB. Determine the force in the cable, the angle x and the forces in the strut and tie. Ignore the weights of the members.

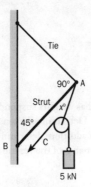

Fig. 2.Q14

15 Determine the forces in the two struts of Fig. 2.Q15 then determine the vertical and horizontal thrusts at each support.

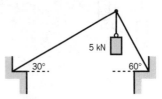

Fig. 2.Q15

16 Figure 2.Q16 shows a simple roof truss consisting of two struts and a tie. Consider the equilibrium of joint

C to determine the forces in the two struts, then, using these values, apply the triangle of forces law to joints A and B respectively to determine the tension in the tie and the vertical reactions at the supports.

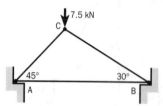

Fig. 2.Q16

17   Determine the forces in the tie and strut of Fig. 2.Q17, then determine the vertical and horizontal components of the reactions at A and B.

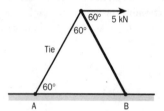

Fig. 2.Q17

18   A continuous string ABCDE has weights suspended from it as shown in Fig. 2.Q18. Determine the tensions in the portions AB and BC. In addition, determine the horizontal and vertical components of the reactions at A and E.

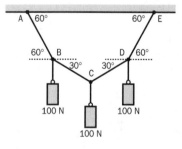

Fig. 2.Q18

19   In Fig. 2.Q19 the masts are vertical and hinged at their bases. Determine the tension in the rope supporting the weight, the tension in the stays and the compression in the masts.

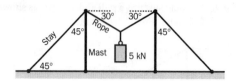

Fig. 2.Q19

20   In Fig. 2.Q20 the rope supporting the weight passes over pulleys at the tops of the masts. Determine the angle A that the masts make with the horizontal and determine the compression in the masts.

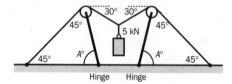

Fig. 2.Q20

21   The block of stone of Fig. 2.Q21 will begin to slide when the horizontal force is 2.5 kN. What is the maximum value of the thrust P without sliding occurring?

Fig. 2.Q21

22   Determine the force in the ropes A and B which will replace the rope C of Fig. 2.Q22.

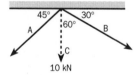

Fig. 2.Q22

23   Two posts support a weight of 1 kN as shown in

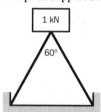

Fig. 2.Q23

Fig. 2.Q23. What horizontal force is each post exerting on the other?

24 A triangular framework of hinged rods is supported by two cables (Fig. 2.Q24). Determine the forces in the framework and in the cables, and determine the vertical components of the reactions at A and B. Neglect the weights of the members.

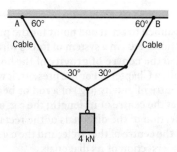

Fig. 2.Q24

25 A framework consisting of five rods hinged at their joints is supported as shown in Fig. 2.Q25. Neglecting the weights of the members, determine the forces in the members of the framework and the reaction at A.

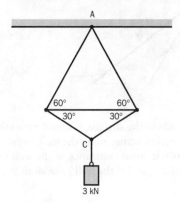

Fig. 2.Q25

26 Figure 2.Q26(a) to (f) shows the free-body diagrams for systems of concurrent forces which are in equilibrium. Determine the magnitude and direction of the unknown force (or forces) marked X and Y.

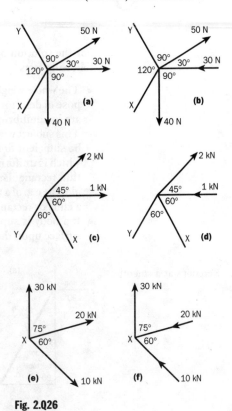

Fig. 2.Q26

27 Three ropes are attached to point O and are pulled with forces as indicated in Fig. 2.Q27. Determine the forces in the strut and tie.

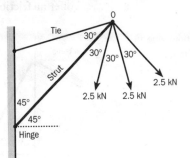

Fig. 2.Q27

3 Non-concurrent coplanar forces

For the solution of the problems in this chapter it is necessary to know the following facts:

- The whole weight of a body can be assumed to act at one point for the purpose of determining supporting reactions, etc., in a system of forces which are in equilibrium. This point is called the **centre of gravity** of the body. This subject will be discussed in detail in Chapter 8. For the present, it will be sufficient to remember that the centre of gravity (c.g.) of a rod or beam which is uniform in cross-section is at the centre of its length; the c.g. of a thin rectangular plate is at the intersection of the diagonals of the rectangle; the c.g. of a thin circular disc is at the centre of the circle; and the c.g. of a cube or rectangular solid is at the intersection of its diagonals.
- If a body is supported by a perfectly smooth surface, the reaction of the surface upon the body acts at right angles to the surface (Fig. 3.1).

Fig. 3.1 Reactions at a smooth surface

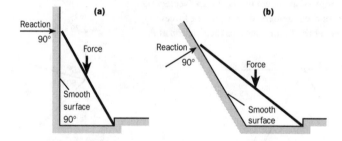

If the reaction were not at right angles to the surface, it would mean that there would be a component of the reaction acting along the wall surface as shown in Fig. 3.2. This shows there is frictional resistance of the wall opposing the downward trend of the ladder, and if the wall is smooth it can offer no frictional resistance.

Fig. 3.2 Reactions due to frictional resistance

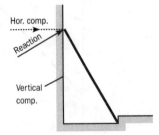

- If two forces (which are not parallel) do not meet at their points of contact with the body, their lines of action can be produced until they do meet.
- If a body is in equilibrium under the action of three forces, and two of these

forces are known to meet at a point, the line of action of the remaining force must pass through the same point.

Example 3.1   A thin rectangular plate weighing 100 N is supported by two strings as indicated in Fig. 3.3(a). Determine the tensions in the strings.

Fig. 3.3   Example 3.1: (a) load diagram; (b) free-body diagram at point O; (c) force diagram

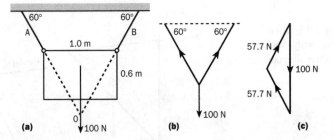

Solution   The whole weight of the plate can be assumed to act at its centre of gravity. The lines of action of the forces in the strings, when produced, meet the line of action of the weight of the plate at O. Consider the equilibrium of this point.

     The tension in string A (and B) is 57.7 N. The principle involved in the solution of this problem can be demonstrated on the apparatus illustrated in Fig. 2.1.

Example 3.2   A ladder rests against a smooth wall and a man weighing 750 N stands on it as indicated in Fig. 3.4. Neglecting the weight of the ladder determine the reactions at the wall and at the ground.

Fig. 3.4   Example 3.2: (a) ladder resting against a smooth surface; (b) free-body diagram at point O; (c) force diagram

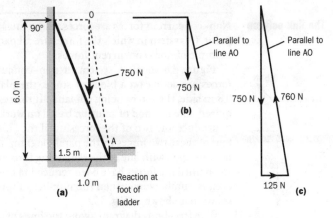

Solution   The ladder is in equilibrium as the result of the action of three forces: the weight of the man, the reaction at the top of the ladder, and the reaction at the foot. Since the wall is smooth the reaction at the top of the ladder is at right angles to the wall. Draw Fig. 3.4(a) to scale. The vertical line representing the weight of the man and the horizontal line representing the reaction at the top of the ladder meet at point O. The line of action of the reaction at the foot of the ladder must therefore pass through the same point. We now have three forces meeting at a point and the solution is shown in Figs. 3.4(b) and (c) which give the reaction at the top of the ladder as 125 N, and the reaction at the foot as 760 N.

Example 3.3 The uniform rod 2 m long shown in Fig. 3.5(a) weighs 250 N. Determine the reactions at A and B.

Fig. 3.5 Example 3.3: (a) load diagram; (b) free-body diagram at point O; (c) force diagram

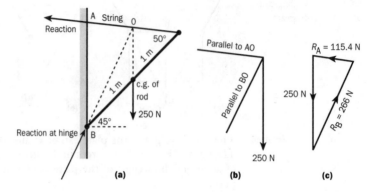

Solution The reaction at A is in line with the string. The weight of the rod can be assumed to act at its centre of gravity and to act vertically downwards. Draw the diagram to scale. Produce the line of action of the force of 250 N to meet at O the line of action of the reaction at A. The line of action of the reaction at B is given by joining B to O. Consider the equilibrium of point O as shown in Fig. 3.5(b). The reactions are 115.4 N at A and 266 N at B.

The link polygon Non-concurrent forces are forces whose lines of action do not all meet at one point. (A system in which the forces are all parallel to one another is a special example of non-concurrent forces.)

Figure 3.6 shows a system of non-concurrent forces. Assume that the forces are acting on a block of stone, the plan view of which is a square of 0.8 m side, the forces being coplanar. If the resultant force on the stone is required, one method of solution based on work in the previous chapter is first to produce any two of the forces until they meet, and then to find the resultant of these two forces by the parallelogram of forces. This resultant can now be combined with one of the remaining forces and a new resultant found, and so on until the forces have been reduced to one resultant. This method, however, is cumbersome when the number of forces is large and a much neater solution is shown in Fig. 3.7.

The free-body diagram giving the lines of action of the forces is drawn accurately and the forces lettered with Bow's notation. These forces are treated as if they all meet at one point and a force polygon is drawn, the sides of which are parallel to the given forces.

The closing line *fa* (direction from *f* to *a*) of the force polygon will give the magnitude and direction of the equilibrant of the five forces and the same line in the direction *a* to *f* will give the resultant. If the five forces met at one point, the answer to the problem would be complete since the resultant must pass

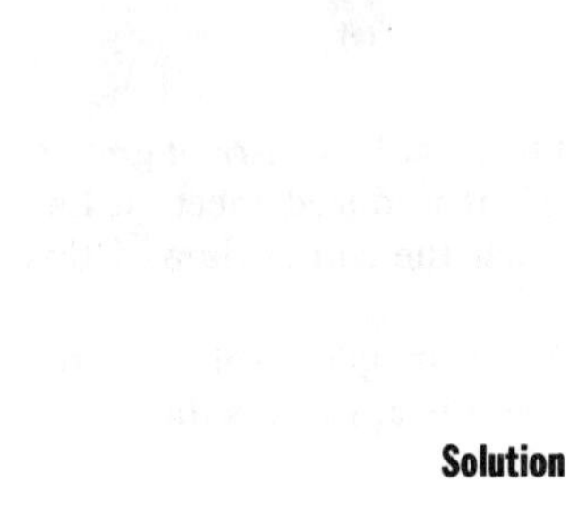

Fig. 3.6 System of non-concurrent forces

Fig. 3.7 Resultant force: (a) free-body diagram with link polygon; (b) force polygon with polar diagram

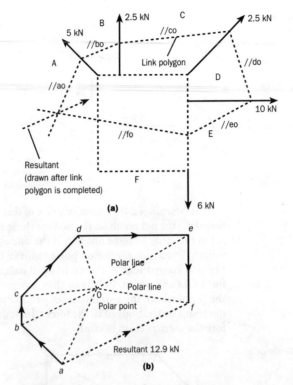

through the same point. In this case, there is no common meeting point of the forces and the position of the resultant on the body must be determined. This is done by placing a point O (the polar point) *anywhere* inside or outside the force polygon and connecting it to the letters *a*, *b*, *c*, etc., by polar lines.

A line parallel to the polar line *ao* is now drawn on the free-body diagram *anywhere across space* A to cut force AB. Then a line parallel to *bo* is drawn across space B to cut force BC and so on. Where the last line (parallel to the polar line *fo*) intersects the first line (parallel to polar line *ao*) is one point on the line of action of the resultant (or equilibrant). The resultant can now be shown in its position on the free-body diagram by drawing a line parallel to the closing line *af* of the force polygon. In this construction the link lines form a polygon on the forces of the free-body diagram and it is known as a **link polygon**.

The graphical conditions of equilibrium for a system of non-concurrent forces are:

- the force polygon must be a closed one
- the link polygon must be a closed one

The validity of the construction of the link polygon can be proved as follows.

Consider the two forces AB and BC of 5 kN and 2.5 kN respectively, which are shown in Fig. 3.8(a) to a larger scale than in Fig. 3.7.

Any two forces *m* and *n* can be substituted for the one force AB without altering the state of equilibrium of the body. The value of these two components *m* and *n* which replace force AB can be obtained from a triangle of forces (Fig. 3.8). The force of 5 kN represented by the line *ab* is the resultant of the two forces. In a similar manner, force BC can be replaced by two forces *p* and *q*; the values of these two forces also being given by a triangle of forces.

Fig. 3.8 Validity of the link polygon: (a) detail of Fig. 3.7; (b) force triangles

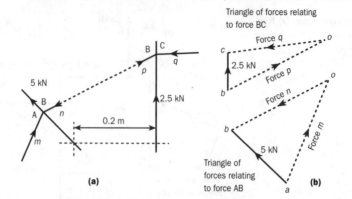

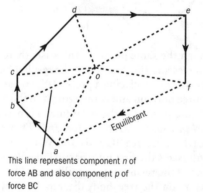

If it is arranged that component *n* of force AB is equal to component *p* of force BC, the net result on the body of these two components is nil, since they act in the same straight line but in the opposite direction (see Fig. 3.8(b)). Similarly, by having a common polar point in the force diagram (given again in Fig. 3.9) component *q* of force BC will balance with an equal component of force CD and so on. It follows that, assuming the equilibrant acts through the point where the two link lines parallel to polar lines *ao* and *fo* meet, the net result on the body of all the forces, including the equilibrant, is nil. Therefore the construction is valid.

Fig. 3.9 Force diagram about a common polar point

Note: the arrow representing the direction of the equilibrant will be in the opposite direction to that given for the resultant in Fig. 3.7.

In brief, this method of dealing with non–concurrent forces consists in replacing each of the given forces by two forces or components and arranging that one component of each force is equal in value to a component of the succeeding force. This can be done by having a common polar point in the force diagram. If the polar point is at any other position different from that shown in Figs. 3.7 and 3.9, the construction is still valid, since a force can be replaced by two components in a multitude of ways. Although the polar point can theoretically be placed anywhere, a neater solution can be obtained by choosing its position so that the link polygon will fit nicely on the free-body diagram. A multitude of link polygons can be drawn for any one problem, by using different positions of the polar point, but the closing lines (such as lines drawn parallel to *ao* and *fo* in Fig. 3.7) will all intersect on points along the line of action of the equilibrant (or resultant).

Application of the link polygon

Examples 3.4–3.9 explore the use of the link polygon construction in a variety of situations.

Example 3.4

A vertical post is subjected to pulls from four cables as shown in Fig. 3.10. The magnitude, direction and position of the resultant pull is obtained as shown.

Note that in Fig. 3.10(a) line *ao* has to be produced beyond space A until it intersects line *eo*.

Example 3.5

A vertical post is subjected to parallel pulls from four cables as shown in Fig. 3.11. Determine the position of the resultant pull.

The resultant pull is 8 kN acting at a point 2.25 m from the foot of the post as shown in Fig. 3.11.

It is usually quicker and easier to obtain the resultant of parallel forces by calculation (see Chapter 4).

The link polygon may also be used to determine the reactions to a beam or truss. It must be realized, however, that the values of two unknown forces in a system of non-concurrent forces which are in equilibrium can only be determined when all the following conditions are satisfied:

- All the forces (apart from the two unknowns) must be known completely, that is, in magnitude, line of action and direction.

Fig. 3.10 Example 3.4: (a) free-body diagram with link polygon; (b) force polygon with polar diagram

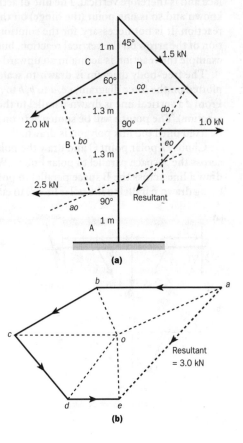

Fig. 3.11 Example 3.5: (a) free-body diagram with link polygon; (b) force and polar diagrams

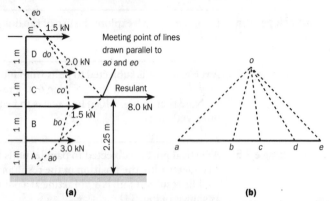

(a)   (b)

- The line of action of one of the unknown reactions (not necessarily its direction) must be known.
- At least one point on the line of action of the remaining reaction must be known.

Example 3.6   A horizontal beam is hinged at one end and supported on a smooth wall at the other end. It is loaded as indicated in Fig. 3.12(a) (the central load including the weight of the beam). Determine the reactions at the supports.

Solution   Since the wall is smooth, the reaction at the wall is at right angles to the surface and is therefore vertical. The line of action of one of the reactions is thus known and so is one point (the hinge) on the line of action of the remaining reaction. It is not necessary, for the solution, to know beforehand the direction of the arrow on the vertical reaction, but it should be obvious that in this example the reaction is acting in an upward direction.

The free-body diagram is drawn to scale and the three known forces are plotted on the force diagram, i.e. *a* to *b*, *b* to *c*, and *c* to *d*, as shown in Fig. 3.12. From *d* a vertical line is drawn parallel to the force DE (the unknown vertical reaction). The point *e* will be somewhere on this line, and its position cannot be fixed until the link polygon is drawn.

Choose a polar point O and draw the polar lines. *From the hinge* draw a line across the A space parallel to polar line *ao*. Where this line cuts the force AB, draw a line across the B space parallel to polar line *bo* and so on, the last line being drawn parallel to the polar line *do* to cut the vertical reaction (force DE).

Fig. 3.12 Example 3.6: (a) load diagram; (b) free–body diagram with link polygon; (c) force and polar diagrams

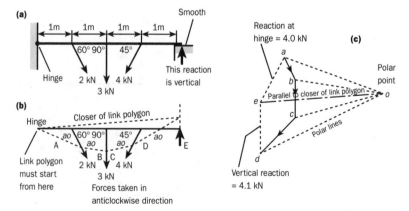

From this point on the force DE draw a line to the hinge. This line closes the link polygon and is known as the **closer**. Draw a line on the force diagram from the polar point O and parallel to the closer. Where this line cuts the vertical line from *d* is the position of *e*. Join *e* to *a*. The length of line *de* gives the magnitude of the vertical reaction at the right-hand support (4.1 kN) and the length of line *ea* gives the magnitude and direction of the reaction at the hinge (4.0 kN).

It is absolutely essential that the link polygon is started by drawing a line from the *only known point* on the reaction at the hinge. Referring to Fig. 3.8 and the procedure necessary to obtain the equilibrant of a system of non-concurrent forces, it was shown that lines drawn parallel to the first and last polar lines intersect on one point of the line of action of the equilibrant. In the procedure just described for obtaining two unknown forces, advantage has been taken of the fact that one point on the unknown reaction is known, and therefore it can be arranged that lines drawn parallel to the first and last polar lines meet at this point (the last polar line being the closer).

Example 3.7   A beam as shown in Fig. 3.13(a) is supported by two walls and loaded as shown, the central load including an allowance for the weight of the beam. Determine the value of the reactions.

Solution   Solving a problem of this type (where all the forces are vertical) by means of the link polygon is equivalent to using a sledgehammer to drive in a tin-tack. Solution by calculation is much quicker and easier (Chapter 5). The graphical

Fig. 3.13 Example 3.7: (a) load diagram; (b) free-body diagram with link polygon; (c) force and polar diagrams

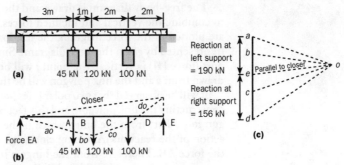

solution is shown in Fig. 3.13. The reaction at the right support is 156 kN (given by *de* on the force diagram) and the reaction at the left support is 109 kN (given by *ea*).

Example 3.8   Figure 3.14(a) represents a roof truss one end of which is supported by a roller bearing. Determine the reactions. The vertical forces are due to the weight of the roof covering, etc., and the inclined forces (which are at right angles to the slope of the roof) are due to wind loads.

Solution   The manner in which the reactions are determined is shown in Fig. 3.14. It should be noted that the right-hand reaction is vertical because the rollers, as indicated, cannot resist horizontal forces, and although both reactions can be determined without combining the pairs of forces by the parallelogram law, the solution given here is probably the most straightforward.

Fig. 3.14 Example 3.8: (a) load diagram; (b) free-body diagram with link polygon; (c) force and polar diagrams

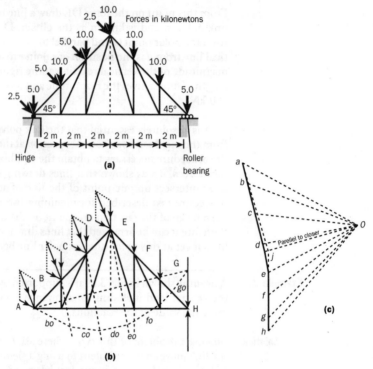

Procedure

The free-body diagram is drawn and the parallelogram of forces law used to combine the vertical and inclined forces. The known loads AB, BC, etc., are plotted in the force diagram, the last load to be plotted being force GH (represented by *gh* on the force diagram). Since the reaction at the right support (force HJ) is vertical, the point *j* will be somewhere on the vertical line drawn from *h* and the link polygon will be the means of fixing this point *j*.

It will be observed that two polar lines (*ao* and *ho*) appear not to be used for constructing the link polygon, and this fact needs explanation. The link polygon must start from the hinge which is the only known point on the line of action of the reaction JA. Going clockwise, the next force to the reaction is the force AB. Therefore from the hinge a line must be drawn parallel to *ao* across the A space to cut force AB. But force AB passes through the hinge. Therefore a line parallel to *ao* must be drawn *from* the hinge *to* the hinge. A line drawn parallel to polar line *ao* may therefore be considered as contained in the hinge.

The next force in order is BC, so a line is drawn from the hinge (which is a point on force AB as well as a point on the reaction) parallel to polar line *bo* to cut force BC and so on, finishing with a line drawn parallel to polar line *go* to cut force GH.

The next force to GH is the vertical reaction HJ so a line parallel to polar line *ho* must be drawn from force GH to cut force HJ. But these two forces are in the same straight line, so, in effect, the line drawn parallel to polar line *ho* from force HG to force HJ is of zero length.

When the point *j* is fixed by drawing a line parallel to the closer of the link polygon, the line *hj* gives the reaction at the right support (34 kN approximately) and the line *ja* gives the direction and magnitude (38 kN approximately) of the reaction at the left support.

Many roof trusses, particularly those of small spans, are equally supported or hinged at both ends so that neither of the reactions is actually vertical. Even so, for the purposes of determining the reactions and the forces in the members of the truss, it is frequently assumed that one reaction is vertical. An alternative assumption with respect to the reactions is that, if the ends of the truss are similarly supported, the horizontal components of the two reactions are equal. This implies that each support resists equally the tendency to horizontal movement due to any inclined loads on the truss.

Example 3.9   A truss as shown in Fig. 3.15 is subjected to wind forces and dead loads. Determine the magnitudes and directions of the reactions at the supports assuming that the total horizontal component of the applied loads is shared equally by the two supports.

Solution   After combining by the parallelogram of forces, the loads are plotted, finishing with *de* on the force diagram. The perpendicular line *ax* gives the sum of the vertical components of all the loads on the truss, and the horizontal line *xe* gives the sum of the horizontal components of the loads. The total hori-

Fig. 3.15   Example 3.9: (a) free-body diagram with link polygon; (b) force and polar diagrams

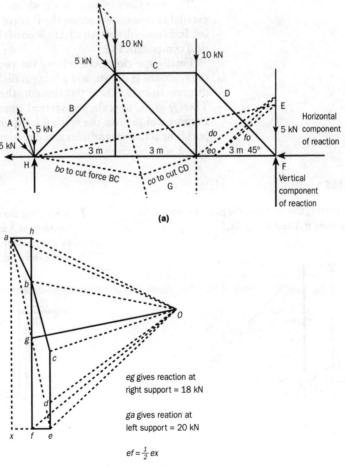

(a)

eg gives reaction at right support = 18 kN

ga gives reation at left support = 20 kN

$ef = \frac{1}{2} ex$

(b)

zontal load on the truss is therefore given by *xe* and, since this is resisted equally by the two supports, the horizontal component of each reaction is given by half the length of line *xe*. Point *f* can now be fixed, *ef* representing the horizontal component EF of the reaction at the right support.

The next force in order is the vertical component FG, so from *f* a vertical line is drawn. The position of *g* on this line is not known but it is known that the vertical component FG plus the vertical component GH must equal the total vertical component of the applied loads. Point *h* must therefore be in a horizontal line from *a* and the force polygon is completed by the line *ha* which represents the horizontal component HA of the reaction at the left support.

The problem is now reduced to determining the vertical components of the two reactions – in other words, to fixing the position of point *g* on the vertical line *fh*.

Starting the link polygon at the left hinge, the first effective link line is the one drawn parallel to polar line *bo* across the B space to cut force BC. (The first imaginary link line is *from* the hinge on the unknown vertical component GH to cut the horizontal component HA *at* the hinge, and the second link line is *from* the hinge on force HA to cut the force AB *at* the hinge.) Link lines parallel to *co* and *do* are now drawn.

The next force in order is EF which is horizontal. Therefore the link line parallel to *eo* is drawn across the E space to cut the horizontal force EF. The last link line is drawn parallel to *fo* across the F space to cut the unknown vertical component FG.

Finally, the closer connecting the two unknown vertical components of the reactions is drawn, and a line parallel to the closer is drawn on the force diagram from O. Where this line cuts the vertical line *fh* is the position of *g*. Then *fg* gives, to scale, the vertical component of the reaction at the right support and *gh* gives the vertical component of the reaction at the left support. The actual magnitudes and directions of the reactions are given by lines *eg* and *ga* respectively.

Exercises

1 Neglecting the weight of the pole, determine the reactions at A and B, Fig. 3.Q1.

2 Neglecting the weights of the members, determine the reactions at A and B of Fig. 3.Q2. (Point A is supported by a horizontal cable.) In addition, determine the forces in the three members of the framework, and the

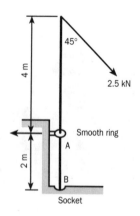

Fig. 3.Q1

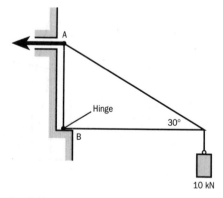

Fig. 3.Q2

horizontal and vertical components of the reaction at B.

3 A gate weighing 500 N is supported by two hinges so that the lower hinge takes the whole weight of the gate (Fig. 3.Q3). Find the reactions at the hinges, and the vertical and horizontal components of the reactions at the bottom hinge.

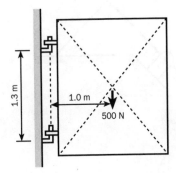

Fig. 3.Q3

4 Determine the tension in the tie-cable and the reaction at the hinge (Fig. 3.Q4).

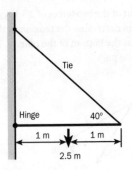

Fig. 3.Q4

5 Determine the tension in the cable and the reaction at the hinge (Fig. 3.Q5).

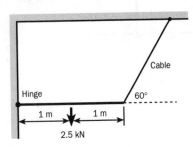

Fig. 3.Q5

6 A ladder rests against a smooth sloping wall, as shown in Fig. 3.Q6. The vertical load includes the weight of the ladder. Determine the reactions at A and B.

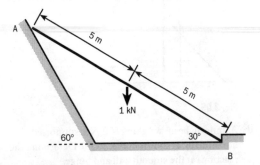

Fig. 3.Q6

7 The structure shown in Fig. 3.Q7 is supported by a smooth ring bolt at A and by a socket at B. Determine the reactions at A and B. Determine also the horizontal and vertical components of the reaction at B.

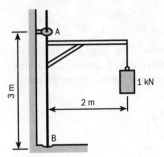

Fig. 3.Q7

8 Determine the tension in the string and the reaction at the hinge (Fig. 3.Q8). Neglect the weight of the members.

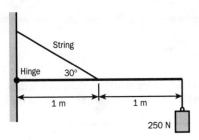

Fig. 3.Q8

9 Determine the tension in the tie-rope and the reaction at the hinge (Fig. 3.Q9).

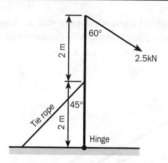

Fig. 3.Q9

10 A uniform beam weighing 500 N is supported horizontally as shown in Fig. 3.Q10. Determine the reactions at the smooth wall and hinge.

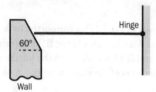

Fig. 3.Q10

11 The horizontal cable of Fig. 3.Q11 is halfway up the rod, the weight of which may be neglected. Determine the tension in the cable and the reaction at the hinge.

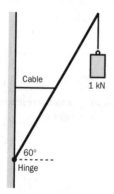

Fig. 3.Q11

12 A uniform rod is hinged halfway along its length as shown in Fig. 3.Q12. Determine the tension in the tie-

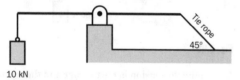

Fig. 3.Q12

rope and the reaction at the hinge, neglecting the weight of the rod.

13 A uniform rod is hinged halfway along its length as shown in Fig. 3.Q13. Determine the tension in the rope and the reaction at the hinge, neglecting the weight of the rod.

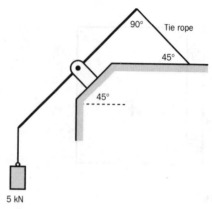

Fig. 3.Q13

14 Determine the resultant of the two forces of 2.5 kN and 1.5 kN (Fig. 3.Q14), then determine the tension in the cable and the reaction at the bottom of the pole. Neglect the weight of the pole.

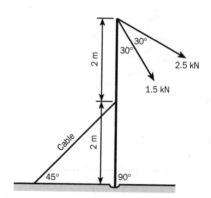

Fig. 3.Q14

15 A framework consisting of three rods hinged at the joints is supported by a smooth wall at B and is hinged to a wall at A (Fig. 3.Q15). Determine the reactions at A and B and the forces in the three members, neglecting their weight.

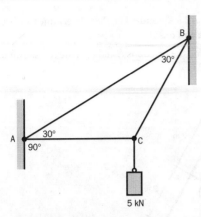

Fig. 3.Q15

16 A bent rigid rod has two equal arms AC and BC as shown in Fig. 3.Q16. The wall at A is smooth. Determine the angle x so that the reaction at B is at right angles to the surface of the wall.

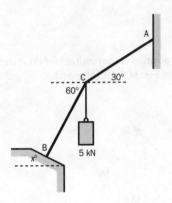

Fig. 3.Q16

17 Determine the tensions in the strings and the reaction at the hook (Fig. 3.Q17).

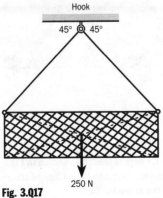

Fig. 3.Q17

18 A horizontal beam, the weight of which may be neglected, is loaded halfway along its length as shown in Fig. 3.Q18. Determine the reactions at A and B.

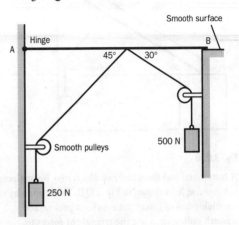

Fig. 3.Q18

19 Neglecting the weight of the beam, determine the reactions at A and B of Fig. 3.Q19.

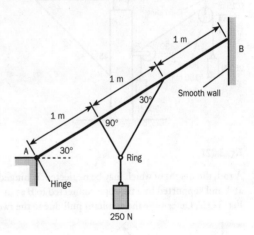

Fig. 3.Q19

20 Neglecting the weight of the rod, determine the reactions at A and B of Fig. 3.Q20.

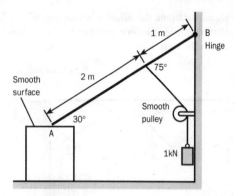

Fig. 3.Q20

21 A horizontal rod the weight of which may be neglected is hinged at X as shown in Fig. 3.Q21. Determine by the parallelogram of forces the resultant pull on the smooth pulley, then use the triangle of forces to determine the tension in the tie-rope and the reaction at the hinge.

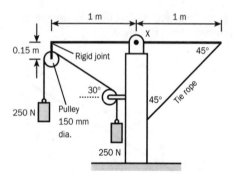

Fig. 3.Q21

22 A rod, the weight of which may be neglected, is hinged at A and supported by a tie-rope connected to B as in Fig. 3.Q22. Determine the resultant pull due to the two

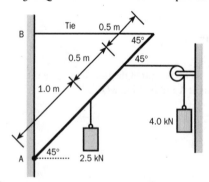

Fig. 3.Q22

weights of 2.5 kN and 4 kN, then determine the forces in the tie and rod.

23 Determine the reactions A and B to the rod shown in Fig. 3.Q23.

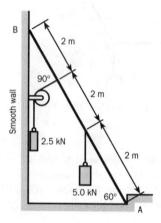

Fig. 3.Q23

24 Determine the position of the resultant for the systems of forces given in Fig. 3.Q24(a), (b) and (c).

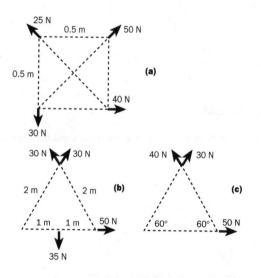

Fig. 3.Q24

25 A mast, fixed at its base is loaded as shown in Fig. 3.Q25. Determine the magnitude and line of action of the resultant of the given loads.

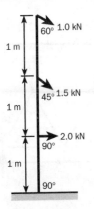

Fig. 3.Q25

26 Determine the magnitude and line of action of the resultant of the four parallel forces acting on the mast of Fig. 3.Q26.

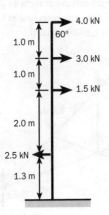

Fig. 3.Q26

27 Determine the magnitude of the vertical reaction at R and the magnitude and line of action of the reaction at L of the system of forces given in Fig. 3.Q27.

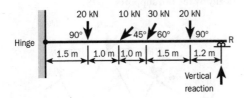

Fig. 3.Q27

28 A beam (Fig. 3.Q28) is hinged at L and supported by means of a cable at R. Determine the reactions at L and R.

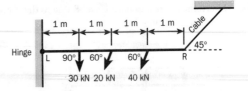

Fig. 3.Q28

29 A vertical mast is supported by means of a smooth ring bolt at X (which means that the reaction at X is horizontal) and by a socket at Y (Fig. 3.Q29). Determine the reactions at X and Y.

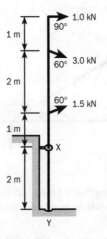

Fig. 3.Q29

30 A tower (Fig. 3.Q30) is supported by roller bearings at R and is hinged at L. Determine the reactions at L and R.

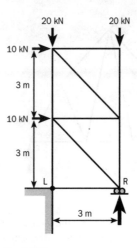

Fig. 3.Q30

31 Determine the reactions at L and R due to the three vertical loads of Fig. 3.Q31.

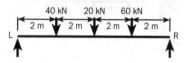

Fig. 3.Q31

32 Assuming the reaction at R of Fig. 3.Q32 to be vertical, determine its magnitude. Also determine the magnitude and direction of the reaction at L.

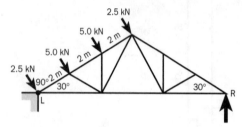

Fig. 3.Q32

33 Repeat Question 32 assuming that the total horizontal component of the applied loads is shared equally by the two supports.

34 The reaction at R to the truss shown in Fig. 3.Q34 is assumed to be vertical. Determine its value due to the wind loading given and also determine the value and line of action of the reaction at L.

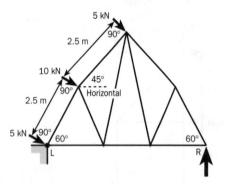

Fig. 3.Q34

35 Repeat Question 34 assuming that the total horizontal component of the applied loads is shared equally by the two supports.

36 Determine the reactions at L and R of Fig. 3.Q36 assuming the reaction at R to be vertical.

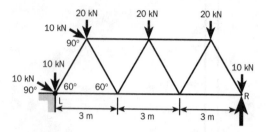

Fig. 3.Q36

37 Repeat Question 36 assuming that the total horizontal component of the applied loads is shared equally by the two supports.

38 The lengths of rafters represented by x and y of Fig. 3.Q38 are equal and the wind loads are at right angles to the slope of the roof. Determine the reactions at L and R assuming the reactions at L to be vertical.

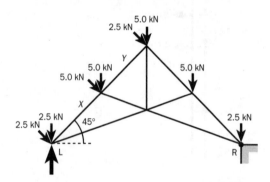

Fig. 3.Q38

39 Fig. 3.Q39 shows a truss which is supported at L and R. Determine the reactions, assuming that the total horizontal component of the applied loads is shared equally by the two supports.

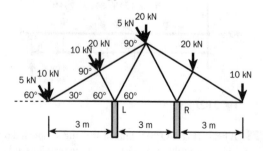

Fig. 3.Q39

4 Moments of forces

The concept of the turning effect or moment of forces is explained in this chapter. The principle of the moment is then applied to the solution of problems connected with the equilibrium of a system of non-concurrent forces, including parallel forces (e.g. beam reactions, couples, etc.).

It is common knowledge that a small force can have a big turning effect or leverage. In mechanics, *moment* is commonly used instead of *turning effect* or *rotational effect*. The word moment is usually thought of as having some connection with time. Moment, however, derives from a Latin word meaning movement and moments of time refer, of course, to movement of time. Similarly, the **moment of a force** can be thought of as the movement of a force, although a more exact definition is 'turning effect of a force'. In many types of problem, there is an obvious **turning point** (or hinge or pivot or fulcrum) about which the body turns or tends to turn as a result of the effect of the force (see Fig. 4.1(a)–(e)). In other cases, where the principle of moments may be used to obtain solutions to problems, there may not be any obvious turning points.

To solve problems by the use of the principle moments, the effect of the forces must be considered in relation to the direction in which the body turns or tends to turn around the given point. In Fig. 4.1(a) the turning effect or

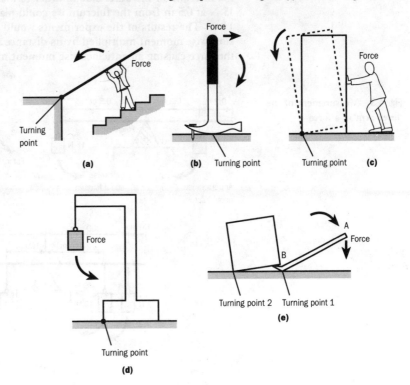

Fig. 4.1 Different forms of moment: (a) anticlockwise; (b) clockwise; (c) and (d) anticlockwise; (e) clockwise at A, anticlockwise at B

moment of the force is anticlockwise; in (b) the moment is clockwise; and in (c) and (d) the moments are anticlockwise. There are two turning points in Fig. 4.1(e). Considering the point where the crowbar touches the ground, (point 1), the moment of force at A is clockwise. At B, the bar applies a force to the stone and causes it to turn about point 2. The moment of force applied to the stone at B is therefore anticlockwise.

If the line of action of a force passes through a point it can have no turning effect (moment) about that point. For example, a door cannot be opened by pushing at the hinge. Further examples are given in Fig. 4.2. The given forces have no moments about points O.

Fig. 4.2 These forces have no moment about point O

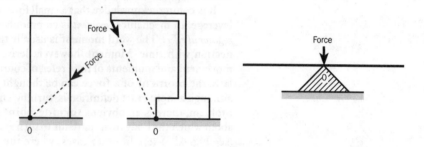

Measurement of moments

Figure 4.3(a) represents a uniform rod which balances on a fulcrum placed halfway along the length of the rod. If a force of 25 N is suspended at 0.3 m from the fulcrum it will be found that a force of 15 N is required at 0.5 m from the fulcrum on the other side in order to maintain equilibrium. Instead of 15 N at 0.5 m from the fulcrum we could place 12.5 N at 0.6 m or 7.5 N at 1.0 m. The results of the experiments would show that the force causing a clockwise moment multiplied by its distance from the turning point equals the force causing an anticlockwise moment multiplied by its distance from

Fig. 4.3 Measurement of the 'lever arm' of a force

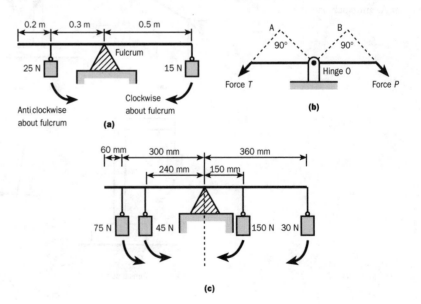

the turning point. For example, referring to Fig. 4.3(a), the clockwise moment of the 15 N force is 15 N × 0.5 m or 7.5 N m.

Moments are always expressed in force-length units such as N m, N mm, kN m, etc. A little consideration should show that this must be so, because the turning effect of a force depends on distance as well as on the magnitude of the force.

The distance from the force to the turning point is often called the **lever arm** of the force. In Fig. 4.3(a) the lever arm of the 15 N force is 0.5 m and the lever arm of the 25 N force is 0.3 m. It should be noted that the lever arm should be measured from the turning point to the point where it intersects the line of action of the force at right angles. In Fig. 4.3(a) the lines of action of the weights are vertical and the lever arms are horizontal. In Fig. 4.3(b) the lever arms of the forces are AO and BO respectively and for equilibrium $T \times AO$ must equal $P \times BO$. The reason for measuring the lever arms in this way will be made clear in Example 4.3 on page 43.

Further demonstrations of the principle of moments can be arranged quite easily with the aid of a fulcrum, a rod and a few weights. In Fig. 4.3(c) the rod itself, since it is supported at its centre of gravity, has no resultant turning effect, clockwise or anticlockwise.

$$\text{anticlockwise moments} = \text{clockwise moments}$$
$$(75 \times 300) + (45 \times 240) = (150 \times 150) + (30 \times 360)$$
$$33\,330 \text{ N mm} = 33\,300 \text{ N mm}$$

Example 4.1   A plank of uniform cross-section 4 m long and weighing 300 N has a support placed under it at 1.3 m from one end as shown in Fig. 4.4(a). Calculate the magnitude of the weight W required to cause the plank to balance.

Solution   It was stated on page 22 that, when considering the equilibrium of forces, the whole weight of a body can be assumed to act at its centre of gravity. In the case of a uniform rod or beam, this is halfway along the rod. Taking moments about the fulcrum or turning point,

$$1.2W \text{ N m} = 300 \times 0.8 \text{ N m}$$
$$W = \frac{240 \text{ N m}}{1.2 \text{ m}} = 200 \text{ N}$$

Fig. 4.4 Example 4.1: (a) moment due to a uniform plank; (b) proof that the whole weight of a body acts at its centre

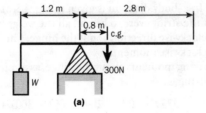

(a)

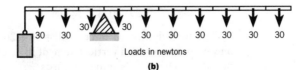

Loads in newtons

(b)

The whole weight of the plank is not, of course, concentrated at its centre of gravity. Each little particle of the plank is being attracted downwards by the earth and the 1.2 m length of the plank to the left of the fulcrum is exerting an anticlockwise moment, whilst the 2.8 m length of the plank to the right of the fulcrum is exerting a clockwise moment. The total net effect is, however, the same as if the whole of the weight were acting at 0.8 m to the right of the fulcrum.

For example, assume the plank is divided into 0.4 m lengths each length weighing 30 N, and assume that each 30 N weight acts halfway along its 0.4 m length (Fig. 4.4(b)).

Taking moments about the fulcrum

$$30 \times (0.2 + 0.6 + 1.0) + 1.2W$$
$$= 30 \times (0.2 + 0.6 + 1.0 + 1.4 + 1.8 + 2.2 + 2.6)$$
$$(54 + 1.2W) \text{ N m} = 294 \text{ N m}$$
$$1.2W \text{ N m} = 240 \text{ N m}$$
$$W = 200 \text{ N}$$

If the plank is divided into a thousand lengths each length weighing 0.3 N, and moments are taken about the fulcrum, the answer will be the same, so there need be no hesitation in assuming the whole weight of a body to act at its centre of gravity.

Example 4.2   A uniform rod weighing 50 N and carrying weights, as shown in Fig. 4.5(a), is hinged at A. The rod is supported in a horizontal position by a vertical string at B. Calculate the tension in the string and the reaction at the hinge.

Fig. 4.5   Example 4.2: (a) load diagram; (b) free-body diagram

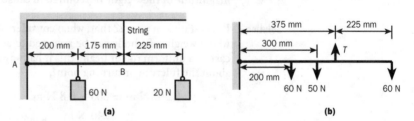

(a)   (b)

Solution   The free-body diagram is given in Fig. 4.5(b).

If the string were to be cut, all the weights would pull the rod downwards in a clockwise direction about the hinge. The string exerts a counterbalancing anticlockwise moment.

Taking moments about the hinge, assuming T newtons to be the tension in the string,

$$375T = (60 \times 200) + (50 \times 300) + (20 \times 600)$$
$$375T = 39\,000$$
$$T = 39\,000/375 = 104 \text{ N}$$

The total downward force is $60 + 50 + 20$, i.e. 130 N. Since the string is only holding up with a vertical force of 104 N, the reaction at the hinge must be $130 - 104$, i.e. 26 N acting upwards.

Example 4.3 A horizontal rod has a weight of 100 N suspended from it as shown in Fig. 4.6(a). Calculate the tension in the string and the reaction at the hinge, ignoring the weight of the rod.

Fig. 4.6 Example 4.3, Solution 1: (a) load diagram; (b) free-body diagram; (c) total reaction at hinge

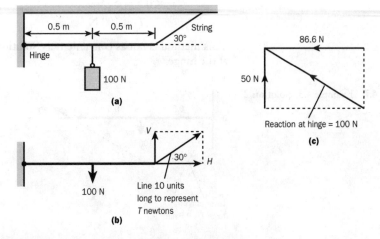

Solution 1 Let the tension in the string be represented by T newtons and resolve this force into its vertical and horizontal components. This may be done by calculation or by drawing a line 10 units long to represent T (Fig. 4.6(b)).

It is found that the vertical component V is 5 units long, i.e. $0.5T$, and that the horizontal component H is 8.66 units long, i.e. $0.866T$. The line of action of the horizontal component of the tension in the string passes through the hinge and therefore has no moment about this point. Taking moments about the hinge

$$\text{anticlockwise moment} = \text{clockwise moment}$$
$$0.5T \times 1 \text{ m} = 100 \times 0.5 \text{ m}$$
$$0.5T = 50$$
$$T = 50/0.5 = 100 \text{ N} = \text{tension in string}$$

Reaction at hinge: the total downward force is 100 N. The vertical component of the tension in the string is $0.5T$, i.e. 50 N. (The horizontal component of the tension in the string is ineffective in holding up the suspended weight.) Since there is 100 N pulling down, there must be 100 N pushing up, therefore the hinge must supply a vertical reaction in the upward direction of 50 N. But the string has also a horizontal component of $0.866T$, i.e. 86.6 N, tending to pull the rod away horizontally from the hinge. The hinge must resist this pull. Therefore there must be a horizontal force acting at the hinge in addition to a vertical force.

The total reaction at the hinge is given in Fig. 4.6(c) and may be found from a scale drawing, or, if R is the reaction at the hinge,

$$R^2 = 50^2 + 86.6^2 = 10\,000$$
$$R = \sqrt{10\,000} = 100 \text{ N}$$

Solution 2   In Fig. 4.7 which may be drawn to scale, produce the line of action of the string and draw a line AO at right angles, then taking moments about O,

$$T \times OA = 100 \times 0.5$$
$$T \times 0.5 = 50$$
$$T = 100 \text{ N}$$

This method is not as convenient as Solution 1 for determining the reaction at the hinge.

Fig. 4.7   Example 4.3, Solution 2

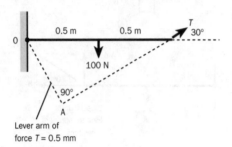

Solution 3   Since there are only three forces in this problem, the triangle of forces law can be applied (see page 8), and this method may be quicker for problems in which the forces are inclined.

Example 4.4   A ladder rests against a smooth vertical wall and a man weighing 750 N stands on it as indicated in Fig. 4.8. Neglecting the weight of the ladder, determine the reactions at the wall and at the ground.

Fig. 4.8   Example 4.4

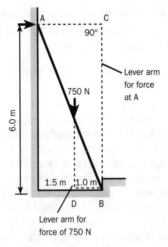

Solution   This problem is identical with that on page 23 but it is now being solved by using the principle of moments. Since the wall is smooth, the reaction at A is horizontal. There may not appear to be any obvious turning point but B can be considered to be a turning point. (If the wall were to collapse, the ladder would swing round B in an anticlockwise direction due to the force of 750 N. If one could imagine the force at A to be increased above that necessary to maintain equilibrium, it would turn the ladder round B in a clockwise direction.) By taking moments about B, the reaction at B can be ignored since it passes through this point and therefore has no moment about it.

$$\text{reaction at A} \times \text{distance CB} = 750 \text{ N} \times \text{distance DB}$$

$$R_A \times 6 = 750 \times 1$$

$$R_A = 750/6 = 125 \text{ N}$$

Reaction at B (R_B): since the wall is smooth, it cannot help at all in holding *up* the ladder, therefore the whole weight of 750 N is taken at B. There is therefore an upward reaction at B of 750 N. There must also be a horizontal reaction to prevent the foot of the ladder from moving outwards. This horizontal reaction equals the force at A, i.e. 125 N, therefore the total reaction at B, which can be obtained graphically or by calculation, is 760 N.

Conditions of equilibrium

The above example is an illustration of the application of the three **laws of equilibrium** which apply to a system of forces acting in one plane. These laws are:

1 The algebraic sum of the vertical forces must equal zero, i.e. if upward forces are called positive and downward forces are called negative, then a force of $+750$ N must be balanced by a force of -750 N and the algebraic sum of the two forces is equal to zero. This can also be expressed by

$$\sum V = 0$$

where Σ is the Greek letter 'sigma' (letter S) and in this connection means 'the sum of', whilst V represents 'the vertical forces', upward forces being plus and downward forces being minus.

2 The algebraic sum of the horizontal forces must equal zero, i.e. the sum of the horizontal forces acting towards the left (plus forces) must equal the sum of the horizontal forces acting towards the right (minus forces) or

$$\sum H = 0$$

3 The algebraic sum of the moments of the forces must equal zero, i.e. if clockwise moments are called plus and anticlockwise moments are called minus then

$$\sum M = 0$$

Example 4.5

A vertical pole (Fig. 4.9(a)) is supported by being hinged in a socket at B and by a smooth ring which can give no vertical reaction, at A. Calculate the reactions at A and B neglecting the weight of the pole.

Solution

First, resolve the pull of 5 kN into its vertical and horizontal components as shown in Fig. 4.9(b) and take moments about B.

Fig. 4.9 Example 4.5: (a) load diagram; (b) free-body diagram; (c) total reaction at point B; (d) total reaction at point A.

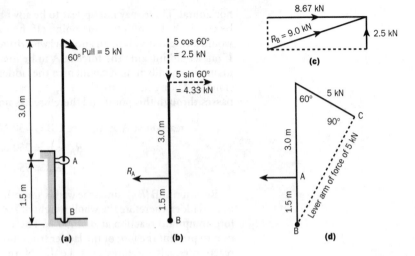

$\sum M = 0$ or clockwise moments equal anticlockwise moments.

$$4.33 \times 4.5 = R_A \times 1.5$$
$$R_A = 13.0 \text{ kN}$$

(Note that the vertical component of 2.5 kN has no moment about B since its line of action passes through B.)

$$\sum V = 0$$

The vertical component of the reaction at B must equal the vertical component of the pull at the top of the pole since the support of A can only *hold back* and cannot *hold up*. The vertical component of the reaction at B therefore equals 2.5 kN and acts upwards.

$$\sum H = 0$$

The horizontal force at the top of the mast is 4.33 kN acting → and the reaction at A is acting ← . There must be, therefore, a horizontal force at B acting → equal to $13.0 - 4.33$, i.e. 8.67 kN. The complete reaction at B is given by Fig. 4.9(c). Note that, if the pole itself weighs, say, 0.5 kN, this weight would be taken entirely at B (since the pole is vertical) and the vertical component of the reaction at B would be 3 kN instead of 2.5 kN as shown in Fig. 4.9(c). The complete reaction at B would then be

$$R_B = \sqrt{(8.67^2 + 3^2)} = 9.2 \text{ kN}$$

Another method of finding the reaction at A (neglecting the weight of the pole) is shown in Fig. 4.9(d).

$$5 \text{ kN} \times \text{distance BC} = R_A \times 1.5 \text{ m}$$

Note also that the triangle of forces principle could be applied to determine both reactions.

Example 4.6 A cantilever truss (Fig. 4.10(a)) is supported at A and B in such a manner that the reaction at A is horizontal. Neglecting the weight of the truss, calculate the reactions at A and B.

Fig. 4.10 Example 4.6: (a) load diagram; (b) total reaction at point B

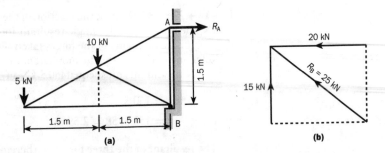

10 kN

5 kN

1.5 m

B

1.5 m 1.5 m

R_A

20 kN

$R_B = 25$ kN

15 kN

(a)   (b)

Solution Take moments about B to determine the reaction at A.

$$\sum M = 0$$
$$1.5R_A = 10 \times 1.5 + 5 \times 3$$
$$R_A = 20 \text{ kN acting towards the right.}$$

$\sum H = 0$, therefore horizontal component of reaction at B is 20 kN, acting towards the left.

$\sum V = 0$, therefore vertical component of reaction at B is 15 kN acting upwards.

Referring to Fig. 4.10(b)

$$R_B = \sqrt{(20^2 + 15^2)} = \sqrt{625} = 25 \text{ kN}$$

Resultant of parallel forces

It was stated in Chapter 2 in connection with concurrent forces that the resultant of a given number of forces is the single force which, when substituted for the given forces, has the same effect on the *state of equilibrium* of the body. This definition applies also to parallel forces.

Consider Fig. 4.11(a) which represents a beam of negligible weight supporting three vertical loads of 13, 18 and 27 kN respectively.

Fig. 4.11 Resultant of parallel forces: (a) load diagram; (b) free-body diagram

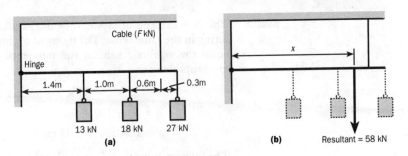

Cable (F kN)

Hinge

1.4m 1.0m 0.6m 0.3m

13 kN 18 kN 27 kN

(a)

x

Resultant = 58 kN

(b)

The force in the vertical cable can be obtained by taking moments about the hinge, i.e.

$$F \times 3 \text{ m} = 13 \times 1.4 + 18 \times 2.4 + 27 \times 3.3$$
$$= 150.5 \text{ kN m}$$
$$F = 150.5/3 = 50.2 \text{ kN}$$

Now, if one force (the resultant) is substituted for the three downward loads without altering the state of equilibrium, its magnitude must be

$13 + 18 + 27 = 58$ kN, and if the reaction in the cable is to remain as before, i.e. 50 kN, the moment of the single resultant force about the hinge must equal the sum of the moments of the forces taken separately.

Referring to Fig. 4.11(b), the moment of the resultant force about the hinge is $58x$ kN m, and this must equal 150.5 kN m, therefore

$$58x = 150.5$$
$$x = 150.5/58 = 2.59 \text{ m}$$

The resultant of the three forces is, therefore, a single force of 58 kN acting at 2.59 m from the hinge. The reactions in the cable and at the hinge will be the same for this one force as for the three forces.

Example 4.7   Calculate the magnitude and position of the resultant of the system of forces shown in Fig. 4.12(a) and determine where a fulcrum must be placed to maintain equilibrium. (Neglect the weight of the rod.)

Fig. 4.12   Example 4.7: (a) load diagram; (b) free-body diagram

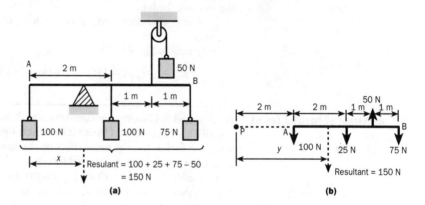

Solution   This is an example of a system of *unlike parallel forces* since all the forces are not acting in the same sense. The moment of this resultant about *any point* must equal the algebraic sum of the moments of the separate forces. Taking moments about end A,

$$150x = 25 \times 2 - 50 \times 3 + 75 \times 4$$
$$= 50 - 150 + 300$$
$$x = 200/150 = 1.33 \text{ m}$$

The fulcrum must therefore be placed at this point in order to maintain the rod in balance (and the reaction at the fulcrum $= 150$ N).

The same answer would be obtained by taking moments about any point, as for example, the point P, 2 m from A (see Fig. 4.12(b)).

$$150y = 100 \times 2 + 25 \times 4 - 50 \times 5 + 75 \times 6$$
$$= 200 + 100 - 250 + 450$$
$$y = 500/150 = 3.33 \text{ m}$$

The position of the resultant is therefore 3.33 m from P or 1.33 m from A.

Problems can often be simplified by working with a resultant instead of with a number of separate forces. This is demonstrated in the following example.

Example 4.8   Calculate the resultant of the three downward forces of Fig. 4.13(a), then use the triangle of forces principle to determine the reaction in the cable and the reaction at the hinge.

Fig. 4.13   Example 4.8: (a) load diagram; (b) free-body diagram; (c) force diagram at point O

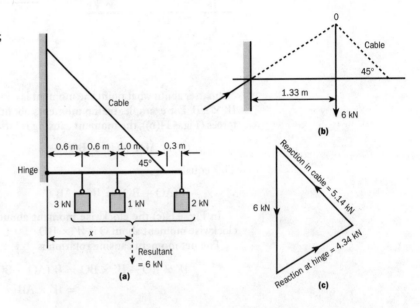

Solution   The resultant of the three downward forces is 6 kN and, taking moments about the hinge,

$$6x = 3 \times 0.6 + 1 \times 1.2 + 2 \times 2.5$$
$$x = 8/6 = 1.33 \text{ m}$$

Now, using the method of Chapter 3, page 23, the reactions can be found as shown in Fig. 4.13(b) and (c).

Couples   Two equal, unlike (i.e. acting in opposite directions) parallel forces are said to form a **couple**.

Imagine a rod resting on a smooth table, as shown in Fig. 4.14(a), and assume that two equal unlike parallel forces of 20 N are applied horizontally as shown. The rod will rotate in a clockwise direction, and the moment causing this rotation is obtained by multiplying *one* of the forces by the distance AB between the two forces. The perpendicular distance AB is called the arm of the couple or in general terms, the **lever arm**, l_a. Therefore

$$\text{moment of the couple} = 20 \text{ N} \times 150 \text{ mm}$$
$$= 3000 \text{ N mm}$$

Fig. 4.14 Couples: (a) load diagrams; (b) and (c) considering moments about different points

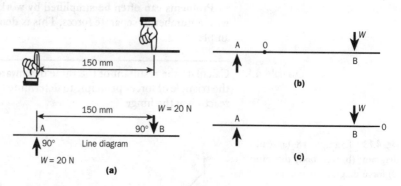

(b)

150 mm

$W = 20$ N

90° B

A

90° Line diagram

$W = 20$ N

(a)

(c)

No matter about what point the moment is considered, the answer is equal to $W \times AB$. For example, taking moments about any point O between the two forces (Fig. 4.14(b)), the moment causing rotation is

$$(W \times AO) + (W \times BO)$$

This equals

$$W(AO + BO) = W \times AB$$

In Fig. 4.14(c) the clockwise moment about O is $W \times AO$, and the anti-clockwise moment about O is $W \times BO$.

The net moment causing rotation is

$$W \times AO - W \times BO = W(AO - BO)$$
$$= W \times AB$$

Note that a couple acting on a body produces rotation and the forces cannot be balanced by a single force. To produce equilibrium another couple of equal and opposite moment is required.

Figure 4.15 represents a mast fixed at its base. The moment tending to break the mast at its base is

$$[2(x + 1) - 2x] \text{ kN m} = 2x + 2 - 2x = 2 \text{ kN m}$$

Fig. 4.15 Fixing moment

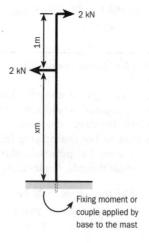

2 kN

1 m

2 kN

x m

Fixing moment or couple applied by base to the mast

No matter what the distance x may be, the moment at the base will be 2 kN m. This moment or couple is resisted by the fixing moment or couple applied by the base to the mast.

A knowledge of the principles of couples is useful in understanding beam problems (see Chapter 9).

Beam reactions

The application of the principle of moments of forces to the calculation of beam reactions can be verified by means of the following experiment.

A beam of wood, a little longer than 1.2 m and weighing 15 N, has weights suspended from it as shown in Fig. 4.16(a). It is required to determine the reactions at supports A and B.

Fig. 4.16 Beam reactions:
(a) balance attached at point B;
(b) balance attached at point A

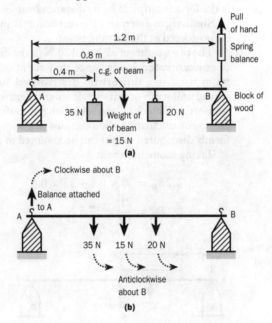

If a spring balance is attached to the beam at B, the beam can be lifted just clear of the support and it will be observed that the reading on the balance is 32.5 N. When the balance is released so that the beam rests again on block B, it follows that the beam must be pushing down on the support with a force of 32.5 N. Similarly, the upward reaction of the support on the beam is 32.5 N.

If the hand holding the balance is moved up and down, the beam will turn round the point where it is supported at A. Taking moments about this point, representing the reaction at B by R_B,

$$R_B \times 1.2 \text{ m (anticlockwise about A)}$$
$$= 35 \times 0.4 + 15 \times 0.6 + 20 \times 0.8$$

therefore

$$1.2 R_B = 39 \quad \text{and} \quad R_B = \frac{39}{1.2} = 32.5 \text{ N}$$

If the balance is transferred to A and the beam lifted off its support, the balance will read 37.5 N (Fig. 4.16(b)).

By calculation, taking B as the turning point,

$$R_A \times 1.2 = 20 \times 0.4 + 15 \times 0.6 + 35 \times 0.8$$

$$\text{clockwise} = \text{anticlockwise}$$

$$R_A = \frac{45}{1.2} = 37.5 \text{ N}$$

Note that the sum of the reactions is 70 N and that this is also the sum of the downward forces ($\sum V = 0$).

Note also that to determine the reaction at A, moments are taken about B, i.e. B is considered as the turning point and the moment of each force or load is the force multiplied by its distance from B.

Similarly, to determine the reaction at B, moments are taken about A, i.e. A is considered as the turning point.

The two weights of 35 N and 20 N in the discussion above are **point loads** or **concentrated loads**, since they act at definite points on the beam. The beam itself is a **uniformly distributed load (UDL)** since its weight is spread uniformly over its whole length. Figure 4.17(a) represents the test beam supporting a block of lead 0.4 m long of uniform cross-section and weighing 30 N. For calculating the reactions at the supports, the whole of this uniformly distributed weight can be assumed to act at its centre of gravity.

Taking moments about A

$$1.2 R_B = 30 \times 0.4 + 15 \times 0.6$$

$$R_B = \frac{21}{1.2} = 17.5 \text{ N}$$

Fig. 4.17   Uniformly distributed loads: (a) load diagram; (b) and (c) representations of a UDL

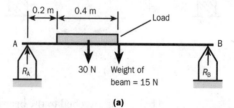

(a)

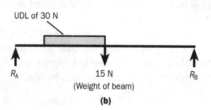

(b)

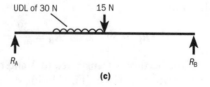

(c)

Similarly, taking moments about B

$$1.2R_A = 15 \times 0.6 + 30 \times 0.8$$

$$R_A = \frac{33}{1.2} = 27.5 \text{ N}$$

$$R_A + R_B = 17.5 + 27.5 = 45 \text{ N} = \text{total load}$$

These calculated values of the reactions can be verified by means of the spring balances.

Instead of drawing the beam every time, a free-body diagram can be drawn as in Fig. 4.17(b) and (c) where two methods of indicating uniformly distributed loads are shown.

Further experiments can be performed with the simple apparatus shown above by placing the supports at different points. Figure 4.18 shows the supports placed 0.8 m apart. In this case, to determine experimentally the reaction at A, the spring balance must be connected to a point on the beam directly above A.

Fig. 4.18 Combination of point and uniformly distributed loads

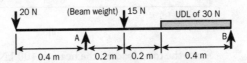

Taking moments about B, the reaction at A produces a clockwise moment and all the loads anticlockwise moments.

$$0.8R_A = 30 \times 0.2 + 15 \times 0.6 + 20 \times 1.2$$

$$R_A = 39/0.8 = 48.75 \text{ N}$$

Taking moments about A, both the reactions at B and the load of 20 N produce anticlockwise moments.

$$0.8R_B + 20 \times 0.4 = 15 \times 0.2 + 30 \times 0.6$$

$$0.8R_B = 3.0 + 18.0 - 8.0$$

$$R_B = \frac{13}{0.8} = 16.25 \text{ N}$$

$$R_A + R_B = 48.75 + 16.25 = 65 \text{ N}$$

$$= \text{sum of the vertical loads}$$

Practical cases   ldings, point loads on beams are usually due to loads from other beams or from columns, and uniformly distributed loads are due to floors, walls and partitions and the weights of the beam themselves.

In Fig. 4.19(a), assume that the beam AB weighs 10 kN and that it supports two beams which transmit loads of 60 kN and 90 kN respectively to beam AB. In addition, a column transmits a load of 240 kN. The reactions are usually taken as acting at the middle of the bearing area supplied by the supporting brickwork or column. Figure 4.19(b) shows the free-body diagram.

Taking moments about B,

$$6R_A = 90 \times 2.0 + 10 \times 3.0 + 240 \times 3.5 + 60 \times 4.0$$

$$= 180 + 30 + 840 + 240$$

$$R_A = \frac{1290}{6} = 215 \text{ kN}$$

Fig. 4.19 Practical example: (a)
loading of beam; (b) free-body
diagram

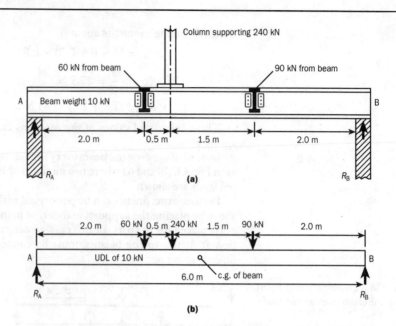

(a)

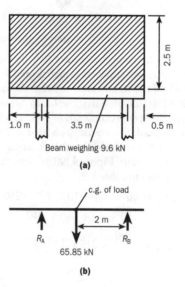

(b)

Since the total load on the beam is 400 kN, the reaction at B is 400 − 215,
i.e. 185 kN. It is advisable, however, to determine the reaction at B by taking
moments about A and thereby to obtain a check on the calculations.

Example 4.9   A beam weighing 9.6 kN supports brickwork 225 mm thick weighing 20
kN/m³ as indicated in Fig. 4.20(a). Calculate the reactions at the supports.

Solution   Both the brickwork and the beam are uniformly distributed loads.

$$\text{volume of brickwork} = 5.0 \times 2.5 \times 0.225 = 2.8125 \text{ m}^3$$
$$\text{weight of brickwork} = 2.8125 \times 20 \quad\quad = 56.25 \text{ kN}$$
$$\text{weight of beam} \quad\quad\quad\quad\quad\quad\quad = 9.60 \text{ kN}$$
$$\text{total weight} = 65.85 \text{ kN}$$

Fig. 4.20   Example 4.9: (a) load
diagram; (b) free-body diagram

The free-body diagram is shown in Fig. 4.20(b).
Taking moments about B,

$$3.5R_A = 65.85 \times 2$$

$$R_A = \frac{131.70}{3.5} = 37.63 \text{ kN}$$

$$R_B = 65.85 - 37.63 = 28.22 \text{ kN}$$

or R_B can be obtained by taking moments about A, thus providing a useful check.

Summary

Moment   The turning effect of a force.

- Moments are considered in relation to a turning point, real or imaginary.
- A moment is always obtained by multiplying a force by a distance and this distance must be measured from the turning point to where it cuts the line of action of the force at right angles.
- Moments are expressed in force-length units such as newton metre (N m), newton millimetre (N mm), etc.
- If the line of action of a force passes through a given point, it can have no moment about that point.

Conditions of equilibrium   For a system of forces acting in one plane, conditions of equilibrium are:

$$\sum V = 0 \qquad \sum H = 0 \qquad \sum M = 0$$

Resultant   For a number of parallel forces the resultant is their algebraic sum. Its position can be found by taking moments about *any point* and moment of resultant equals the algebraic sum of moments of the individual forces.

Couple   Two equal unlike parallel forces. The distance between the forces is called the arm of the couple, and the moment of a couple equals one of the forces multiplied by the arm of the couple.

Exercises

1 A uniform rod is in equilibrium under the action of weights as shown in Fig. 4.Q1. Calculate the value of W and the reaction at the fulcrum, ignoring the weight of the rod.

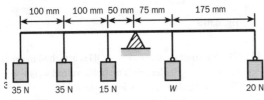

100 mm   100 mm   50 mm   75 mm   175 mm

35 N   35 N   15 N   W   20 N

Fig. 4.Q1

2 Calculate the value of x metres so that the uniform rod of Fig. 4.Q2 will balance.

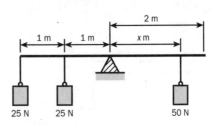

2 m

1 m   1 m   x m

25 N   25 N   50 N

Fig. 4.Q2

3 A uniform rod weighing 100 N and supporting 500 N is hinged to a wall and kept horizontal by a vertical rope (Fig. 4.Q3). Calculate the tension in the rope and the reaction at the hinge.

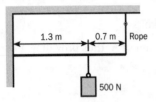

Fig. 4.Q3

4 A uniform rod weighing 40 N is maintained in equilibrium as shown in Fig. 4.Q4. Calculate the distance x and the reaction at the fulcrum.

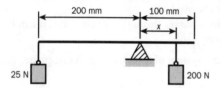

Fig. 4.Q4

5 A uniform rod 0.4 m long weighing 10 N supports loads as shown in Fig. 4.Q5. Calculate the distance x if the rod is in equilibrium.

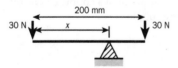

Fig. 4.Q5

6 A uniform horizontal beam 1.3 m long and weighing 200 N is hinged to a wall and supported by a vertical prop as shown in Fig. 4.Q6. Calculate the reactions at the prop and hinge.

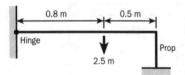

Fig. 4.Q6

7 Calculate the reaction at the prop and hinge in Fig. 4.Q7. The uniform beam weighs 200 N.

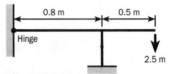

Fig. 4.Q7

8 A compound lever is shown in Fig. 4.Q8. Each of the horizontal rods weighs 50 N and the vertical rod weighs 40 N. Calculate the value of the force W which is required for equilibrium, and calculate the reactions at the fulcrums. (First take moments about A to find the force in the vertical rod, then take moments about B.)

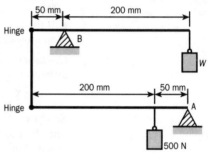

Fig. 4.Q8

9 Solve Exercises 1 to 13 of Chapter 3, with the aid of the principle of moments.

10 Determine the tension in the chain and the reaction at the hinge (Fig. 4.Q10). The uniform rod weighs 500 N.

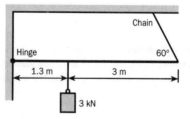

Fig. 4.Q10

11 The horizontal cable in Fig. 4.Q11 is attached to the rod halfway along its length. The uniform rod weighs 250 N. Calculate the tension in the cable and the reaction at the hinge.

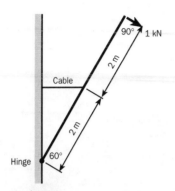

Fig. 4.Q11

12 Calculate the tension in the guy rope of Fig. 4.Q12, then use the triangle of forces law to determine the forces in the mast, tie and jib. Neglect the weight of the members.

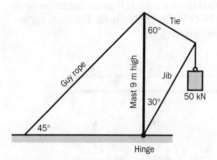

Fig. 4.Q12

13 A uniform rod weighing 50 N is loaded as shown in Fig. 4.Q13. Calculate the tension in the cable and the reaction at the hinge.

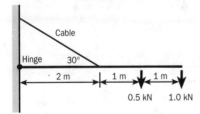

Fig. 4.Q13

14 Two uniform rods each 3 m long and weighing 250 N are connected by a link bar (Fig. 4.Q14). Calculate the tensions in the link bar and chain neglecting their weights.

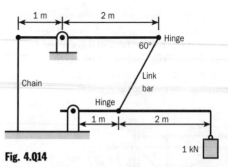

Fig. 4.Q14

15 For each of the structures shown in Fig. 4.Q15(a), (b) and (c) determine the position of the resultant of the vertical loads; then, by using the triangle of forces, determine the tension in the cable and the reaction at the hinge. Neglect the weight of the rod.

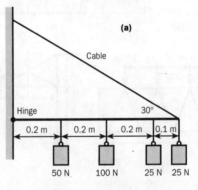

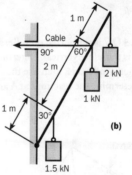

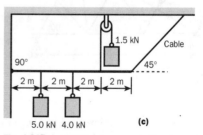

Fig. 4.Q15

16 The weights of the structures shown in Fig. 4.Q16(a),
(b), (c) and (d) may be neglected. In each case the
reaction at A is horizontal, and B is a hinge. Determine
the position of the resultant of the vertical loads; then,
by applying the principle of the triangle of forces,
determine the reactions at A and B.

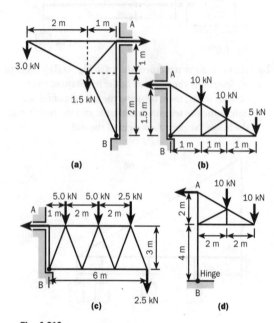

(a)   (b)

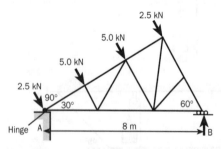

(c)   (d)

Fig. 4.Q16

17 The roof truss of Fig. 4.Q17 is supported on rollers at B
and the reaction at B is therefore vertical. Determine
the resultant of the four wind loads, then use the
triangle of forces to determine the reactions at A and B.

Fig. 4.Q17

18 Reduce the system of forces shown in Fig. 4.Q18 to a
couple. Determine the arm and moment of the couple.

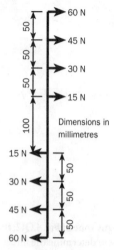

Fig. 4.Q18

19 Calculate the reactions at A and B due to the given
point loads in Fig. 4.Q19(a) to (e). The loads are in
kilonewtons (kN).

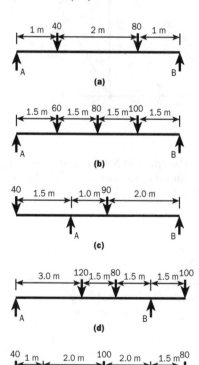

Fig. 4.Q19

20 Calculate the reactions at A and B due to the given system of loads in Fig. 4.Q20(a) to (g).

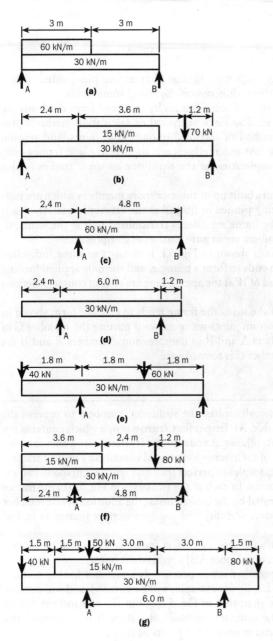

(a) 3 m | 3 m, 60 kN/m, 30 kN/m, A, B

(b) 2.4 m | 3.6 m | 1.2 m, 15 kN/m, 30 kN/m, 70 kN, A, B

(c) 2.4 m | 4.8 m, 60 kN/m, A, B

(d) 2.4 m | 6.0 m | 1.2 m, 30 kN/m, A, B

(e) 1.8 m | 1.8 m | 1.8 m, 40 kN, 60 kN, 30 kN/m, A, B

(f) 3.6 m | 2.4 m | 1.2 m, 15 kN/m, 30 kN/m, 80 kN, 2.4 m | 4.8 m, A, B

(g) 1.5 m | 1.5 m | 50 kN | 3.0 m | 3.0 m | 1.5 m, 40 kN, 15 kN/m, 80 kN, 30 kN/m, 6.0 m, A, B

Fig. 4.Q20

21 A beam AB carries two point loads and is supported by two beams CD and EF as shown in plan in Fig. 4.Q21. Neglecting the weights of the beams, calculate the reactions at C, D, E and F.

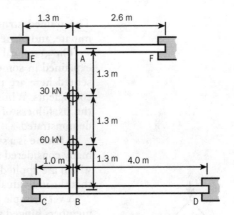

Fig. 4.Q21

22 A reinforced concrete bridge between two buildings spans 8 m. The cross-section of the bridge is shown in Fig. 4.Q22. The brickwork weighs 20 kN/m^3, the reinforced concrete weighs 23 kN/m^3, and the stone coping weighs 0.5 kN/m. The floor of the bridge has to carry a uniformly distributed load of 1.5 kN/m^2 in addition to its own weight. Calculate the force on the supporting buildings from one end of each beam.

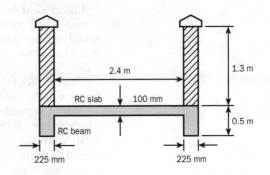

Fig. 4.Q22

5 Framed structures

The framed structures dealt with in this chapter are pin-jointed, determinate, and plane and three-dimensional (or space) frameworks.

The graphical solutions of plane frames by means of force diagrams are explained in some detail. The three analytical or calculation methods presented here are the method of sections, resolution at joints, and tension coefficients. Whilst the first two methods are applied to plane frames only, the usefulness of the application of the third one to space frames is also demonstrated.

A **frame** is a structure built up of three or more members which are normally considered as being pinned or hinged at the various joints. Any loads which are applied to the frame are usually transmitted to it at the joints, so that the individual members are in pure tension or compression.

A very simple frame is shown in Fig. 5.1. It consists of three individual members hinged at the ends to form a triangle, and the only applied loading consists of a vertical load of W at the apex. There are also, of course, reactions at the lower corners.

Under the action of the loads the frame tends to take the form shown in broken lines, i.e. the bottom joints move outward putting the member C in tension, and the members A and B in compression. Members A and B are termed **struts** and member C is termed a **tie**.

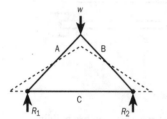

Fig. 5.1 A simple frame

Perfect, imperfect and redundant pin-jointed frames

A **perfect frame** is one which has just sufficient members to prevent the frame from being unstable. An **imperfect frame** is one which contains too few members to prevent collapse. A **redundant frame** is one which contains more than the number of members which would constitute a perfect frame.

Figure 5.2(a) shows examples of perfect frames. Figure 5.2(b) shows two examples of imperfect frames. In each of the two cases in Fig. 5.2(b) the frames would collapse as suggested by the broken lines; the addition of one member in each case would restore stability and produce perfect frames as in Fig. 5.2(a).

Figure 5.2(c), for example, shows a perfect frame, and Fig. 5.2(d) (the same frame with the addition of member AB) a redundant frame.

Note: A redundant member is not necessarily a member having no load. The member AB (Fig. 5.2(d)), for example, will be stressed, and will serve to create a stronger frame than that in Fig. 5.2(c), but the **redundant frame** cannot be solved by the ordinary methods of statics. It is, therefore, also called a **statically indeterminate** or hyperstatic frame.

Number of members in a perfect frame

The simplest perfect frame consists of three members in the form of a triangle as in Fig. 5.3(a). The frame also has three joints or node points, A, B and C.

If *two* more members are added to form another triangle as in Fig. 5.3(b), then *one* more joint has been added.

Fig. 5.2 Perfect, imperfect and redundant frames: (a) and (c) perfect; (b) imperfect; (d) redundant

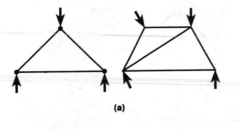

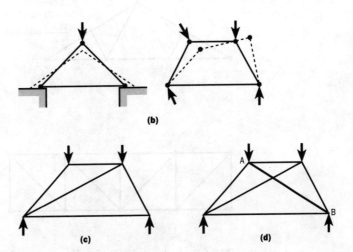

Fig. 5.3 Number of members in a perfect frame: (a) simplest form − 3 members and 3 joints; (b) adding a triangle to a perfect frame

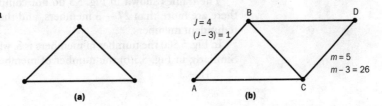

So long as triangles only are added, then the frame will remain a perfect frame.

If J = number of joints, and m = number of members, then

$$2(J - 3) = m - 3$$
$$2J - 6 = m - 3$$
$$2J = m - 3 + 6$$
$$2J = m + 3 \quad \text{or} \quad m = 2J - 3$$

i.e.

Number of members to form a perfect frame
= (twice the number of joints) − 3

or

$$\text{Number of joints} = \frac{\text{number of members} + 3}{2}$$

Fig. 5.4 Examples of perfect frames showing the relationship between numbers of members and joints

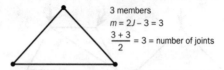

3 members
$m = 2J - 3 = 3$
$\dfrac{3 + 3}{2} = 3$ = number of joints

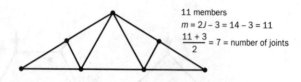

11 members
$m = 2J - 3 = 14 - 3 = 11$
$\dfrac{11 + 3}{2} = 7$ = number of joints

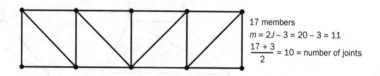

17 members
$m = 2J - 3 = 20 - 3 = 11$
$\dfrac{17 + 3}{2} = 10$ = number of joints

The frames shown in Fig. 5.4 all comply with this requirement.

The frames shown in Fig. 5.5 do not comply with this requirement, i.e. there are more than $2J - 3$ members, and the frames contain one or more redundant members.

In Fig. 5.5(a) the number of members is 8, which is more than $(2 \times 5) - 3$. Similarly, in Fig. 5.5(b) the number of members is 6, which is greater than $(2 \times 4) - 3$.

Fig. 5.5 These frames contain one or more redundant members

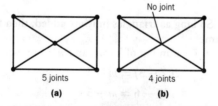

No joint

5 joints   4 joints

(a)   **(b)**

Graphical solutions for frames – force diagrams

Consider the simple perfect frame shown in Fig. 5.6(a). The known forces are the load at the apex and the two equal reactions of 10 kN.

In describing the loads, it is convenient to use Bow's notation (see Chapter 2). Reading clockwise, the 20 kN load at the apex is 'load AB', the reaction at the right support is 'load BC', and the reaction at the left support is 'load CA'. Similarly, the spaces *inside* the frame are usually denoted by numbers so that the force in the horizontal member is 'force C1' or 'force 1C'.

As explained in Chapter 2, the forces in the two inclined members can be obtained by considering the equilibrium of the joint at the apex (Fig. 5.6(b) and (c)).

Fig. 5.6 Frame force diagrams:
(a) load diagram for a simple
frame; (b) reading clockwise the
forces are AB, Bl and 1A;
(c) force diagram (note that the
arrows follow each other round
the triangle)

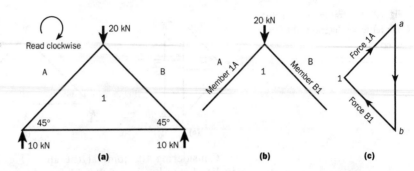

(a)   (b)   (c)

The force in the horizontal tie member can be obtained by considering either the joint at the left reaction (Fig. 5.7) or the joint at the right reaction (Fig. 5.8).

The three triangles of forces drawn separately for the three joints may, for convenience, be superimposed upon each other to form in one diagram the means of determining all unknown forces. This resultant diagram, which is called a **force diagram**, is shown in Fig. 5.9(a).

Note that no arrows are shown on the combined force diagram. The direc-

Fig. 5.7 Reaction at left-hand
joint: (a) load diagram; (b) force
diagram

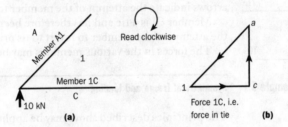

(a)   (b)

tions of the arrows in the frame diagram are obtained by considering each joint as in Figs. 5.6–5.8. One must consider each joint to be the centre of a clock, and the letters are read clockwise round this centre. Thus, having drawn the combined diagram, consider, say, the joint at the left reaction (Fig. 5.7). Reading clockwise round this joint, the inclined member is A1. On

Fig. 5.8 Reaction at right-hand
joint: (a) load diagram; (b) force
diagram

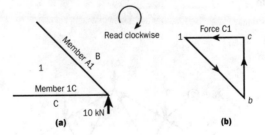

(a)   (b)

the force diagram, the direction from a to 1 is downwards to the left, therefore the arrow is placed in the frame diagram as shown, near the joint. The tie is member 1C and 1 to c on the force diagram is a direction left to right, so the arrow is placed in this direction on the frame (or free-body or space) diagram.

Fig. 5.9 For frame given in Fig. 5.6: (a) force diagram; (b) frame diagram

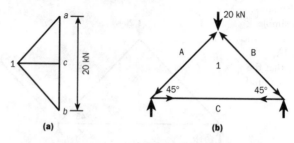

(a)   (b)

Considering the joint at the apex (Fig. 5.6(b) and (c)), the left-hand inclined member is member 1A (not A1 as was the case when the joint at the left-hand reaction was being considered). From the force diagram 1 to *a* is upwards to the right and this fixes the direction of the arrow on the frame diagram. Similarly, for the same joint, the other inclined member is B1. From the force diagram *b* to 1 is upwards towards the left. (Study the force diagrams and the directions of the arrows in Figs. 5.6–5.8.)

Note the directions of the arrows indicating compression (strut) and tension (tie) respectively (Fig. 5.9(b)). These arrows represent the directions of the internal resistances in the members. Member A1 is a strut, which means that it has shortened as a result of the force in it. If the force were removed, the member would revert to its original length (i.e. it would lengthen) and the arrows indicate the attempt of the member to revert to its original length.

Member C1 is a tie and has therefore been stretched. The arrows indicate the attempt of the member to revert to its original (shorter) length.

The forces in the various members may be entered as shown in Table 5.1.

Table 5.1

Member	Force (kN) in:	
	Strut	Tie
A1	14	
B1	14	
C1		10

Example 5.1   **Symmetrical frame and loading**

The principles described above may be applied to frames having any number of members. The work of determining the values of all the unknown forces may be enormously simplified by drawing *one* combined force diagram as shown in Fig. 5.10(b); for this example the operations are explained in detail.

1 Starting from force AB (Fig. 5.10(a)) the *known* forces AB, BC, CD, DE, EF, FG and GA, working clockwise round the frame, are set down in order and

Fig. 5.10 Example 5.1: (a) free-body, space or frame diagram; (b) force diagram

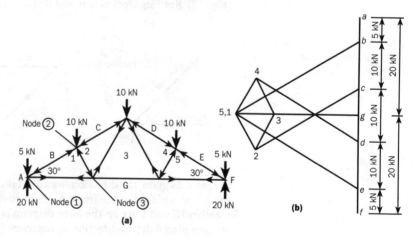

(a)

(b)

to scale as *ab, bc, cd, de, ef, fg* and *ga* in Fig. 5.10(b). (Since the loading is symmetrical each of the reactions FG and GA is 20 kN.)

2 Point 1 on the force diagram is found by drawing *b*–1 parallel to B1, and *g*–1 parallel to G1. Point 1 lies at the intersection of these lines.

3 Point 2 on the force diagram is found by drawing *c*–2 parallel to C2 and 1–2 parallel to 1–2 from the frame (node 2). The point 2 is the intersection of these two lines.

4 Point 3 on the force diagram is found in the same manner by drawing 2–3 parallel to 2–3, and *g*–3 parallel to G3 from the frame, point 3 being the intersection.

5 The remaining points 4 and 5 are found in the same way as above, though in this case, as the frame is symmetrical and forces E5 and 4–5 are the same as B1 and 1–2, etc., only half the diagram need have been drawn.

6 The directions of the arrows may now be transferred to the frame diagram (Fig. 5.10(a)) by working clockwise around each node point.

Taking node 1 for example: moving clockwise; force AB is known to be downward, and *ab* (5 kN) is read downward on the force diagram. The next force in order (still moving clockwise round the joint) is force B1, which as *b*–1 on the force diagram acts to the left and downwards. Thus force B1 must act towards node 1 as indicated in Fig. 5.10(a). The next force in order (still moving clockwise) is 1G. 1–*g* on the force diagram acts from left to right, and thus the arrow is put in on the frame diagram acting to the right from node 1.

7 Node 2: BC is known to act downward. The next force (moving clockwise) is C2, which, as *c*–2 in the force diagram acts downward to the left, thus must act towards node 2. The arrow is shown accordingly. The next force (still moving clockwise) is 2–1, and as 2–1 on the force diagram acts upward to the left, this must act toward the node 2, as shown by the arrow.

8 When all nodes have been considered in this way, it will be seen that each member of the frame in Fig. 5.10(a) has two arrows, one at each end of the member. As explained earlier, these arrows indicate not what is being done to the member, but what the member is doing to the nodes at each end of it.

Hence a member with arrows thus ↔ is in compression, i.e. it is being compressed and is reacting by pressing outward. This member is called a **strut**.

Similarly, two arrows thus →← indicate tension in the member. This member is called a **tie**.

9 The *amounts* of all the forces B1, C2, 1–2, etc., are now scaled from the force diagram, and the amounts and types of force can be tabulated as in Table 5.2.

Table 5.2

Member		Forces (kN) in:	
		Strut	Tie
B1	E5	30.0	
C2	D4	25.0	
G1	G5		26.0
–	G3		17.5
1–2	4–5	9.0	
2–3	3–4		9.0

It should be noted that constructing a force diagram involves considering each joint of the frame in turn and fixing the position of one figure on the force diagram by drawing two lines which intersect. No joint can be dealt with unless *all except two* of the forces meeting at the joint are known. This is why node 1 was taken as the first joint since, of the four forces meeting at this joint, only two B1 and 1G are unknown. Solving this joint gives the forces in the members B1 and 1G. Node 2 can now be dealt with since only two forces, C2 and 2–1, are unknown. Node 3 cannot be dealt with before node 2 because, although the force in the member G1 is known, there are still three unknown forces, 1–2, 2–3, and 3G.

Example 5.2 **Frame containing a member having no force**

The external loads and the two equal reactions are set off to scale as shown in Fig. 5.11(b).

Fig. 5.11 Example 5.2: (a) frame diagram; (b) force diagram

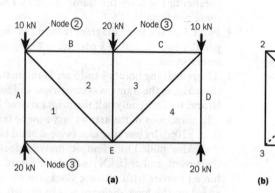

(a)

(b)

Starting with node 1 the two unknown forces are A1 and 1E.

From *a* on the force diagram a vertical line must be drawn (parallel to A1) and point 1 must lie on this vertical line where it meets a horizontal line drawn parallel to member 1E. The point 1 must, therefore, be on the same point as *e* and force 1E = 0.

Considering node 2, point 2 is found by drawing 1–2 to intersect *b*–2.

Point 3 is found by drawing *c*–3 and 3–2 on the force diagram.

Since the frame and loading are symmetrical there is no need to draw further lines, but in this simple example, one more line, shown dotted, completes the force diagram.

Example 5.3 **Frame with unsymmetrical loading**

When the load and/or the frame is unsymmetrical, *every* joint must be considered when drawing the force diagram. Figure 5.12 is an example. The reactions may be determined graphically using the link polygon (Chapter 3) or by calculation (Chapter 4).

Fig. 5.12 Example 5.3: (a) frame diagram; (b) force diagram

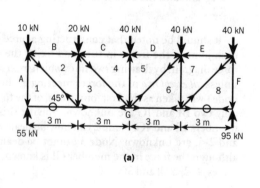

(a)

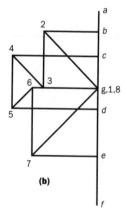

(b)

Example 5.4   **Frame with loads suspended from the bottom chord of the frame in addition to loads on the top chord**

The reactions in Fig. 5.13 are determined in the usual manner and are 40 kN at the left support and 50 kN at the right support. The vertical load line is

Fig. 5.13 Example 5.4: (a) frame diagram; (b) force diagram

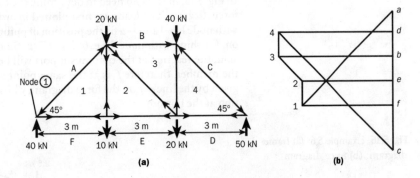

(a)   (b)

then drawn as follows: *a–b* is 20 kN down; *b–c* is 40 kN down. The next force is the reaction CD of 50 kN, therefore from *c* to *d* is measured upwards a distance equal to 50 kN. The next load is DE and is 20 kN down, therefore *d–e* represents this load on the force diagram. Similarly, *e–f* represents the load of 10 kN (EF) and finally *f–a* represents the vertical reaction FA of 40 kN.

The remaining force diagram can now be drawn in the usual manner, e.g. starting with node 1, two lines *a–1* and *f–1* are drawn to intersect at the point 1, and so on, as explained in Example 5.1.

Example 5.5   **Frame cantilevering over one support**

Figure 5.14 shows a frame cantilevering 2 m over the left support. By calculation, the reactions are found to be 7.5 kN and 4.0 kN respectively. The load line is drawn as explained in the previous example, *a–b, b–c, c–d, d–e* (all

Fig. 5.14 Example 5.5: (a) frame diagram; (b) force diagram

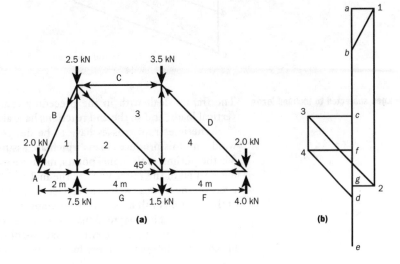

(a)   (b)

downwards), *e–f* upwards (4.0 kN), *f–g* downwards (1.5 kN), *g–a* upwards (7.5 kN).

Example 5.6 Cantilever frames

In Fig. 5.15, there is no need to determine the reactions before starting on the force diagram. The loads are first plotted in order, *a–b, b–c, c–d, d–e*. Starting with node 1, *b–1* and *1–a* fix the position of point 1 on the force diagram, and so on, finishing with node 7, i.e. *6–7* and *7–a*. Since member 7A is in compression, the reaction at the bottom support will be equal in value to the force in the member, therefore *f* is at the same point as 7. The reaction at the top is given by the line *e–f* on the force diagram, whilst line *a–f* represents the reaction at the bottom.

Fig. 5.15 Example 5.6: (a) frame diagram; (b) force diagram

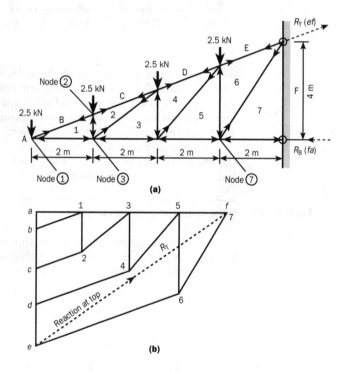

Frames subjected to inclined loads

The frames dealt with in the preceding examples have all carried purely vertical loads, and so the end reactions have also been vertical.

Sometimes roof trusses have to be designed to withstand the effect of wind, and this produces loads which are assumed to be applied to the truss (via the purlins) at the panel points, in such a way that the panel loads are acting normal to the rafter line. As the inclined loads have a horizontal thrust effect on the truss as well as a vertical thrust effect, it follows that at least one of the reactions will have to resist the horizontal thrust effect.

It was stated in Chapter 3 that, when a truss is subjected to inclined loads, the usual assumptions are either that one of the reactions is vertical, or the horizontal components of the two reactions are equal. The method of determining such reactions by the use of the link polygon has been described in Chapter 3.

When only symmetrical inclined loads have to be considered, and one of the reactions is assumed to be vertical as in Fig. 5.16(a), the reactions can be found more easily by the method described below than by the link polygon.

Fig. 5.16 Inclined loads: (a) load diagram; (b) force diagram; (c) taking moments about A

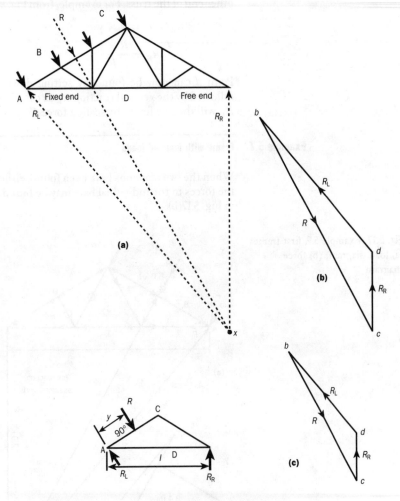

If the four loads are considered to be replaced temporarily by the resultant R, then only three loads act on the truss, i.e. R, R_L and R_R. The three loads must act through one point for equilibrium, and the loads R and R_R obviously act through point x, the point of intersection of the vertical reaction R_R and the resultant inclined load R. As R_L must act through point x, the direction of R_L is found by joining point A to point x.

The direction of all three loads is thus known, together with the magnitude of the resultant R, so the amounts of R_L and R_R may be found graphically as in Fig. 5.16(b).

The resultant R is set down to scale (b to c). The point d is found by drawing in the direction of R_R from c and the direction of R_L from b to their intersection point d.

This vector diagram *bcd* is, of course, the triangle of forces for the loads, and *cd* = reaction R_R to scale, and *db* = reaction R_L to scale. From these three points, *b*, *c*, and *d*, the force diagram for the frame can be drawn in the usual way.

The vertical reaction can also be calculated by taking moments about the other end of the truss. For example, from Fig. 5.16(c), taking moments about A

$$R_R \times L = R \times y$$

$$R_R = \frac{R \times y}{L}$$

Point *d* can now be found by setting down the resultant load *R* (line *bc*) followed by the vertical reaction R_R as calculated to give the point *d*. Reaction R_L will then be found by scaling force *db*.

Example 5.7   **Frame with inclined loads**

When the two reactions have been found, either graphically or by calculation, the forces in the individual bars may be found by drawing a force diagram, as in Fig. 5.17(b).

Fig. 5.17 Example 5.7, first frame: (a) load diagram; (b) force diagram

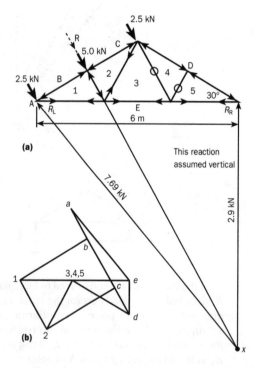

The resultant load *R* (inclined) is produced to intersect the vertical reaction R_R at the point *x*. The direction of the left-hand reaction R_L is then found by joining *x* to A.

The two reactions and the three inclined loads are thus now all known, and may be set down as in Fig. 5.17(b) in the order *ab*, *bc*, *cd* (inclined), *de* (vertical), and *ea* parallel to R_L.

The points 1, 2, 3, 4, 5 on the force diagram are then found in the usual way, as shown in Fig. 5.17(b).

Note that there is no force in members 3–4 and 4–5 from this system of loading (see Table 5.3, first example).

Another example is shown in Fig. 5.18 and Table 5.3, second example.

Fig. 5.18 Example 5.7, second frame: (a) load diagram; (b) force diagram

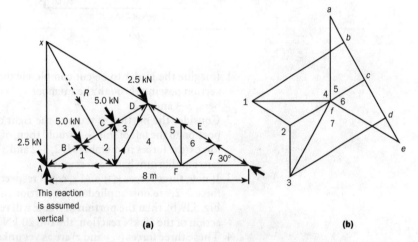

Table 5.3

First example			Second example		
Member	Force (kN) in:		Member	Force (kN) in:	
	Strut	Tie		Strut	Tie
B1	7.0		B1	12.7	
C2	7.0		C2	10.0	
D4	5.8		D3	13.1	
D5	5.8		E5	8.7	
E1		9.8	E6	8.7	
E3		5.0	E7	8.7	
E5		5.0	F1		9.7
1–2	5.0		1–2	5.7	
2–3		5.0	2–3	6.0	
			3–4		10.0
No force in 3–4 and 4–5			No force in 4–5, 5–6, 6–7, F4 and F7		

Calculation methods for frames

When the forces in *all* the members have to be found, the force diagram is usually the most convenient method, but on occasions it is necessary to determine the force in a single individual member. In cases of this type it is normally more convenient to calculate the amount of the force.

Method of sections

Consider the N girder shown in Fig. 5.19(a). It is required to calculate the force in the member 1–2 of the bottom boom.

Fig. 5.19 Section method: (a) load diagram; (b) load diagram of left-hand section

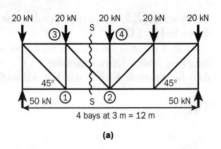 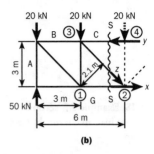

(a)

(b)

1 Imagine the girder to be cut completely through, along the section S–S, a section passing through the member 1–2 concerned and two other members, 3–4 and 3–2.

2 Consider the portion of truss to the right of line S–S to be removed. The portion to the left of line S–S would then, of course, collapse, because three forces (the forces in members 1–2, 3–2 and 3–4) which were necessary to retain equilibrium had been removed.

3 If now, three forces x, y and z, equal respectively to force 1–2, force 3–4 and force 3–2, are now applied to the portion of frame concerned, as shown in Fig. 5.19(b), then the portion of frame will remain in equilibrium under the action of the 50 kN reaction, the two 20 kN loads, and the forces x, y and z.

4 These three forces (x, y and z) are as yet unknown in amount and direction, but the force in member 1–2, i.e. force x, is to be determined at this stage, and it will be seen that the other two unknown forces (y and z) intersect at and pass through the point 3 so that they have no moment about that point.

5 Thus, taking moments about the point 3, the portion of frame to the left of S–S is in equilibrium under the moments of the two applied loads, the moment of the reaction, and the moment of the force x.

The moment of the reaction is a clockwise one of (50×3), i.e. 150 kN m.

The moment of the applied load is anticlockwise about point 3 and equals (20×3), i.e. 60 kN m.

For equilibrium the moment of force x about point 3 (i.e. $3x$) must equal $(150 - 60)$, i.e. 90 kN m anticlockwise.

Therefore $3x = 90$, whence $x = 30$ kN and the arrow must act in the direction shown in Fig. 5.19(b) (in order that the moment of force x about point 3 is anticlockwise).

The member 1–2 is therefore in tension (pulling away from joint 1).

Note: It will be seen that, in general, the rule must be to take a cut through three members (which include the one whose force is to be determined) and to take moments about the point through which the lines of action of the other two forces intersect.

To determine the force in the member 3–4:

Take moments about the point 2 where forces x and z intersect.

$$\text{Moment of reaction} = (50 \times 6)$$
$$= 300 \text{ kN m clockwise}$$
$$\text{Moments of applied loads} = (20 \times 6) + (20 \times 3)$$
$$= 180 \text{ kN m anticlockwise}$$

Therefore

$$\text{Moment of force } y = 3y = 300 - 180$$
$$= 120 \text{ kN m anticlockwise}$$
$$y = 40 \text{ kN}$$

and, for the moment to be anticlockwise about point 2, the arrow must act towards the joint 3, therefore member 3–4 is a strut.

It should be noted that this particular method has its limitations. It would not appear to apply, for example, to member 3–2 (force z) for no point exists through which forces x and y intersect, as x and y are, in fact, parallel.

However, when force y has been calculated, the force z may be found by taking moments about point 1, including the known value of force y (member 3–4).

Note: force y is 40 kN and is anticlockwise about point 1.

$$\text{Moment of the reaction about point 1}$$
$$= (50 \times 3) = 150 \text{ kN m}$$

$$\text{Moments of the applied loads and the force in member 3–4}$$
$$= (20 \times 3) + (40 \times 3) = 180 \text{ kN m}$$

Therefore

$$\text{Moment of force } z \text{ about point 1}$$
$$= 180 - 150 = 30 \text{ kN m clockwise}$$

Hence

$$2.1z = 30$$
$$z = 14.2 \text{ kN}$$

The arrow must act as shown in Fig. 5.19(b), therefore member 3–2 is a tie.

It should be noted that these arrows must be considered in relation to the nearest joints of that portion of the frame which remains after the imaginary cut through the three members has been made. These results should be checked by drawing a force diagram for the entire frame.

Example 5.8   **Roof truss with vertical loads**

Calculate the forces in the members marked 2–3, 1–6 and 2–6 of the frame shown in Fig. 5.20.

Fig. 5.20 Example 5.8

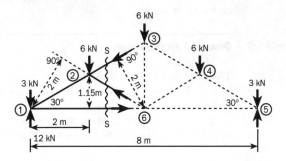

Solution   Member 2–3: take moments about point 6 (section at S–S).

$$\text{Moment of reaction} = 12 \times 4$$
$$= 48 \text{ kN m clockwise}$$
$$\text{Moment of applied loads} = (3 \times 4) + (6 \times 2)$$
$$= 24 \text{ kN m anticlockwise}$$
$$\text{Moment of force in member } 2–3 = 48–24$$
$$= 24 \text{ kN m anticlockwise}$$

Therefore

$$\text{Force } 2–3 \times 2 \text{ m} = 24 \text{ kN m}$$
$$\text{Force } 2–3 = 12 \text{ kN (strut)}$$

Member 1–6: take moments about point 2

$$\text{Moment of reaction} = 12 \times 2$$
$$= 24 \text{ kN m clockwise}$$
$$\text{Moment of loads} = 3 \times 2$$
$$= 6 \text{ kN m anticlockwise}$$
$$\text{Force } 1–6 \times 1.15 \text{ m (anticlockwise)} = 24 – 6$$
$$\text{Force } 1–6 = 18/1.15$$
$$= 15.7 \text{ kN (tie)}$$

Member 2–6: take moments about point 1

$$\text{Moment of reaction} = 0$$
$$\text{Moment of loads} = 6 \times 2$$
$$= 12 \text{ kN m clockwise}$$
$$\text{Force } 2–6 \times 2 \text{ m (anticlockwise)} = 12 \text{ kN m}$$
$$\text{Force } 2–6 = 6 \text{ kN (strut)}$$

Remember that, for determining whether members are struts or ties, the arrows as shown in Fig. 5.20 must be considered as acting towards or away from the nearest joint in that portion of the frame which remains after the *cut* has been made.

Example 5.9   **Girder with parallel flanges**

The application of this method may be simplified considerably when the frame concerned has parallel flanges. Consider, for example, the Warren girder shown in Fig. 5.21(a).

In using the method of sections to find the force in member EF, the cut would be made through the section X–X, and moments taken about the point C.

Taking moments about C to the right,

$$(\text{Force in EF}) \times h = \text{moments of forces } R_R \text{ and } W_3$$
$$= (R_R \times x_2) - (W_3 \times x_1)$$
$$\text{Force in EF} = \frac{(R_R \times x_2) - (W_3 \times x_1)}{h}$$

But the value $(R_R \times x_2) - (W_3 \times x_1)$ is obviously the bending moment at C, considering the girder as an ordinary beam as shown in Fig. 5.21(b).

Fig. 5.21 Example 5.9: (a) load diagram; (b) girder considered as a simple beam

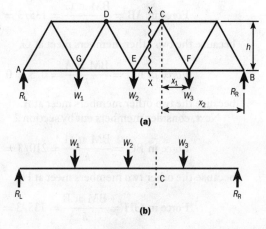

(a)

(b)

So, to calculate the force in any top or bottom boom member of a girder of this type, calculate the bending moment at the node point opposite to the member and divide by the vertical height of the girder, e.g.

$$\text{Force in EF} = \text{BM at point C} \div h$$
$$\text{Force in DC} = \text{BM at point E} \div h$$
$$\text{Force in GE} = \text{BM at point D} \div h$$

Note: This simplification of the method is sometimes known as the **method of moments**.

Example 5.10   In Fig. 5.22, by calculation, the reactions are

$$R_L = 45 \text{ kN} \quad \text{and} \quad R_R = 55 \text{ kN}$$
$$\text{BM at B and G} = (45 \times 3) = 135 \text{ kN m}$$
$$\text{BM at C and H} = (45 \times 6) - (20 \times 3) = 210 \text{ kN m}$$
$$\text{BM at D and J} = (55 \times 3) = 165 \text{ kN m}$$

Fig. 5.22 Example 5.10

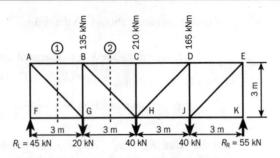

Therefore, considering members cut by section 1,

$$\text{Force in AB} = \frac{\text{BM at G}}{h} = 135/3 = 45 \text{ kN}$$

because the two other members meet at G,

$$\text{Force in FG} = \frac{\text{BM at A}}{h} = 0/3 = 0$$

because the two other members meet at A.
    Next, consider members cut by section 2.

$$\text{Force in BC} = \frac{\text{BM at H}}{h} = 210/3 = 70 \text{ kN}$$

because the other two members meet at H.

$$\text{Force in GH} = \frac{\text{BM at B}}{h} = 135/3 = 45 \text{ kN}$$

because the other two members meet at B.
    Hence

$$\text{Force in CD} = \frac{\text{BM at H}}{h} = 210/3 = 70 \text{ kN}$$

$$\text{Force in HJ} = \frac{\text{BM at D}}{h} = 165/3 = 55 \text{ kN}$$

$$\text{Force in DE} = \frac{\text{BM at J}}{h} = 165/3 = 55 \text{ kN}$$

$$\text{Force in JK} = \frac{\text{BM at E}}{h} = 0/3 = 0$$

Method of resolution of forces at joints

The forces in the individual members of loaded frames may also be determined by considering the various forces acting at each node point. This method is particularly useful for dealing with those members not so easily tackled by the method of sections, e.g. the vertical members of the frame dealt with in the previous example.

Example 5.11

Consider the node F in Figure 5.23(a).
    There is a load of 45 kN *upward* at F. Obviously, member AF must itself counteract this (as FG, a horizontal member, cannot resist vertical force). Hence the force in member AF must equal 45 kN, and it must, at the node

point F, be acting downwards in opposition to the reaction R_L, as shown by the arrowhead 1. This shows member AF to be a strut, and the other arrowhead 2 may be put in, acting towards A.

Consider the node A.

It is now known that a force of 45 kN acts upwards at this point (arrowhead 2). Again, member AB, being horizontal, cannot resist a vertical load, so member AG must resist the upward 45 kN at A. Thus the force in AG must act downwards at A, and the arrow 3 may be placed in position. The amount of force AG must be such that its vertical component equals 45 kN as in Fig. 5.23(b). Thus

$$\frac{45}{\text{force AG}} = \sin 45° = 0.7071$$

Therefore

$$\text{Force in AG} = 45/0.7071 = 63.6 \text{ kN}$$

and AG is in tension, so the arrow 4 may be drawn in.

The load of 63.6 kN acting down and to the right at A tends to pull point A to the right, horizontally. Thus the force in member AB at A must *push* point A to the left for equilibrium (arrow 5, showing member AB to be a strut); thus arrow 6 may be placed as shown in Fig. 5.23(c). The amount of force in AB is such that it counteracts the horizontal component of force 63.6 kN (arrow 3).

Thus, as in Fig. 5.23(d),

$$\frac{\text{Force in AB}}{63.6} = \cos 45° = 0.7071$$

Fig. 5.23 Example 5.11: (a) load diagram; (b) force AG; (c) force in AB; (d) force AB; (e) force in BG

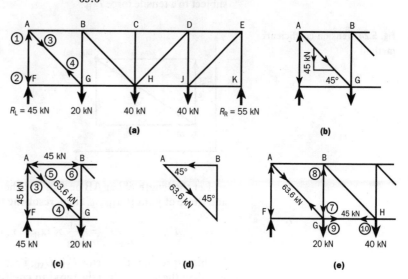

Thus force $AB = 63.6 \times 0.7071 = 45$ kN compression, as was seen by the method of moments.

Consider the node G.

It has already been seen that there is no force in member FG, since the bending moment is zero at the opposite point A. Thus, as there is a force of 63.6 kN at G, acting upwards and to the left, the force in BG must act in such a way that it counteracts the resultant of the vertical forces at G. These vertical forces are:

- the vertical resultant upward of the 63.6 kN (i.e. 45 kN↑)
- the downward applied load of 20 kN↓

Their resultant is obviously 25 kN upward, so member BG must act downwards to the extent of 25 kN (arrows 7 and 8 may thus be placed) and BG is thus a strut with a force of 25 kN (Fig. 5.23(e)). Also, as AG pulls to the *left* at G to the extent of 45 kN (the horizontal component of the force AG) then member GH must pull to the *right* to the extent of 45 kN (arrow 9), and member GH is seen to be a tie with a force of 45 kN. This again agrees with the result obtained by the method of moments.

This process may be continued from each node to the next, and although its application is admittedly somewhat more laborious than the drawing of a force diagram, this method is strongly recommended. Its use most definitely results in a better and more complete understanding of the forces acting at each node, and by its use struts and ties are more easily distinguished.

Method of tension coefficients   This is a modified version of the previous method and it is particularly useful in the case of pin-jointed, three-dimensional or space frames.

The basis of this method is the resolution into components of the external and internal forces acting on each joint of the frame, using the lengths of members and the coordinates of joints. All the members of the frame are initially assumed to be in tension.

Figure 5.24 shows bar AB, a member of a pin-jointed frame, assumed to be subject to a tensile force T_{AB}.

Fig. 5.24 Tension coefficients method

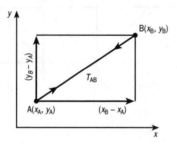

Let the length of bar AB be L_{AB} and the coordinates of joint A be x_A, y_A and those of joint B, x_B, y_B. Then, resolving the force T_{AB} in the x direction

$$T_{AB} \times \frac{x_B - x_A}{L_{AB}} = \frac{T_{AB}}{L_{AB}} \times (x_B - x_A)$$

In this expression the factor T_{AB}/L_{AB}, i.e. tension in bar AB divided by the length of the bar, is called the **tension coefficient** for the bar and is denoted by the symbol t_{AB}.

Hence the component in the x direction is

$$t_{AB}(x_B - x_A)$$

and, similarly, the component in the y direction is

$$t_{AB}(y_B - y_A)$$

Where there are a number of members and applied loads acting at a joint, which is in equilibrium under the action of the internal and external forces, two sets of equations can be formed, one for each of the directions x and y.

Example 5.12   A load of 7.2 kN is suspended from a soffit by two ropes PQ and QR (Fig. 5.25). Determine the forces in the ropes.

Fig. 5.25 Example 5.12

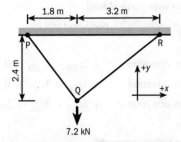

Solution   The equations for joint Q (i.e. assuming the origin of coordinates is at Q) in terms of tension coefficients are

$$\text{In direction } x \quad -1.8t_{PQ} + 3.2t_{QR} = 0 \tag{1}$$

$$\text{In direction } y \quad +2.4t_{PQ} + 2.4t_{QR} - 7.2 = 0 \tag{2}$$

Then from (2)

$$t_{PQ} = 3.0 - t_{QR}$$

and substituting into (1)

$$-1.8(3.0 - t_{QR}) + 3.2t_{QR} = 0$$

$$t_{QR} = 5.4/5.0 = 1.08$$

and

$$t_{PQ} = 3.0 - 1.08 = 1.92$$

But $t_{PQ} = T_{PQ}/L_{PQ}$ and thus $T_{PQ} = t_{PQ} \times L_{PQ}$ Therefore

$$T_{PQ} = 1.92 \times \surd(2.4^2 + 1.8^2)$$

$$= 1.92 \times 3.0 = 5.76 \text{ kN}$$

and

$$T_{QR} = 1.08 \times \surd(2.4^2 + 3.2^2)$$

$$= 1.08 \times 4.0 = 4.32 \text{ kN}$$

It should be noted that both T_{PQ} and T_{QR} have positive values and, therefore, they are ties. A negative solution denotes a compression member, i.e. a strut.

From the above examples it follows that, for a frame having J joints, $2J$ equations can be formed. This, incidentally, is the same number of equations as that used in the method of resolution at joints, which may explain the lack of popularity of the method of tension coefficients for plane frames. It is, however, very useful for computer application, particularly in the solution of space frames.

Space frames   A space frame has to be considered in three dimensions, hence the components of the forces at each joint are resolved in three directions x, y and z.

Consequently, there will be $3\mathcal{J}$ equations for the frame and the number of members in a perfect pin-jointed space frame is $m = 3\mathcal{J} - 6$. Most space frames, however, will have pinned supports, so the number of members in such a frame is $m = 3\mathcal{J}_f$, where $\mathcal{J}_f$ denotes a free joint (i.e. a joint other than at a support).

Example 5.13   Space frame with vertical load

The three equations for joint A of the shear legs shown in Fig. 5.26 are formed as follows:

$$\text{In direction } x \quad + 2t_{AB} - 2t_{AD} = 0 \tag{1}$$

$$\text{In direction } y \quad + 3t_{AB} + 3t_{AC} + 3t_{AD} + 21 = 0 \tag{2}$$

$$\text{In direction } z \quad + 2t_{AB} + 4t_{AC} + 2t_{AD} = 0 \tag{3}$$

Fig. 5.26 Example 5.13: (a) plan; (b) elevation

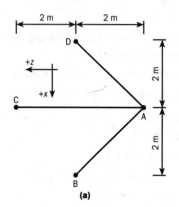
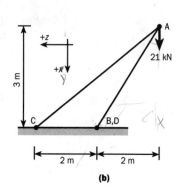

Then from (1), $t_{AB} = t_{AD}$, and adding (1) and (3) gives

$$t_{AB} = -t_{AC}$$

and substituting both into (2),

$$- 3t_{AC} + 3t_{AC} - 3t_{AC} = -21$$

$$t_{AC} = 7 \quad \text{and} \quad t_{AB} = -7 = t_{AD}$$

Therefore

$$T_{AB} = -7 \times \sqrt{(2^2 + 2^2 + 3^2)}$$
$$= 28.86 \text{ kN (strut)}$$
$$T_{AC} = +7 \times \sqrt{(4^2 + 3^2)}$$
$$= 35.00 \text{ KN (tie)}$$
$$T_{AD} = 28.86 \text{ kN (strut)}$$

Example 5.14   Space frame with inclined load

In the case of Fig. 5.27 the simple space frame is supporting an inclined load. It is, therefore, necessary to resolve the load into components in directions x, y and z, before writing down the usual equations.

Fig. 5.27 Example 5.14: (a) plan; (b) elevation

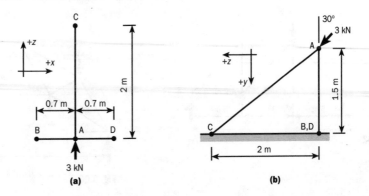

(a)   (b)

Let the load be F and its components F_x, F_y and F_z, respectively. Therefore

$$F_x = 0$$
$$F_y = 3 \times \cos 30° = 2.6 \text{ kN}$$
$$F_z = 3 \times \sin 30° = 1.5 \text{ kN}$$

It is usually helpful, particularly for a many-jointed frame, to present the equations and the resulting tension coefficients in tabular form as shown in Table 5.4.

Hence

$$T_{AB} = T_{AD} = -0.49 \times \sqrt{(0.7^2 + 1.5^2)}$$
$$= -0.811 \text{ kN (negative, i.e. strut)}$$

and

$$T_{AC} = -0.75 \times \sqrt{(2.0^2 + 1.5^2)}$$
$$= -1.875 \text{ kN (also a strut)}$$

Table 5.4

Joint	Direction	Equation	Tension coefficient
A	x	$-0.7t_{AB} + 0.7t_{AD} = 0$	$t_{AB} = t_{AD}$
	y	$+2.6 + 1.5t_{AB} + 1.5t_{AC} + 1.5t_{AD} = 0$	$t_{AB} = -0.49$
	z	$1.5 + 2.0t_{AC} = 0$	$t_{AC} = -0.75$

Exercises

Note: All loads are in kN.

1–15 Determine the reactions and the type and magnitude of the forces in the members of the frames shown in Figs. 5.Q1–5.Q15. Show the results in tabular form.

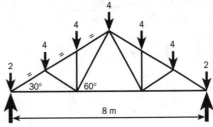

Fig. 5.Q1

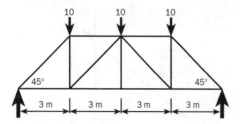

Fig. 5.Q2

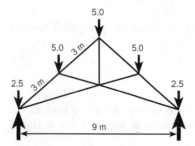

Fig. 5.Q3

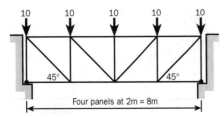

Fig. 5.Q4

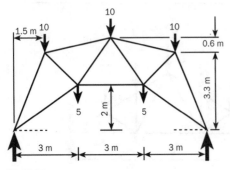

Fig. 5.Q5

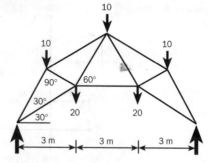

Fig. 5.Q6

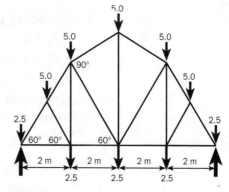

Fig. 5.Q7

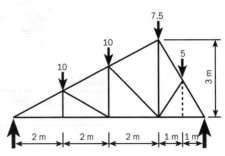

Fig. 5.Q8

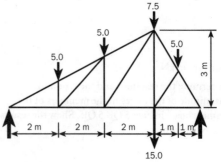

Fig. 5.Q9

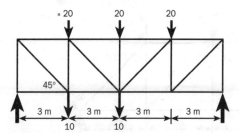

Fig. 5.Q10

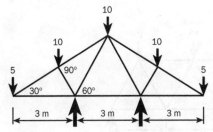

Fig. 5.Q11

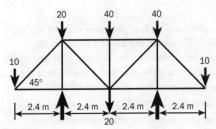

Fig. 5.Q12

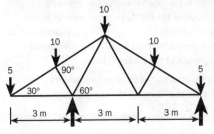

Fig. 5.Q13

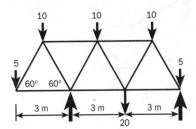

Fig. 5.Q14

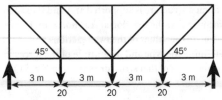

Fig. 5.Q15

16–18 For the frames shown in Figs. 5.Q16–5.Q18 determine:
- the directions and values of the reactions
- the type and magnitude of the forces in all the members

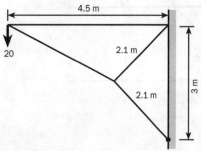

Fig. 5.Q16

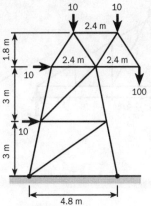

Fig. 5.Q17

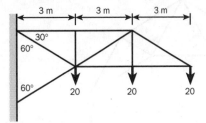

Fig. 5.Q18

19   A framed structure is supported at A and B in such
a manner that the reaction at A is horizontal.
Calculate the values of the two reactions and
determine the type and magnitude of the forces in
the members of that structure shown in Fig. 5.Q19.

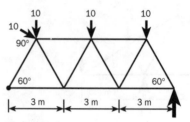

Fig. 5.Q22

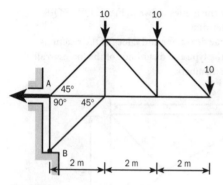

Fig. 5.Q19

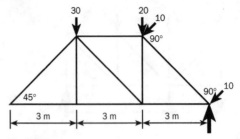

20–23 The right-hand ends of the frames shown in Figs.
5.Q20–5.Q23 are supported on rollers, i.e. the
reaction there is vertical as indicated. Determine
the reactions and the type and magnitude of the
forces in the members of the frames.

24–27 Repeat Questions 20–23 assuming that the roller
supports are now at the left-hand ends of the
frames.

28–31 Determine the reactions and the type and
magnitude of the forces in the members in the
frames shown in Figs. 5.Q20–5.Q23, assuming that
the two supports resist equally the horizontal
components of the inclined loads.

32–34 Using the method of tension coefficients
determine the forces in the space frames shown in
Figs. 5.Q32– 5.Q34.

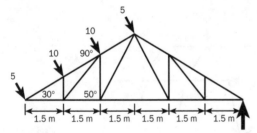

Fig. 5.Q20

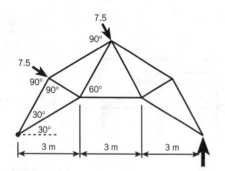

Fig. 5.Q21

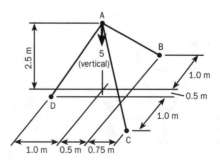

Fig. 5.Q32

Fig. 5.Q23

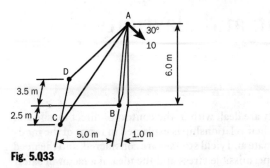

Fig. 5.Q33

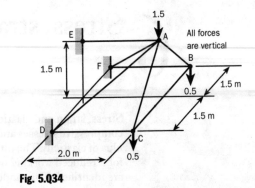

Fig. 5.Q34

6 Stress, strain and elasticity

Stress, strain and elasticity are dealt with in the context of direct tensile and compressive stresses and their relationship is explained in terms of the modulus of elasticity. The ultimate and yield stresses are considered, and the need for a reduced working or permissible stress and the idea of a factor of safety are identified. The modular ratio for composite elements is introduced.

Stress   When a member is subjected to a load of any type, the many fibres or particles, of which the member is made up, transmit the load throughout the length and section of the member, and the fibres doing this work are said to be in a state of stress.

There are different types of stress, but the principal kinds are tensile and compressive stresses.

Tensile stress   Consider the steel bar shown in Fig. 6.1(a) having a cross-sectional area of A mm^2 and pulled out at each end by forces W. Note that the total force in the member is W kN (not $2W$ kN). There would be no force at all in the member if only one load W was present as the member would not be in equilibrium.

Fig. 6.1 Stress in a steel bar: (a) tensile; (b) compressive

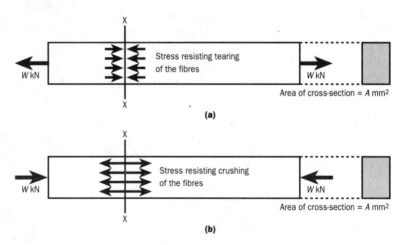

At *any* plane such as X–X taken across the section, there exists a state of stress between the fibres on one side of the plane and those on the other. Here the stress is tensile by nature – the fibres on one side exerting stress to resist the tendency of being pulled away from the fibres on the other side of the plane. This type of resistance to the external loads is set up along the whole length of the bar and not at one plane only, just as resisting forces are set up by every link in a chain.

In this particular case (Fig. 6.1(a)), if the loads act through the centroid of the shape, the stress is provided equally by the many fibres and each square millimetre of cross-section provides the same resistance to the 'pulling apart' tendency. This is known as **direct** or **axial stress**.

Figure 6.2 represents a bar of cast steel which is thinner at the middle of its length than elsewhere, and which is subjected to an axial pull of 45 kN. If the

Fig. 6.2 Tensile stress in a non-uniform steel bar

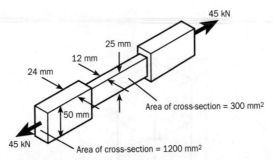

bar were to fail in tension, it would be due to the bar snapping where the amount of material is a minimum. The total force tending to cause the bar to fracture is 45 kN at all cross-sections, but whereas 45 kN is being resisted by a cross-sectional area of 1200 mm² for part of its length it is being resisted by only 300 mm² at the middle portion of the bar. The intensity of load is greatest at this middle section and is at the rate of 150 N/mm² of cross-section. At other points along the bar, 45 kN is resisted by 1200 mm² and the stress is equal to 37.5 N/mm².

In cases of direct tension, therefore

$$\text{stress} = \frac{\text{applied load}}{\text{area of cross-section of member}} = \frac{W}{A}$$

The SI unit for stress is the pascal, Pa = 1 N/m². However, since the square metre is a rather inconvenient unit of area when measuring stress, British Standards have adopted the multiple N/mm² as a stress unit for materials.

It should be noted that 1 N/mm² = 1 MN/m² = 1 MPa.

Example 6.1 A bar of steel 2000 mm² in cross-sectional area is being pulled with an axial force of 180 kN. Find the stress in the steel.

Solution Since 2000 mm² of cross-section is resisting 180 kN or 180 000 N it means that each mm² is resisting 90 N. In other words,

$$\text{stress} = \frac{W}{A} = \frac{180\,000 \text{ N}}{2000 \text{ mm}^2} = 90 \text{ N/mm}^2$$

Compressive stress Figure 6.1(b) shows a similar member to that of Fig. 6.1(a) but with two axial forces of W kN each acting inwards towards each other and thus putting the bar into a state of compression.

Again, at any plane section such as X–X there is a state of stress between the fibres, but this time the stress which is generated is resisting the tendency of the fibres to be crushed. Once more the stress is shared equally beween the fibres, and the stress is

$$\frac{\text{load}}{\text{area}} \quad \text{i.e.} \quad \frac{W}{A} \text{ MN/m}^2 \quad \text{or} \quad \text{N/mm}^2$$

Example 6.2   A brick pier is 0.7 m square and 3 m high and weighs 19 kN/m^3. It is supporting an axial load from a column of 490 kN (Fig. 6.3). The load is uniformly spread over the top of the pier so the arrow shown merely represents the resultant of the load. Calculate the stress in the brickwork immediately under the column and the stress at the bottom of the pier.

Fig. 6.3 Example 6.2

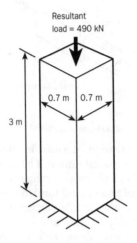

Resultant
load = 490 kN

0.7 m | 0.7 m

3 m

Solution   Area of cross-section $= 0.49$ m^2. 490 kN on 0.49 m^2 is equivalent to 1000 kN on each square metre. Immediately under the column, therefore

$$\text{Stress} = 490 \text{ kN}/0.49 \text{ m}^2 = 1 \text{ MN/m}^2 \quad \text{or} \quad 1 \text{ N/mm}^2$$
$$\text{Weight of pier} = 0.7 \times 0.7 \times 3.0 \times 19 = 28 \text{ kN, hence}$$
$$\text{Total load} = 490 + 28 = 518 \text{ kN}$$

and at the bottom of the pier

$$\text{Stress} = 518 \text{ kN}/0.49 \text{ m}^2 = 1057 \text{ kN/m}^2 \quad \text{or} \quad 1.06 \text{ N/mm}^2$$

Strain   All materials slightly alter their shape when they are stressed. A member which is subjected to tensile stress increases in length and its cross-section becomes slightly smaller. Similarly, a compression member becomes shorter and slightly larger in cross-section. The very slight alteration in cross-section of building members is not as a rule important and only the alterations in length will be considered in this book.

When a tension force is applied to a rubber band it becomes longer, and when the force is removed, the material reverts to its original length. This property is common to some building materials such as steel, although under its normal working loads the amount of elongation is usually too small to be detected by the unaided eye. Another property of elastic materials is that the alteration in length is directly proportional to the load. For example, if in a given member

10 kN produces an extension of 2 mm

20 kN will produce an extension of 4 mm

30 kN will produce an extension of 6 mm

and so on, provided the *elastic limit* of the material is not exceeded.

This law:

Change in length is proportional to the force.

was first stated by Robert Hooke (1635–1703) and is therefore known as **Hooke's law**. Hooke was a mathematician and scientist and a contemporary of Sir Christopher Wren.

Although all materials alter in length when stressed, materials which are not elastic do not obey Hooke's law. That is, the changes in length are not directly proportional to the load. Structural steel and timber are almost perfect elastic materials.

Just as it is convenient to represent the stress in a member as force per unit area, so it is convenient to represent the change in length of a member in terms of change per unit length, i.e.

$$\text{strain} = \frac{\text{change in length}}{\text{original length}} = \frac{\delta l}{l}$$

where δ is the Greek letter delta, which is frequently used to denote small changes.

Since strain is the direct outcome of stress it, too, is classified as tensile and compressive.

Elasticity
Suppose that a general formula is required for determining the amount of elongation (or shortening) in any member composed of an elastic material. Let that material be rubber. By carrying out experiments on pieces of rubber of different lengths, different areas of cross-section and different qualities, the following facts would be demonstrated:

- Increase in the load W produces proportionate increase of elongation (Hooke's law), therefore W is one term of the required formula.
- Increase in length l of the member means increase in elongation. It will be found that the elongation is directly proportional to the length of the member, i.e. for a given load and a given area of cross-section a member 500 mm long will stretch twice as much as a member of the same material 250 mm long. l is therefore a term in the formula.
- It is harder to stretch a member of large cross-sectional area A than a member of small cross-sectional area. Experiments will show that the elongation is inversely proportional to the area, i.e. for a given load and a given length a member 10 mm^2 in cross-sectional area will stretch only half as much as a member 5 mm^2 in area. A is therefore a term in the formula.
- Rubber can be obtained of different qualities, i.e. of different stiffnesses. Other conditions being constant (i.e. length, area, etc.) it is more difficult to stretch a 'stiff' rubber than a more flexible rubber. Let this quality of stiffness be represented by the symbol E, then the greater the value of E, i.e. the greater the stiffness of the material, the smaller will be the elongation; other conditions (length, load and area) being constant.

The formula for elongation (or shortening) of an elastic material can now be expressed as

$$\delta l = \frac{Wl}{AE}$$

W and l are in the numerator of the fraction, because the greater their values, the greater the elongation. A and E are in the denominator of the fraction because the greater their values (i.e. the greater the area and the greater the stiffness) the less will be the elongation.

The symbol E in the above formula represents the stiffness of a given material, i.e. it is a measure of its elasticity and is called the **modulus of elasticity**. If it can be expressed in certain definite units, the formula can be used to determine elongations of members of structures. Now if $\delta l = Wl/AE$, by transposing the formula,

$$E = Wl/A\,\delta l$$

and the value of E for different materials can be obtained experimentally by compression or tension tests.

The foregoing explanation is lengthy and the usual more concise treatment of the subject is as follows. In an elastic material stress is proportional to strain (Hooke's law), i.e.

$$\frac{\text{stress}}{\text{strain}} = \text{a constant value}$$

This value is called the modulus (i.e. measure) of elasticity of the material and is denoted by E.

Therefore

$$\frac{\text{stress}}{\text{strain}} = E$$

but

$$\text{stress} = \frac{\text{load}}{\text{area}} = \frac{W}{A}$$

and

$$\text{strain} = \frac{\text{change in length}}{\text{original length}} = \frac{\delta l}{l}$$

Thus

$$\frac{W}{A} \bigg/ \frac{\delta l}{l} = E \quad \text{or} \quad E = \frac{Wl}{A\,\delta l}$$

The modulus of elasticity is often called Young's modulus, after the scientist Thomas Young (1773–1829). Note that modulus of elasticity is expressed in the same units as stress. This is because stress is load per unit area and strain has no dimensions, it is just a number and therefore

$$\frac{\text{stress}}{\text{strain}} \left(\frac{\text{MN/m}^2 \text{ or N/mm}^2}{\text{a number}} \right) = E\,\text{N/mm}^2$$

Figure 6.4(a) indicates in essentials the manner of testing a block of timber 75 mm × 100 mm in cross-section and 300 mm high. In the actual experiment four gauges (extensometers) were used, one at each corner, and the average of the four readings was taken for each increment of load. A gauge was used at each corner because possible small inequalities in the top and bottom surfaces of the timber might result in unequal shortening of the timber block. The following readings (Table 6.1), which have been slightly

Fig. 6.4 Testing a block under compression: (a) experimental method; (b) graph of results given in Table 6.1

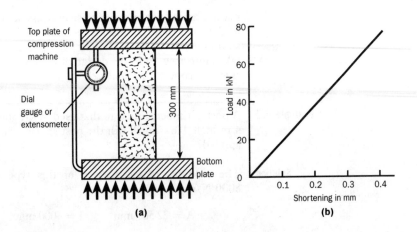

Top plate of compression machine

Dial gauge or extensometer

300 mm

Bottom plate

Load in kN

Shortening in mm

(a)

(b)

Table 6.1

Reading	Load on specimen W (kN)	Shortening of specimen δl (mm)
1	0	0.0
2	20	0.1
3	40	0.2
4	60	0.3
5	80	0.4

adjusted for purposes of clearer explanation, were recorded as the compressive load was gradually increased.

Now

$$E = \frac{Wl}{A\,\delta l}, \quad \text{where} \quad l = 300 \text{ mm} \quad \text{and} \quad A = 7500 \text{ mm}^2.$$

Hence

From reading 2 $\quad E = \dfrac{20\,000 \times 300}{7500 \times 0.1} = 8000 \text{ N/mm}^2$

From reading 3 $\quad E = \dfrac{40\,000 \times 300}{7500 \times 0.2} = 8000 \text{ N/mm}^2$

From reading 4 $\quad E = \dfrac{60\,000 \times 300}{7500 \times 0.3} = 8000 \text{ N/mm}^2$

From reading 5 $\quad E = \dfrac{80\,000 \times 300}{7500 \times 0.4} = 8000 \text{ N/mm}^2$

The results of the experiment can be plotted as shown in Fig. 6.4(b). In an actual experiment results might not be as uniform as given in Table 6.1, but the resultant graph would approximate very closely to a straight line.

This indicates that E is a constant value, and it is this value which is taken as the modulus of elasticity for the timber.

It has to be noted, however, that materials such as timber and concrete are subject to the phenomenon of **creep**. This is a time-dependent strain deformation caused not by an increase in stress but by the duration of the applied load. Under these conditions the value of E is not constant since the stress/strain diagram is not a straight line. These strain deformations have to be taken into account in the design of structural elements but the topic is outside the scope of this book.

Example 6.3

A post of timber similar to that used in the above test is 150 mm square and 4 m high. How much will the post shorten when an axial load of 108 kN is applied?

Solution

The modulus of elasticity E of this type of timber is known to be 8000 N/mm².

$$A = 22\,500 \text{ mm}^2 \qquad l = 4000 \text{ mm}$$

Therefore,

$$\delta l = \frac{Wl}{AE} = \frac{108\,000 \times 4000}{22\,500 \times 8000} = 2.4 \text{ mm}$$

This problem demonstrates that if the modulus of elasticity of a material is known, the amount a given member will shorten (or lengthen) under a given load can be calculated.

Behaviour of steel in tension – yield point

A steel bar, 12 mm in diameter, was gripped in the jaws of a testing machine and subjected to a gradually increasing pull (Fig. 6.5). A gauge was attached to two points on the bar 250 mm apart (gauge length = 250 mm) and the following readings were recorded (Table 6.2). (The elongations given in the table have been slightly adjusted.)

The area A of a 12 mm diameter circle is 113.1 mm².

$$l = 250 \text{ mm}$$

From reading 2,

$$E = \frac{Wl}{A\,\delta l} = \frac{3800 \times 250}{113.1 \times 0.04} = 210\,000 \text{ N/mm}^2$$

From reading 3,

$$E = \frac{Wl}{A\,\delta l} = \frac{7600 \times 250}{113.1 \times 0.08} = 210\,000 \text{ N/mm}^2$$

and so on.

The constant value of E in this case is 210 000 N/mm².

Note that the modulus of elasticity of steel is much greater than that of timber. Steel, of course, is much more difficult to stretch than is timber.

Within the range of the above experiment, where the highest recorded load was 19 kN, the steel behaved as an elastic material. This means that, when the load is taken off, the bar will revert to its original length; on reloading, the bar will again elongate; and on unloading the elongation will disappear, and so on.

If the load is gradually increased beyond 19 kN, the bar will continue to stretch proportionately to the applied load until a loading of between 30 and 32 kN is reached. (These loads on a bar 12 mm in diameter are equivalent to

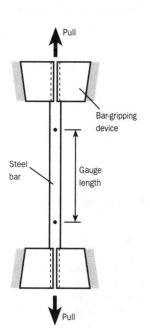

Fig. 6.5 Testing a steel bar in tension

Table 6.2

Reading	Load (kN)	Elongation (mm)
1	0.0	0.00
2	3.8	0.04
3	7.6	0.08
4	11.4	0.12
5	15.2	0.16
6	19.0	0.20

stresses of 265–280 N/mm^2.) At about this point the steel reaches what is called its **elastic limit** (i.e. it ceases to behave as an elastic material) and begins to stretch a great amount compared with the previous small elongations. This stretching takes place without the application of any further load and the steel is said to have reached its **yield point**.

At this point the **plastic or ductile behaviour** of the steel begins. If the load is removed from the steel after it has been loaded beyond its yield point, the steel will not revert to its original length. The elongation which remains is called *permanent set*.

After a short time the steel recovers a little and ceases to stretch. Additional load can now be applied, but the steel has been considerably weakened and stretches a great deal for each small increment of load. Finally, at about a stress of 450 N/mm^2 the bar breaks, but just before it breaks, it *waists* at the point of failure as indicated in Fig. 6.6(a). If the two fractured ends are placed together and the distance between the original gauge points is measured it will be found that the total elongation is 20 per cent or more of the original length.

Fig. 6.6 Results of a tensile test of steel: (a) waisting of bar at failure point; (b) typical graph of results

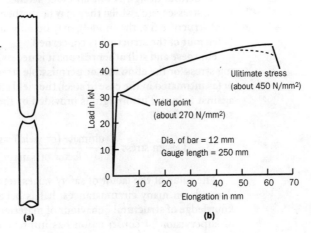

Figure 6.6(b) shows a graph which has been constructed from the results of a test on a 12 mm diameter bar, the gauge length being 250 mm.

Actually, the limit of proportionality (stress proportional to strain) is reached a little before the yield point, but for most practical purposes it is quite justifiable to consider the limit of proportionality, the elastic limit and the yield point to be identical. In addition, referring to Fig. 6.6(a), the bar reduces considerably in cross-section just before it fails, and some load can be taken off the bar so that the actual load which causes failure is less than the maximum recorded load. This is indicated on the graph by the dotted line. The ultimate or failing stress, i.e. the tensile strength of the steel, is calculated on the original cross-sectional area of the bar and not on the final reduced cross-section.

Strength of materials and factor of safety

In Examples 6.1 and 6.2 the stress in the steel was calculated as 90 N/mm^2 and the stress in the brickwork as 1.0 N/mm^2 (or 1.06 N/mm^2) respectively. Two questions now arise:

- Is it safe to allow these materials to be stressed to this extent?

• Can higher stresses be allowed so that smaller members may be used, thus economizing in material?

From the above discussion of the behaviour of steel in tension it can be deduced that the bar of 2000 mm^2 (Example 6.1) would require about 900 kN to cause it to fail in tension. The bar, however, is only supporting 180 kN, therefore it is amply strong and a smaller bar would suffice. The question now is: How small can the bar be? It would be, surely, unwise to make it of such dimensions that the steel in the bar would be stressed beyond its yield point because the steel would be in an unstable state. Various other factors also argue against using too high a stress for design purposes. The actual stress in the bar might be more (or even less) than the calculated value, due to assumptions made during the calculation stage. For example, the calculations may be based on the presence of a perfect hinge at a certain point in the structure. If the construction is such that there is a certain amount of fixity at the so-called hinge, the actual stresses in the members might be somewhat different from the calculated stresses.

Structural design is not an exact science, and calculated values of reactions, stresses, etc., whilst they may be mathematically correct for the theoretical structure (i.e. the model), may be only approximate as far as the actual behaviour of the structure is concerned.

For these and still other reasons it is necessary to make the design or working stress (or the allowable or permissible stress) less than the ultimate stress or (as intimated in the case of steel) the yield stress to allow for a safety margin against failure. This margin is provided by the introduction of the **factor of safety**, so that

$$\text{Design stress} = \frac{\text{ultimate (or yield) stress}}{\text{factor of safety}}$$

The value of the factor of safety varies between 1.4 and 2.4 at present and depends on many circumstances. It has been progressively reduced as the knowledge of structural behaviour of materials has increased and the quality of supervision of construction has improved. Its application to structural calculations is explained in later chapters.

Stresses in composite members – modular ratio

The knowledge of the modulus of elasticity E is particularly useful in determining stresses in composite members. These are structural elements made up of two (or more) materials (e.g. steel and timber in flitch beams or steel and concrete in reinforced concrete beams), in which the materials are rigidly fixed together so that any changes in the length of the element are the same in each of the constituent materials.

Consider an element consisting of materials A and B. Then

$$E_A = \frac{\text{stress}_A}{\text{strain}_A} \quad \text{and} \quad E_B = \frac{\text{stress}_B}{\text{strain}_B}$$

but strain$_A$ = strain$_B$, from the above definition of composite member, so taking the ratio of the moduli of elasticity

$$\frac{E_A}{E_B} = \frac{\text{stress}_A}{\text{strain}_B}$$

The ratio $E_A A/E_B$, being the ratio of two moduli, is called the **modular ratio** and is denoted by the letter m.

$$\text{Modular ratio } m = \frac{E_A}{E_B}$$

Example 6.4 Two 150 mm × 75 mm × 4 m long timber members are reinforced with a steel plate 150 mm × 6 mm × 4 m long (Fig. 6.7), the three members being adequately bolted together.

Fig. 6.7 Example 6.4

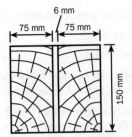

The permissible stresses for the timber and the steel are 6 N/mm² and 130 N/mm², respectively, and E for timber is 8200 N/mm² and for steel is 205 000 N/m².

Calculate the permissible tensile load for this composite member and the amount of elongation due to this load.

Solution

$$\text{Area of timber} = 2 \times 150 \times 75 = 22\,500 \text{ mm}^2$$

$$\text{Area of steel} = 150 \times 6 = 900 \text{ mm}^2$$

The stresses in the composite parts will be in the ratio

$$\frac{E_s}{E_t} = \frac{205\,000}{8200} = 25$$

so that if the stress in timber is 6 N/mm², the stress in steel would have to be $6 \times 25 = 150$ N/mm² which exceeds the permissible stress for steel of 130 N/mm².

It follows, therefore, that the timber may not be fully stressed and the steel stress is the critical one.

Thus the stress in timber will be $130/25 = 5.2$ N/mm².

Hence

$$\text{Safe load for timber} = 5.2 \times 22\,500 = 117 \text{ kN}$$

$$\text{Safe load for steel} = 130 \times 900 \quad = \underline{117 \text{ kN}}$$
$$\overline{234 \text{ kN}}$$

and the elongation

$$\delta l = \frac{Wl}{AE} = \frac{117\,000 \times 4000}{900 \times 205\,000} = 2.6 \text{ mm} \quad \text{or}$$

$$\delta l = \frac{117\,000 \times 4000}{22\,500 \times 8200} = 2.6 \text{ mm}$$

Summary   *Stress*   Obtained by dividing the applied load by the area of cross-section of the member, i.e.

$$\text{Tensile or compressive stress} = \frac{W}{A}$$

The design, permissible, working or allowable stress for a material depends on the nature of the material, the type of stress and the use of the material in the building, for example, whether it is used in a long column or in a short column.

Factor of safety   This is the failing or ultimate stress of the material divided by the design, permissible, working or allowable stress.

Strain   In tension or compression, strain is given by

$$\frac{\text{change in length}}{\text{original length}} = \frac{\delta l}{l}$$

   In elastic materials obeying Hooke's law stress is proportional to strain provided the elastic limit of the material is not exceeded.

Modulus of elasticity   Also called Young's modulus, E, it is a measure of the resistance of an elastic material to being stretched or shortened. The greater the value of E, the more difficult it is to cause shortening or lengthening of the material.

$$E = \frac{\text{stress}}{\text{strain}} = \frac{Wl}{A\,\delta l}$$

and is measured in MN/m^2, N/mm^2, etc.

Modular ratio   When two materials A and B are combined,

$$\frac{\text{stress}_A}{\text{stress}_B} = \frac{E_A}{E_B} = m$$

where m denotes the modular ratio.

Exercises

1 A steel tie-bar 100 mm × 10 mm in cross-section is transmitting a pull of 135 kN. Calculate the stress in the bar.

2 Calculate the safe tension load for a steel bar 75 mm × 6 mm in cross-section, the working stress being 155 N/mm^2.

3 A tie-bar is 75 mm wide and it has to sustain a pull of 100 kN. Calculate the required thickness of the bar if the permissible stress is 150 N/mm^2.

4 A bar of steel circular in cross-section is 25 mm in diameter. It sustains a pull of 60 kN. Calculate the stress in the bar.

5 Calculate the safe load for a bar of steel 36 mm in diameter if the working stress is 155 N/mm^2.

6 A bar of steel, circular in cross-section, is required to transmit a pull of 40 kN. If the permissible stress is 150 N/mm^2 calculate the required diameter of the bar.

7 A timber tension member is 100 mm square in cross-section. Calculate the safe load for the timber if the permissible stress is 8 N/mm^2. Calculate the diameter of a steel bar which would be of equal strength to the timber member. Permissible stress for the steel is 150 N/mm^2.

8 A tie-bar of the shape shown in Fig. 6.Q8 has a uniform thickness of 12 mm and has two holes 20 mm diameter each. Calculate the width x so that the bar is equally strong throughout its length. Calculate the safe pull for the bar if the permissible stress is 150 N/mm^2.

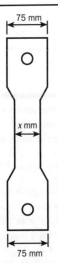

75 mm

x mm

75 mm

Fig. 6.Q8

9 A tie-bar of steel 150 mm wide is connected to a gusset plate by six rivets as shown in Fig. 6.Q9. The diameter of each rivet hole is 22 mm. Calculate the required thickness of the bar if the working stress is 155 N/mm^2.

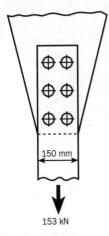

150 mm

153 kN

Fig. 6.Q9

10 Calculate the cross-sectional dimensions of a square brick pier to support an axial load of 360 kN, if the permissible stress for the brickwork is 1.7 N/mm^2.

11 A short specimen of deal timber 50 mm square in cross-section failed in a compression machine at a load of 70 kN. The permissible stress for such timber is 5.6 N/mm^2. Calculate the factor of safety.

12 A steel stanchion carrying a load of 877.5 kN is to be provided with a square steel base plate to spread the load on to a concrete foundation block. Calculate the minimum length of side (in mm) of the base plate if the stress on the concrete must not exceed 4.5 N/mm^2.

13 A steel column circular in cross-section is 150 mm in diameter and carries a load of 1.2 MN. Calculate (a) the compressive stress in the column; (b) the length of side of a square steel plate to transmit the column load to a concrete foundation block; the permissible stress on the concrete is 4.5 N/mm^2; (c) assuming the concrete base to weigh 150 kN, calculate the plan dimensions of the concrete foundation so that the stress on the soil does not exceed 200 kN/m^2.

14 The following data were recorded during a tensile steel test:

Diameter of bar = 20 mm
Distance between gauge points = 200 mm
Elongation due to load of 50 kN = 0.18 mm
Load at yield point = 79 kN
Failing or ultimate load = 127 kN

Calculate, in N/mm^2, (a) the stress at yield point; (b) the ultimate stress; (c) the modulus of elasticity of the steel.

15 A steel bar 100 mm × 12 mm in cross-section and 3 m long is subjected to an axial pull of 130 kN. How much will it increase in length if the modulus of elasticity of the steel is 210 000 N/mm^2?

16 A hollow steel tube 100 mm external diameter and 80 mm internal diameter and 3 m long is subjected to a tensile load of 400 kN. Calculate the stress in the material and the amount the tube stretches if Young's modulus is 200 000 N/mm^2.

17 During an experiment on a timber specimen 75 mm × 75 mm in cross-section, a shortening of 0.22 mm was recorded on a gauge length of 300 mm when a load of 36 kN was applied. Calculate the modulus of elasticity of the timber. Using this value of E, determine the amount of shortening of a timber post 150 mm square and 2.4 m high due to an axial load of 130 kN.

18 Assuming the permissible stress for a timber post 150 mm square and 2.7 m high is 6.5 N/mm^2, calculate the safe axial load for the post. How much will the post shorten under this load, assuming E to be 11 200 N/mm^2?

19 Three separate members of steel, copper and brass are of identical dimensions and are equally loaded. Young's moduli for the materials are steel, 205 000 N/mm^2; copper, 100 000 N/mm^2; brass, 95 000 N/mm^2. If the steel member stretches 0.13 mm calculate the amount of elongation in the copper and brass members.

20 During a compression test, a block of concrete 100 mm square and 200 mm long (gauge length = 200 mm) shortened 0.2 mm when a load of 155 kN was applied. Calculate the stress and strain and Young's modulus for the concrete.

21 A tension member is made of timber and steel firmly fixed together side by side. The cross-sectional area of the steel is 1300 mm^2 and that of the timber is 4000 mm^2 and the length of the member is 3 m. If the maximum permissible stresses for the steel and timber when used separately are 140 N/mm^2 and 8 N/mm^2 respectively, calculate the safe load which the member can carry and the increase in length due to the load. Young's modulus for steel is 205 000 N/mm^2 and for timber is 8200 N/mm^2.

22 A structural member made of timber is 125 mm × 100 mm in cross-section. It is required to carry a tensile force of 300 kN and is to be strengthened by two steel plates 125 mm wide bolted to the 125 mm sides of the timber. Calculate the thickness of steel required if the permissible stresses for the steel and timber are 140 N/mm^2 and 7 N/mm^2 respectively. Assume that Young's modulus for the steel is 25 times that for the timber.

23 A timber post 150 mm square has two steel plates 150 mm × 6 mm bolted to it on opposite sides along the entire length of the post. Calculate the stresses in the timber and steel due to a vertical axial load of 350 kN. If the post is 3 m high calculate the amount of shortening under the load. E for steel is 205 000 N/mm^2 and E for timber is 8200 N/mm^2.

24 A metal bar consists of a flat strip of steel rigidly fixed alongside a flat strip of brass. The brass has a cross-sectional area of 900 mm^2 and the steel 300 mm^2. The compound bar was placed in a tensile testing machine and the extension measured by means of an extensometer fixed over a 250 mm gauge length. The extension was recorded as 0.13 mm. Calculate the load applied to the bar and the stress in each material.

E for brass = 80 000 N/mm^2

E for steel = 205 000 N/mm^2

7 Shear force and bending moment

In this chapter the twin effects of the action of applied loads on beams are investigated in terms of the shear force and bending moment. The plotting of shear force diagrams (SFDs) and bending moment diagrams (BMDs) is demonstrated, and the significance of the position of the points of zero shear (maximum bending moment M_{max}) and zero bending (contraflexure) is explained.

When a beam is loaded, the applied loads have a tendency to cause failure of the beam, and whether or not the beam *actually* does fail depends obviously upon the extent or amount of loading and on the size and strength of the beam in question.

It is necessary to provide a beam that will safely carry the estimated loading with a reasonable factor of safety, and which will at the same time be light enough for economy and shallow enough to avoid unnecessary encroachment upon headroom.

In order to calculate the stresses that loading will induce into the fibres of a beam's cross-section, and to compare them with the known safe allowable stress for the material of which the beam is made, it is necessary to study the ways in which loading *punishes* a beam, and to assess the degree of *punishment*.

Loading tends to cause failure in two main ways:

- By 'shearing' the beam across its cross-section, as shown in Fig. 7.1(a).
- By bending the beam to an excessive amount, as shown in Fig. 7.1(b).

These two tendencies to failure or collapse do occur simultaneously, but for a clearer understanding of each they will be examined separately.

Fig. 7.1 Failure caused by loading: (a) shearing; (b) bending

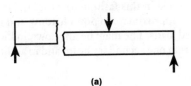
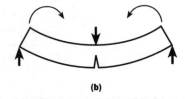

(a) (b)

Shear force Consider the portion of beam shown in Fig. 7.2(a). For simplicity of explanation the weight of the beam itself has been ignored.

Figure 7.2(b) shows how the beam would tend to shear at point A. The only load to the left of A is the l.h. reaction (acting upwards) of 55 kN and there is therefore a resultant force of 55 kN tending to shear the portion of beam to the left of A upwards as shown. Note that the loads to the right of A are 70 kN downward and reaction 15 kN upward, so that there is also a resultant of 55 kN tending to shear the portion of beam to the right of A downward.

This 55 kN upward to the left, and 55 kN downward to the right, constitutes the shearing force at the point A.

Fig. 7.2 Shear force in a beam

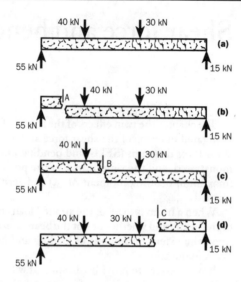

It follows from the above that **shear force** may be defined as the algebraic sum of the loads to the left or to the right of a point.

This type of shearing force, where the resultant shear is upwards to the left and downwards to the right, may be called **positive** shear.

Consider now the point B as shown in Fig. 7.2(c). The resultant shear to the left is seen to be $55 - 40 = 15$ kN upward to the left, and $30 - 15 = 15$ kN downward to the right (again positive shear). At point C in Fig. 7.2(d), the shear to the left is $30 + 40 - 55 = 15$ kN down, and to the right equals the r.h. reaction of 15 kN upward. This type of shear, down to the left and up to the right, is called **negative** shear.

The above example of shear force ignored the weight of the beam and the only loading consisted of point loads – that is to say the loading was considered to be applied at a definite point along the span. Loads in actual fact are rarely applied in this fashion to structural members, but many loads applied to beams approximate to point loading and in design are considered as concentrated loads. The main beam shown in Fig. 7.3, for example, carries the reaction from one secondary beam and in addition has sitting on its top flange a

Fig. 7.3 Shear forces considered as point loads

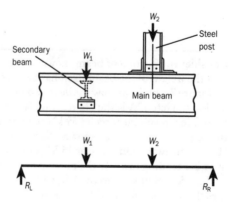

short steel post; both these loads are applied to such a short length of the beam that they may be considered as point loads.

On the other hand, there are many cases where the loading is applied at a more or less uniform rate to the span of the beam. Such an example is found where a brick wall is carried on the top flange of a beam, or when a reinforced concrete slab sits upon a steel or reinforced concrete beam. Now, shear force has been described as the algebraic summation of loads to the left (or to the right) of a point. Therefore, the shear alters in such cases only where another point load occurs. When the loading is uniformly distributed, however, as in Fig. 7.4, then the shear force will vary at a uniform rate also and a sudden jump in the value will only occur at the point of application of the one point load. The shear at point A (Fig. 7.4(a)) upward to the left will be

$$140 \text{ kN up} - 30 \text{ kN down} = +110 \text{ kN}$$

Similarly the shear at point B (Fig. 7.4(b)) upward to the left is

$$140 \text{ kN up} - 60 \text{ kN down} = 80 \text{ kN (again positive)}$$

The shear just to the left of C (Fig. 7.4(c)) is

$$140 - 90 = 50 \text{ kN positive}$$

whilst just to the right of C when the 40 kN point load is included it is

$$140 - (90 + 40) = 10 \text{ kN positive}$$

Thus it will be seen that uniform loads cause gradual and uniform change of shear, whilst point loads bring about a sudden change in the value of the shear force.

Fig. 7.4 Shear force due to uniform loading

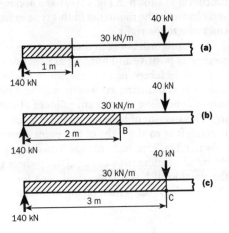

Bending moment   The degree of punishment in bending is measured as bending moment, and the amount of bending tendency is dependent upon the loads and upon the distance between them.

For example, the beam shown in Fig. 7.5(a) tends to split in bending under the 30 kN load because the l.h. reaction of 15 kN acting 1.5 m to the left has a

Fig. 7.5 Bending moments:
(a) sagging of beam; (b) hogging
of beam

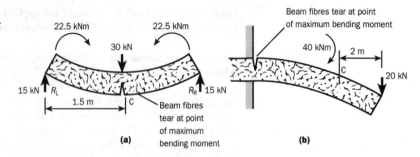

(a)

(b)

clockwise bending tendency of $15 \times 1.5 = 22.5$ kN m, at the point where splitting of this type would most easily occur in this case.

If the beam had been of 10 m span instead of only 3 m the l.h. reaction (ignoring for a while the self-weight of the beam) would still be only 15 kN, but this time the bending tendency at the point of maximum stress would be $15 \times 5 = 75$ kN m.

Reverting to the beam shown in Fig. 7.5(a), which we have seen has a clockwise bending tendency to the left of point C of 22.5 kN m, there is also an anticlockwise bending tendency to the right of C of $15 \times 1.5 = 22.5$ kN m caused by the action of the r.h. reaction. At *any point* along the span of a simple beam of this type supported at its ends, the bending tendency to the left will always be clockwise, and that to the right anticlockwise but of the same amount. Thus at any point of such a beam the bending will be of the *sagging* type as shown in Fig. 7.1(b), and the fibres towards the lower face of the section will be subjected to tension.

There are types of beams that bend in the opposite way. For example, a cantilever as shown in Fig. 7.5(b) has a *hogging* rather than a sagging tendency and obviously the moments in this type must be anticlockwise to the left and clockwise to the right.

To distinguish between these two types of bending it is normal to describe sagging as positive and hogging as negative.

From the foregoing example, which has been of the very simplest nature, it follows that **bending moment** may be described as the factor which measures the bending effect at any point of a beam's span due to a system of loading. The amount of bending moment is found by taking the moments acting to the left *or* to the right of the point concerned. The beam discussed and shown in Fig. 7.5(a) had only one load acting to the left of the point C at which the bending moment was calculated; where there are more than one load on the portion of beam concerned – as for example in the portion of beam shown in Fig. 7.6(a) – then taking moments to the left of point C, the reaction

Fig. 7.6 Example of bending
moments

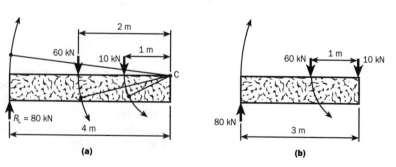

(a)

(b)

has a clockwise bending effect and the two downward loads have anticlockwise bending tendency. The net bending moment at C will be the difference between these two types of moment, i.e. their algebraic sum. The bending moment at any point is, therefore, defined as the algebraic sum of the moments caused by the forces acting to the left or to the right of that point.

In Fig. 7.6(a), the l.h. reaction exerts a clockwise moment of $80 \text{ kN} \times 4 \text{ m} = 320 \text{ kN m}$ about point C and the downward loads exert moments in an anticlockwise direction equal to $60 \text{ kN} \times 2 \text{ m} = 120 \text{ kN m}$ and $10 \text{ kN} \times 1 \text{ m} = 10 \text{ kN m}$. Thus the bending moment at C is

$$+80 \times 4 - 60 \times 2 - 10 \times 1 = 320 - 120 - 10 = 190 \text{ kN m}$$

The bending moment at the point of the 10 kN load in the above beam would be

$$80 \text{ kN} \times 3 \text{ m} - 60 \text{ kN} \times 1 \text{ m} = 240 - 60 = 180 \text{ kN m}$$

as shown in Fig. 7.6(b). Note that in this case the 10 kN load is ignored, as, passing through the point concerned, it does not exert a moment about that point.

When, on the other hand, a uniformly distributed load is applied to the beam, the punishment caused is less than that exerted by a comparable point load. Therefore, it would be most uneconomical in design to treat such loads as being concentrated at their midpoint.

Where such uniform rate of loading occurs, as in the case of the portion of beam shown in Fig. 7.7 for example, only that portion of loading which lies to the left of the point C (shown shaded) need be considered when the moment at that point to the left is being calculated. The shaded portion of load is $30 \text{ kN/m} \times 4 \text{ m} = 120 \text{ kN}$, and its resultant lies halfway along its length (2 m from C). Therefore the moment about C of the uniform load is $120 \text{ kN} \times 2 \text{ m} = 240 \text{ kN}$ and the total bending moment at point C is

$$200 \times 4 - 120 \times 2 = 800 - 240 = 560 \text{ kN m}$$

Fig. 7.7 Bending moments due to uniform loading

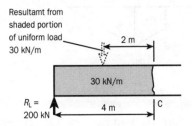

Resultamt from shaded portion of uniform load
30 kN/m
2 m
30 kN/m
$R_L =$ 200 kN
4 m
C

Shear force and bending moment diagrams

Shear force and bending moment have been described in general terms and it will have been seen that the values of both vary at different points along the span.

It is often desirable to show this variation by means of diagrams, which are really graphs, and these diagrams are called shear force diagrams and bending moment diagrams.

The following examples will serve to show how the diagrams of this type may be constructed for simple and more complex cases.

Simply supported beam with point load

Figure 7.8(a)(i) shows a simply supported beam of span l carrying one point load of W kN at the centre of the span. Since the loading is symmetrical the reactions must be equal to each other, and each reaction will be $\frac{1}{2}W$ kN. Ignoring the self-weight of the beam, at any point C at x m from the left-hand reaction, the shear force to the left will be simply the value of the l.h. reaction, i.e. $\frac{1}{2}W$ kN upward to the left, and wherever point C lies between the l.h. end and the load, the shear will still be $+\frac{1}{2}W$ kN.

Similarly at any point to the right of the load as at point D, the shear to the left of D is W kN downward and $\frac{1}{2}W$ kN upward, which is equal to $\frac{1}{2}W$ kN down to the left or $\frac{1}{2}W$ kN up to the right (negative shear). These variations of the shear values are shown to scale on the **shear force diagram (SFD)** (Fig. 7.8(a)(ii)). The vertical ordinate of the diagram at any point along the span shows the shear force at that point, and the diagram is drawn to two scales: a vertical scale of 1 mm = a suitable number of kilonewtons, and a horizontal scale of 1 mm = a suitable number of metres, being the same scale as that used in showing the span of the beam. Figure 7.8(a)(iii) shows the variation of bending moment and constitutes a **bending moment diagram (BMD)**.

Fig. 7.8 Shear force and bending moments: (a)(i) beam with a single central load point, (ii) shear force diagram (SFD), (iii) bending moment diagram (BMD); (b)(i) beam with a single non-central load point, (ii) SFD, (iii) BMD

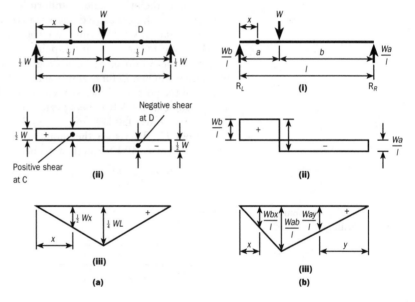

Obviously the moment at each end is zero, and at any point C at x m from the l.h. end the bending moment (summation of moments to the left) is simply $\frac{1}{2}W \times x = \frac{1}{2}Wx$ kN m. This moment increases as x increases, and will reach a maximum amount of $\frac{1}{2}W \times \frac{1}{2}l = \frac{1}{4}Wl$ at the centre of span as shown.

Again the diagram is drawn horizontally to the same scale as the span of the beam, but this time vertically to a scale of 1 mm = a suitable number of kN m or N mm, and the vertical ordinate at any point along the span represents the bending moment at that point.

Comparing the two diagrams, it will be seen that the bending moment everywhere on the span is positive, and that the shear force changes its type from positive to negative at the point along the span where the bending moment reaches its maximum amount.

Figure 7.8(b)(i) shows a simply supported beam loaded with one single non-central point load, at distance a from one end and at b from the other.

Taking moments about the l.h. end,

$$\text{r.h. reaction} \times l = W \times a$$

$$\text{r.h. reaction} = Wa/l$$

Similarly, taking moments about the r.h. end,

$$\text{l.h. reaction} \times l = W \times b$$

$$\text{l.h. reaction} = Wb/l$$

At any point between the l.h. end and the point load,

$$\text{shear to the left} = \text{l.h. reaction} = Wb/l \quad \text{(positive)}$$

Similarly, at any point between the load and the r.h. end,

$$\text{Shear to the right} = \text{r.h. reaction of } Wa/l$$
$$\text{(up to the right, thus negative shear)}$$

Thus the shear changes sign from positive to negative at the point load as in the previous example, and the SFD is as shown in Fig. 7.8(b)(ii).

As before, the bending moment is zero at the l.h. end, and at any point between that end and the load is equal to

$$R_L \times x = (Wb/l) \times x = Wbx/l$$

This reaches a maximum value of Wab/l at the point load (where $x = a$).

Also, at any point between the load and the r.h. end, at a distance of y from the r.h. reaction,

$$\text{Bending moment} = \text{r.h. reaction} \times y = Way/l$$

and this also reaches a maximum at the point load (where $y = b$) of Wab/l.

The full bending moment diagram (Fig. 7.8(b)(iii)) is thus a triangle with a maximum vertical height of Wab/l at the load, and the bending moment at any point along the span may thus be scaled from the diagram to the same scale which was used in setting up the maximum ordinate of Wab/l.

Simply supported beam with uniformly distributed load

The next case will deal with a simply supported beam of span l carrying a uniformly distributed load of intensity w kN/metre run (Fig. 7.9(a)). The span consists of l m of load, and thus the total load is wl kN, and, as the beam is symmetrical, each reaction will be half of the total load

$$R_L = R_R = \tfrac{1}{2}wl \text{ kN}$$

The shear (up to the left) at a point just in the span and very very near to R_L is quite obviously simply the l.h. reaction of $\tfrac{1}{2}wl$.

When the point concerned is, say, 1 m from R_L however, the shear to the left (summation of loads to the left of the point) is then

$$\tfrac{1}{2}wl \text{ upwards} + 1 \text{ m of load } w \text{ downwards}$$
$$= \tfrac{1}{2}wl - w$$

Fig. 7.9 Simply supported beam
carrying a uniform load: (a) load
diagram; (b) SFD; (c) BMD

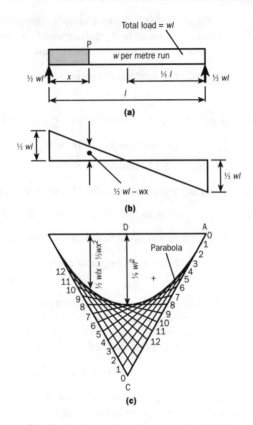

Similarly when the point concerned is 2 m from R_{L}, the shear up to the left is

$$\tfrac{1}{2} wl \text{ upwards} - 2w \text{ (2 m of load) downwards}$$
$$= \tfrac{1}{2} wl - 2w$$

Putting this in general terms, the shear at any point P on the span at distance x from R_{L} is

$$\tfrac{1}{2} wl - x \text{ metres of uniform load} = \tfrac{1}{2} wl - wx$$

This will give a positive result wherever x is less than $\tfrac{1}{2}l$, and a negative result where x exceeds $\tfrac{1}{2}l$, so the shear will 'change sign' where $x = \tfrac{1}{2}l$ (at the point of midspan), and the SFD will be as shown in Fig. 7.9(b).

Referring again to Fig. 7.9(a), the bending moment at the l.h. end is again zero, and at a point 1 m from R_{L}, the bending moment (summation of moment to the left) is simply the algebraic sum of:

- the clockwise moment of the l.h. reaction ($\tfrac{1}{2} wl \times 1$)
- the anticlockwise moment from 1 m of downward load ($w \times 1 \times \tfrac{1}{2} = \tfrac{1}{2} w$)

Thus, the bending moment $= \tfrac{1}{2} wl - \tfrac{1}{2} w$.

If the bending moment is required at a point P, at x m from R_{L}, then the bending moment is the algebraic sum again of:

- the clockwise moment of the l.h. reaction ($\tfrac{1}{2} wl \times x$)
- the anticlockwise moment of x m of load (wx) $\times \tfrac{1}{2} x$

Thus, the bending moment $= \frac{1}{2}wlx - \frac{1}{2}wx^2$.

This bending moment will be positive for any value of x, and will reach a maximum value of

$$\frac{1}{2}wl \times \frac{1}{2}l) - \frac{1}{2}w(\frac{1}{2}l)^2 = \frac{1}{4}wl^2 - \frac{1}{8}wl^2$$
$$= \frac{1}{8}wl^2$$

It should be most carefully noted that the maximum bending moment (M_{max}) is $\frac{1}{8}wl^2$, where w is the amount of uniform load per metre.

Sometimes it is more convenient to think in terms of the *total load W* (i.e. capital W), and in this case $W = wl$, and M_{max} in terms of the total load will then be

$$\frac{1}{8}wl \times l = \frac{1}{8}Wl$$

If the values of this bending moment at points along the span are plotted as a graph, the resulting BMD will be a parabola with a maximum ordinate of $\frac{1}{8}wl^2$ or $\frac{1}{8}Wl$ as shown in Fig. 7.9(c). Where the diagram has to be drawn, it will be necessary only to draw a parabola having a central height of $\frac{1}{8}Wl$, and any other ordinates at points away from the centre may be scaled or calculated as required.

Shear force diagrams
It is advisable, at this stage, to study carefully the shear force diagrams already drawn and to note that these diagrams are drawn on a horizontal base. The upward loads (reactions) are projected upward on this base, and at the point loads, the SFD drops vertically by the amount of the point load.

Where a uniform load occurs, however, the SFD slopes down at a uniform rate, dropping for each metre of span an amount equal to the amount of uniform load per metre as shown in Fig. 7.10. It should be observed that in every case, the value of the bending moment has been a maximum at the point where the shear force changes its sign (generally referred to as the point of zero shear).

Fig. 7.10 Slope of SFD due to uniform loading

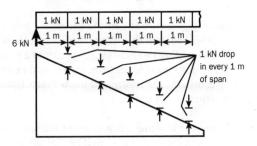

Bearing this in mind, it is common practice to draw the shear force diagram simply to discover where the maximum bending moment occurs. This is quite unnecessary, as will be seen from consideration of the two cases shown in Fig. 7.11.

In Fig. 7.11(a), the l.h. reaction is 130 kN, and the shear force will change sign at the point of M_{max} where the downward loads starting from the l.h. end just equal 130 kN.

Fig. 7.11 SFDs for beams carrying uniform loads

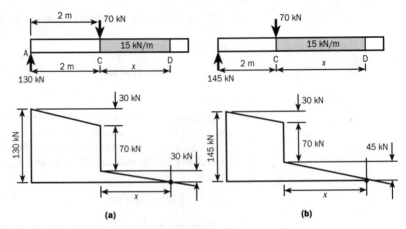

(a)   (b)

From A to C, the downward uniform load is 2 m at 15 kN/m $= 30$ kN. Then at the point load, there is a further drop of 70 kN, making a drop in the shear diagram from A to C of 100 kN.

Hence, if D is the point of M_{max} (zero shear) then the portion of load shown shaded must equal $130 - 100 = 30$ kN, so that the downward loads up to D equal the l.h. reaction.

Thus, if the distance from C to D is x, then x m at 15 kN/m must equal 30 kN so $15x = 30$, i.e. $x = 2$ m and the maximum bending moment occurs at $2 + 2 = 4$ m from R_L.

Similarly in the case shown in Fig. 7.11(b) the load from the l.h. end up to and including the 70 kN point load is $2 \times 15 + 70 = 100$ kN.

Therefore, a further $145 - 100 = 45$ kN of load is required beyond C to D, the point of maximum bending moment. Therefore

$$15 \times x = 45 \text{ kN}$$

$$x = \frac{45}{15} = 3 \text{ m}$$

and M_{max} occurs at point D which is $2 + 3 = 5$ m from the l.h. end.

It will be seen that where the loading is uniform, the position of the maximum bending moment (M_{max}) occurs at a point on the span such that the sum of the downward loads from the l.h. end exactly equals the l.h. reaction.

Similarly, the downward loads taken from the r.h. end will equal the r.h. reaction at this point.

Uniformly distributed and point loads

When the loading includes point loading, however, as it does in the example shown in Fig. 7.12, it may well be that the point at which the bending moment is a maximum coincides with the position of a point load – and in the beam illustrated in Fig. 7.12 the maximum value occurs at the 40 kN load.

Summing up the loads (starting from the l.h. end) it will be seen that just to the left of the 40 kN load they are $30 \times 1.7 + 10 = 61$ kN, that is just less than R_L.

Immediately to the right of the 40 kN load they add up to $30 \times 1.7 + 10 + 40 = 101$ kN which is more than R_L.

The rule for finding the position of the **maximum bending moment** may therefore be stated as follows: Add the downward loads together, starting

Fig. 7.12 SFD for a beam carrying uniform and point loads

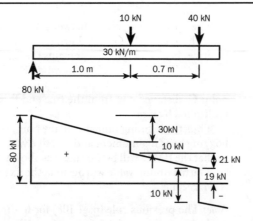

from one reaction to the point where they equal (or suddenly become greater than) that reaction. This is the point of zero shear, and the point of maximum bending moment.

Cantilever
A beam which is supported at one end only by being firmly built into a wall or which is held horizontally at one end only by other means, is called a **cantilever**. Figure 7.13(a)(i) shows such a cantilever AB having a length of l m and loaded with one point load of W kN at the free end B.

The only downward load is W kN, so for equilibrium the l.h. upward reaction will also be W kN, but in addition, to prevent the beam from rotating, the wall or other form of restraint at the fixed end A must exert a moment of Wl in an anticlockwise direction.

Fig. 7.13 SFDs and BMDs for cantilever beams: (a)(i) point load, (ii) SFD; (iii) BMD; (b)(i) uniformly distributed load, (ii) SFD, (iii) BMD

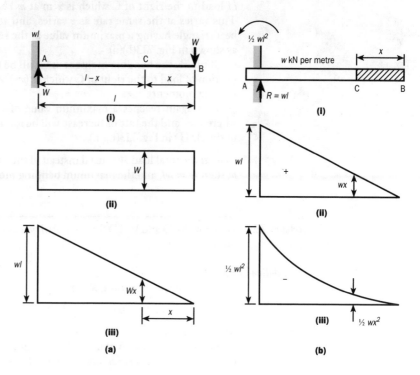

The shear at any point C between A and B (sum of the loads to the left or to the right of C) will be W kN down to the right, or W kN up to the left. This is positive shear, and the value remains the same at all points along the span, so that the SFD is a rectangle as shown in Fig. 7.13(a)(ii).

The bending moment (taken as the sum of moments to the right) at any point C, distance x m from the free end B, will be the downward load of W multiplied by x m.

Thus the bending moment at any point is Wx kN m. This is a negative or hogging bending moment and its value varies directly as the value of x varies, so that the BMD will be a triangle as shown in Fig. 7.13(a)(iii).

The maximum value will occur at the fixed end when $x = l$, and its amount is Wl kN m.

Note: The previous rule (page 108) for finding the position of M_{max} will not apply to cantilevers. In the case of cantilevers M_{max} will always occur at the fixed end.

Cantilever with uniformly distributed load

Figure 7.13(b)(i) shows a cantilever with a uniformly distributed load of w kN/m. The total load will be $w \times l = wl$ kN and the upward reaction at A will also be wl kN.

The moment of the downward load wl about the support A is

$$wl \times \tfrac{1}{2}l = \tfrac{1}{2}wl^2 \text{ kN m}$$

Thus the wall (or other form of restraint) at A must exert an anticlockwise moment of $\tfrac{1}{2}wl^2$ on the beam to prevent its rotating under the couple formed by the upward reaction and the resultant of the downward load.

The shear at any point C at distance x from the free end will be the portion of load to the right of C which is x m at w kN/m $= wx$ kN (positive shear). This varies at the same rate as x varies, and so the shear force diagram will be a triangle having a maximum value at the reaction (where $x = l$) of wl kN, as shown in Fig. 7.13(b)(ii).

Similarly the bending moment at C will be the moment (clockwise) of the portion of load to the right of C which is $(wx) \times \tfrac{1}{2}x = \tfrac{1}{2}wx^2$ kN m (negative or sagging moment).

This again reaches a maximum value of $\tfrac{1}{2}wl^2$ kN m at the fixed end A where $x = l$ and the rate of increase will be found to form a parabola as shown in the BMD in Fig. 7.13(b)(iii).

Note: If the total load W is used instead of the value of the load per metre run, w, then $W = wl$, and the maximum bending moment will be $\tfrac{1}{2}Wl$ kN m.

Example 7.1

Draw the SFD and BMD for the cantilever shown in Fig. 7.14(a) indicating all important values.

Solution

Reaction

There is only one reaction, R_L at A

$$R_L = 30 + 20 + 10 = 60 \text{ kN}$$

Shear force

At any point between A and C, the shear downward to the right is the sum of the three downward loads which is 60 kN positive.

Fig. 7.14 Example 7.1: (a) load
diagram; (b) SFD; (c) BMD

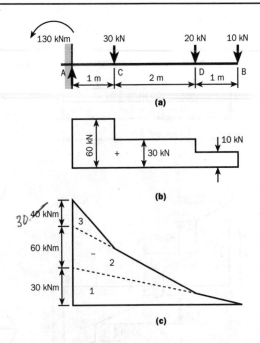

Similarly between C and D the shear is the sum of the two loads to the
right which is 30 kN positive. Between D and the free end B the shear is
the single load of 10 kN positive.

This is shown in the SFD constructed in Fig. 7.14(b).

Bending moment

The BMD is most easily drawn by treating the three loads separately: the
10 kN load produces a bending moment of $10 \times 4 = 40$ kN m at A and the
BMD for this load alone is the triangle 1 shown in Fig. 7.20(c).

The 20 kN load causes a bending moment at A of $20 \times 3 = 60$ kN m and
the BMD for this load alone is the triangle 2.

Finally, the 30 kN load causes a $30 \times 1 = 30$ kN m bending moment at A
and the BMD for this load is the triangle 3.

As in fact all three loads are on the beam at the same time, then the final
BMD is the sum of the three triangles 1, 2 and 3 as shown in Fig. 7.14(c).

Example 7.2   Draw the SFD and BMD for the beam loaded as shown in Fig. 7.15(a)(i) indicating all important values.

Solution   *Reactions*
Taking moments about A

$$R_R \times 4 = 20 \times 4 \times \tfrac{1}{2} \times 4 + 40 \times 0.8 + 30 \times 1.6 + 20 \times 3.2$$
$$= 160 + 32 + 48 + 64 = 304$$
$$R_R = 76 \text{ kN}$$

and

$$R_L = 80 + 40 + 30 + 20 - 76 = 94 \text{ kN}$$

Fig. 7.15 Example 7.2 (a)(i) shear
force load diagram, (ii) construction of SFD, (iii) final SFD;
(b)(i) BMD for uniformly
distributed loads, (ii) BMD for
point loads; (iii) addition of
BMDs, (iv) final BMD

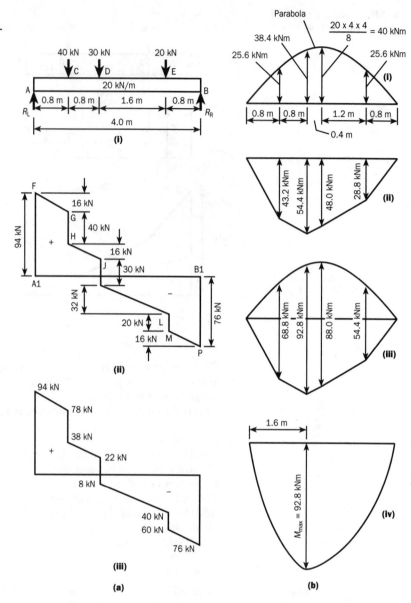

Shear force

Note: It should be obvious by now that SFDs represent all point loads as vertical lines – upward in the case of reactions, and downward in the case of
downward loads.

Downward uniform loads are shown as sloping lines (gradual change in
the value of the shear force) and therefore the SFD may be plotted quickly
from a horizontal base by merely plotting the loads in this way as they occur.

The horizontal base is shown as line A1–B1 in Fig. 7.15(a)(ii). The procedure
is as follows:

1 Draw up from A1 the vertical reaction R_L of 94 kN (A1–F).

2 Draw the gradual change of (20×0.8) kN between A and C as the sloping line F–G.
3 Draw G–H of 40 kN vertically downward to represent the sudden change of shear at the 40 kN load at point C.
4 Draw the gradual change of 16 kN between C and D as the sloping line H–J.
5 Draw J–K of 30 kN vertically downward to represent the sudden change of shear at the 30 kN load at point D.
6 Draw the gradual change of (20×1.6) kN between D and E as the sloping line K–L.
7 Draw L–M of 20 kN vertically downward to represent the sudden change of shear at the 20 kN point load at E.
8 Draw the gradual change of 16 kN between E and B as the sloping line M–P.
9 Finally draw the vertical right hand reaction of 76 kN upward from P to join the horizontal base at Bl.

Figure 7.15(a)(ii) shows the construction of this SFD, and the final diagram with its important values is shown in Fig. 7.15(a)(iii).

Bending moment

The bending moment at any point or the maximum bending moment may be calculated in the usual way by taking moments to the left or to the right of that point. But if the final BMD is required (as it is in this example) then it may be drawn by constructing separately:

- the BMD for the uniformly distributed load (above baseline — Fig. 7.15(b)(i)
- the BMD for the point loads (below baseline — Fig. 7.15(b)(ii)

Then, by adding the calculated ordinates (Fig. 7.15(b)(iii)), the final BMD may be drawn, as shown in Fig. 7.15(b)(iv).

The value of the bending moment at any point may also be obtained by computing the area of the SFD to the left or to the right of that point.

Consider the bending moment at the centre of the span. The positive shear force area is

$$\tfrac{1}{2}(94 + 78) \times 0.8 + \tfrac{1}{2}(38 + 22) \times 0.8$$
$$= 68.8 + 24.0 = 92.8 \text{ kN m}$$

To obtain the negative shear force area, it is necessary to determine the value of the shear force at the centre of the span which is 0.4 m to the right of the 30 kN point load.

So, the shear force at the centre of span is

$$8 + 20 \times 0.4 = 16 \text{ kN}$$

Hence the negative shear force area to the left of the centre of the span is

$$\tfrac{1}{2}(8 + 16) \times 0.4 = 4.8 \text{ kN m}$$

Thus the bending moment at the centre of the span is

$$92.8 - 4.8 = 88.0 \text{ kN m}$$

Example 7.3 Draw the SFD for the beam shown in Fig. 7.16, and calculate the value of $M_{\max}$. Determine also the bending moment at a point C at 2 m from the l.h. reaction.

Fig. 7.16 Example 7.3: (a) load diagram; (b) SFD

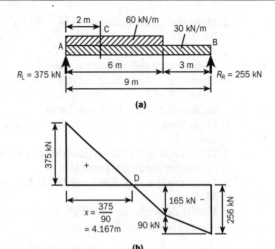

(a)

(b)

Solution   Taking moments about A to determine reactions:

$$R_R \times 9 = 30 \times 9 \times (\tfrac{1}{2} \times 9) + 60 \times 6 \times (\tfrac{1}{2} \times 6)$$
$$= 1215 + 1080 = 2295 \text{ kN}$$
$$R_R = 255 \text{ kN}$$

and

$$R_L = 60 \times 6 + 30 \times 9 - 255 = 375 \text{ kN}$$

M_{max} will occur at x m from A where

$$x \text{ m} \times (60 \times 30) \text{ kN/m} = R_L = 375 \text{ kN}$$

$$x = \frac{375}{60 + 30} = 4.167 \text{ m}$$

The value of M_{max} will, therefore, be

$$M_{max} = 375 \times 4.167 - (60 + 30) \times 4.167 \times (\tfrac{1}{2} \times 4.167)$$
$$= 1562.50 - 781.25 = 781.25 \text{ kN m}$$

or by computing the areas of the SFD

$$M_{max} = 375 \times (\tfrac{1}{2} \times 4.167) = 781.25 \text{ kN m} \quad \text{(as before)}$$

Bending moment at C, 2 m from R_L is

$$375 \times 2 - (30 + 60) \times 2 \times (\tfrac{1}{2} \times 2) = 750 - 180 = 570 \text{ kN m}$$

If a BMD is required, bending moment values may be calculated at several points of the span and these values drawn to scale vertically from a horizontal line representing the span of the beam. The BMD is obtained by joining the tops of these lines.

Example 7.4   Draw the SFD for the loaded beam shown in Fig. 7.17, and determine the position and amount of the maximum bending moment (M_{max}).

Fig. 7.17 Example 7.4: (a) load diagram; (b) SFD

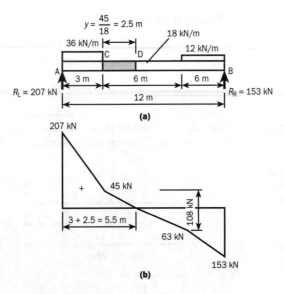

(a)

(b)

Solution   Taking moments about A to determine reactions,

$$R_R \times 12 = 36 \times 3 \times 1.5 + 18 \times 12 \times 6 + 12 \times 3 \times 10.5$$
$$= 162 + 1296 + 378 = 1836$$
$$R_R = 153 \text{ kN}$$

and

$$R_L = 36 \times 3 + 18 \times 12 + 12 \times 3 - 153$$
$$= 108 + 216 + 36 - 153$$
$$= 207 \text{ kN}$$

Position of M_{max}
At point C, 3 m from l.h. end, downward load from l.h. end is

$$(36 + 18) \times 3 = 162 \text{ kN}$$

M_{max} will occur where this has increased to 207 kN. Thus the M_{max} will occur at point D where the amount of load between C and D (shown shaded) is

$$207 - 162 = 45 \text{ kN}$$

Thus

$$\text{Distance } y \times 18 \text{ kN/m} = 45 \text{ kN}$$
$$y = 45/18 = 2.5 \text{ m}$$

Therefore M_{max} occurs at D at $(3.0 + 2.5) = 5.5$ m from the l.h. end:

$$M_{max} = 207 \times 5.5 - 162 \times (2.5 + \tfrac{1}{2} \times 3.0) - 18 \times 2.5 \times \tfrac{1}{2} \times 2.5$$
$$= 1138.5 - 648 - 56.25$$
$$= 434.25 \text{ kN m}$$

Check by shear force areas:

$$M_{max} = \tfrac{1}{2}(207 + 45) \times 3 + 45 \times \tfrac{1}{2} \times 2.5$$
$$= 378 + 56.25$$
$$= 434.25 \text{ kN m}$$

Example 7.5 Draw the SFD and BMD for the beam shown in Fig. 7.18(a), and determine the position and amount of the maximum bending moment.

Fig. 7.18 Example 7.5: (a) load diagram; (b) SFD; (c) BMD; (d) deflected shape of beam

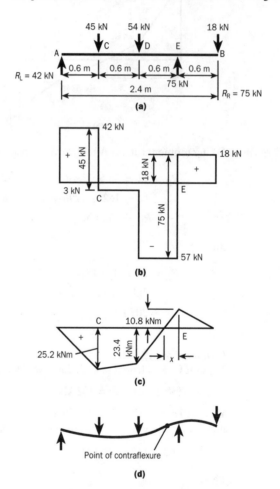

Solution Taking moments about A

$$R_R \times 1.8 = 45 \times 0.6 + 54 \times 1.2 + 18 \times 2.4$$
$$= 27.0 + 64.8 + 43.2$$
$$= 135$$
$$R_R = 75 \text{ kN}$$

and

$$R_{\mathrm{L}} = 45 + 54 + 18 - 75 = 42 \text{ kN}$$

The shear force diagram is as shown in Fig. 7.18(b). The values of bending moments at

$$45 \text{ kN load} = 42 \times 0.6 = 25.2 \text{ kN m} \qquad (+)$$

$$54 \text{ kN load} = 42 \times 1.2 - 45 \times 0.6 = 23.4 \text{ kN m} \quad (+)$$

$$\text{r.h. support} = 42 \times 1.8 - 45 \times 1.2 - 59 \times 0.6$$

$$= 10.8 \text{ kN m} \qquad (-)$$

The bending moment diagram is as shown in Fig. 7.18(c).

Note that, due to the overhanging of the r.h. end, part of the beam is subjected to negative bending moment.

There are, therefore, two points of zero shear: one at C which marks the point of maximum positive moment, and one at E where the maximum negative moment occurs. The absolute maximum bending moment is thus 25.2 kN m at C, and the shape of the deflected beam is as shown in Fig. 7.18(d).

The point at which the negative bending moment changes to positive (and vice versa) is called the **point of contraflexure**. The value of the bending moment at that point is zero.

Example 7.6   Draw the SFD and BMD for the beam shown in Fig. 7.19(a) and determine the position and amount of the maximum bending moment.

Solution   Since the loading and the beam are symmetrical, the reactions are each equal to $\frac{1}{2}(20 + 70 + 20) = 55$ kN.

Fig. 7.19 Example 7.6: (a) load diagram; (b) SFD; (c) BMD; (d) deflected shape of beam

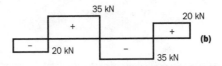

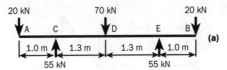

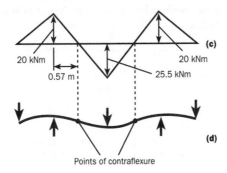

Bending moment at $D = 55 \times 1.3 - 20 \times 2.3$
$$= 71.5 - 46.0$$
$$= 25.5 \text{ kN m} \quad (+)$$
Bending moment at $C = -20 + 1.0$
$$= -20.0 \text{ kN m} \quad (-)$$

Example 7.7 A beam 4 m long carrying a uniformly distributed load of 60 kN/m cantilevers over both supports as shown in Fig. 7.20(a). Sketch the SFD and BMD and determine the position of the point of contraflexure.

Fig. 7.20 Example 7.7: (a) load diagram; (b) SFD; (c) BMD

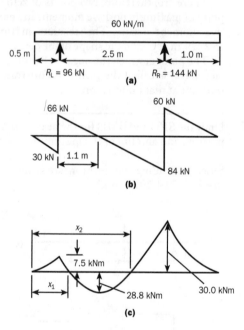

Solution The shear force and bending moment diagrams are given in Fig. 7.20(b) and (c). Check that the reactions are 96 kN and 144 kN respectively. The maximum positive bending moment occurs at 1.6 m from the left end of the beam and is

$$M_{max} = \tfrac{1}{2}(66 \times 1.1 - 30 \times 0.5)$$
$$= \tfrac{1}{2}(72.6 - 15.0)$$
$$= 28.8 \text{ kN m}$$

The maximum negative bending moment occurs over the right-hand support and is $60 \times 0.5 = 30$ kN m.

By calculating several other bending moment values the diagram can be constructed as shown in Fig. 7.20(c).

Bending moment at point of contraflexure is

$$96(x - 0.5) - 60 \times x \times \tfrac{1}{2}x = 0$$

$$5x^2 - 16x + 8 = 0$$

$$x = \frac{16 \pm \sqrt{(256 - 160)}}{10}$$

$$x_1 = 0.62 \text{ m} \qquad x_2 = 2.58 \text{ m}$$

Summary

Shear force At a point in the span of a beam, shear force is defined as the algebraic sum of the loads to the left or to the right of the point.

Bending moment At a point in the span of the beam, bending moment is defined as the algebraic sum of all the moments caused by the loads acting to the left or to the right of that point.

Maximum bending moment (M_{max}) *The algebraic sum of the moments of the loads and reaction acting to* one side *of the point of maximum bending moment.*
  To determine the position of the maximum bending moment in a beam simply supported at its two ends, add together the downward loads starting from one reaction to the point where they equal (or suddenly become greater than) that reaction. This is the point of zero shear and the point of maximum bending moment.
  The maximum bending moment in cantilevers will always occur at the fixed end and equals the sum of the moments of all the loads about the fixed end.

Point of contraflexure This occurs where the bending moments change sign and their value is zero.

Exercises

1–14 Calculate the reactions, determine the position and amount of the maximum bending moment for the beams shown in Figs. 7.Q1 to 7.Q14.

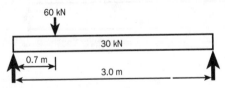

Fig. 7.Q1

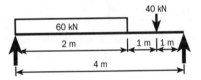

Fig. 7.Q3

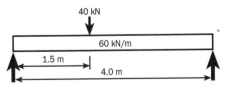

Fig. 7.Q2

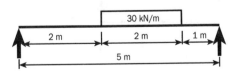

Fig. 7.Q4

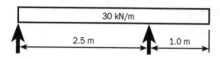

Fig. 7.Q5

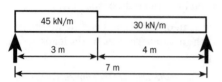

Fig. 7.Q6

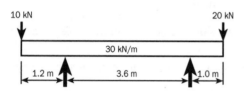

Fig. 7.Q7

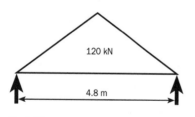

Fig. 7.Q8

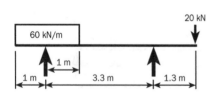

Fig. 7.Q9

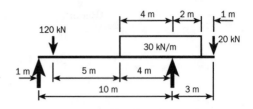

Fig. 7.Q10

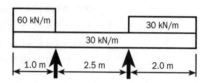

Fig. 7.Q11

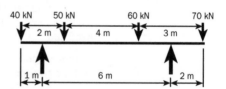

Fig. 7.Q12

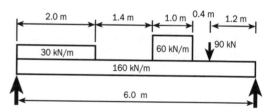

Fig. 7.Q13

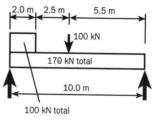

Fig. 7.Q14

15 Fig. 7.Q15 shows the shear force diagram for a loaded beam. Sketch the beam, showing the loading conditions, and calculate the maximum positive and negative bending moments.

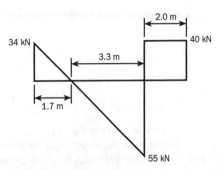

Fig. 7.Q15

16 A steel beam spans 4.8 m and carries on its whole length a brick wall 220 mm thick and 2.7 m high. The brickwork weighs 20 kN/m³ and the self-weight of beam and casing is estimated as being 5.7 kN. What is the maximum bending moment on the beam?

17 A steel beam is as shown in Fig. 7.Q17. The portion between supports A and B carries a uniform load of 30 kN/m and there are point loads at the free ends as shown. What is the length *l* in metres between A and B if the bending moment at a point C midway between these supports is just zero?

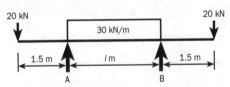

Fig. 7.Q17

18 For the beam loaded as shown in Fig. 7.Q18 (a) calculate the reactions; (b) determine the position and amount of the maximum positive and maximum negative bending moments.

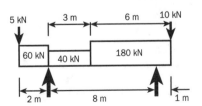

Fig. 7.Q18

19 Referring to Fig. 7.Q19, calculate (a) end reactions; (b) position and amount of M_{max}.

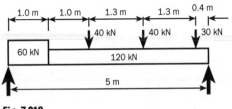

Fig. 7.Q19

20 Calculate the maximum bending moment in Fig. 7.Q20.

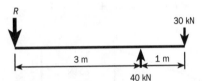

Fig. 7.Q20

21 Determine the position and amount of the maximum shear force and bending moment in Fig. 7.Q21.

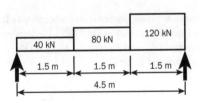

Fig. 7.Q21

22 Referring to Fig. 7.Q22 (a) calculate the end reactions; (b) determine the position and amount of the maximum bending moment, M_{max}.

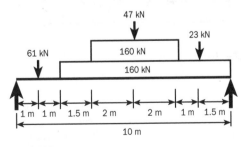

Fig. 7.Q22

23 For the cantilever loaded as shown in Fig. 7.Q23, calculate (a) the bending moment at the 40 kN load; (b) the bending moment at the point D; (c) the M_{max}.

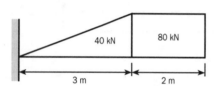

Fig. 7.Q23

24 Calculate the maximum bending moment for the cantilever shown in Fig. 7.Q24.

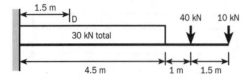

Fig. 7.Q24

25 A steel beam simply supported at its ends carries a load of varying intensity as shown (Fig. 7.Q25). Determine (a) the end reactions R_L and R_R; (b) the M_{max}; (c) the bending moment at point C.

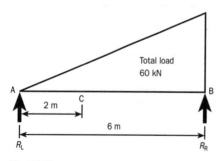

Fig. 7.Q25

26 A steel post cantilevers vertically carrying two point loads as shown in Fig. 7.Q26. Calculate the bending moments at (a) the 35 kN load; (b) the base; (c) 1 m from the base.

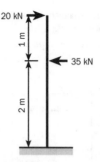

Fig. 7.Q26

27 A steel beam loaded as shown in Fig. 7.Q27 has a maximum bending moment (occurring at the point load) of 135 kN m. What is the value W kN of the point load?

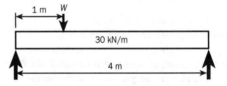

Fig. 7.Q27

28 In the beam shown in Fig. 7.Q28, the maximum bending moment occurs at the supports R_L and R_R and the bending moment at the central 20 kN load is zero. What is the length in metres of span l?

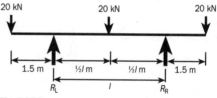

Fig. 7.Q28

29 In the beam shown in Fig. 7.Q29, the maximum negative bending moment is twice the amount of the maximum positive bending moment. What is the value in kN of the central load W?

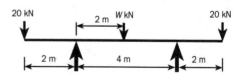

Fig. 7.Q29

30 Derive an expression for the beam as shown in Fig.
7.Q30 for (a) the maximum bending moment; (b) the
bending moment at C.

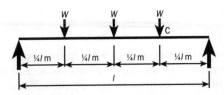

Fig. 7.Q30

8 Properties of sections

This chapter considers the effect of the shape or profile of a beam's section on the beam's resistance to the punishing forces and moments induced by the loading. The beam itself must, of course, be made just strong enough to withstand these punishing effects with a reasonable factor of safety, and the strength of the beam or its degree of resistance to bending moment and shear force is built up in terms of:

- the shape and size of the beam's section
- the strength of the particular material of which the beam is made

The final degree of 'defence' or resistance will be measured in units which take into account both of these two factors, but for a clearer understanding they will be treated separately to begin with. The present chapter will show how the shape or profile of a beam's section affects its strength.

The properties which various sections have by virtue of their shape alone are:

- cross-sectional area
- position of centre of gravity or area (centroid)
- moment of inertia or the second moment of area
- section modulus or modulus of section
- radius of gyration

Section modulus and radius of gyration will be considered in Chapters 9 and 12 respectively.

The cross-sectional area should need no description and may be calculated with ease for most structural members.

The centre of gravity or centroid

The **centre of gravity** of a body is a point in or near the body through which the resultant attraction of the earth, i.e. the weight of the body, acts for all positions of the body.

It should be noted, however, that the section of a beam is a plane figure without weight, and therefore the term **centre of area** or **centroid** is more appropriate and is frequently used in this case. The determination of the position of the centre of gravity of a body or centroid of a section is equivalent to determining the resultant of a number of like, parallel forces.

Figure 8.1(a) shows a thin sheet of tinplate with several small holes drilled in it. To one of these holes (hole A) is attached a string. It should be obvious that the position shown for the sheet in Fig. 8.1(a) is an impossible one. The sheet will swing round in a clockwise direction and come to rest as indicated in Fig. 8.1(b). Each particle of the sheet in Fig. 8.1(a) is attracted vertically downwards by the force of gravity and the parallel lines indicate the direction of the gravity forces. The resultant of these parallel forces is the total weight of the sheet. The sheet comes to rest in such a position that the line of the resultant weight and the line of the vertical reaction in the string form one continuous line. When the string is attached to point B the sheet will hang

Fig. 8.1 Determining the centre of gravity of a metal sheet

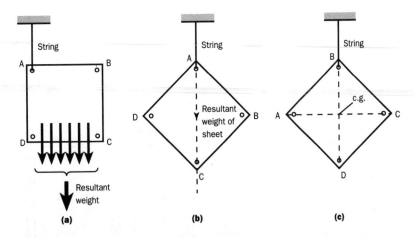

as in Fig. 8.1(c) and the intersection of the two lines AC and BD is called the centre of gravity of the body. The actual position of the centre of gravity is in the middle of the thickness of the metal immediately behind the intersection of the two lines AC and BD. If the thickness is infinitely reduced, as in the case of a beam section, the position of the centre of gravity will coincide with that of the centroid (centre of area).

The position of the centre of gravity of sheets of metal of various shapes can be obtained by the method shown in Fig. 8.2. The sheet is suspended

Fig. 8.2 Experimental determination of the centre of gravity of an irregular shape: (a) front view; (b) side view

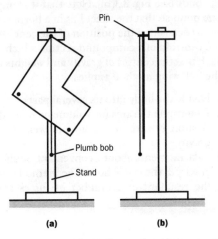

from a pin and from the same pin is suspended a plumb-line. The line of the string forming the plumb-line can be marked on the sheet behind it with pencil or chalk. The sheet can now be suspended in a different position and another line marked. The intersection of these two lines gives the position of the centre of gravity of the sheet.

The position of the centre of gravity of certain simple shapes is obvious by inspection (Fig. 8.3(a)), e.g. the centre of gravity of a line or uniform rod is halfway along its length and the centre of gravity of a circle is at its centre.

If it were required to balance any of these shapes in a horizontal position by placing a pencil underneath, the point of application of the pencil would

Fig. 8.3 The centre of gravity of
simple shapes

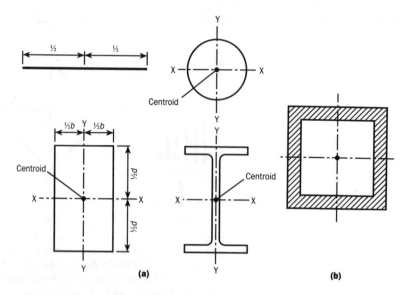

(a)

(b)

be the centre of gravity of the sheet, and the reaction of the pencil would be
equal to the weight of the sheet. Each of the shapes shown in Fig. 8.3 has at
least two axes of symmetry and whenever a body has two axes of symmetry
the centre of gravity is at the intersection of the axes.

The centre of gravity of a body need not necessarily be in the material of
the body (see Fig. 8.3(b)). Note that it is impossible to balance this shape on
one point so that the sheet lies in a horizontal plane.

To determine the position of the centre of gravity of a compound body or
the centroid of a compound section which can be divided into several parts,
such that the centres of gravity and weights of the individual parts are known,
the following method applies.

1 Divide the body into its several parts.
2 Determine the area (or volume or weight) of each part.
3 Assume the area (or volume or weight) of each part to act at its centre of
 gravity.
4 Take moments about a convenient point or axis to determine the centre of
 gravity of the whole body. The method is identical with that of determining
 the resultant of a number of forces and is explained in the following
 example.

Example 8.1   A thin uniform sheet of material weighs w newtons for each mm^2 of its surface (Fig. 8.4). Determine the position of its centre of gravity.

Solution   Since the figure is symmetrical about line AB, its centre of gravity must lie on
this line. The figure can be divided into three rectangles.

$$
\begin{array}{llll}
\text{Area (1)} = 60 \times 150 = & 9000 \text{ mm}^2 & \text{Weight} = & 9w \text{ kN} \\
\text{Area (2)} = 200 \times 20 = & 4000 \text{ mm}^2 & \text{Weight} = & 4w \text{ kN} \\
\text{Area (3)} = 20 \times 100 = & \underline{2000 \text{ mm}^2} & \text{Weight} = & \underline{2w \text{ kN}} \\
& \text{Total area} = 15\,000 \text{ mm}^2 & \text{Total weight} = & \overline{15w} \text{ kN}
\end{array}
$$

Fig. 8.4 Example 8.1

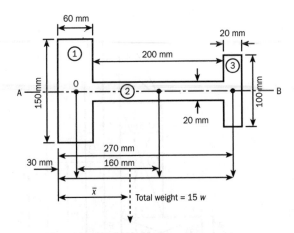

Let $\bar{x}$ be the distance of the centre of gravity of the whole figure from O. Take moments about this point.

$$15w\bar{x} = 9w \times 30 + 4w \times 160 + 2w \times 270$$
$$= 270w + 640w + 540w$$
$$= 1450w$$
$$\bar{x} = 1450w/15w = 97 \text{ mm}$$

The centre of gravity is therefore on the line AB at 97 mm from O at the extreme left edge of the figure.

Note that the centre of gravity has been determined without knowing the actual weight of the material, and when a body is of uniform density throughout, its weight may be ignored and moments of areas or moments of volumes can be taken. The position of the centroid of a section is of great importance in beam design for, as will be seen later, the portion of section above this centroid performs a different function to that below the centroid.

Example 8.2 Determine the position of the centre of gravity of the body shown in Fig. 8.5(a). The body has a uniform thickness of 100 mm and weighs 10 N/m³.

Solution Since the body is homogeneous (i.e. of uniform density) and is of uniform thickness throughout, its weight and thickness may be ignored and moments taken of areas.

$$\begin{aligned}
\text{Area (1)} &= \tfrac{1}{2}(3.0 \times 1.5) = & 2.25 \text{ m}^2 \\
\text{Area (2)} &= 4.0 \times 1.5 &= 6.00 \text{ m}^2 \\
\text{Area (3)} &= 1.0 \times 3.0 &= \underline{3.00 \text{ m}^2} \\
&\text{Total area} &= \overline{11.25 \text{ m}^2}
\end{aligned}$$

In the centre of gravity problems it is usually convenient to choose two axes A–A and B–B at one extreme edge of the figure.

Let $\bar{x}$ be the horizontal distance of the centre of gravity of the whole figure from A–A and $\bar{y}$ be the vertical distance of the centre of gravity from axis B–B. The positions of the centre of gravity of the three separate parts of the figure are shown in Fig. 8.5(a).

Fig. 8.5 Example 8.2

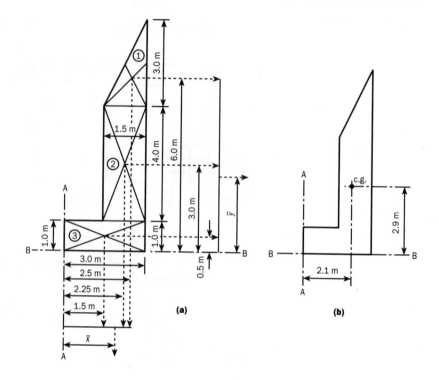

(a)   (b)

Taking moments about axis A–A

$$11.25\bar{x} = 3.00 \times 1.5 + 6.00 \times 2.25 + 2.25 \times 2.5$$
$$= 4.5 + 13.5 + 5.6 = 23.6$$
$$\bar{x} = 23.6/11.25 = 2.1 \text{ m}$$

Taking moments about axis B–B

$$11.25\bar{y} = 3.00 \times 0.5 + 6.00 \times 3.0 + 2.25 \times 6.0$$
$$= 1.5 + 18.0 + 13.5 = 33.0$$
$$\bar{y} = 33.0/11.25 = 2.9 \text{ m}$$

The centre of gravity is in the position indicated by Fig. 8.5(b) and is within the thickness of the body halfway between the front and back faces.

Use of the link polygon   The above problem may also be solved graphically by means of the principle of the link polygon as shown in Fig. 8.6, where the body has to be drawn to scale.

Example 8.3   A compound girder is built up of a 254 × 146 UB37 steel beam (i.e. a universal beam of 254 mm × 146 mm at 37 kg/m) with one 250 mm × 24 mm plate on

Fig. 8.6 Solving Example 8.2 using the link polygon method: (a) free-body diagram with link polygon; (b) force and polar diagrams (horizontal); (c) force and polar diagrams (vertical)

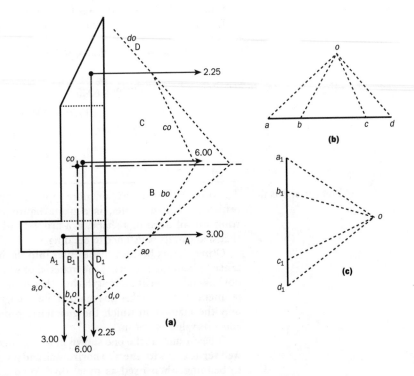

Fig. 8.7 Example 8.3

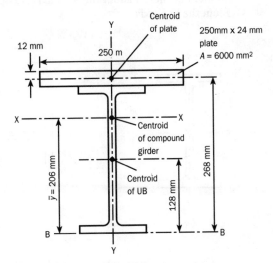

the top flange only. The area of the steel beam alone is 4720 mm². Determine the position of the centroid of the compound girder (Fig. 8.7).

Solution
- The compound section is symmetrical about the vertical axis Y–Y, so the centroid must lie on this line.
- The addition of a plate to the top flange has the effect of moving the

centroid (the X–X axis) towards that plate, and taking moments about line B–B at the lower edge of the section,

$$\text{Total area of section} \times \bar{y} = (\text{area of UB} \times 128\ \text{mm})$$
$$+ (\text{area of plate} \times 268\ \text{mm})$$
$$(4720 + 6000)\bar{y} = 4720 \times 128 + 6000 \times 268$$
$$10\,720 = 604\,160 + 1\,608\,000$$
$$= 2\,212\,160$$
$$\bar{y} = 2\,216\,000/10\,720 = 206\ \text{mm}$$

Moment of inertia The **moment of inertia** or, more appropriately in the case of a beam section (which, as stated earlier, is a plane figure without weight), the **second moment of area** of a shape, is a property which measures the *efficiency* of that shape in its resistance to bending.

Other factors besides shape enter into the building up of a beam's resistance to bending moment; the material of which a beam is made has a very obvious effect on its strength, but this is allowed for in other ways. The moment of inertia takes no regard of the strength of the material; it measures only the manner in which the geometric properties or shape of a section affect its value as a beam.

A beam such as the one shown in Fig. 8.8 may, in theory, be used with the web vertical or with the web horizontal, and it will offer much more resistance to bending when fixed as in (a) than when as in (b). This is because the moment of inertia about the X–X axis is larger than the moment of inertia about the Y–Y axis.

Fig. 8.8 The universal beam (UB) section

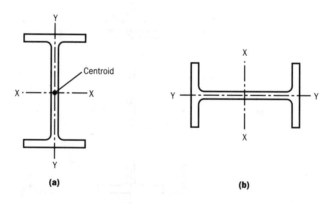

(a)   (b)

These axes of symmetry X–X and Y–Y are termed the **principal axes** and they are in fact axes which intersect at the centroid of the section concerned.

It will normally be necessary to calculate the moments of inertia about both these principal axes, and these are usually described as I_{xx} and I_{yy}, the moments of inertia about the X–X and Y–Y axes respectively.

The manner in which the 'build–up' of a shape affects its strength against bending must – to be understood completely – involve the use of calculus, but it may still be made reasonably clear from fairly simple considerations.

The term *moment of inertia* is itself responsible for a certain amount of confusion. Inertia suggests laziness in some ways, whereas in fact the true meaning of inertia may be described as the resistance which a body makes to any forces or moments which attempt to change its shape (or its motion in the case of a moving body). A beam tends to change its shape when loaded, and inertia is the internal resistance with which the beam opposes this change of shape. The moment of inertia is a measure of the resistance which the section can supply in terms of its shape alone.

Good and bad beam shapes

In order to make the best use of the beam material, the 'shape' of the section has to be chosen with care. Certain shapes are better able to resist bending than others, and in general it may be stated that a shape will be more efficient when the greater part of its area is as far away as possible from its centroid.

A universal beam (UB) section (Fig. 8.8(a)), for example, has its flanges (which comprise the greater part of its area) well away from the centroid and the X–X axis, and is in consequence an excellent shape for resisting bending.

Figure 8.9(a) shows a shape built up by riveting together four angles. The bulk of its area (shown shaded) is situated *near to* the centroid, and the section is not a good one from the point of view of resistance to bending about the X–X axis.

If the same four angles are arranged as shown in Fig. 8.9(b), however, then the bulk of the area has been moved away from the X–X axis, and the section – which now resembles roughly a UB form – is a good one, with a high resistance to bending.

When the same four angles are used with a plate placed between them as in Fig. 8.9(c), the flanges are moved even further from the axis which passes

Fig. 8.9 Good and bad beam shapes

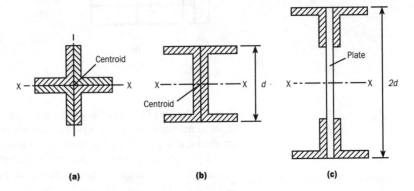

(a)   (b)   (c)

through the centroid and is called the **neutral axis** or the axis of bending, and the efficiency of the shape is much greater. Large deep girders are built up in this way when very heavy loads have to be supported.

When a shape such as that shown in Fig. 8.9(b) is being designed, it will be found that doubling the area of the *flange* will approximately double the flange's efficiency, but doubling the depth d as in Fig. 8.9(c) will increase the efficiency by 4, i.e. 2^2. The efficiency of the flange, then, varies directly as its area and as the square of its distance from the neutral axis.

Calculation of the moment of inertia

In order to measure the increase in efficiency of a section as its area and depth increase, the moment of inertia (second moment of area) is calculated, and this property does in fact measure both the direct increase in area and the square of the increase in depth.

To determine the moment of inertia of the rectangle shown in Fig. 8.10(a), it will be necessary to divide the shape into a number of strips of equal area as

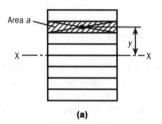

Fig. 8.10 Calculating the moment of inertia of a rectangle by division into a number of strips

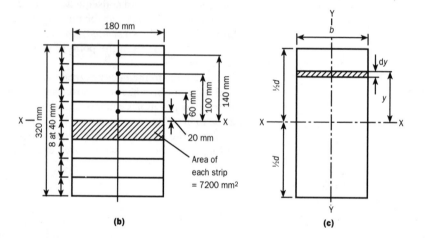

shown. The area of each strip will be multiplied by the square of the distance of its centroid from the centroid of the whole section. The sum of all such products, $a \times y^2$, will be the moment of inertia (second moment of area) of the whole shape about the X–X axis, i.e. the total I_{xx} of section.

Assume a rectangle to be as shown in Fig. 8.10(b), 180 mm wide and 320 mm in depth, divided into 8 strips each 40 mm deep, each strip having an area of 7200 mm^2.

The sum of all the products

$$\sum ay^2 = \sum 2 \times a(y_1^2 + y_2^2 + y_3^2 + \ldots)$$
$$= 2 \times 7200(20^2 + 60^2 + 100^2 + 140^2)$$
$$= 14\,440 \times 33\,600$$
$$= 483.84 \times 10^6 \text{ mm}^4$$

This is an approximate value of the I_{xx}, or moment of inertia about the X–X axis, but the exact value of this factor will, of course, depend upon the number of strips into which the shape is divided. It can be shown that if b is the width of the rectangle and d is the depth, then the exact value of I_{xx} is given by

$$I_{xx} = \tfrac{1}{12} bd^3$$

In the case above this is

$$\tfrac{1}{12} \times 180 \times 320^3 \quad \text{or} \quad 491.52 \times 10^6 \text{ mm}^4$$

The units of the answer are important, and it should be appreciated that, as an area (mm^2) has been multiplied by a distance squared (y^2), the answer is the second moment of area or more simply the moment of inertia of the shape, and is measured in mm^4.

For those who have an elementary knowledge of calculus, the foregoing approximate derivation of $\tfrac{1}{12} bd^3$ will appear somewhat lengthy, and the exact value is obtained by integrating as follows.

Consider a small strip of area of breadth b and depth dy at a distance of y from the neutral axis as shown in Fig. 8.10(c). The moment of inertia of this strip (shown shaded) is

$$(\text{its area}) \times (y)^2 = b \times dy \times y^2 = b \times y^2 \times dy$$

The second moment of half of the rectangle is the sum of all such quantities $by^2 \, dy$ between the limits of $y = 0$ and $y = \tfrac{1}{12} d$

$$I_{xx} \text{ of half rectangle} = \int_0^{d/2} by^2 \, dy = \left[\frac{by^3}{3} \right]_0^{d/2} = \frac{bd^3}{24}$$

Thus,

$$I_{xx} \text{ of the complete rectangle} = \frac{bd^3}{24} \times 2 = \frac{bd^3}{12}$$

Similarly,

$$I_{yy} \text{ of the rectangle} = \frac{db^3}{12}$$

The values of the moments of inertia of a number of common shapes are listed below for reference (Fig. 8.11).

Rectangle, about neutral axes:

$$I_{xx} = \tfrac{1}{12} bd^3$$
$$I_{yy} = \tfrac{1}{12} db^3$$

Rectangle, about one edge:

$$I_{uu} = \tfrac{1}{3} bd^3$$
$$I_{vv} = \tfrac{1}{3} db^3$$

Hollow rectangular shape:

$$I_{xx} = \tfrac{1}{12} (BD^3 - bd^3)$$
$$I_{yy} = \tfrac{1}{12} DB^3 - db^3)$$

Fig. 8.11 Moments of inertia of common shapes

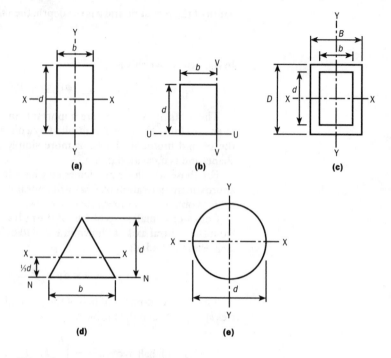

(a)   (b)   (c)

(d)   (e)

Triangle:

$$I_{xx} \text{ about neutral axis} = bd^3/36$$

$$I_{nn} \text{ about base} = bd^3/12$$

Circle:

$$I_{xx} = I_{yy} = \pi d^4/64$$

Principle of parallel axes   Figure 8.12(a) shows a simple rectangular section of size $b \times d$. The I_{xx} of the rectangle is $\frac{1}{12}bd^3$. It will be seen, however, that there are times when the moment of inertia of the rectangle about some other parallel axis such as Z–Z is required, and the I_{zz} will obviously be greater than the I_{xx}, since larger distances are involved in the summation AH^2.

The rule for use in these cases may be stated as follows:

To find the moment of inertia of any shape about an axis Z–Z, parallel to the neutral axis (X–X) and at a perpendicular distance of H away from the

Fig. 8.12 Principle of parallel axes

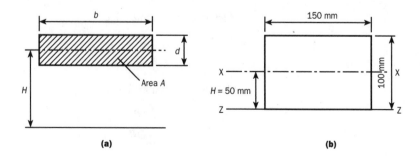

(a)   (b)

neutral axis, the amount AH^2 (area of shape × distance H squared) must be added to I_{xx}.

For example, in the case of the rectangle shown in Fig. 8.12(b),

$$I_{xx} = \tfrac{1}{12} bd^3 = \frac{150 \times 100^3}{12} = 12.5 \times 10^6 \text{ mm}^4$$

The I_{zz} about the base (where $H = 50$ mm) is

$$I_{zz} = I_{xx} + AH^2 = 12.5 \times 10^6 + 15\,000 \times 50^2$$

$$= (12.5 + 37.5) \times 10^6$$

$$= 50 \times 10^6 \text{ mm}^4 \text{ units}$$

Note: This particular case could have been derived directly from the formula of $\tfrac{1}{3} bd^3$ given in Fig. 8.11(b).

$\tfrac{1}{3} bd^3$ in this example is $\tfrac{1}{3} \times 150 \times 100^3 = 50 \times 10^6 \text{ mm}^4$

Example 8.4   The principle of parallel axes may be used to calculate the values of the moments of inertia of structural sections like the I-beam shown in Fig. 8.13(a).

Fig. 8.13 Example 8.4

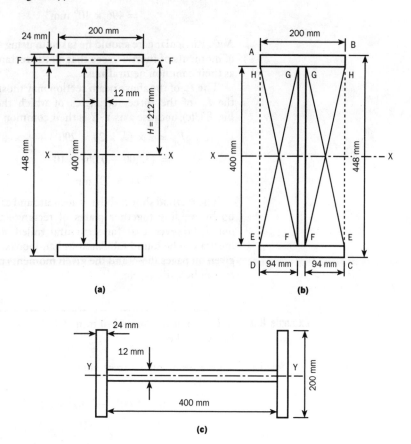

(a)

(b)

(c)

The web has its centroid on the X–X axis of the beam and the I_{xx} of the web thus equals

$$\tfrac{1}{12} \times 12 \times 400^3 = 64 \times 10^6 \text{ mm}^4$$

The moment of inertia of one flange about *its own axis* (F–F) is

$$\tfrac{1}{12} \times 200 \times 24^3 = 0.23 \times 10^6 \text{ mm}^4$$

and from the principle of parallel axis, the I_{xx} of the one flange is

$$I_{xx} = 0.23 \times 10^6 + 200 \times 24 \times 212^2$$
$$= 215.96 \times 10^6 \text{ mm}^4$$

The total I_{xx} of the two flanges plus the web is

$$\text{Total } I_{xx} = (64 + 2 \times 216) \times 10^6 \text{ mm}^4$$
$$= 496 \times 10^6 \text{ mm}^4$$

An alternative method of calculating the moment of inertia about X–X is to calculate the I_{xx} of rectangle ABCD and to subtract the I_{xx} of the two rectangles EFGH (Fig. 8.13(b)). Thus

$$\text{Total } I_{xx} = \tfrac{1}{12} \times 200 \times 448^3 - \tfrac{1}{12} \times 2(94 \times 400^3)$$
$$= (1498.59 - 1002.66) \times 10^6$$
$$= 496 \times 10^6 \text{ mm}^4$$

Note: Particular care should be taken in using this method of the subtraction of moments of inertia to see that all the rectangles concerned have axis X–X as their common neutral axis.

The I_{yy} of the above beam section may most easily be calculated by adding the I_{yy} of the three rectangles of which the joist consists, as shown in Fig. 8.13(c), because axis Y–Y is their common neutral axis.

$$I_{yy} = 2 \times (\tfrac{1}{12} \times 24 \times 200^3) + \tfrac{1}{12} \times 400 \times 12^3$$
$$= (32.00 + 0.06) \times 10^6$$
$$= 32.06 \times 10^6 \text{ mm}^4$$

The method shown above is accurate and can be used for steel girders built up by welding together plates of rectangular cross-section. It should be noted, however, that for structural rolled steel sections the moments of inertia can be found tabulated in handbooks. An example of such a table is given on pages 166–9 and the given moments of inertia take into account the root radius, fillets, etc.

Example 8.5   A T-section measures 140 mm × 140 mm × 20 mm as shown in Fig. 8.14. Calculate the I_{xx}.

Solution   Taking moments of areas about the base to determine the position of the neutral axis X–X:

$$(2400 + 2800)\bar{x} = 2400 \times 80 + 2800 \times 10$$
$$\bar{x} = \frac{(192 + 28) \times 10^3}{5.2 \times 10^3} = 42.3 \text{ mm}$$

Fig. 8.14 Example 8.5

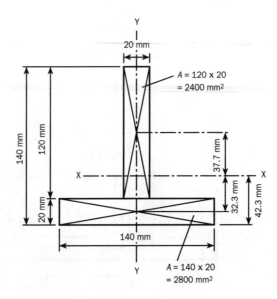

I_{xx} of vertical rectangle

$$\tfrac{1}{12} \times 20 \times 120^3 + 2400 \times 37.7^2$$
$$= (2.88 + 3.41) \times 10^6$$
$$= 6.29 \times 10^6 \text{ mm}^4$$

I_{xx} of horizontal rectangle

$$\tfrac{1}{12} \times 140 \times 20^3 + 2800 \times 32.2^2$$
$$= (0.09 + 2.92) \times 10^6$$
$$= 3.01 \times 10^6 \text{ mm}^4$$

Total I_{xx}

$$(6.29 + 3.01) \times 10^6 = 9.30 \times 10^6 \text{ mm}^4$$

Example 8.6   Calculate the I_{xx} and I_{yy} of the section shown in Fig. 8.15.

Solution   Taking moments of area about the base to find the distance $\bar{x}$ to the neutral axis X–X:

$$9600\bar{x} = 4800 \times 12 + 2400 \times 124 + 2400 \times 236$$
$$= 57\,600 + 297\,600 + 566\,400$$
$$\bar{x} = 921\,600/9600 = 96 \text{ mm}$$

Fig. 8.15 Example 8.6

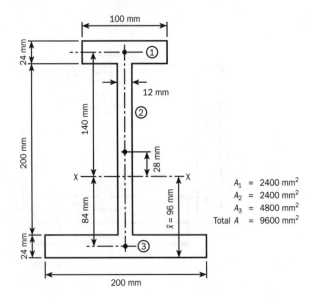

$$A_1 = 2400 \text{ mm}^2$$
$$A_2 = 2400 \text{ mm}^2$$
$$A_3 = 4800 \text{ mm}^2$$
$$\text{Total } A = 9600 \text{ mm}^2$$

The figure is divided into three rectangles as shown in Fig. 8.15 and the distances of the centroids of each from the centroid of the whole section are also given.

$$I_{xx} \text{ (top flange)} = \tfrac{1}{12} \times 100 \times 24^3 + 2400 \times 140^2$$
$$= (0.11 + 47.04) \times 10^6$$
$$= 47.15 \times 10^6 \text{ mm}^4$$
$$I_{xx} \text{ (web)} = \tfrac{1}{12} \times 12 \times 200^3 + 2400 \times 28^2$$
$$= (8.00 + 1.88) \times 10^6$$
$$= 9.88 \times 10^6 \text{ mm}^4$$
$$I_{xx} \text{ (bottom flange)} = \tfrac{1}{12} \times 200 \times 24^3 + 4800 \times 84^2$$
$$= (0.23 + 53.87) \times 10^6$$
$$= 34.10 \times 10^6 \text{ mm}^4$$

Total I_{xx} of the whole section
$$= (47.15 + 9.88 + 34.10) \times 10^6$$
$$= 91.13 \times 10^6 \text{ mm}^4$$

Total $I_{yy} = \tfrac{1}{12} \times 24 \times 100^3 + \tfrac{1}{12} \times 24 \times 200^3 + \tfrac{1}{12} \times 200 \times 12^3$
$$= (2.00 + 16.00 + 0.03) \times 10^6$$
$$= 18.03 \times 10^6 \text{ mm}^4$$

Example 8.7   Calculate the I_{xx} and I_{yy} of the channel section shown in Fig. 8.16.

Fig. 8.16 Example 8.7

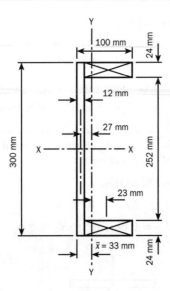

Solution   X–X axis:

$$I_{xx} = \tfrac{1}{12} \times 100 \times 300^3 - \tfrac{1}{12} \times 88 \times 252^3$$
$$= (225.00 - 117.36) \times 10^6$$
$$= 107.64 \times 10^6 \text{ mm}^4$$

Check by addition:

$$I_{xx} = \tfrac{1}{12} \times 12 \times 300^3 + 2 \times (\tfrac{1}{12} \times 88 \times 24^3 + 88 \times 24 \times 138^2)$$
$$= 27 \times 10^6 + 2(0.10 + 40.22) \times 10^6$$
$$= 107.64 \times 10^6 \text{ mm}^4 \quad \text{(as before)}$$

Y–Y axis:
  Taking moments of areas about back of channel

$$\text{Area} \times \text{distance } \bar{x} = 300 \times 12 \times 6 + 2(24 \times 88 \times 56)$$
$$= 21\,600 + 236\,544$$
$$= 258\,144 \text{ mm}^3$$

$$\text{Distance } \bar{x} = \frac{258\,144}{3600 + 2 \times 2112} = 33 \text{ mm}$$

$$I_{yy} = \tfrac{1}{12} \times 300 \times 12^3 + 3600 \times 27^2$$
$$+ 2(\tfrac{1}{12} \times 24 \times 88^3 + 2112 \times 23^2)$$
$$= [0.04 + 2.63 + 2(1.36 + 1.12)] \times 10^6$$
$$= 7.63 \times 10^6 \text{ mm}^4$$

Check by subtraction:

$$I_{yy} = (\tfrac{1}{12} \times 300 \times 100^3 + 30\,000 \times 17^2)$$
$$- (\tfrac{1}{12} \times 252 \times 88^3 + 22\,176 \times 23^2)$$
$$= (25.00 + 8.67 - 14.31 - 11.73) \times 10^6$$
$$= 7.63 \times 10^6 \text{ mm}^4 \quad \text{(as before)}$$

Example 8.8   Calculate the I_{xx} and I_{yy} of the compound girder shown in Fig. 8.17. The properties of the UB alone are

$$\text{Area} = 2800 \text{ mm}^2$$
$$I_{xx} = 28.41 \times 10^6 \text{ mm}^4$$
$$I_{yy} = 1.19 \times 10^6 \text{ mm}^4$$

Fig. 8.17 Example 8.8

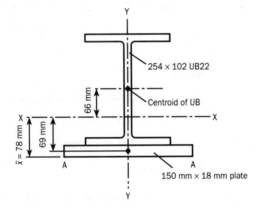

Solution   *Note*: The moment of inertia given above for the UB (without the plate) is of course about the X–X axis of the UB.

The addition of a single plate renders the compound section unsymmetrical about the X–X axis of the compound, and the AH^2 of the UB must be added to its own I_{xx}.

Taking moments about the line A–A:

$$\text{Distance } \bar{x} = \frac{2800 \times 145 + 2700 \times 9}{2800 + 2700}$$
$$= 430\,300/5500$$
$$= 78 \text{ mm}$$
$$I_{xx} = 28.41 \times 10^6 + 2800 \times 66^2 + \tfrac{1}{12} \times 150 \times 18^3 + 2700 \times 69^2$$
$$= (28.41 + 12.20 + 0.07 + 12.85) \times 10^6$$
$$= 53.53 \times 10^6 \text{ mm}^4$$
$$I_{yy} = 1.19 \times 10^6 + \tfrac{1}{12} \times 18 \times 150^3$$
$$= (1.19 + 5.06) \times 10^6$$
$$= 6.25 \times 10^6 \text{ mm}^4$$

Example 8.9   Calculate the I_{xx} and I_{yy} of the compound girder shown in Fig. 8.18. The properties of the UB alone are

$$Area = 2800 \text{ mm}^2$$

$$I_{xx} = 28.41 \times 10^6 \text{ mm}^4$$

$$I_{yy} = 1.19 \times 10^6 \text{ mm}^4$$

Fig. 8.18 Example 8.9

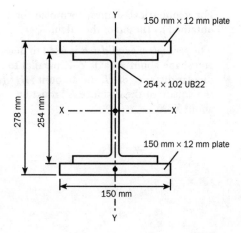

Solution   *Note*: The addition of a 150 mm × 12 mm plate to each flange does not move the neutral axis of the UB from its original position, so that

$$I_{xx} = 28.41 \times 10^6 + \tfrac{1}{12} \times 150 \times 278^3 - \tfrac{1}{12} \times 150 \times 254^3$$

$$= (28.41 + 268.56 - 204.84) \times 10^6$$

$$= 92.13 \times 10^6 \text{ mm}^4$$

Check I_{xx} by addition of the I_{xx} of plates:

$$I_{xx} = 28.41 \times 10^6 + 2(\tfrac{1}{12} \times 150 \times 12^3 + 150 \times 12 \times 133^2)$$

$$= (28.41 + 2 \times 31.86) \times 10^6$$

$$= 92.13 \times 10^6 \text{ mm}^4 \quad \text{(as before)}$$

$$I_{yy} = 1.19 \times 10^6 + 2(\tfrac{1}{12} \times 12 \times 150^3)$$

$$= (1.19 + 6.75) \times 10^6$$

$$= 7.94 \times 10^6 \text{ mm}^4$$

Summary   The *centre of gravity* of a body is a point in or near the body through which the resultant weight of the body acts. When dealing with areas, the terms *centroid*, *centre of area* and *neutral axis* are frequently used.

With respect to beam sections, which are plane figures without weight, the *moment of inertia* about an axis is the sum of the *second moments of area* about that axis, i.e.

$$I = \sum ay^2$$

If a section is divided into an infinite number of strips parallel to the axis in question, and the area of each strip is multiplied by the square of its distance from the axis, and then all these quantities are added together, the result is the moment of inertia of the section.

For geometrical shapes, formulae for the moments of inertia can be obtained by the aid of the calculus.

The *principle of parallel axes* is used to determine the moment of inertia of any shape about an axis Z–Z parallel to another axis X–X at a perpendicular distance of H, the amount AH^2 (i.e. area of shape multiplied by the square of the distance H) must be added to the moments of inertia about X–X.

Exercises

1 Figure 8.Q1 shows a system of three weights connected together by rigid bars. Determine the position of the centre of gravity of the system with respect to point A, ignoring the weight of the bars.

2 Each of the shapes shown in Fig. 8.Q2(a), (b), (c), (d) is symmetrical about a vertical axis. Calculate the distance of the centre of gravity from the base. (All dimensions are in millimetres.)

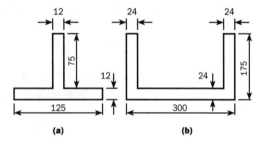
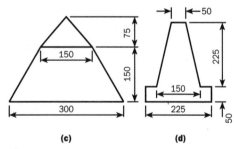

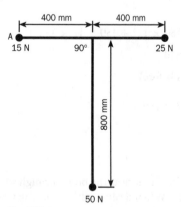

Fig. 8.Q1

Fig. 8.Q2

3 In Fig. 8.Q3(a) to (e) take point P as the intersection of the A–A and B–B axes and calculate the position of the centre of gravity with respect to these axes. ($\bar{x}$ is the distance of the c.g. from the vertical axis A–A and $\bar{y}$ is the distance of the c.g. from the horizontal axis B–B. All dimensions are in millimetres.)

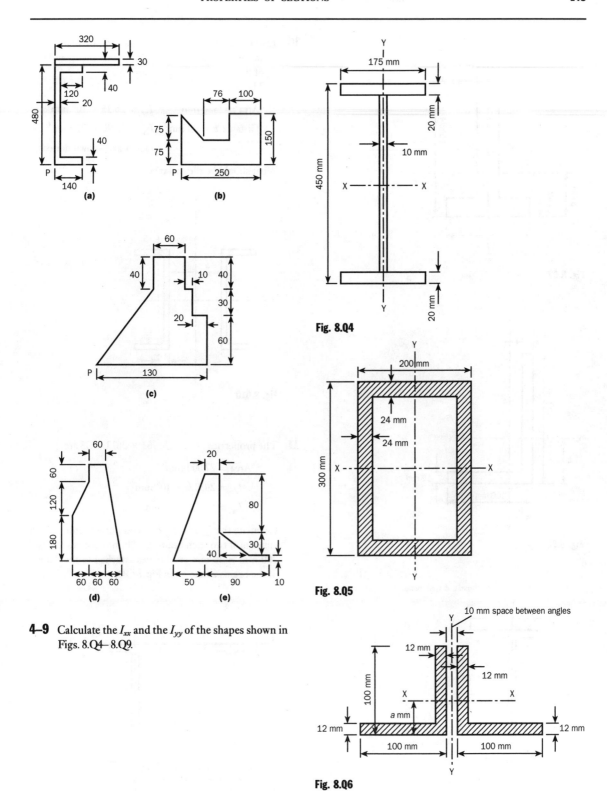

Fig. 8.Q4

Fig. 8.Q5

Fig. 8.Q6

4–9 Calculate the I_{xx} and the I_{yy} of the shapes shown in Figs. 8.Q4– 8.Q9.

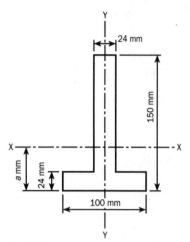

Fig. 8.Q7

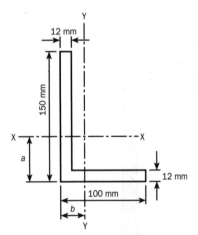

Fig. 8.Q8

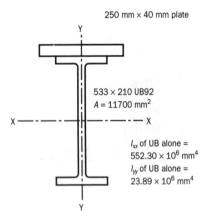

250 mm × 40 mm plate

533 × 210 UB92
A = 11700 mm²

I_{xx} of UB alone =
552.30 × 10⁶ mm⁴
I_{yy} of UB alone =
23.89 × 10⁶ mm⁴

Fig. 8.Q9

10 Two steel channels, 229 mm × 76 mm, are to be arranged as shown in Fig. 8.Q10, so that the I_{xx} and the I_{yy} of the compound section are equal.

The properties of one single channel are

Area = 3320 mm² I_{xx} = 26.15 × 10⁶ mm⁴

Distance b = 20 mm I_{yy} = 15.90 × 10⁶ mm⁴

(about axis shown dotted)

What should be the distance a?

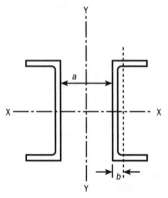

Fig. 8.Q10

11 The properties of a single 762 × 267 UB173 are

$$\text{Area} = 22\,000 \text{ mm}^2$$
$$I_{xx} = 2053.0 \times 10^6 \text{ mm}^4$$
$$I_{yy} = 68.5 \times 10^6 \text{ mm}^4$$

Calculate the I_{xx} and I_{yy} of a compound girder which consists of two such beams at 300 mm centres and two 700 mm × 36 mm steel plates attached to the flanges of the beams as shown in Fig. 8.Q11.

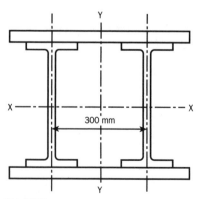

Fig. 8.Q11

12–14 Determine the position of the centre of area of the shapes shown in Figs. 8.Q12– 8.Q14 and calculate the values of their I_{xx}.

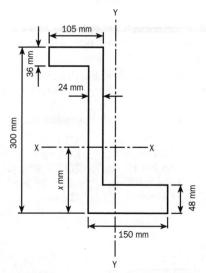

Fig. 8.Q12

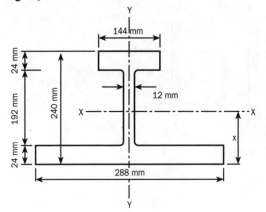

Fig. 8.Q13

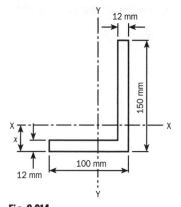

Fig. 8.Q14

15 A strut is made up by inserting a 178 mm × 24 mm steel plate between two channels as shown in Fig. 8.Q15. The properties of a single 178 mm × 76 mm steel channel are

$$\text{Area} = 2660 \text{ mm}^2$$
$$I_{xx} = 13.38 \times 10^6 \text{ mm}^4$$
$$I_{yy} = 1.34 \times 10^6 \text{ mm}^4$$

Calculate the values of I_{xx} and I_{yy} for the strut.

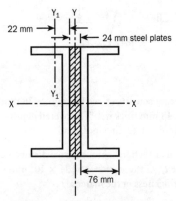

Fig. 8.Q15

16 Calculate the I_{xx} and I_{yy} of the square section shown in Fig. 8.Q16.

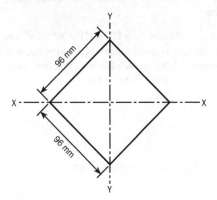

Fig. 8.Q16

17 A special stanchion section is built up using two UBs fixed together as shown in Fig. 8.Q17.
The properties of each individual UB are

	356 × 171 UB57	254 × 102 UB22
Area	7260 mm²	2800 mm²
I_{xx}	160.40 × 10⁶ mm⁴	28.41 × 10⁶ mm⁴
I_{yy}	11.08 × 10⁶ mm⁴	1.19 × 10⁶ mm⁴

Calculate the dimension x to the centroid of the compound section and the values of I_{xx} and I_{yy}.

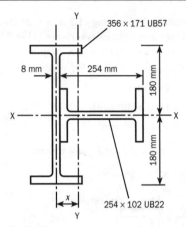

Fig. 8.Q17

18 A timber I-beam consists of two 120 mm × 36 mm flanges and a 48 mm thick web. The overall depth of the beam is 276 mm. Determine its I_{xx} and I_{yy}.

19 A hollow circular section has an external diameter of 200 mm. The I_{xx} of the section is 52.34 × 10^6 mm^4. What is the thickness of the material?

20 The overall depth of a hollow rectangular section is 300 mm and the thickness of the material is 20 mm throughout. The I_{xx} about the horizontal axis is 184.24 × 10^6 mm^4. What is the overall width of the section?

21 The I_{xx} of a 352 mm deep UB is 121 × 10^6 mm^4. The beam has to be increased in strength by the addition of a 220 mm wide plate to each flange so that the total I_{xx} of the compound girder is brought up to 495 × 10^6 mm^4.

What thickness plate will be required? Bolt or rivet holes need not be deducted.

22 The UB mentioned in Question 21 ($I_{xx} = 121 \times 10^6$ mm^4) is to be strengthened by the addition of 30 mm thick plate to each flange so that the I_{xx} is increased to 605 × 10^6 mm^4. What is the required width of the plates? Bolt or rivet holes may be ignored.

23 Calculate the second moment of area about the X–X (horizontal) axis of the crane gantry girder shown in Fig. 8.Q23.

The properties of the individual components are

	610 × 305 UB238	432 × 102 [
Area	30 300 mm^2	8340 mm^2
I_{xx}	2095.0 × 10^6 mm^4	213.7 × 10^6 mm^4
I_{yy}	158.4 × 10^6 mm^4	6.27 × 10^6 mm^4
Web thickness	18.4 mm	12.2 mm

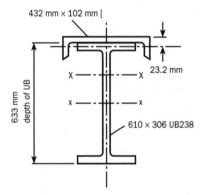

Fig. 8.Q23

9 Simple beam design

In the previous two chapters it has been shown that

- When a beam is loaded, the beam has a bending tendency which is measured in newton millimetres (N mm), etc., and is known as the *bending moment* on the beam. There is, for every loaded beam, a certain critical point at which this bending moment has a maximum value (M_{max}).
- The shape of the beam's cross-section has an effect upon its strength and this shape effect is measured in mm^4 units and is called the *moment of inertia* (second moment of area).
- The material of which the beam is constructed also affects the beam's strength and this factor is measured in terms of the material's *safe allowable stress* (in tension or compression) and is measured in newtons per square millimetre (N/mm^2), etc.

The maximum bending moment, M_{max} depends on the length of the beam and on the nature and disposition of the applied loads.

The factor measuring the strength of the section by virtue of its shape may be varied by the designer so that, when the strength of the material is also taken into account, a beam may be designed of just sufficient strength to take the calculated bending moment.

The bending moment may be thought of as the *punishment factor*.

A suitable combination of the moment of inertia and the safe allowable stress may be described as the beam's *resistance factor* and, indeed, this factor is termed the *moment of resistance* M_r of the beam.

The general theory of bending

It has been seen that, in general, most beams tend to bend in the form shown in Fig. 9.1(a), and if X–X is the neutral axis of such a beam, the fibres above X–X are stressed in compression and those below X–X in tension.

Fig. 9.1 General theory of bending

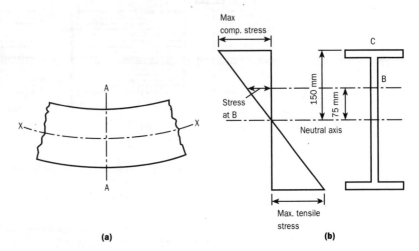

(a)   (b)

Furthermore, fibres far away from the neutral axis are stressed more heavily than those near to the neutral axis. Fibres lying on the neutral axis are neither in tension or compression, and are, in fact, quite unstressed.

The distribution of stress across a plane section such as A–A in Fig. 9.1(a) may thus be seen to vary from zero at the neutral axis to maximum tension and maximum compression at the extreme fibres, as shown in the stress distribution diagram (Fig. 9.1(b)).

The stress varies directly as the distance from the neutral axis, i.e. the stress at C is twice that at B.

It should be appreciated by now that material far from the neutral axis is more useful than material near the neutral axis. This is why the universal beam (UB) is such a popular beam section; the mass of its material (i.e. the two flanges) is positioned as far as possible from the centroid, that is, from the neutral axis (NA).

A simple analogy will help to show how the compression and tension generated within the beam combine to resist the external bending moment.

Consider, for example, the cantilever shown in Fig. 9.2(a), and let a load of W kN be suspended from the free end B, the self-weight of the beam being neglected. If the beam were to be sawn through at section C–C then the portion to the right of the cut would fall, but it could be prevented from falling by an arrangement as shown in Fig. 9.2(b).

Fig. 9.2 Bending of a cantilever: (a) load diagram; (b) theoretical representation at section C–C; (c) representation preventing rotation

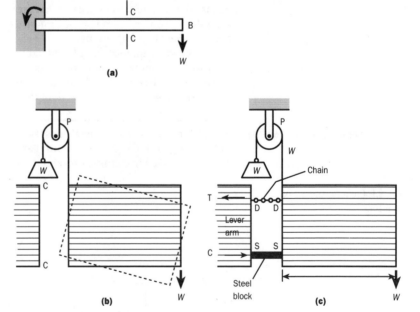

A load of W kN, represented by the weight W, is attached to a wire which, in turn, passes over a pulley P and is connected to the small portion of cantilever. (The weight of the cantilever itself is ignored for the purpose of this discussion.)

However, although this arrangement would prevent the small portion from falling, the portion would still not be in equilibrium, for it would now twist or rotate as shown by the broken lines. This rotation could also be prevented by a small length of chain D–D and a steel block at S–S, as shown in Fig. 9.2(c).

This is because the chain exerts a pull (tension) on the portion of cantilever to the right, and the steel block exerts a push (compression) on the portion.

These forces T and C are equal to each other, and they form an *internal couple* whose moment is equal to and resists the bending moment Wx. This couple provides the moment of resistance of the section which equals $T \times$ lever arm or $C \times$ lever arm. In a real beam these forces T and C are, of course, provided by the actual tension and compression in the many fibres of the cross–section.

Similarly, in a real beam, the force W which prevents the portion from falling is provided by the shear stress generated within the beam itself.

A cantilever has been chosen to illustrate the nature of these internal forces and moments, but the principle remains the same in the case of the simply supported beam, except that the tension will now occur in the layers below the NA and compression in those above the NA.

Figure 9.3(a) shows a short portion of beam (of any shape) before loading, with two plane sections A–B and C–D taken very close together. The layer E–F lies on the NA and layer G–H is one at a distance of y below the NA and having a small cross–sectional area of a.

Fig. 9.3 Bending of a simple beam: (a) cross–section and longitudinal section; (b) the beam after bending

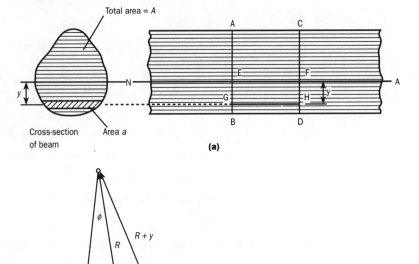

Figure 9.3(b) shows the same length of beam after bending. The sections A–B and C–D remain straight but are rotated to the positions shown, A_1–B_1 and C_1–D_1. The small portion has bent through an angle of ϕ radians, at a radius of curvature (measured to the NA) of R.

The fibre AC has decreased in length to A_1–C_1, whilst the fibre BD has increased in length to B_1–D_1. G–H has also increased to G_1–H_1 but the increase is less than in the case of the fibre B–D. The fibre E–F, being on the NA, neither increases nor decreases and E–F = E_1–F_1.

This straining (alteration in length) of the various fibres in the depth of the section is shown in Fig. 9.4(b), and as stress and strain are proportional to each other (stress/strain = E, see Chapter 6) a similar diagram, Fig. 9.4(c), shows the distribution of stress over the depth of the section.

Fig. 9.4 Stress and strain within the beam: (a) cross-section; (b) strain profile; (c) stress profile

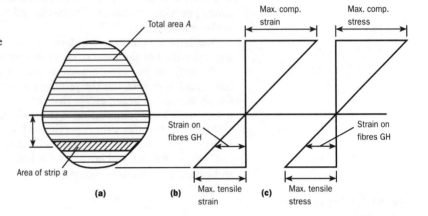

Referring again to Fig. 9.3(b), the fibre G–H increases in length to G_1–H_1. Therefore

$$\text{Strain in fibre G–H} = \frac{\text{increase in length}}{\text{original length}}$$

$$= \frac{(G_1\text{–}H_1) - (G\text{–}H)}{G\text{–}H}$$

But length of arc = radius × angle in radians, therefore

$$G_1 - H_1 = (R + y)\phi$$

Now E_1–F_1 remains the same length and thus E_1–F_1 = G–H. Also E_1–$F_1 = R\phi$.

Thus

$$\frac{(G_1\text{–}H_1) - (G\text{–}H)}{G\text{–}H} = \text{strain in fibres G–H} = \frac{(R + y)\phi - R\phi}{R\phi}$$

$$= \frac{R\phi + y\phi - R\phi}{R\phi} = \frac{y}{R} \tag{1}$$

and, since stress/strain = E, then stress = strain × E, therefore,

$$\text{Stress in fibres G–H} = f = \frac{y}{R} \times E = \frac{Ey}{R} \tag{2}$$

i.e.

$$\frac{f}{y} = \frac{E}{R} \tag{3}$$

The total load over the area a = stress on the area × the area

$$= \frac{Ey}{R} \times a = \frac{Eay}{R}$$

The first moment of the load about the NA is

$$\text{Load} \times y = \frac{Eay^2}{R}$$

The summation of all such moments over the total area A (i.e. the sum of the moments of all such areas as a), including areas above the neutral axis as well as below, is

$$\sum \frac{Eay^2}{R}$$

and this is the total moment of resistance of the section.

But E and R are constants (i.e. only a and y^2 vary) and, since in design the moment of resistance is made equal to the bending moment (usually the maximum bending moment, M_{max}),

$$M_{max} = M_r = \frac{E}{R} \sum ay^2$$

i.e.

$$\text{Safe bending moment} = \frac{E}{R} \times \text{the sum of all the } ay^2$$

But as was already stated in Chapter 8, the sum of all the ay^2, i.e. $\sum ay^2$, is the moment of inertia or second moment of area of the shape, I. So it may now be written

$$M_r = \frac{EI}{R} \tag{4}$$

Referring again to equation (3), $f/y = E/R$ and substituting f/y for E/R in equation (4),

$$M_r = \frac{fI}{y} = f\left(\frac{I}{y}\right) \tag{5}$$

Finally, the main formulae arising from this theory may be stated as follows:

$$\frac{M_r}{I} = \frac{f}{y} = \frac{E}{R} \tag{6}$$

and the most important portion consists of the first two terms

$$\frac{M_r}{I} = \frac{f}{y} \quad \text{or} \quad M_r = f\left(\frac{I}{y}\right) \quad \text{i.e. equation (5)}$$

In the above formula f is the permissible bending stress for the material of the beam. For the most commonly used steel (Grade 43), f may be taken as 165 N/mm^2, whereas the value of f for timber ranges between 3 N/mm^2 and 12 N/mm^2 depending on the species.

These values, however, apply only in those cases where the compression flange of the beam is adequately restrained laterally to prevent sideways buckling. For beams which are not restrained laterally, the above values have to be appropriately reduced. This is outside the scope of this textbook and it can be assumed that all the beams dealt with in this chapter are adequately supported laterally.

Elastic section modulus – symmetrical sections

In equation (5) the factor I/y is called the **section modulus** (or modulus of section) and is usually denoted by the letter Z,

$$\frac{I}{y} = Z$$

Where a beam is symmetrical about its X–X axis, as in the case of those shown in Fig. 9.5, the X–X axis is also the neutral axis (NA) of the beam section, and so the distance y to the top flange is the same as that to the bottom flange, hence

$$Z = \frac{I_{xx}}{y} = \frac{I_{xx}}{\frac{1}{2}d}$$

Fig. 9.5 Beams with symmetrical sections about the X–X axis

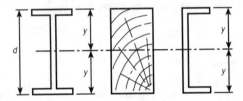

Now, for a rectangular section, $I_{xx} = \frac{1}{12}bd^3$, and so

$$Z = \frac{I}{y} = \frac{bd^3}{12} \div \frac{d}{2} = \frac{bd^2}{6}$$

In the case of rolled steel sections the values of Z (i.e. I/y) are normally taken directly from tables similar to those on pages 166–9 (elastic modulus column) but the following examples include some in which the values are calculated.

Example 9.1   A timber beam of rectangular cross-section is 150 mm wide and 300 mm deep. The maximum allowable bending stress in tension and compression must not exceed 6 N/mm². What maximum bending moment in N mm can the beam safely carry?

Solution   Section modulus $Z = \frac{1}{6}bd^2 = \frac{1}{6} \times 150 \times 300^2$
$$= 2.25 \times 10^6 \text{ mm}^3 \text{ units}$$

Safe allowable bending moment $M_{max} = fZ = 6 \times 2.25 \times 10^6$
$$= 13.5 \times 10^6 \text{ N mm}$$

Example 9.2   A timber beam has to support loading which will cause a maximum bending moment of 25×10^6 N mm. The safe bending stress must not exceed 7 N/mm². What section modulus will be required in choosing a suitable size section?

Solution

$$M_r = f Z \text{ and, therefore, the required } Z = M_r/f, \text{ i.e.}$$

$$Z = \frac{25 \times 10^6}{7} = 3.57 \times 10^6 \text{ mm}^3$$

Example 9.3 A timber beam is required to span 4 m carrying a total uniform load (inclusive of the beam's self-weight) of 40 kN. The safe allowable bending stress is 8 N/mm². Choose a suitable depth for the beam if the width is to be 120 mm.

Solution

$$M_{max} = \tfrac{1}{8} W l = \tfrac{1}{8} \times 40 \times 4 \times 10^6 = 20 \times 10^6 \text{ N mm}$$

but $M_{max} = M_r = f \times Z$, and it follows that

$$\text{required } Z = \frac{M_{max}}{f} = \frac{20 \times 10^6}{8} = 2.5 \times 10^6 \text{ mm}^3$$

The Z of the section is $\tfrac{1}{6} b d^2$, where b is given as 120 mm, therefore

$$\tfrac{1}{6} \times 120 \times d^2 = 2.5 \times 10^6 \text{ mm}^3$$

$$d^2 = \frac{25 \times 10^6 \times 6}{120}$$

$$d = \sqrt{125\,000} = 353 \text{ mm, say } 360 \text{ mm}$$

Example 9.4 The section of floor shown in Fig. 9.6 is to be carried by 125 mm × 75 mm timber joists spanning the 3 m length. The bending stress must not exceed 4.6 N/mm² and the total inclusive load per m² of floor is estimated to be 2.0 kN. At what cross-centres x in mm must the timber beams be fixed?

Fig. 9.6 Example 9.4

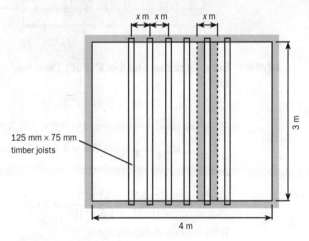

125 mm × 75 mm timber joists

Solution Total load carried by *one* timber joist is

$$3 \times x \times 2.0 = 6x \text{ kN}$$

$$Z \text{ of joist} = \tfrac{1}{6} b d^2 = \tfrac{1}{6} \times 75 \times 125^2 = 195 \times 10^3 \text{ mm}^3$$

Moment of resistance of one joist = bending moment on one joist

$$M_r = f Z = 4.6 \times 195 \times 10^3 = 0.9 \times 10^6 \text{ N mm}$$

Thus

$$M_{max} = \tfrac{1}{8} Wl = \tfrac{1}{8} \times 6x \times 3 \times 10^6 = 0.9 \times 10^6 \text{ N mm}$$

and

$$x = \frac{0.9 \times 10^6 \times 8}{6 \times 3 \times 10^6} = 0.4 \text{ m}$$

Example 9.5   If the floor mentioned in Example 9.4 has its 125 mm × 75 mm timber joists at 450 mm centres and the span of the timber joists is halved by the introduction of a main timber beam 150 mm wide, as shown in Fig. 9.7, what load in kN/m² will the floor now safely carry? And what will be the required depth of the 150 mm wide main timber beam if $f = 5.25$ N/mm²?

Fig. 9.7 Example 9.5

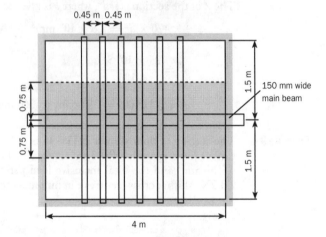

Solution   Let w be the safe load in kN/m². Then load carried by one secondary timber joist is

$$1.5 \times 0.45 \times w = 0.675w \text{ kN}$$

M_r of the joists is unchanged at 0.9×10^6 N mm, therefore

$$M_{max} = \tfrac{1}{8} \times 0.675w \times 1.5 \times 10^6 = 0.9 \times 10^6 \text{ N mm}$$

and

$$w = \frac{0.9 \times 10^6 \times 8}{0.675 \times 1.5 \times 10^6} = 7 \text{ kN/m}^2$$

Total load on the main beam
Each secondary timber joist transfers half of its load to the main beam in the form of an end reaction, and the other half to the wall.

Thus, the total area of floor load carried by the main beam is the area shown shaded in Fig. 9.7.

The total load carried by the main beam is thus $4.0 \times 1.5 \times 7 = 42$ kN and, although this in fact consists of a large number of small point loads from the secondary beams, it is normal practice to treat this in design as a uniformly distributed load. Therefore

$$M_{max} \text{ on main beam} = \tfrac{1}{8} \times 42 \times 4 \times 10^6 = 21 \times 10^6 \text{ N mm}$$

and

$$Z \text{ required} = \frac{M_r}{f} = \frac{21 \times 10^6}{5.25} = 4 \times 10^6 \text{ mm}^3$$

but for a rectangular section, $Z = \tfrac{1}{6}bd^2$, so

$$d = \sqrt{\frac{4 \times 10^6 \times 6}{150}} = 400 \text{ mm}$$

Example 9.6 A welded steel girder is made up of plates as shown in Fig. 9.8.

Fig. 9.8 Example 9.6

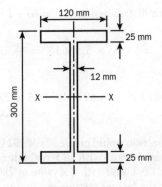

What safe, uniformly distributed load can this girder carry on a simply supported span of 4.0 m if the permissible bending stress is limited to 165 N/mm²?

Solution

$$
\begin{aligned}
I_{xx} \text{ of section} &= \tfrac{1}{12} \times 120 \times 300^3 - \tfrac{1}{12} \times 108 \times 250^3 \\
&= 270 \times 10^6 - 140.625 \times 10^6 \\
&= 129 \times 10^6 \text{ mm}^4
\end{aligned}
$$

$$Z_{xx} \text{ of section} = \frac{I}{y} = \frac{129 \times 10^6}{150} = 0.86 \times 10^6 \text{ mm}^3$$

Hence

$$M_r = 165 \times 0.86 \times 10^6 = 142 \times 10^6 \text{ N mm}$$

Let W be the total safe, uniformly distributed load the girder can carry (in kN), then

$$\tfrac{1}{8} W \times 4 \times 10^6 = 142 \times 10^6 \text{ N mm}$$

$$W = \frac{142 \times 8}{4} = 284 \text{ kN}$$

Example 9.7 A steel beam is required to span 5.5 m between centres of simple supports carrying a 220 mm thick brick wall as detailed in Fig. 9.9.

Choose from the table of properties (pages 166–9) a suitable beam section given that the permissible stress in bending is 165 mm².

Fig. 9.9 Example 9.7

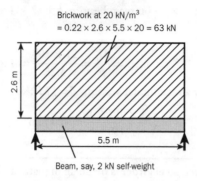

Brickwork at 20 kN/m³
= 0.22 × 2.6 × 5.5 × 20 = 63 kN

2.6 m

5.5 m

Beam, say, 2 kN self-weight

Solution   Total uniformly distributed load $W = 63 + 2 = 65$ kN.
Therefore,

$$M_{max} = \tfrac{1}{8} \times 65 \times 5.5 \times 10^6 = 44.69 \times 10^6 \text{ N mm}$$

and

$$\text{required } Z = 44.69 \times 10^6/165 = 270\,800 \text{ mm}^3$$

$$= 270.8 \text{ cm}^3 \text{ (as shown in tables)}$$

A suitable section would be the 305 × 102 UB25 (i.e. 305 mm × 102 mm at 25 kg/m universal beam) with an actual value of Z at 292 cm³.

The 203 × 133 UB30 with a Z value of 280 cm³ is nearer to the required 270.5 cm³, but the former section was chosen because of its smaller weight.

Example 9.8   Figure 9.10 shows a portion of floor of a steel-framed building. The floor slab, which weighs 3 kN/m², spans in the direction of the arrows, carrying a super (live) load of 5 kN/m² and transferring it to the secondary beams marked A. These in turn pass the loads to the main beams marked B and to the columns.

Fig. 9.10 Example 9.8

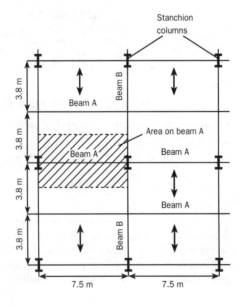

Stanchion columns

3.8 m

Beam A

Beam B

3.8 m

Area on beam A

Beam A

Beam A

3.8 m

Beam A

3.8 m

Beam A

Beam B

7.5 m   7.5 m

Choose suitable sizes for beams A and B, using a permissible bending stress of 165 N/mm².

Solution *Secondary beam, A*

$$Total\ load\ W = 3.8 \times 7.5 \times (3 + 5) = 228\ kN$$
$$M_{max} = \tfrac{1}{8} \times 228 \times 7.5 \times 10^6 = 213.75 \times 10^6\ N\ mm$$

and

$$required\ Z = \frac{213.75 \times 10^6}{165} = 1.295 \times 10^6\ mm^3$$
$$= 1295\ cm^3$$

Use 457 × 191 UB67 with a Z value of 1296 cm³.

Main beam, B
Reaction from beams A at midspan = 228 kN (point load). Assumed weight of beam (UDL), say, 3 kN/m

$$M_{max} = (\tfrac{1}{4} \times 228 + \tfrac{1}{8} \times 3 \times 7.6) \times 7.6 \times 10^6$$
$$= (57 + 2.85) \times 7.6 \times 10^6$$
$$= 454.86 \times 10^6\ N\ mm$$

and

$$required\ Z = \frac{454.86 \times 10^6}{165} = 2.757 \times 10^6\ mm^3$$
$$= 2757\ cm^3$$

Use 610 × 229 UB113 with a Z value of 2874 cm³.

Example 9.9 The timber box beam shown in Fig. 9.11 spans 6 m and carries a uniformly distributed load of 2 kN/m and a centrally placed point load of 3 kN. Determine the maximum stresses due to bending.

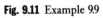
Fig. 9.11 Example 9.9

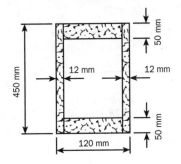

Solution   Because of the symmetry of the loading, the maximum bending moment due to the combined action of the two loads will occur at midspan.

$$M_{max} = (\tfrac{1}{8} \times 2 \times 6 + \tfrac{1}{4} \times 3) \times 6 \times 10^6 = 13.5 \times 10^6 \text{ N mm}$$

$$I_{xx} = \tfrac{1}{12} \times 120 \times 450^3 - \tfrac{1}{12} \times 96 \times 350^3$$

$$= (911.25 - 343.00) \times 10^6 = 568.25 \times 10^6 \text{ mm}^4$$

$$Z_{xx} = \frac{568.25 \times 10^6}{225} = 2.525 \times 10^6 \text{ mm}^3$$

Now, from $M = f \times Z, f = M/Z$, therefore

$$f = \frac{13.5 \times 10^6}{2.525 \times 10^6} = 5.35 \text{ N/mm}^2$$

Elastic section modulus – non-symmetrical sections

Figure 9.12 shows two examples of beam sections which are not symmetrical about the X–X axis. In such cases the distance y between the neutral axis and

Fig. 9.12 Beams with non-symmetrical sections about the X–X axis

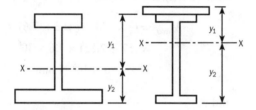

the top and bottom fibres will have two values, y_1 and y_2, and, consequently, there will also be two values of Z_{xx} – one obtained by dividing I_{xx} by the distance y_1, and the other obtained by dividing I_{xx} by the value of y_2.

If in these unsymmetrical shapes the permissible stress is the same for both top and bottom flanges, as it is, for example, in the case of steel, then

$$\text{Safe bending moment} = f \times \text{least } Z$$

If, as in the case of the now obsolete cast iron joist shown in Fig. 9.12, the permissible stresses in tension and compression are not the same, then the safe bending moment is the lesser of the following:

$$M_{rc} = \frac{I}{y_1} \times \text{compression stress}$$

$$M_{rt} = \frac{I}{y_2} \times \text{tension stress}$$

Example 9.10   A 254 × 102 UB22 has one 200 mm × 12 mm steel plate welded to the top flange only as shown in Fig. 9.13. The area of the universal beam alone is 2800 mm² and $I_{xx} = 28.41 \times 10^6$ mm⁴. What safe uniform load can such a beam carry on a 5.0 m simply supported span if $f = 165$ N/mm² in tension or compression?

Fig. 9.13 Example 9.10

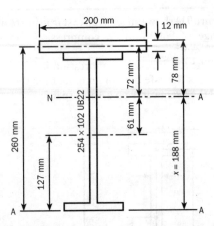

Solution   Taking moments about line A–A to find the neutral axis which passes through the centroid of the section:

$$(2800 + 2400)x = 2800 \times 127 + 2400 \times 260 = 979\,600$$
$$x = 979\,600/5200 = 188 \text{ mm}$$

Applying the principle of parallel axis (Chapter 8) to determine the second moment of area of the section about NA:

$$I_{NA} = 28.41 \times 10^6 + 2800 \times 61^2 + \tfrac{1}{12} \times 200 \times 12^3 + 2400 \times 72^2$$
$$= (28.41 + 10.45 + 0.03 + 12.44) \times 10^6$$
$$= 51.30 \times 10^6 \text{ mm}^4$$
$$Z_{top} = (51.30 \times 10^6)/78 = 657\,700 \text{ mm}^3$$
$$Z_{bot} = (51.30 \times 10^6)/188 = 272\,800 \text{ mm}^3$$

Therefore

$$\text{Critical } M_r = 165 \times 272\,800 = 45.01 \times 10^6 \text{ N mm}$$

For a UDL, $M_{max} = \tfrac{1}{8}Wl$

$$W = 45.01 \times 8/5 = 72.0 \text{ kN}$$

This load will cause the bottom fibres of the beam to be stressed in tension at 165 N/mm^2, while the maximum compressive stress in the top fibres will be much less, namely

$$165 \times \frac{272\,800}{657\,700} \quad \text{or, more simply,}$$
$$165 \times \frac{78}{188} = 68.5 \text{ N/mm}^2$$

Example 9.11   Figure 9.14 shows an old type cast iron joist (dimensions in mm) with a tension flange of 9600 mm², a compression flange of 2880 mm² and a web of 7200 mm².

   The safe stress in compression is 5 N/mm², and in tension 2.5 N/mm². What is the safe bending moment for the section? What safe uniform load will the beam carry on a 4.8 m span?

Fig. 9.14 Example 9.11

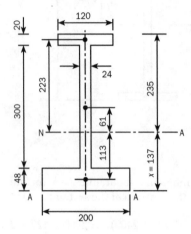

Solution   Taking moments about the lower edge A–A to find the neutral axis,

$$2880 \times 360 = 1\,036\,800$$
$$7200 \times 198 = 1\,425\,600$$
$$9600 \times \ \ 24 = \ \ \ 230\,400$$
$$\overline{19\,680} \qquad = \overline{2\,692\,800}$$

Therefore $x = 2\,692\,800/19\,680 = 137$ mm
   Total I_{NA} is

$$\tfrac{1}{12} \times 120 \times 24^3 + 2880 \times 223^2 = 143.36 \times 10^6$$
$$\tfrac{1}{12} \times 24 \times 300^3 + 7200 \times 61^2 \ \ = \ \ 80.79 \times 10^6$$
$$\tfrac{1}{12} \times 200 \times 48^3 + 960 \times 113^2 \ = 124.42 \times 10^6$$
$$\overline{348.57 \times 10^6 \ \text{mm}^4}$$

Safe bending moment in tension = tension $f \times$ tension Z

$$= 2.5 \times \frac{348.57 \times 10^6}{137}$$
$$= 6.36 \times 10^6 \ \text{N mm}$$

Safe bending moment in compression
$$= \text{compression } f \times \text{compression } Z$$

$$= 5.0 \times \frac{348.57 \times 10^6}{235}$$
$$= 7.42 \times 10^6 \ \text{N mm}$$

- Thus tension is the critical stress, and safe bending moment $= 6.36 \times 10^6$ N mm
- Therefore, since $\frac{1}{8}Wl = 6.36 \times 10^6$ N mm the safe uniformly distributed load, $W = (6.36 \times 8)/4.8 = 10.6$ kN

Example 9.12 A timber joist has a 48 mm thick flange of the same timber fixed to it as shown below in Fig. 9.15, so that the resulting section may be considered to act as a tee beam.

Fig. 9.15 Example 9.12

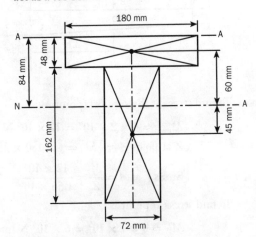

Calculate the safe distributed load on a span of 3.6 m with an extreme fibre stress of 7 N/mm^2.

Solution Taking moments about line A–A to find the NA:

$$8640 \times 24 \qquad\qquad = \ \ 207\,360$$
$$11\,664 \times (48 + \tfrac{1}{2} \times 162) = 1\,504\,656$$
$$\overline{20\,304} \qquad\qquad\qquad \overline{1\,712\,016}$$

Therefore $x = 1\,712\,016/20\,304 = 84$ mm
Total I_{NA} is

$$\tfrac{1}{12} \times 180 \times 48^3 + 8640 \times 60^2 + \tfrac{1}{12} \times 72 \times 162^3 + 11\,664 \times 45^2$$
$$= (1.66 + 31.10 + 25.51 + 23.63) \times 10^6$$
$$= 81.9 \times 10^6 \text{ mm}^4$$

$$\text{Least } Z_{xx} = \frac{I_{xx}}{y} = \frac{81.9 \times 10^6}{210 - 84} = 650\,000 \text{ mm}^3$$

Therefore

$$M_r = fZ = 7 \times 650\,000 = 4.55 \times 10^6 \text{ N mm}$$
If W = safe distributed load in kN,
$$\tfrac{1}{8}Wl = 4.55 \times 10^6 \text{ N mm}$$
$$W = \frac{4.55 \times 10^6 \times 8}{3600} = 10\,000 \text{ N} = 10 \text{ kN}$$

Example 9.13   A timber cantilever beam projects 2 m and carries a 6 kN point load at the free end. The beam is 150 mm wide throughout, but varies in depth from 150 mm to 250 mm, as shown in Fig. 9.16. Calculate the stress in the extreme fibres at a the support and at point 1 m from the support. Ignore the weight of the beam.

Fig. 9.16 Example 9.13

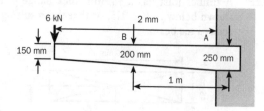

Solution   To find stress at the support (point A):

$$M_{max} = 6 \times 2 \times 10^6 = 12 \times 10^6 \text{ N mm}$$
$$Z \text{ at point } A = \tfrac{1}{6} bd^2 = \tfrac{1}{6} \times 150 \times 250^2 = 1.56 \times 10^6 \text{ mm}^3$$
$$\text{Stress } f \text{ at } A = \frac{M}{Z} = \frac{12 \times 10^6}{1.56 \times 10^6} = 7.68 \text{ N/mm}^2$$

To find stress at point B

$$M_B = 6 \times 1 \times 10^6 = 6 \times 10^6 \text{ N mm}$$
$$Z \text{ at point } B = \tfrac{1}{6} bd^2 = \tfrac{1}{6} \times 150 \times 200^2 = 1.0 \times 10^6 \text{ mm}^3$$
$$\text{Stress } f \text{ at } B = \frac{M}{Z} = \frac{6 \times 10^6}{1 \times 10^6} = 6.0 \text{ N/mm}^2$$

Example 9.14   A hollow steel pipe of 150 mm external and 100 mm internal diameter, as shown in Fig. 9.17 is to span between two buildings. What is the greatest permissible span in metres if the stresses in tension and compression must not exceed 150 N/mm²?

Fig. 9.17 Example 9.14

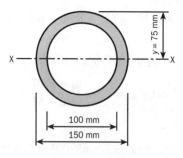

Note: The unit weight of steel is 78 kN/m³

Solution
$$\text{Area} = \tfrac{1}{4} \pi (150^2 - 100^2) = 0.7854 \times (22\,500 - 10\,000)$$
$$= 9.8 \times 10^3 \text{ mm}^2$$

If $l =$ length of span in millimetres, then

$$\text{Total volume of pipe} = l \times 9800 \text{ mm}^3$$

$$\text{Total weight of pipe} = \frac{l \times 9800 \times 78\,000}{1\,000\,000\,000} = 0.77 \times lN$$

But I_{xx} of pipe $= \frac{1}{64}\pi(150^4 - 100^4) = \frac{1}{64}\pi(406.25 \times 10^6) = 19.94 \times 10^6$ mm^4, therefore

$$Z_{xx} \text{ of pipe} = \frac{I}{y} = \frac{19.94 \times 10^6}{75} = 0.27 \times 10^6 \text{ mm}^3$$

$$\text{Permissible } M_{\max} = M_r = \tfrac{1}{8} Wl = f\, Z = 150 \times 0.27 \times 10^6$$

$$= 40.5 \times 10^6 \text{ N mm}$$

therefore

$$\tfrac{1}{8} \times 0.77l \times l = 40.5 \times 10^6 \text{ N mm}$$

$$l = \sqrt{\frac{40.5 \times 10^6 \times 8}{0.77 \times 10^6}} = 20.5 \text{ m}$$

Limit state design

A limit state may be defined as that state of a structure at which it becomes unfit for the use for which it was designed.

To satisfy the object of this method of design, all relevant limit states should be considered in order to ensure an adequate degree of safety and serviceability. The codes therefore distinguish between the **ultimate limit states**, which apply to the safety of the structure, and the **serviceability limit states**, which deal with factors such as deflection, vibration, local damage and cracking of concrete.

Whereas the permissible stress design methods relied on a single factor of safety (or load factor), the limit state design codes introduce two partial safety factors, one applying to the strength of the materials, γ_m, and the other to loads, γ_f. This method, therefore, enables the designer to vary the degree of risk by choosing different partial safety factors. In this way the limit state design ensures an acceptable probability that the limit states will not be reached, and so provides a safe and serviceable structure economically.

Part 1 of BS5950 *Structural use of steelwork in building*, 'Code of practice for design in simple and continuous construction: hot rolled sections', first published in 1985, gives some examples of limit states. These are listed in Table 9.1.

Table 9.1 Limit states

Ultimate		Serviceability	
1	Strength (including general yielding, rupture, buckling and transformation into a mechanism)	5	Deflection
		6	Vibration (e.g. wind-induced oscillation)
2	Stability against overturning and sway	7	Repairable damage due to fatigue
3	Fatigue fracture	8	Corrosion and durability
4	Brittle fracture		

In the design of a steel beam for the ultimate limit state of strength, the following points should be considered.

Loads For the purpose of checking the strength and stability of a structure the specified loads, which are based on the values given in BS 6399: Part 1: 1984 for dead and imposed loads and in CP3: Chapter V: Part 2: 1972 for wind loads, have to be multiplied by the partial factor γ_f. The value of γ_f depends on the combination of loads: for dead loads only it is 1.4, for imposed (or live) loads it is 1.6 and for wind loads it is 1.4. However, for dead loads combined with wind and live loads it is 1.2.

So, the **factored load**, also called the **design load**, for a combination of dead and live loads will be 1.4 × dead + 1.6 × live load and for the combination of dead, live and wind loads it will be 1.2 × (dead + live + wind load).

Strength of steel The design strength, p_y, is based on the minimum yield strength as specified in BS 4360: 1990 '*Specification for weldable steels*' (see Table 9.2). The material factor, γ_m, for steel is 1.0 and the modulus of elasticity, E, is 205 000 N/mm^2.

Table 9.2 Design strength, p_y, for sections, plates and hollow sections

Thickness ≤ (mm)	Design strength (N/mm^2) for BS 4360 steel grade:		
	43	50	55
16	275	355	450
25			430
40	265	345	
63	255	335	400
80	245	325	
100	235	315	

(*Source*: Adapted from Table 6 BS 5950: Part 1: 1990)
Note: For rolled sections the thickness is that of the flange as given in BS 4: Part 1: 1993.

Limit state of strength The **moment capacity** (or moment of resistance), M_c, of a steel section depends on lateral restraint and the cross-section of the beam. Clause 3.5.2 of BS 5950: Part 1: 1990 distinguishes four classes of cross-section, according to their width-to-thickness ratio. The universal beam sections are Class 1 or plastic cross-sections, and as such may be used for plastic (limit state) design.

For these sections the moment capacity should be taken as follows (see Clause 4.2.5 of BS 5950):

$$M_c = p_y S \leqslant 1.2 p_y Z$$

provided that the shear force, F_v, does not exceed

$$0.36 p_y s h$$

where

S is the plastic modulus of section

s is the web thickness of the section

h is the depth of the section

Example 9.15 Using the data given in Example 9.8, choose suitable sections for beams A and B, given that $p_y = 265$ N/mm^2 and the beams have full lateral restraint.

Solution *Secondary beam, A*

Loading $= 3.8 \times 7.5 \times (1.4 \times 3 + 1.6 \times 5) = 347.7$ kN

$$\text{Applied moment, } M_{max} = \tfrac{1}{8} \times 347.7 \times 7.5 \times 10^6$$

$$= 325.97 \times 10^6 \text{ N mm}$$

$$\text{So required } S = \frac{325.97 \times 10^6}{265}$$

$$= 1.23 \times 10^6 \text{ mm}^3 = 1230 \text{ cm}^3$$

Use a 457×152 UB60 ($S = 1287$ cm$^3 < 1.2 \times 1122$ cm^2)

Check for shear:

$$F_v = \frac{347.7}{2} \text{ kN} < \frac{0.36 \times 265 \times 8.1 \times 454.6}{1000} \text{ kN}$$

i.e. the chosen section is satisfactory.

Main beam, B

Loading: own weight of beam, say, 3 kN/m UDL, point load at midspan from beams A $= 347.7$ kN.

$$\text{So } M_{max} = (\tfrac{1}{4} \times 347.7 + \tfrac{1}{8} \times 1.4 \times 3 \times 7.6) \times 7.6 \times 10^6$$

$$= 690.95 \times 10^6 \text{ N mm}$$

$$\text{and required } S = \frac{690.95 \times 10^6}{265} = 2.607 \times 10^6 \text{ mm}^3 = 2607 \text{ cm}^3$$

Use a 533×210 UB101 ($S = 2612$ cm$^3 < 1.2 \times 2292$ cm^3)

Check for shear:

$$F_v = \frac{347.7 + 31.92}{2} \text{ kN} < \frac{0.36 \times 265 \times 10.8 \times 536.7}{1000} \text{ kN}$$

It has to be pointed out that this very brief introduction to limit state design of steelwork treats the topic in an extremely simplified manner. For more detailed application of the principles it is imperative that specialist textbooks be consulted.

Summary The following variables are used in the design of timber and steel beams by the *elastic method*:

$$M_r = f \times \frac{I}{y} = fZ$$

$\qquad$ = moment of resistance

$\qquad$ = maximum bending moment, M_{max} in N mm, kN m, etc.

$\quad f$ = maximum permissible bending stress in N/mm^2

$\quad I$ = moment of inertia about the axis of bending (mm^4)

$\quad y$ = the distance from the neutral axis to the extreme beam fibres

$\quad Z$ = elastic section modulus (mm^3) $= \dfrac{I}{y}$

Table 9.3 Dimensions and properties of universal beams

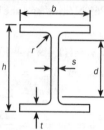

Designation	Area of section (cm²)	Mass per metre (kg/m)	Dimensions (mm)						Ratios for local buckling	
			h Section depth	b Section width	s Web thickness	t Flange thickness	r Root radius	d Depth between fillets	b/2t Flange	d/s Web
914 × 419 × 388	494	388.0	921.0	420.5	21.4	36.6	24.1	799.6	5.74	37.4
914 × 419 × 343	437	343.3	911.8	418.5	19.4	32.0	24.1	799.6	6.54	41.2
914 × 305 × 289	368	289.1	926.6	307.7	19.5	32.0	19.1	824.4	4.81	42.3
914 × 305 × 253	323	253.4	918.4	305.5	17.3	27.9	19.1	824.4	5.47	47.7
914 × 305 × 224	286	224.2	910.4	304.1	15.9	23.9	19.1	824.4	6.36	51.8
914 × 305 × 201	256	200.9	903.0	303.3	15.1	20.2	19.1	824.4	7.51	54.6
838 × 292 × 226	289	226.5	850.9	293.8	16.1	26.8	17.8	761.7	5.48	47.3
838 × 292 × 194	247	193.8	840.7	292.4	14.7	21.7	17.8	761.7	6.74	51.8
838 × 292 × 176	224	175.9	834.9	291.7	14.0	18.8	17.8	761.7	7.76	54.4
762 × 267 × 197	251	196.8	769.8	268.0	15.6	25.4	16.5	686.0	5.28	44.0
762 × 267 × 173	220	173.0	762.2	266.7	14.3	21.6	16.5	686.0	6.17	48.0
762 × 267 × 147	187	146.9	754.0	265.2	12.8	17.5	16.5	686.0	7.58	53.6
762 × 267 × 134	171	133.9	750.0	264.4	12.0	15.5	16.5	686.0	8.53	57.2
686 × 254 × 170	217	170.2	692.9	255.8	14.5	23.7	15.2	615.1	5.40	42.4
686 × 254 × 152	194	152.4	687.5	254.5	13.2	21.0	15.2	615.1	6.06	46.6
686 × 254 × 140	178	140.1	683.5	253.7	12.4	19.0	15.2	615.1	6.68	49.6
686 × 254 × 125	159	125.2	677.9	253.0	11.7	16.2	15.2	615.1	7.81	52.6
610 × 305 × 238	303	238.1	635.8	311.4	18.4	31.4	16.5	540.0	4.96	29.3
610 × 305 × 179	228	179.0	620.2	307.1	14.1	23.6	16.5	540.0	6.51	38.3
610 × 305 × 149	190	149.1	612.4	304.8	11.8	19.7	16.5	540.0	7.74	45.8
610 × 229 × 140	178	139.9	617.2	230.2	13.1	22.1	12.7	547.6	5.21	41.8
610 × 229 × 125	159	125.1	612.2	229.0	11.9	19.6	12.7	547.6	5.84	46.0
610 × 229 × 113	144	113.0	607.6	228.2	11.1	17.3	12.7	547.6	6.60	49.3
610 × 229 × 101	129	101.2	602.6	227.6	10.5	14.8	12.7	547.6	7.69	52.2
533 × 210 × 122	155	122.0	544.5	211.9	12.7	21.3	12.7	476.5	4.97	37.5
533 × 210 × 109	139	109.0	539.5	210.8	11.6	18.8	12.7	476.5	5.61	41.1
533 × 210 × 101	129	101.0	536.7	210.0	10.8	17.4	12.7	476.5	6.03	44.1
533 × 210 × 92	117	92.1	533.1	209.3	10.1	15.6	12.7	476.5	6.71	47.2
533 × 210 × 82	105	82.2	528.3	208.8	9.6	13.2	12.7	476.5	7.91	49.6
457 × 191 × 98	125	98.3	467.2	192.8	11.4	19.6	10.2	407.6	4.92	35.8
457 × 191 × 89	114	89.3	463.4	191.9	10.5	17.7	10.2	407.6	5.42	38.8
457 × 191 × 82	104	82.0	460.0	191.3	9.9	16.0	10.2	407.6	5.98	41.2
457 × 191 × 74	94.6	74.3	457.0	190.4	9.0	14.5	10.2	407.6	6.57	45.3
457 × 191 × 67	85.5	67.1	453.4	189.9	8.5	12.7	10.2	407.6	7.48	48.0
457 × 152 × 82	105	82.1	465.8	155.3	10.5	18.9	10.2	407.6	4.11	38.8
457 × 152 × 74	94.5	74.2	462.0	154.4	9.6	17.0	10.2	407.6	4.54	42.5
457 × 152 × 67	85.6	67.2	458.0	153.8	9.0	15.0	10.2	407.6	5.13	45.3
457 × 152 × 60	76.2	59.8	454.6	152.9	8.1	13.3	10.2	407.6	5.75	50.3
457 × 152 × 52	66.6	52.3	449.8	152.4	7.6	10.9	10.2	407.6	6.99	53.6

(*Source:* Structural sections to BS 4: Part 1 and BS 4848: Part 4 British Steel)

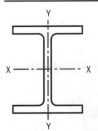

Second moment of area (cm⁴)		Radius of gyration (cm)		Elastic modulus (cm³)		Plastic modulus (cm³)		Parameters and constants				Designation
X–X axis	Y–Y axis	X–X axis	Y–Y axis	X–X axis	Y–Y axis	X–X axis	Y–Y axis	u Buckling parameter	x Torsional index	H Warping constant (dm⁶)	J Torsional constant (cm⁴)	
719600	45440	38.2	9.59	15630	2161	17607	3341	0.885	26.7	88.9	1734	914 × 419 × 388
625800	39160	37.8	9.46	13730	1871	15480	2890	0.883	30.1	75.8	1193	914 × 419 × 343
504200	15600	37.0	6.51	10880	1014	12570	1601	0.867	31.9	31.2	926	914 × 305 × 289
436300	13300	36.8	6.42	9501	871	10940	1371	0.866	36.2	26.4	626	914 × 305 × 253
376400	11240	36.3	6.27	8269	739	9535	1163	0.861	41.3	22.1	422	914 × 305 × 224
325300	9423	35.7	6.07	7204	621	8351	982	0.854	46.8	18.4	291	914 × 305 × 201
339700	11360	34.3	6.27	7985	773	9155	1212	0.870	35.0	19.3	514	838 × 292 × 226
279200	9066	33.6	6.06	6641	620	7640	974	0.862	41.6	15.2	306	838 × 292 × 194
246000	7799	33.1	5.90	5893	535	6808	842	0.856	46.5	13.0	221	838 × 292 × 176
240000	8175	30.9	5.71	6234	610	7176	959	0.869	33.2	11.3	404	762 × 267 × 197
205300	6850	30.5	5.58	5387	514	6198	807	0.864	38.1	9.39	267	762 × 267 × 173
168500	5455	30.0	5.40	4470	411	5156	647	0.858	45.2	7.40	159	762 × 267 × 147
150700	4788	29.7	5.30	4018	362	4644	570	0.854	49.8	6.46	119	762 × 267 × 134
170300	6630	28.0	5.53	4916	518	5631	811	0.872	31.8	7.42	308	686 × 254 × 170
150400	5784	27.8	5.46	4374	455	5000	710	0.871	35.5	6.42	220	686 × 254 × 152
136300	5183	27.6	5.39	3987	409	4558	638	0.868	38.7	5.72	169	686 × 254 × 140
118000	4383	27.2	5.24	3481	346	3994	542	0.862	43.9	4.80	116	686 × 254 × 125
209500	15840	26.3	7.23	6589	1017	7486	1574	0.886	21.3	14.5	875	610 × 305 × 238
153000	11410	25.9	7.07	4935	743	5547	1144	0.886	27.7	10.2	340	610 × 305 × 179
125900	9308	25.7	7.00	4111	611	4594	937	0.886	32.7	8.17	200	610 × 305 × 149
111800	4505	25.0	5.03	3622	391	4142	611	0.875	30.6	3.99	216	610 × 229 × 140
98610	3932	24.9	4.97	3221	343	3676	535	0.873	34.1	3.45	154	610 × 229 × 125
87320	3434	24.6	4.88	2874	301	3281	469	0.870	38.0	2.99	111	610 × 229 × 113
75780	2915	24.2	4.75	2515	256	2881	400	0.864	43.1	2.52	77.0	610 × 229 × 101
76040	3388	22.1	4.67	2793	320	3196	500	0.877	27.6	2.32	178	533 × 210 × 122
66820	2943	21.9	4.60	2477	279	2828	436	0.875	30.9	1.99	126	533 × 210 × 109
61520	2692	21.9	4.57	2292	256	2612	399	0.874	33.2	1.81	101	533 × 210 × 101
55230	2389	21.7	4.51	2072	228	2360	356	0.872	36.5	1.60	75.7	533 × 210 × 92
47540	2007	21.3	4.38	1800	192	2059	300	0.864	41.6	1.33	51.5	533 × 210 × 82
45730	2347	19.1	4.33	1957	243	2232	379	0.881	25.7	1.18	121	457 × 191 × 98
41020	2089	19.0	4.29	1770	218	2014	338	0.880	28.3	1.04	90.7	457 × 191 × 89
37050	1871	18.8	4.23	1611	196	1831	304	0.877	30.9	0.922	69.2	457 × 191 × 82
33320	1671	18.8	4.20	1458	176	1653	272	0.877	33.9	0.818	51.8	457 × 191 × 74
29380	1452	18.5	4.12	1296	153	1471	237	0.872	37.9	0.705	37.1	457 × 191 × 67
36590	1185	18.7	3.37	1571	153	1811	240	0.873	27.4	0.591	89.2	457 × 152 × 82
32670	1047	18.6	3.33	1414	136	1627	213	0.873	30.1	0.518	65.9	457 × 152 × 74
28930	913	18.4	3.27	1263	119	1453	187	0.869	33.6	0.448	47.7	457 × 152 × 67
25500	795	18.3	3.23	1122	104	1287	163	0.868	37.5	0.387	33.8	457 × 152 × 60
21370	645	17.9	3.11	950	84.6	1096	133	0.859	43.9	0.311	21.4	457 × 152 × 52

Continued

Table 9.3 *Continued*

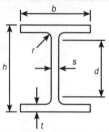

Designation	Area of section (cm²)	Mass per metre (kg/m)	Dimensions (mm)						Ratios for local buckling	
			h Section depth	b Section width	s Web thickness	t Flange thickness	r Root radius	d Depth between fillets	b/2t Flange	d/s Web
406 × 178 × 74	94.5	74.2	412.8	179.5	9.5	16.0	10.2	360.4	5.61	37.9
406 × 178 × 67	85.5	67.1	409.4	178.8	8.8	14.3	10.2	360.4	6.25	41.0
406 × 178 × 60	76.5	60.1	406.4	177.9	7.9	12.8	10.2	360.4	6.95	45.6
406 × 178 × 54	69.0	54.1	402.6	177.7	7.7	10.9	10.2	360.4	8.15	46.8
406 × 140 × 46	58.6	46.0	403.2	142.2	6.8	11.2	10.2	360.4	6.35	53.0
406 × 140 × 39	49.7	39.0	398.0	141.8	6.4	8.6	10.2	360.4	8.24	56.3
356 × 171 × 67	85.5	67.1	363.4	173.2	9.1	15.7	10.2	311.6	5.52	34.2
356 × 171 × 57	72.6	57.0	358.0	172.2	8.1	13.0	10.2	311.6	6.62	38.5
356 × 171 × 51	64.9	51.0	355.0	171.5	7.4	11.5	10.2	311.6	7.46	42.1
356 × 171 × 45	57.3	45.0	351.4	171.1	7.0	9.7	10.2	311.6	8.82	44.5
356 × 127 × 39	49.8	39.1	353.4	126.0	6.6	10.7	10.2	311.6	5.89	47.2
356 × 127 × 33	42.1	33.1	349.0	125.4	6.0	8.5	10.2	311.6	7.38	51.9
305 × 165 × 54	68.8	54.0	310.4	166.9	7.9	13.7	8.9	265.2	6.09	33.6
305 × 165 × 46	58.7	46.1	306.6	165.7	6.7	11.8	8.9	265.2	7.02	39.6
305 × 165 × 40	51.3	40.3	303.4	165.0	6.0	10.2	8.9	265.2	8.09	44.2
305 × 127 × 48	61.2	48.1	311.0	125.3	9.0	14.0	8.9	265.2	4.47	29.5
305 × 127 × 42	53.4	41.9	307.2	124.3	8.0	12.1	8.9	265.2	5.14	33.2
305 × 127 × 37	47.2	37.0	304.4	123.3	7.1	10.7	8.9	265.2	5.77	37.4
305 × 102 × 33	41.8	32.8	312.7	102.4	6.6	10.8	7.6	275.9	4.74	41.8
305 × 102 × 28	35.9	28.2	308.7	101.8	6.0	8.8	7.6	275.9	5.78	46.0
305 × 102 × 25	31.6	24.8	305.1	101.6	5.8	7.0	7.6	275.9	7.26	47.6
254 × 146 × 43	54.8	43.0	259.6	147.3	7.2	12.7	7.6	219.0	5.80	30.4
254 × 146 × 37	47.2	37.0	256.0	146.4	6.3	10.9	7.6	219.0	6.72	34.8
254 × 146 × 31	39.7	31.1	251.4	146.1	6.0	8.6	7.6	219.0	8.49	36.5
254 × 102 × 28	36.1	28.3	260.4	102.2	6.3	10.0	7.6	225.2	5.11	35.7
254 × 102 × 25	32.0	25.2	257.2	101.9	6.0	8.4	7.6	225.2	6.07	37.5
254 × 102 × 22	28.0	22.0	254.0	101.6	5.7	6.8	7.6	225.2	7.47	39.5
203 × 133 × 30	38.2	30.0	206.8	133.9	6.4	9.6	7.6	172.4	6.97	26.9
203 × 133 × 25	32.0	25.1	203.2	133.2	5.7	7.8	7.6	172.4	8.54	30.2
203 × 102 × 23	29.4	23.1	203.2	101.8	5.4	9.3	7.6	169.4	5.47	31.4
178 × 102 × 19	24.3	19.0	177.8	101.2	4.8	7.9	7.6	146.8	6.41	30.6
152 × 89 × 16	20.3	16.0	152.4	88.7	4.5	7.7	7.6	121.8	5.76	27.1
127 × 76 × 13	16.5	13.0	127.0	76.0	4.0	7.6	7.6	96.6	5.00	24.1

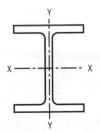

Second moment of area (cm⁴)		Radius of gyration (cm)		Elastic modulus (cm³)		Plastic modulus (cm³)		Parameters and constants				Designation
								u	x	H	J	
X–X axis	Y–Y axis	X–X axis	Y–Y axis	X–X axis	Y–Y axis	X–X axis	Y–Y axis	Buckling parameter	Torsional index	Warping constant (dm⁶)	Torsional constant (cm⁴)	
27310	1545	17.0	4.04	1323	172	1501	267	0.882	27.6	0.608	62.8	406 × 178 × 74
24330	1365	16.9	3.99	1189	153	1346	237	0.880	30.5	0.533	46.1	406 × 178 × 67
21600	1203	16.8	3.97	1063	135	1199	209	0.880	33.8	0.466	33.3	406 × 178 × 60
18720	1021	16.5	3.85	930	115	1055	178	0.871	38.3	0.392	23.1	406 × 178 × 54
15690	538	16.4	3.03	778	75.7	888	118	0.871	38.9	0.207	19.0	406 × 140 × 46
12510	410	15.9	2.87	629	57.8	724	90.8	0.858	47.5	0.155	10.7	406 × 140 × 39
19460	1362	15.1	3.99	1071	157	1211	243	0.886	24.4	0.412	55.7	356 × 171 × 67
16040	1108	14.9	3.91	896	129	1010	199	0.882	28.8	0.330	33.4	356 × 171 × 57
14140	968	14.8	3.86	796	113	896	174	0.881	32.1	0.286	23.8	356 × 171 × 51
12070	811	14.5	3.76	687	94.8	775	147	0.874	36.8	0.237	15.8	356 × 171 × 45
10170	358	14.3	2.68	576	56.8	659	89.1	0.871	35.2	0.105	15.1	356 × 127 × 39
8249	280	14.0	2.58	473	44.7	543	70.3	0.863	42.2	0.0812	8.79	356 × 127 × 33
11700	1063	13.0	3.93	754	127	846	196	0.889	23.6	0.234	34.8	305 × 165 × 54
9899	896	13.0	3.90	646	108	720	166	0.891	27.1	0.195	22.2	305 × 165 × 46
8503	764	12.9	3.86	560	92.6	623	142	0.889	31.0	0.164	14.7	305 × 165 × 40
9575	461	12.5	2.74	616	73.6	711	116	0.873	23.3	0.102	31.8	305 × 127 × 48
8196	389	12.4	2.70	534	62.6	614	98.4	0.872	26.5	0.0846	21.1	305 × 127 × 42
7171	336	12.3	2.67	471	54.5	539	85.4	0.872	29.7	0.0725	14.8	305 × 127 × 37
6501	194	12.5	2.15	416	37.9	481	60.0	0.866	31.6	0.0442	12.2	305 × 102 × 33
5366	155	12.2	2.08	348	30.5	403	48.5	0.859	37.4	0.0349	7.40	305 × 102 × 28
4455	123	11.9	1.97	292	24.2	342	38.8	0.846	43.4	0.0273	4.77	305 × 102 × 25
6544	677	10.9	3.52	504	92.0	566	141	0.891	21.2	0.103	23.9	254 × 146 × 43
5537	571	10.8	3.48	433	78.0	483	119	0.890	24.3	0.0857	15.3	254 × 146 × 37
4413	448	10.5	3.36	351	61.3	393	94.1	0.880	29.6	0.0660	8.55	254 × 146 × 31
4005	179	10.5	2.22	308	34.9	353	54.8	0.874	27.5	0.0280	9.57	254 × 102 × 28
3415	149	10.3	2.15	266	29.2	306	46.0	0.866	31.5	0.0230	6.42	254 × 102 × 25
2841	119	10.1	2.06	224	23.5	259	37.3	0.856	36.4	0.0182	4.15	254 × 102 × 22
2896	385	8.71	3.17	280	57.5	314	88.2	0.881	21.5	0.0374	10.3	203 × 133 × 30
2340	308	8.56	3.10	230	46.2	258	70.9	0.877	25.6	0.0294	5.96	203 × 133 × 25
2105	164	8.46	2.36	207	32.2	234	49.8	0.888	22.5	0.0154	7.02	203 × 102 × 23
1356	137	7.48	2.37	153	27.0	171	41.6	0.888	22.6	0.00987	4.41	178 × 102 × 19
834	89.8	6.41	2.10	109	20.2	123	31.2	0.890	19.6	0.00470	3.56	152 × 89 × 16
473	55.7	5.35	1.84	74.6	14.7	84.2	22.6	0.895	16.3	0.00199	2.85	127 × 76 × 13

For rectangular cross-section beams, $Z = \frac{1}{6}bd^2$.

For circular cross-section beams, $Z = \pi D^3/32$.

For rolled steel sections, values of Z can be found from Table 9.3 (pages 166–9) under the heading Elastic modulus.

For sections built up with rectangular plates, values of Z can be calculated from first principles, having first obtained the moments of inertia (see Chapter 8).

For limit state design of steel beams see BS 5950: Part 1.

The code for the limit state design in timber is BS 5268: Part 1.

Exercises

1   A 250 mm × 75 mm timber beam with its longer edge vertical spans 2 m between simple supports. What safe uniformly distributed load W can the beam carry if the permissible bending stress is 8 N/mm²?

2   A timber joist 75 mm wide has to carry a uniform load of 10 kN on a span of 4 m. The bending stress is to be 6 N/mm². What depth should the joist be?

3   A girder shaped as shown in Fig. 9.Q3 spans 3 m carrying a uniform load over the entire span. The bending stress is to be 150 N/mm². What may be the value of the uniform load?

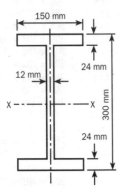

Fig. 9.Q3

4   The properties of a 356 × 171 UB45 are as follows:

$$I_{xx} = 120.7 \times 10^6 \text{ mm}^4 \quad I_{yy} = 8.11 \times 10^6 \text{ mm}^4$$

What safe uniform load will the beam carry on a span of 4 m, if the stress is to be 125 N/mm²?

5   The universal beam used in Question 4, now spans 4.8 m carrying a uniform load of 170 kN. What is the maximum bending stress in N/mm²?

6   Two 254 mm × 89 mm steel channels, arranged back to back with 20 mm space between them, act as a beam on a span of 4.8 m. Each channel has a section modulus (Z) about axis X–X of 350 000 mm³. Calculate the maximum point load that the beam can carry at midspan if the safe stress is 165 N/mm² and the beam's self-weight is ignored.

7   A timber T-beam is formed by rigidly fixing together two joists, as shown in Fig. 9.Q7. The resulting section is used (with the 216 mm rectangle placed vertically) as a beam spanning 3.6 m between simple supports. Calculate the safe uniformly distributed load if the permissible fibre stress is 5.6 N/mm².

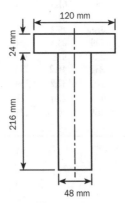

Fig. 9.Q7

8 A 152 mm × 76 mm@19 kg/m steel tee section, as shown in Fig. 9.Q8, may be stressed to not more than 155 N/mm². What safe inclusive uniform load can the section carry as a beam spanning 2.0 m between simple supports?

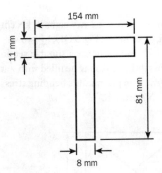

Fig. 9.Q8

9 Timber floor joists 200 mm × 75 mm at 300 mm centres span 3 m between centres of simple supports (Fig. 9.Q9). What safe inclusive load, in kN/m², may the floor carry if the timber stress is not to exceed 6.0 N/mm²?

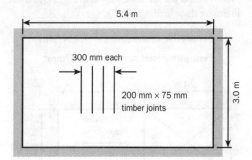

Fig. 9.Q9

10 A small floor 4.8 m × 4.2 m is to be supported by one main beam and 150 mm × 48 mm joists spanning between the wall and the beam, as shown in Fig. 9.Q10.
(a) Calculate the safe inclusive floor load if the stress in the timber has a maximum value of 7.0 N/mm².
(b) Choose a suitable section modulus for the main steel beam if the stress is not to exceed 165 N/mm². Ignore the weights of the joists and the beam.

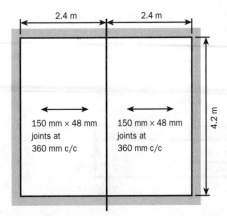

Fig. 9.Q10

11 (a) Calculate the dimension x to the centroid of the section shown in Fig. 9.Q11. (b) Determine I_{xx} and the two values of Z_{xx} for the section. (c) What safe inclusive uniformly distributed load can a beam of this section carry on a span of 3.6 m if the tension stress must not exceed 20 N/mm² and the compression stress 100 N/mm²?

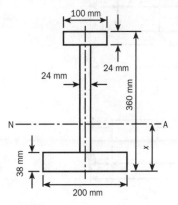

Fig. 9.Q11

12 How deep would a 150 mm wide timber beam need to be to carry the same load as the beam investigated in Question 11 if the maximum flexural stress equals 8 N/mm²?

13 Choose suitable section moduli and select appropriate UBs for the conditions shown in Fig. 9.Q13(a) to (f). $f = 165$ N/mm^2.

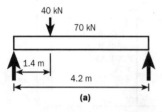

(a)

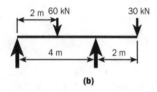

(b)

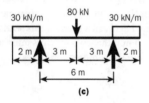

(c)

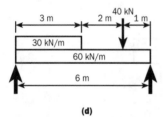

(d)

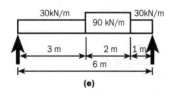

(e)

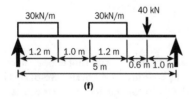

(f)

Fig. 9.Q13

14 A UB 620 mm deep has a section modulus (Z_{xx}) of 4.9×10^6 mm^3. The beam spans 10 m carrying an inclusive uniform load of 100 kN and a central point load of 50 kN; (a) calculate the maximum stress due to bending; (b) what is the intensity of flexural stress at a point 150 mm above the neutral axis and 3.0 m from l.h. reaction?

15 A steel tank 1.8 m $\times$ 1.5 m $\times$ 1.2 m weighs 15 kN empty and is supported, as shown in Fig. 9.Q15, by two steel beams weighing 1 kN each. Choose a suitable section for the steel beams if the tank may be filled with water weighing 10 kN/m^3. The permissible bending stress on the section is 165 N/mm^2.

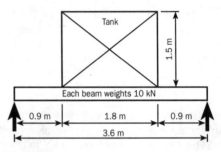

Fig. 9.Q15

16 Choose suitable timber joists and calculate sizes for UBs A and B for the floor shown in Fig. 9.Q16 if the inclusive floor loading is 8.4 kN/m^2.

Permissible timber stress $= 7.2$ N/mm^2

Permissible steel stress $= 165$ N/mm^2

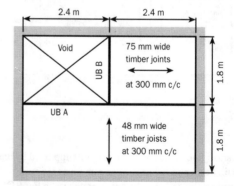

Fig. 9.Q16

17 The symmetrically loaded beam, shown in Fig. 9.Q17, carries three loads, and the internal span l has to be such that the negative bending moment at each support equals the positive bending moment at C. What is the span *l*? If each load *W* is 100 kN, choose a suitable UB ($f = 165$ N/mm^2).

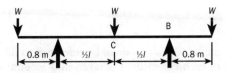

Fig. 9.Q17

18 A 610 × 305 UB149, 610 mm deep ($I_{xx} = 1.25 \times 10^9$ mm^4) is simply supported on a span of *l* m. What will be the maximum permissible span if the stress in the beam under its own weight reaches 22 N/mm^2?

19 The plan of a floor of a steel-framed building is as shown in Fig. 9.Q19. Reinforced concrete slabs spanning as indicated (↔) are supported by steel beams AB. Each beam AB carries a stanchion at C, and the point load from each stanchion is 90 kN. The total inclusive loading will be 9 kN/m^2. Select a suitable beam AB for the floor, using a safe bending stress of 160 N/mm^2.

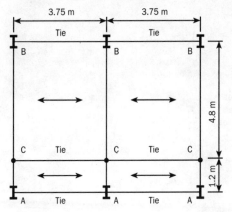

Fig. 9.Q19

20 A steel beam carries loads, as shown in Fig. 9.Q20. Calculate the position and amount of the maximum bending moment and draw the shear force and bending moment diagrams. Choose a suitable steel section for the beam, using a safe bending stress of 150 N/mm^2.

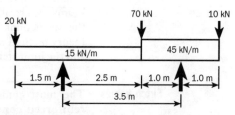

Fig. 9.Q20

21 A rectangular timber beam 300 mm deep and 250 mm wide, freely supported on a span of 6 m carries a uniform load of 3 kN and a triangular load of 12 kN as shown in Fig. 9.Q21.

What is the greatest central point load that can be added to this beam if the maximum bending stress is 8 N/mm^2?

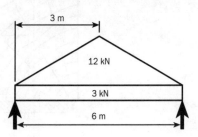

Fig. 9.Q21

10 Beams of two materials

The basis of design of composite beams is explained by reference to flitch beams, where all the material of each constituent is taken into account in the calculations. The reinforced concrete beam is then dealt with, the elastic theory method receiving a fairly detailed treatment. The load factor method and the limit state design method are treated in their simplified form.

Flitch beams

The most common case of beams composed of two materials is the **reinforced concrete beam**, where steel and concrete combine with each other in taking stress. A special case and, therefore, not dealt with in this book is the use of steel and concrete in **composite construction**, where steel beams and concrete slabs are fastened together by means of **shear connectors**.

The simplest case, however, is that of the **flitch beam** which consists of timber and steel acting together, as shown in Fig. 10.1. A timber joist is strengthened by the addition of a steel plate on the top and bottom, the three members being securely bolted together at intervals.

Fig. 10.1 Flitch beam

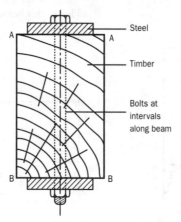

- Steel
- Timber
- Bolts at intervals along beam

The bolting together ensures that there is no slip between the steel and the timber at the sections A–A and B–B. Thus the steel and timber at these sections alter in length by the same amount. That is to say,

Strain in the steel at A–A = strain in timber at A–A

But, as shown in Chapter 6,

$$E = \frac{\text{stress}}{\text{strain}} \quad \text{and stress} = \text{strain} \times E$$

Stress in steel at A–A = E steel × strain

Stress in timber at A–A = E timber × (same strain)

Thus

$$\frac{\text{Stress in steel}}{\text{Stress in timber}} = \frac{E \text{ steel}}{E \text{ timber}}$$

This ratio of the moduli of elasticity is called the **modular ratio** of the two materials and is usually denoted by the letter m. Therefore,

Stress in steel $= m \times$ stress in timber

If, for example, E for steel is 210 000 N/mm^2 and E for timber is 7000 N/mm^2, then

$$m = \frac{210\,000 \text{ N/mm}^2}{7000 \text{ N/mm}^2} = 30$$

and, in effect, the steel plate can thus carry the same load as a timber member of equal thickness but m times the width of the steel.

The beam of the two materials can therefore be considered as an *all-timber beam or equivalent timber beam* as shown in Fig. 10.2(b).

Fig. 10.2 Equivalent beams: (a) composite beam of timber and steel; (b) equivalent timber beam; (c) equivalent steel beam

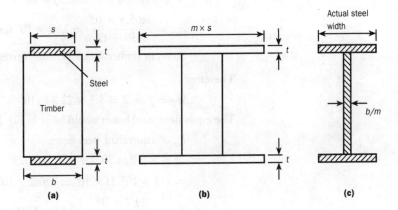

Similarly, an imaginary *equivalent steel beam* could be formed by substituting for the timber a steel plate having the same depth but only $1/m$ times the thickness of the timber, as in Fig. 10.2(c).

It should be noted that, in forming these equivalent beams, the width only has been altered. Any alteration in the vertical dimensions of the timber or steel would affect the value of the strain and therefore only horizontal dimensions may be altered in forming the equivalent sections.

The strength of the *real* beam may now be calculated by determining the strength of the equivalent timber beam or the equivalent steel beam in the usual manner – treating the section as a normal homogeneous one, but making certain that neither of the maximum permissible stresses (timber and steel) is exceeded.

Example 10.1 A composite beam consists of a 300 mm × 200 mm timber joist strengthened by the addition of two steel plates 180 mm × 12 mm, as in Fig. 10.3(a). The safe stress in the timber is 5.5 N/mm^2, the safe stress in the steel is 165 N/mm^2 and $m = 30$. Calculate the moment of resistance in N mm.

Fig. 10.3 Example 10.1:
(a) composite beam;
(b) equivalent timber beam;
(c) equivalent steel beam

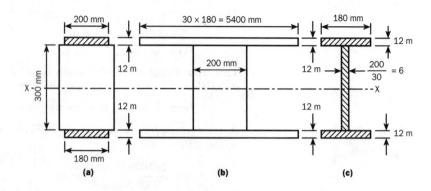

(a)   (b)   (c)

Solution   The equivalent timber beam would be as shown in Fig. 10.3(b).

I_{xx} of equivalent timber beam is

$\frac{1}{12} \times 200 \times 300^3 + 2(\frac{1}{12} \times 5400 \times 12^3 + 5400 \times 12 \times 156^2)$

$= (450.0 + 1.6 + 3153.9) \times 10^6 = 3605.5 \times 10^6 \text{ mm}^4$

$Z_{xx} = \dfrac{3605.5 \times 10^6}{150 + 12} = 22.3 \times 10^6 \text{ mm}^3$

Stress in timber $= 5.5 \text{ N/mm}^2$ (given)

Therefore

$M_T = f \times Z = 5.5 \times 22.3 \times 10^6 = 122.4 \times 10^6 \text{ N mm}$

The equivalent steel beam would be as in Fig. 10.3(c).

I_{xx} of equivalent steel beam

$= \frac{1}{12} \times 6.7 \times 300^3 + \frac{1}{12} \times 180(324^3 - 300^3)$

$= (15 + 105.2) \times 10^6 = 120.2 \times 10^6 \text{ mm}^4$

$Z_{xx} = \dfrac{120.2 \times 10^6}{162} = 0.74 \times 10^6 \text{ mm}^3$

Stress in steel $= 165 \text{ N/mm}^2$ (given)

$\text{or} = 30 \times \text{stress in timber} = 30 \times 5.5$

$= 165 \text{ N/mm}^2$

Therefore

$M_S = f \times Z = 165 \times 0.74 \times 10^6 = 122.4 \times 10^6 \text{ N mm}$

Note: In this example, the ratio of the permissible stresses is equal to the modular ratio.

Example 10.2   A composite beam, shown in Fig. 10.4(a), consists of a 300 mm × 150 mm timber joist strengthened by the addition of two steel plates each 24 mm thick. The stress in the steel must not exceed 150 N/mm²; the stress in the timber must not exceed 8 N/mm²; the modular ratio for the materials is 30. Calculate the safe moment of resistance of the section in N mm and the safe uniform load in kN on a span of 4.0 m.

Fig. 10.4 Example 10.2: (a) composite beam; (b) steel stressed to 150 N/mm² (c) timber stressed to 8 N/mm²; (d) equivalent timber beam

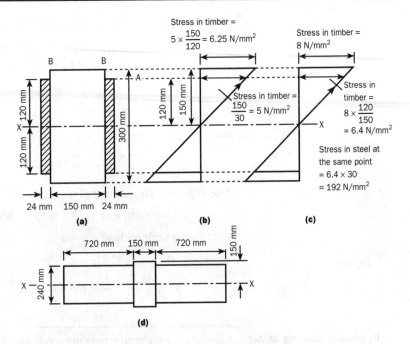

Solution This case has one additional complication which was not present in Example 10.1. The stresses in steel and timber are not to exceed 150 and 8 N/mm² (Fig. 10.4(b) and (c)), respectively, but it may be that stressing the steel to 150 N/mm² would result in a stress of more than 8 N/mm² in the timber. Alternatively, a stress of 8 N/mm² in the timber may result in causing more than 150 N/mm² in the steel.

The critical case may be reasoned as follows. If the steel at A is stressed to 150 N/mm², then

$$\text{Stress in the timber at A} = 150/30 = 5 \text{ N/mm}^2$$

From the stress distribution triangles it follows that

$$\text{Stress in the timber at B} = 5 \times 150/120 = 6.25 \text{ N/mm}^2$$

Both these stresses (150 and 6.25 N/mm²) are permissible.

If, on the other hand, the timber at B is stressed to 8 N/mm², then (from the triangles of stress distribution)

$$\text{Stress in the timber at A} = 8 \times 120/150 = 6.4 \text{ N/mm}^2$$
$$\text{Stress in the steel at A} = 30 \times 6.4 = 192 \text{ N/mm}^2$$

This stress would be exceeding the safe allowable stress in the steel.

Thus it follows that, in calculating the strength of the composite beam, the stress in the steel may be kept at 150 N/mm², but the maximum permissible stress in the equivalent timber section must be reduced to 6.25 N/mm².

Figure 10.4(d) shows the equivalent timber section.

I_{xx} of equivalent section

$$= 2(\tfrac{1}{12} \times 720 \times 240^3) + \tfrac{1}{12} \times 150 \times 300^3$$

$$= (1658.9 + 337.5) \times 10^6$$

$$= 1996.4 \times 10^6 \text{ mm}^4$$

and

$$Z_{xx} = \frac{1996.4 \times 10^6}{150} = 13.3 \times 10^6 \text{ mm}^3$$

Therefore the safe moment of resistance is

$$M_T = fZ = 6.25 \times 13.3 = 83.13 \text{ kN m}$$

and the safe uniform load is given by

$$\tfrac{1}{8} W \times 4 = 83.13$$

$$W = \frac{83.13 \times 8}{4} = 166.4 \text{ kN}$$

Reinforced concrete beams

Concrete is a material strong in its resistance to compression, but very weak indeed in tension. A good concrete will safely take a stress upward of 7 N/mm² in compression, but the safe stress in tension is usually limited to no more than $\tfrac{1}{10}$th of its compressive stress. It will be remembered that in a homogeneous beam the stress distribution is as shown in Fig. 10.5(a), and in

Fig. 10.5 Concrete beams: (a) homogeneous beam: (b) beam with steel bars

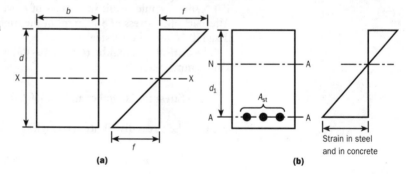

the case of a section symmetrical about the X–X axis, the actual stress in tension equals the actual stress in compression. If such a beam was of ordinary unreinforced concrete, then the stresses in tension and in compression would of necessity have to be limited to avoid overstressing in the tension face, whilst the compressive fibres would be taking only $\tfrac{1}{10}$th of their safe allowable stress. The result would be most uneconomical since the compression concrete would be understressed.

In order to increase the safe load–carrying capacity of the beam, and allow the compression concrete to use to the full its compressive resistance, steel bars are introduced in the tension zone of the beam to carry the whole of the tensile forces.

This may, at first, seem to be a little unreasonable – for the concrete is capable of carrying some tension, even though its safe tensile stress is very low. In fact it will be seen from Fig. 10.5(b) that the strain in the steel and in the concrete at A–A are the same. Therefore, if the stress in steel is 140 N/mm^2, the concrete would be apparently stressed to $1/m$ times this and, since m, the ratio of the elastic moduli of steel and concrete, is usually taken as 15, the tensile stress in concrete would be in excess of 9 N/mm^2. The concrete would crack through failure in tension at a stress very much lower than this and thus its resistance to tension is disregarded.

Basis of design – elastic theory method

In the design of reinforced concrete beams the following assumptions are made:

- Plane sections through the beam before bending remain plane after bending.
- The concrete above the neutral axis carries *all* the compression.
- The tensile steel carries *all* the tension.
- Concrete is an elastic material, and therefore stress is proportional to strain.
- The concrete shrinks in settling and thus grips the steel bars firmly so that there is no slip between the steel and the surrounding concrete.

As the concrete below the neutral axis (NA) is ignored from the point of view of its tensile strength, the actual beam shown in Fig. 10.6(a) is, in fact, repre-

Fig. 10.6 Elastic theory method: (a) beam cross-section; (b) representation of tensile strength

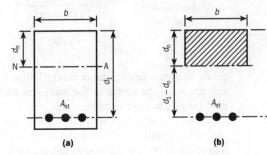

sented by an area of concrete above the neutral axis carrying compression, and a relatively small area of steel taking tension as in Fig. 10.6(b). Thus the position of the neutral axis will not normally be at $d_1/2$ from the top of the beam, but will vary in position according to the relation between the amount of concrete above the NA and the area of steel.

The distance from the top face of the beam to the NA which indicates the depth of concrete in compression in a beam is denoted by d_n.

Critical area of steel

In a beam of a given size, say 200 mm wide with a 300 mm effective depth, the area of steel could be varied – it could be 1000 mm^2 in area, or 2000 mm^2, etc. – and each addition of steel would make for a stronger beam, but not necessarily for an economical one.

If, for example, only a very small steel area is included, as in Fig. 10.7(a), and this small steel content is stressed up to its safe allowable value, then the concrete will probably be very much understressed, so that the concrete is not used to its full advantage. From this point of view, an uneconomical beam results.

If, on the other hand, an overgenerous amount of steel is included, as in Fig. 10.7(b), then stressing this steel to its safe allowable value will probably

Fig. 10.7 Critical area of steel: (a) small steel area; (b) large steel area; (c) normal reinforced concrete beam; (d) normal strain diagram

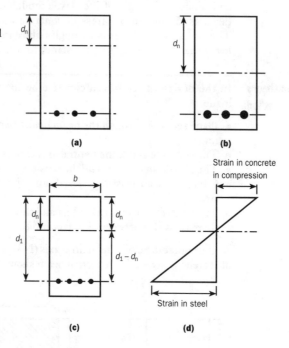

result in overstressing the concrete. Thus, to keep the concrete stress to its safe amount, the stress in the steel will have to be reduced, and again the result is an uneconomical section.

Given that $m = 15$, there will be only one critical area of steel for a given size of beam which will allow the stress in the concrete to reach its permissible value at the same time as the stress in the steel reaches its maximum allowable value. In reinforced concrete beam calculations the designer is nearly always seeking to choose a depth of beam and an area of steel which will allow these stresses to be attained together thus producing what is called a 'balanced' or 'economical' section. The breadth of the beam is normally chosen from practical considerations.

If the two given safe stresses are reached together, then the required area of steel will always remain a constant fraction of the beam's area (bd_1) and the factor d_n/d_1 will also have a constant value.

Figure 10.7(c) shows a normal reinforced concrete (RC) beam and Fig. 10.7(d) the strain diagram for the beam.

From similar triangles it will be seen that

$$\frac{\text{strain in concrete}}{\text{strain in steel}} = \frac{d_n}{d_1 - d_n} \tag{1}$$

But stress/strain for any material is E (Young's modulus) for the material.

Therefore strain = stress/E

Let p_{cb} represent the permissible compressive stress for concrete in bending in N/mm^2 and let p_{st} represent the permissible tensile stress in the steel in N/mm^2.

Thus from equation (1)

$$\frac{d_n}{d_1 - d_n} = \frac{\text{stress in concrete}}{E_c} \div \frac{\text{stress in steel}}{E_s}$$

$$\frac{d_n}{d_1 - d_n} = \frac{p_{cb}}{E_c} \div \frac{p_{st}}{E_s} = \frac{p_{cb}}{E_c} \times \frac{E_s}{p_{st}}$$

But $E_s/E_c = m$, modular ratio, therefore

$$\frac{d_n}{d_1 - d_n} = \frac{m p_{cb}}{p_{st}} \tag{2}$$

Equation (2) shows that d_n and d_1 will always be proportional to each other when m, p_{cb}, and p_{st} are constant.

From equation (2)

$$d_1 - d_n = \frac{d_n p_{st}}{m p_{cb}}$$

$$d_1 = d_n \left(1 + \frac{p_{st}}{m p_{cb}}\right)$$

So

$$d_n = \frac{d_1}{1 + (p_{st}/m p_{cb})} \tag{3}$$

Note: d_n is the distance down to the neutral axis from the compression face of the beam, and is expressed in millimetres.

As d_n is, however, seen to be a constant fraction of d_1 for fixed values of m, p_{cb} and p_{st}, it is frequently useful to know the value of the constant d_n/d_1, and this constant is usually denoted by n, i.e.

$$d_n/d_1 = n$$

As d_n is equal to $\dfrac{d_1}{1 + (p_{st}/m p_{cb})}$ from equation (3), then

$$n = \frac{d_n}{d_1} = \frac{1}{1 + (p_{st}/m p_{cb})} \tag{4}$$

For the values $m = 15$, $p_{cb} = 7$ N/mm^2, $p_{st} = 140$ N/mm^2

$$n = \frac{1}{1 + \dfrac{140}{15 \times 7}} = 0.43$$

and $d_n = 0.43 d_1$

Figure 10.8 shows that the stress in the concrete from the top of the section to the NA follows the normal distribution for ordinary homogeneous sections, and the total compression force $C = \frac{1}{2} p_{cb} b d_n$, i.e. an area $b d_n$ of concrete, taking stress at an average rate of $\frac{1}{2} p_{cb}$ N/mm^2.

The tension, however, will now all be taken by the steel and will be $T = A_{st} \times p_{st}$, i.e. an area of steel A_{st} taking stress at p_{st} N/mm^2. These two internal forces T and C are equal and opposite in their direction, and they

Fig. 10.8 Stress within concrete

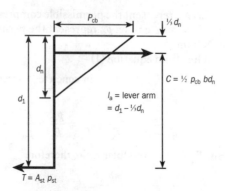

form an internal couple acting at $d_1 - \frac{1}{3}d_n$ apart, where $d_1 - \frac{1}{3}d_n$ is the lever arm, denoted by l_a.

This internal couple or moment of resistance may thus be written as follows:

$$\left.\begin{array}{l} \text{In terms of concrete } \tfrac{1}{2}p_{cb}bd_n(d_1 - \tfrac{1}{3}d_n) = M_{rc} \\[2mm] \text{In terms of steel } A_{st} \times p_{st} \times (d_1 - \tfrac{1}{3}d_n) = M_{rt} \end{array}\right\} M_{max} \qquad \begin{array}{c}(5)\\[4mm](6)\end{array}$$

But just as d_n/d_1 is a 'constant' n, similarly $(d_1 - \frac{1}{3}d_n)/d_1$ is also a constant, say, j and so the lever arm l_a may be expressed as jd_1.

For the values $m = 15$, $p_{cb} = 7$ N/mm^2, $p_{st} = 140$ N/mm^2,

$$j = \frac{d_1 - \frac{1}{3}d_n}{d_1} = \frac{1 - \frac{1}{3} \times 0.43}{1} = 0.86$$

and $l_a = 0.86d_1$

Substituting the constants n and j in (5),

$$M_{rc} = \tfrac{1}{2}p_{cb}bd_n(d_1 - \tfrac{1}{3}dn) = \tfrac{1}{2}p_{cb} \times b \times nd_1 \times jd_1$$
$$= \tfrac{1}{2}p_{cb} \times n \times jbd_1^2$$

As $\frac{1}{2}p_{cb}$, n and j are all constants, then $\frac{1}{2}p_{cb} \times n \times j$ has a constant value which may be called R, and therefore

$$M_{rc} = Rbd_1^2 \qquad (7)$$

Using the values $m = 15$, $p_{cb} = 7$ N/mm^2, $p_{st} = 140$ N/mm^2

$$R = \tfrac{1}{2}p_{cb} \times n \times j = \tfrac{1}{2} \times 7 \times 0.43 \times 0.86 = 1.29$$

and

$$M_{rc} = 1.29bd_1^2$$

Hence

$$d_1^2 = \frac{M}{1.29b} \quad \text{and} \quad d_1 = \sqrt{\frac{M}{1.29b}} \qquad (8)$$

As $M_{rt} = A_{st} \times p_{st} \times jd_1$ then the required area of steel is

$$A_{st} = \frac{M}{p_{st} \times jd_1} = \frac{M}{p_{st}l_a} \qquad (9)$$

Note: As the area of steel is a constant fraction of the area $b \times d_1$ (i.e. $A_{st} = bd_1 \times$ constant), this relationship may be specified as the percentage area of steel, p, where

$$p = \frac{A_{st}}{bd_1} \times 100\% \quad \text{or} \quad A_{st} = bd_1 \times \frac{p}{100}$$

From equation (9)

$$A_{st} = \frac{M}{p_{st}jd_1}$$

but

$$M = Rbd_1^2 = \tfrac{1}{2}p_{cb} \times n \times j \times bd_1^2$$

therefore

$$A_{st} = \frac{p_{cb} \times n \times j \times bd_1^2}{2 \times p_{st} \times jd_1} = \frac{p_{cb} \times n \times bd_1}{2 \times p_{st}}$$

$$\frac{A_{st}}{bd_1} = \frac{p_{cb}n}{2p_{st}}$$

So

$$p = \text{percentage of steel} = \frac{A_{st}}{bd_1} \times 100 = \frac{p_{cb} \times n \times 100}{2p_{st}}$$

$$p = \frac{50 \times p_{cb} \times n}{p_{st}} \tag{10}$$

Using the values $p_{cb} = 7 \text{ N/mm}^2$, $p_{st} = 140 \text{ N/mm}^2$, $m = 15$,

$$p = \frac{50 \times 7 \times 0.43}{140} = 1.07\%$$

Hence

$$A_{st} = bd_1 \times (1.07/100) = 0.011bd_1$$

It must be remembered, however, that there is a wide range of concrete strengths, hence it is not advisable to memorize the numerical values of these design constants.

They may be derived, quite easily, from

$$n = \frac{1}{1 + (p_{st}/mp_{cb})} \qquad j = 1 - \tfrac{1}{3}n$$

$$R = \tfrac{1}{2}p_{cb} \times n \times j \qquad p = \frac{50 \times p_{cb} \times n}{p_{st}}$$

From all the above it will be seen that in designing simple reinforced concrete beams the procedure is as follows:

1 Calculate maximum bending moment, M_{max}, in N mm.
2 Choose a suitable breadth of beam, b.
3 Calculate effective depth to centre of steel,

$$d_1 = \sqrt{\frac{M_{max}}{R \times b}}$$

4 Determine area of steel required from

$$A_{st} = \frac{M_{max}}{p_{st} \times j \times d_1} \quad \text{or} \quad A_{st} = \frac{pbd_1}{100}$$

Example 10.3 Design a simple reinforced concrete beam 150 mm wide to withstand a maximum bending moment of 20×10^6 N mm using the following permissible stresses: $p_{cb} = 7$ N/mm^2, $p_{st} = 140$ N/mm^2 and $m = 15$.

Solution Solution by design constants:

$$n = \frac{1}{1 + (p_{st}/mp_{cb})} = \frac{7}{7 + 140/5} = 0.43$$

$$j = 1 - \tfrac{1}{3}n = 1 - \tfrac{1}{3} \times 0.43 = 0.86$$

$$R = \tfrac{1}{2}p_{cb} \times n \times j = \tfrac{1}{2} \times 7 \times 0.43 \times 0.86 = 1.29$$

$$p = \frac{50 \times p_{cb} \times n}{p_{st}} = \frac{50 \times 0.43 \times 7}{140} = 1.07\%$$

$$d_1 = \sqrt{\frac{20\,000\,000}{1.29 \times 150}} = \sqrt{103\,352} = 321 \text{ mm}$$

$$A_{st} = 0.0107 \times 150 \times 321 = 516 \text{ mm}^2$$

Alternative solution:

From equation (3)

$$d_n = \frac{d_1}{1 + \dfrac{140}{15 \times 7}} = \tfrac{3}{7}d_1 = 0.43d_1$$

From equation (5) and substituting $0.43d_1$ for d_n

$$M_{rc} = \tfrac{1}{2}p_{cb} \times b \times 0.43d_1 \times (d_1 - \tfrac{1}{3} \times 0.43d_1)$$

$$20 \times 10^6 = \tfrac{1}{2} \times 7 \times 150 \times 0.43 \times 0.86 \times d_1^2$$

$$d_1 = \sqrt{\frac{20\,000\,000 \times 2}{7 \times 150 \times 0.43 \times 0.86}} = 321 \text{ mm}$$

From equation (6)

$$A_{st} = \frac{M}{p_{st} \times 0.86 \times d_1} = \frac{20\,000\,000}{140 \times 0.86 \times 321} = 516 \text{ mm}^2$$

Use five 12 mm diameter bars (area $= 5 \times 113.1 = 566$ mm^2). The overall depth d is determined as follows.

The effective depth d_1 is measured from the top face to the centre of the steel bars. The bars themselves need an effective cover of at least 25 mm, so to determine the overall depth d it is necessary to add (half the bar diameter + 25 mm) to the effective depth d_1 (see Fig. 10.9).

In this case overall depth $d = 321 + 6 + 25 = 352$ mm. Say 150 mm $\times$ 355 mm overall with five 12 mm diameter bars.

BS 5328: 1991, in its 'Guide to specifying concrete', lists four types of concrete mix (see Part 1, Clause 7.2) and gives detailed methods for specifying concrete mixes in Part 2.

Fig. 10.9 Example 10.3

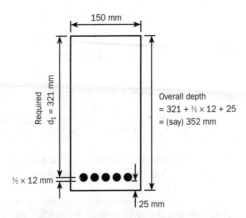

150 mm

Required
$d_1 = 321$ mm

Overall depth
$= 321 + \frac{1}{2} \times 12 + 25$
$= $ (say) 352 mm

$\frac{1}{2} \times 12$ mm

25 mm

For the purpose of exercises in this chapter the permissible stresses in compression due to bending, p_{cb}, may be taken as follows:

$$10.0 \text{ N/mm}^2 \text{ for } 1:1:2 \text{ mix (or C30 grade)}$$

$$8.5 \text{ N/mm}^2 \text{ for } 1:1\tfrac{1}{2}:3 \text{ mix (or C25 grade)}$$

$$7.0 \text{ N/mm}^2 \text{ for } 1:2:4 \text{ mix (or C20 grade)}$$

The permissible tensile stress, p_{st}, for steel reinforcement may be taken as $0.55 \, f_y$, which for hot rolled mild steel bars is generally accepted to be 140 N/mm^2.

Example 10.4

A simply supported reinforced concrete beam is to span 5 m carrying a total uniform load of 60 kN inclusive of self-weight, and a point load of 90 kN from a secondary beam as shown in Fig. 10.10. The beam is to be 200 mm

Fig. 10.10 Example 10.4

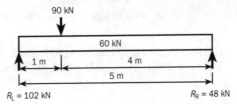

90 kN

60 kN

1 m

4 m

5 m

$R_L = 102$ kN

$R_R = 48$ kN

wide. Choose a suitable overall depth and area of tensile steel reinforcement for the maximum bending moment. Assume: $p_{cb} = 8.5$ N/mm^2, $p_{st} = 140$ N/mm^2, $m = 15$.

Solution

$$R_R = \tfrac{1}{2} \times 60 + \tfrac{1}{5} \times 90 \times 1 = 30 + 18 = 48 \text{ kN}$$

$$R_L = 60 + 90 - 48 = 102 \text{ kN}$$

M_{max} occurs at a distance

$$x = \frac{(102 - 90) \times 5}{60} = 1 \text{ m}$$

from R_R, i.e. at the point load.

$$M_{\max} = 102 \times 1 - \tfrac{1}{5} \times 60 \times \tfrac{1}{2} = 96 \times 10^6 \text{ N mm}$$

$$d_n = \frac{8.5}{8.5 + (140/15)} = 0.48d_1 \quad \text{and} \quad l_a = 0.84d_1$$

$$d_1 = \sqrt{\frac{2 \times 96\,000\,000}{8.5 \times 200 \times 0.48 \times 0.84}} = 530 \text{ mm}$$

$$A_{st} = \frac{96\,000\,000}{140 \times 0.84 \times 530} = 1540 \text{ mm}^2$$

Use five 20 mm diameter bars (1570 mm^2).
Then $d = 530 + \tfrac{1}{2} \times 20 + 25 = 565$ mm.

Example 10.5   A simply supported beam is to span 3.6 m carrying a uniform load of 60 kN inclusive of self-weight. The beam is to be 150 mm wide, and the stresses in steel and concrete respectively are to be 140 N/mm^2 and 10 N/mm^2 ($m = 15$).
Determine the constants n, j, R and p, and choose a suitable effective depth, overall depth and area of steel for the beam.

Solution

$$n = \frac{1}{1 + (p_{st}/mp_{cb})} = \frac{1}{1 + (140/15 \times 10)} = 0.517$$

$$j = 1 - \tfrac{1}{3}n = 1 - \tfrac{1}{3} \times 0.517 = 0.828$$

$$R = \tfrac{1}{2}p_{cb} \times n \times j = \tfrac{1}{2} \times 10 \times 0.517 \times 0.828 = 2.140$$

$$p = \frac{50 \times p_{cb} \times n}{p_{st}} = \frac{50 \times 10 \times 0.517}{140} = 1.846\%$$

$$M_{\max} = \tfrac{1}{8}Wl = \tfrac{1}{8} \times 60 \times 3.6 \times 10^6 = 27 \times 10^6 \text{ N mm}$$

$$d_1 = \sqrt{\frac{27\,000\,000}{2.140 \times 150}} = 290 \text{ mm}$$

$$A_{st} = 0.01846 \times 150 \times 290 = 803 \text{ mm}^2$$

Use four 16 mm diameter bars (804 mm^2).
Then $d = 290 + \tfrac{1}{2} \times 16 + 25 = 323$, say 325 mm.

Basis of design – load factor method   The introduction of this method alongside the elastic theory method to CP114 in 1957 was the result of tests to destruction. They have shown that, at failure, the compressive stresses adjust themselves to give a compressive resistance greater than that obtained by the elastic theory method.

The load factor method, therefore, does not use the modular ratio nor does it assume a proportionality between stress and strain in concrete. It requires the knowledge of plastic behaviour, which is outside the scope of this textbook. CP114, however, introduced simplified formulae for rectangular beams and slabs which give acceptable results. They are based on the assumption that the compressive stress in the concrete is two-thirds of the permissible compressive stress in the concrete in bending (i.e. $\tfrac{2}{3}p_{cb}$). The stress is considered to be uniform over a depth d_n, not exceeding one-half of the effective depth ($\tfrac{1}{2}d_1$).

Figure 10.11 shows the stress distribution diagram based on the above assumptions.

Total compressive force $C = \frac{2}{3} p_{cb} \times b \times d_n$

Total tensile force $T = p_{st} \times A_{st}$

For equilibrium $C = T$ and therefore

$$d_n = \frac{3}{2} \times \frac{p_{st} \times A_{st}}{p_{cb} \times b}$$

Fig. 10.11 Load factor method

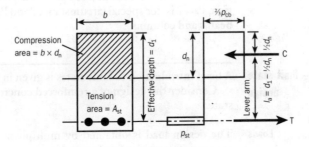

Since the compressive stress is uniform

Lever arm $l_a = d_1 - \frac{1}{2} d_n$

$$= d_1 - \frac{3 \times p_{st} \times A_{st}}{4 \times p_{cb} \times b}$$

but, for an 'economical' or balanced section, the code recommends $d_n = \frac{1}{2} d_1$

$$l_a = d_1 - \frac{1}{2} \times \frac{1}{2} d_1 = \frac{3}{4} d_1$$

Hence, the moment of resistance, based on concrete in compression, is

$$M_{rc} = \frac{2}{3} p_{cb} \times b \times \frac{1}{2} d_1 \times \frac{3}{4} d_1$$

$$= \frac{1}{4} p_{cb} \times b \times (d_1)^2$$

and the moment of resistance, based on tensile reinforcement, is

$$M_{rt} = A_{st} \times p_{st} \times l_a$$

Example 10.6 Consider the beam in Example 10.5 and, using the same concrete mix, design the beam by the load factor method.

Solution

$$M_{max} = 27 \times 10^6 \text{ N mm}$$

$$= M_{rc} = \frac{1}{4} \times 10 \times 150 \times (d_1)^2$$

$$d_1 = \sqrt{\frac{27\,000\,000 \times 4}{10 \times 150}} = 268 \text{ mm}$$

and

$$A_{st} = \frac{27\,000\,000 \times 4}{140 \times 3 \times 268} = 960 \text{ mm}^2$$

It may be correctly deduced from the above that the load factor method of design results in shallower beams with more reinforcement.

The load factor method gained fairly general acceptance in a relatively short time, so that, when in 1972 CP110, the code of practice for the structural use of concrete, appeared, it did not contain any explicit reference to the modular ratio or elastic theory. This code introduced yet another approach to structural design in what it called 'strict conformity with the theory of limit states'.

CP110 has been superseded by BS 8110: 1985, with 'no major changes in principle'. The new code in three parts: Part 1 deals with design and construction, Part 2 is 'for special circumstances' and Part 3 contains design charts for beams and columns.

Basis of design – limit state design

A brief introduction to limit states is given in Chapter 9.

Consider the design of a reinforced concrete beam for the ultimate limit state.

Loads

The design load is obtained by multiplying the characteristic load by an appropriate partial safety factor, γ_f.

The characteristic loads (G_k = dead load, Q_k = imposed load, and W_k = wind load) are based on values given in BS 6399: Part 1: 1984 for dead and imposed (live) loads, and CP3: Chapter V: Part 2: 1972 for wind loads.

The partial safety factors γ_f depend on the combination of loads. For example, for dead + live loads, the design load would be

$$(1.4 \times G_k + 1.6 \times Q_k)$$

and for dead + live + wind loads, the design load would be

$$1.2 \times (G_k + Q_k + W_k).$$

Strength of materials

The design strength is obtained by dividing the characteristic strength f_k, by the appropriate partial safety factor γ_m.

Characteristic strength is defined as that value of the cube strength of concrete (f_{cu}) or the yield (or proof) stress of reinforcement (f_y) below which not more than 5 per cent of the test results will fall.

The partial safety factors γ_m for ultimate limit state are 1.5 for concrete and 1.15 for the reinforcement.

Ultimate moment of resistance

In their simplest form, as given in Clause 3.4.4.4 of BS8110: Part 1: 1985, the calculations are based on assumptions similar to the load factor method, i.e. rectangular stress block with the stress value of $0.477f_{cu}$ over $0.9 \times$ distance to NA.

Therefore

$$M_{uc} = 0.477f_{cu} \times b \times 0.446d \times 0.777d$$
$$= 0.156f_{cu} \times b \times d^2$$

and

$$M_{ut} = 0.87f_y \times A_s \times z$$

where M_{uc} = ultimate resistance moment (compression)

M_{ut} = ultimate resistance moment (tension)

b = width of beam section

d = effective depth to tension reinforcement

A_s = area of tension reinforcement

z = lever arm (in this case $0.777d$)

Example 10.7   Take the beam in Example 10.5 and adapt the data to limit state design method for the ultimate limit state.

Solution   Assume the 60 kN load to consist of 60 per cent dead and 40 per cent live loads and

$$f_y = 250 \text{ N/mm}^2 \qquad f_{cu} = 30 \text{ N/mm}^2$$

Therefore,

$$\text{Design load} = (1.4 \times 36 + 1.6 \times 24) = 88.8 \text{ kN}$$

and

$$M_{max} = \tfrac{1}{8} \times 88.8 \times 3.6 \times 10^6 = 40 \times 10^6 \text{ N mm}$$

Hence

$$M_{uc} = 0.156 \times 30 \times 150 \times d^2 = 40 \times 10^6 \text{ N mm}$$

$$d = \sqrt{\frac{40\,000\,000}{0.156 \times 30 \times 150}} = 238 \text{ mm}$$

and

$$M_{ut} = 0.87 \times 250 \times A_s \times 0.777 \times 238 = M_{max}$$

$$A_s = \frac{40\,000\,000}{0.87 \times 259 \times 0.777 \times 238} = 995 \text{ mm}^2$$

It must be emphasized that the above exposition of the reinforced concrete beam design methods is very much simplified. It is essential to consult text-books on concrete design for more detailed discussion of both principles and their application.

Summary   *Timber and steel composite (flitch) beams*   Replace the steel by its equivalent area of timber by multiplying the width of the steel (parallel to the axis of bending) by the modular ratio E_s/E_t. Calculate the value of Z for the equivalent timber section. Determine the maximum permissible stress f in the extreme fibres of the beam such that neither the maximum per-missible steel stress nor the maximum permissible timber stress is ex-ceeded. Then

$$M_r = fZ$$

Reinforced concrete beams

- Elastic theory method

$$d_n = \frac{1}{1 + p_{st}/mp_{cb}} \times d_1$$

$$M_{rc} = \tfrac{1}{2}p_{cb} \times b \times d_n \times l_a$$

$$M_{rt} = p_{st} \times A_{st} \times l_a$$

where $l_a = d_1 - \tfrac{1}{3}d_n$.

- Load factor method

$$M_{rc} = \tfrac{1}{4}p_{cb} \times b \times (d_1)^2$$

$$M_{rt} = p_{st} \times A_{st} \times \tfrac{3}{4}d_1$$

- Limit state design method

$$M_{uc} = 0.156f_{cu} \times b \times d^2$$

$$M_{ut} = 0.87f_y \times A_s \times z$$

where $z = 0.777d$.

Exercises

1 A composite beam is formed using a 400 mm × 180 mm timber beam with a 300 mm × 12 mm steel plate securely fixed to each side as shown in Fig. 10.Q1. The maximum stresses in the steel and timber respectively must not exceed 140 and 8 N/mm², and the modular ratio is 20. (a) What will be the actual stresses used for the steel and for the timber? (b) What is the safe moment of resistance in N mm for the beam section?

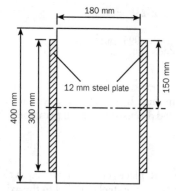

Fig. 10.Q1

2 A timber flitch beam is composed of two 300 mm × 150 mm timber beams and one 250 mm × 20 mm steel plate placed between the timbers so that, when properly bolted together, the centre lines of all three members coincide. Calculate the maximum safe uniformly distributed load in kilonewtons that this beam could carry over a span of 4.5 m if the stress in the steel is not to exceed 125 N/mm² and that in the timber 7 N/mm², and given that the modular ratio $E_s/E_t = 20$.

3 A timber beam 150 mm × 300 mm deep has two steel plates, each 125 mm × 12 mm bolted to it as shown in Fig. 10.Q3. Assuming the safe steel stress is 140 N/mm², the safe timber stress is 7 N/mm², E for steel is 205 000 N/mm² and E for timber is 8200 N/mm², calculate the moment of resistance of the beam. (Ignore bolt holes.)

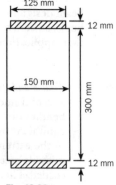

Fig. 10.Q3

4 A timber beam in an existing building is 200 mm wide and 380 mm deep and is simply supported at the ends of a 6 m span. (a) Calculate the maximum safe uniformly distributed load for the timber alone if the bending stress must not exceed 7 N/mm². (b) It is proposed to strengthen the beam to enable it to carry a uniformly distributed load of 150 kN by bolting two steel plates to the beam as indicated in Fig. 10.Q4. Calculate the required thickness t of the plates if the maximum permissible stress for the steel is 140 N/mm² and the modular ratio is 24.

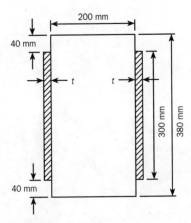

Fig. 10.Q4

5 A short concrete beam is to be constructed without any steel reinforcement to span 2.8 m, carrying a total inclusive uniform load of 20 kN. If the concrete has a safe tensile stress of only 0.6 N/mm², state what depth would be needed for a suitable beam 200 mm wide.

6 Design a reinforced concrete section for the loading conditions as in Question 5 if the beam remains 200 mm wide, using the following stresses: $p_{cb} = 10$ N/mm², $p_{st} = 140$ N/mm², $m = 15$.

7 A simply supported reinforced concrete beam, 240 mm wide, carries inclusive loads as shown in Fig. 10.Q7.

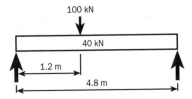

Fig. 10.Q7

Determine (a) the effective depth d_1 in mm, (b) the required steel area if $p_{cb} = 7$ N/mm², $p_{st} = 140$ N/mm², $m = 15$. The weight of the beam may be assumed to be included in the given loads.

8 Referring to Example 9.5 design a reinforced concrete beam as an alternative to the 400 mm × 150 mm timber beam. Assume that the floor load is 7 kN/m² and allow 12 kN for the weight of the reinforced concrete beam. Take the breadth of the beam as 250 mm and $p_{cb} = 7$ N/mm², $p_{st} = 140$ N/mm², $m = 15$.

9 Referring to Example 9.7 for loading conditions but substituting 18 kN for the weight of the beam, design a reinforced concrete beam 225 mm wide. When the design is complete, check the assumed weight of beam, taking the weight of reinforced concrete as 24 kN/m³. Use $1 : 1\frac{1}{2} : 3$ mix of concrete.

10 A small floor is to be supported as shown in Fig. 10.Q10. The total floor load is 5 kN/m², and 12 kN can be assumed as the weight of the beam. Design the beam assuming a breadth of 250 mm and $1 : 2 : 4$ mix concrete.

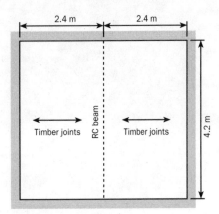

Fig. 10.Q10

11 Referring to Exercise 15 in Chapter 9, but taking 8 kN as the weight of each beam instead of the value given, design reinforced concrete beams assuming a breadth of 225 mm and $1 : 1 : 2$ mix concrete.

12 Referring to Exercise 16 in Chapter 9 design beams A and B in reinforced concrete ($1 : 2 : 4$ mix). Take the weight of beam B as 1.8 kN and the weight of beam A as 16.0 kN (in addition to the floor load given).

13 A reinforced concrete beam 300 mm wide simply supported on a span of 6 m carries a triangular load of 80 kN in addition to its own weight, which may be assumed to be 20 kN (Fig. 10.Q13). Design the beam in $1 : 1\frac{1}{2} : 3$ mix concrete.

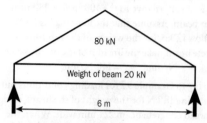

Fig. 10.Q13

11 Deflection of beams

So far in this book the beam has been considered from the point of view of its safety and strength in its resistance to bending. This chapter investigates the deformation of beams as the direct effect of that bending tendency, which affects their serviceability and stability, and does so in terms of their deflection.

A beam may be strong enough to resist safely the bending moments due to the applied loading and yet not be suitable because its deflection is too great. Excessive deflection might not only impair the strength and stability of the structure but also give rise to minor troubles such as cracking of plaster ceilings, partitions and other finishes, as well as adversely affecting the functional needs and aesthetic requirements or simply being unsightly.

The relevant BS specifications and codes of practice stipulate that the deflection of a beam shall be restricted within limits appropriate to the type of structure. In the case of structural steelwork the maximum deflection due to unfactored imposed loads for beams carrying plaster or other brittle finish must not exceed 1/360 of the span, but for all other beams it may be span/200 (Clause 2.5.1 of BS 5950: Part 1: 1990). For timber beams, on the other hand, the figure is 0.003 of the span when the supporting member is fully loaded (Clause 14.7 of BS 5268: Part 2: 1991). In reinforced concrete the deflection is generally governed by the span/depth ratio (Clause 3.4.6.3 of BS 8110: Part 1: 1985).

Factors affecting deflection

For many beams in most types of buildings, e.g. flats, offices, warehouses, it will usually be found that, if the beams are made big enough to resist the bending stresses, the deflections will not exceed the permitted values. In beams of long spans, however, it may be necessary to calculate deflections to ensure that they are not excessive. The derivation of formulae for calculating deflections usually involves the calculus. In this chapter, therefore, only a general treatment will be attempted and deflection formulae for a few common cases of beam loadings will be given without proof. General methods of calculation of deflections are given in standard books on theory of structures or strength of materials.

Load

AB (Fig. 11.1) represents a beam of span l metres supported simply at its ends and carrying a point load of W kN at midspan. Let us assume that the deflection due to the load is 5 mm. It is obvious that, if the load is increased, the deflection will increase. It can be proved that the deflection is directly proportional to the load, i.e. a load of $2W$ will cause a deflection of 10 mm, $3W$ will produce a deflection of 15 mm and so on. W must therefore be a term in any formula for calculating deflection.

Fig. 11.1 Deflection of a beam under loading

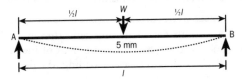

Span

In Fig. 11.2(a) and (b) the loads W are equal and the weights of the beams, which are assumed to be equal in cross-section, are ignored for purposes of

Fig. 11.2 Effect of span upon deflection: the span of (a) is twice that of (b)

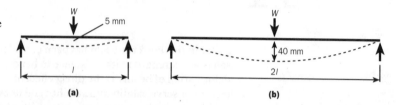

this discussion. The span of beam (b) is twice that of beam (a). It is obvious that the deflection of beam (b) will be greater than that of beam (a), but the interesting fact (which can be demonstrated experimentally or proved by mathematics) is that, instead of the deflection of (b) being twice that of (a), it is 8 times (e.g. 40 mm in this example). If the span of beam (b) were $3l$, its deflection would be 27 times that of beam (a). In other words, the deflection of a beam is proportional to the cube of the span, therefore l^3 is a term in the deflection formula.

Size and shape of beam

Figure 11.3(a) and (b) represents two beams (their weights being ignored) of equal spans and loading but the moment of inertia of beam (b) is twice that

Fig. 11.3 Effect of size and shape upon deflection: (a) moment of inertia of beam = 1 unit; (b) moment of inertia of beam = 2 units

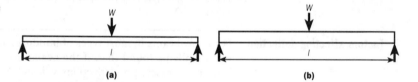

of beam (a). Obviously, the greater the size of the beam, the less the deflection (other conditions being equal). It can be proved that the deflection is inversely proportional to the moment of inertia, e.g. the deflection of beam (b) will be one-half that of beam (a). Moment of inertia I is therefore a term in the denominator of the deflection formula. (It may be noted that, since the moment of inertia of a rectangular cross-section beam is $bd^3/12$, doubling the breadth of a rectangular beam decreases the deflection by one-half, whereas doubling the depth of a beam decreases the deflection to one-eighth of the previous value.)

'Stiffness' of material

The stiffer the material of a beam, i.e. the greater its resistance to bending, the less will be the deflection, other conditions such as span, load, etc., remaining constant. The measure of the 'stiffness' of a material is its modulus of elasticity E and deflection is inversely proportional to the value of E.

Derivation of deflection formulae

A formula for calculating deflection must therefore contain the load W, the cube of the span l^3, the moment of inertia I, and the modulus of elasticity E. For standard cases of loading, the deflection formula can be expressed in the form cWl^3/EI, where c is a numerical coefficient depending on the disposition of the load and also on the manner in which the beam is supported, that is, whether the ends of the beam are simply supported or fixed, etc. For Fig. 11.4 the values of c are respectively $1/48$ and $5/384$. W and l^3 are in the numera-

Fig. 11.4 Deflection formulae

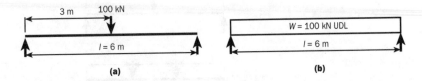

tor of the formula because increase in their values means increase of deflection, whereas E and I are in the denominator because increase in their values means decrease of deflection.

Referring to Fig. 11.4 it should be obvious (neglecting the weights of the beams) that although the beams are equally loaded, the deflection of beam (b) will be less than that of beam (a). In fact, the maximum deflection of beam (a) is

$$\frac{1}{48}\frac{Wl^3}{EI} = \frac{8}{384}\frac{Wl^3}{EI}$$

and the maximum deflection of beam (b) is

$$\frac{5}{384}\frac{Wl^3}{EI}$$

Table 11.1 gives the values of c for some common types of loading, etc. When the load system is complicated, e.g. several point loads of different magnitudes, or various combinations of point loads and uniformly distributed loads, the deflections must be calculated from first principles.

In certain simple cases it is possible to derive deflection formulae mathematically without using the calculus, and the following example is given for the more mathematically minded student. Neglecting the weight of the beam, shear force and bending moment diagrams are given in Fig. 11.5(b) and (c) for the beam loaded as shown. The maximum bending moment is $Wa/2$ and this moment is constant along the length AB, the shear force being zero. Since the moment is constant, the portion AB of the beam bends into the arc of a circle with a radius of curvature R. In triangle OBC (Fig. 11.6),

$$R^2 = (R - \delta)^2 + (\tfrac{1}{2}l)^2$$
$$= R^2 - 2R\delta + \delta^2 + \tfrac{1}{4}l^2$$
$$R^2 - R^2 + 2R\delta - \delta^2 = \tfrac{1}{4}l^2$$
$$2R\delta - \delta^2 = \tfrac{1}{4}l^2 \tag{1}$$

δ^2 is the square of a small quantity and can be ignored. Therefore

$$2R\delta = \tfrac{1}{4}l^2 \tag{2}$$

Table 11.1 Values of coefficient c for deflection formula $\delta = cWl^3/EI$

Condition of loading	Value of c (δ_{max} at A)
	$\dfrac{1}{48} = 0.02083$
	$\dfrac{23}{1296} = 0.01775$
	$\dfrac{11}{768} = 0.01432$
	$\dfrac{5}{384} = 0.01302$
	$\dfrac{1}{192} = 0.00521$
	$\dfrac{1}{384} = 0.00260$
	$\dfrac{1}{3} = 0.33333$
	$\dfrac{1}{8} = 0.12500$

For example, if $l = 3000$ mm and $\delta = 30$ mm, which is a big deflection for such a small span, then from equation (1)

$$2R \times 30 - 900 = \tfrac{1}{4} \times 9 \times 10^6$$

$$60R - 900 = 2.25 \times 10^6$$

$$60R = 2\,249\,100$$

$$R = 37\,485 \text{ mm}$$

Fig. 11.5 Derivation of the deflection formulae: (a) loading diagram; (b) shear force diagram; (c) bending moment diagram

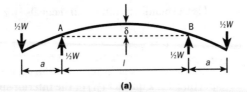

(a)

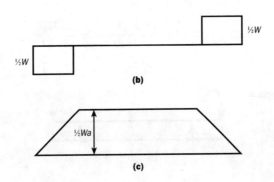

(b)

(c)

Fig. 11.6 Beam bending into an arc of a circle

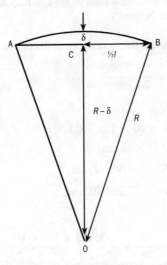

If we ignore the term δ^2, $R = 37\,500$ mm, so the inaccuracy is very small, even with such a comparatively large deflection.

It was shown in Chapter 9, page 151, that

$$\frac{M}{I} = \frac{E}{R} \quad \text{or} \quad R = \frac{EI}{M}$$

Substituting this value of R in equation (2)

$$2 \times \frac{EI}{M} \times \delta = \tfrac{1}{4} l^2$$

$$\delta = \frac{Ml^2}{8EI}$$

M is the bending moment and equals $\frac{1}{2}Wa$. Therefore

$$\delta = \frac{Wal^2}{16EI} \tag{3}$$

a can be expressed in terms of the span l. If, for example, the overhanging portion a is a quarter ($\frac{1}{4}l$) of the interior span (Fig. 11.7) then, substituting in equation (3)

$$\delta = \frac{W \times \frac{1}{4}l \times l^2}{16EI} = \frac{1}{64}\frac{Wl^3}{EI}$$

Fig. 11.7 Effect of beam overhang upon deflection

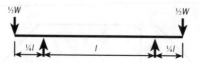

Example 11.1 A 406 × 178 UB54 simply supported at the ends of a span of 5 m carries a uniformly distributed load of 60 kN/m. Calculate the maximum deflection ($E = 205\,000$ N/mm^2).

Solution The formula for the maximum deflection is (from Table 11.1)

$$\delta = \frac{5}{384}\frac{Wl^3}{EI}$$

where

$$W = 60 \times 5 = 300 \text{ kN} = 0.3 \times 10^6 \text{ N}$$
$$l = 5000 \text{ mm}$$
$$E = 205\,000 \text{ N/mm}^2$$
$$I = 187.2 \times 10^6 \text{ mm}^4$$

Therefore

$$\delta = \frac{5}{384} \times \frac{300\,000 \times 125 \times 10^9}{205\,000 \times 187.2 \times 10^6}$$
$$= 13 \text{ mm}$$

Example 11.2 Calculate the safe inclusive uniformly distributed load for a 457 × 152 UB52 simply supported at its ends if the span is 6 m and if the span is 12 m. The maximum permissible bending stress is 165 N/mm^2 and the maximum permissible deflection is 1/360 of the span. E is 205 000 N/mm^2.

Solution
$$Z = 950\,000 \text{ mm}^3 \text{ (from Table 11.1)}$$
$$M_r = f \times Z = 165 \times 950\,000 = 156.75 \times 10^6 \text{ N mm}$$
$$M_{max} = \frac{1}{8}Wl = \frac{1}{8}W \times 6000 = 156.75 \times 10^6 \text{ N mm}$$

Therefore

$$W = \frac{156.75 \times 10^6 \times 8}{6000} = 209 \text{ kN}$$

Maximum deflection $\delta = \dfrac{5}{384} \dfrac{Wl^3}{EI}$

where $\quad W = 209\,000 \text{ N}$

$l = 6000 \text{ mm}$

$E = 205\,000 \text{ N/mm}^2$

$I = 213.7 \times 10^6 \text{ mm}^4$

Therefore

$$\delta = \frac{5}{384} \times \frac{209 \times 6^3 \times 10^{12}}{0.205 \times 213.7 \times 10^{12}} = 13.4 \text{ mm}$$

Maximum permissible deflection $= \dfrac{6000}{360} = 17 \text{ mm}$

Therefore

Safe load $= 209.0 \text{ kN}$

For a 12 m span $M_r = 156.75 \times 10^6 \text{ N mm}$ as before.
Therefore

$$W = \frac{156.75 \times 10^6 \times 8}{12\,000} = 104.5 \text{ kN}$$

Maximum actual deflection due to this load

$$\delta = \frac{5}{384} \times \frac{104.5 \times 12^3 \times 10^{12}}{0.205 \times 213.7 \times 10^{12}} = 53.7 \text{ mm}$$

Maximum permissible deflection $= \dfrac{12\,000}{360} = 33 \text{ mm}$

This means that, although the beam is quite satisfactory from the strength point of view, the deflection is too great, therefore the load must be reduced.

Now $\quad \delta = \dfrac{5}{384} \dfrac{Wl^3}{EI}$

Therefore

$$33 = \frac{5}{384} \times \frac{W \times (12\,000)^3}{205\,000 \times 213.7 \times 10^6}$$

giving $W = 64.2 \text{ } kN$ and this is the maximum permitted load for the beam. (Instead of using the deflection formula, W can also be obtained from $W = (33/53.7) \times 104.5 = 64.2 \text{ kN.}$)

Example 11.3 Calculate the safe inclusive uniformly distributed load for a 200 mm × 75 mm timber joist, simply supported at its ends, if the span is 4 m and if the span is 8 m. The maximum permissible bending stress is 6 N/mm^2 and the maximum permissible deflection is 0.003 of the span. E is 9500 N/mm^2.

Solution   From $M_r = \frac{1}{6} f b d^2 = M_{max} = \frac{1}{8} W l$

$$\frac{1}{6} \times 6 \times 75 \times 200^2 = \frac{1}{8} \times W \times 4000$$

and so $W = 6.0$ kN

Maximum actual deflection $= \dfrac{5}{384} \dfrac{W l^3}{E I}$

where $W = 6000$ N

$l = 4000$ mm

$E = 9500$ N/mm^2

$I = \frac{1}{12} \times 75 \times 200^3 = 50 \times 10^6$ mm^4

$$\delta = \frac{5}{384} \times \frac{6 \times 4^3 \times 10^{12}}{9.5 \times 50 \times 10^9} = 11 \text{ mm}$$

Maximum permissible deflection $= 0.003 \times 4000$

$= 12$ mm

The safe UDL for the 4 m span $= 6.0$ kN

When the span is doubled only half the previous load will be applied: $W = 3.0$ kN.

Maximum actual deflection due to this load

$$= \frac{5}{384} \times \frac{3 \times 8^3 \times 10^{12}}{9.5 \times 50 \times 10^9} = 42 \text{ mm}$$

Maximum permissible deflection $= 0.003 \times 8000 = 24$ mm

The load must be reduced so that $\dfrac{5}{384} \times \dfrac{W l^3}{E I} = 24$ mm

or the answer can be obtained by multiplying the load of 3 kN by 24/42 which gives 1.7 kN. Therefore, for this span, the safe load is 1.7 kN.

Span/depth ratios   It appears from the above examples that, before a beam can be passed as suitable, deflection calculations must be made in addition to bending calculations. Fortunately, it is possible to derive simple rules which replace deflection calculations in many cases. For example, if having designed a UB simply supported at its ends and carrying a UDL over its full length it is found that the span of the beam does not exceed 17 times its depth, the beam will be suitable from the deflection point of view. This rule is derived as follows.

The maximum deflection for a simply supported beam with a UDL is

$$\delta = \frac{5}{384} \times \frac{W l^3}{E I}$$

This formula can be rearranged as follows:

$$\delta = \frac{5}{48} \times \frac{W l}{8} \times \frac{l^2}{E I}$$

Now $\frac{1}{8} W l$ is the maximum bending moment, which is M. Thus

$$\delta = \frac{5}{48} \times \frac{M}{I} \times \frac{l^2}{E}$$

But $M/I = f/y$ (see page 151)

$$\delta = \frac{5}{48} \times \frac{f}{y} \times \frac{l^2}{E}$$

where $y = \frac{1}{2}d$ (see page 152)

$$f = 0.6 \times p_y = 165 \text{ N/mm}^2$$
$$E = 205\,000 \text{ N/mm}^2$$

Therefore

$$\delta = \frac{5}{48} \times \frac{165 \times 2}{d} \times \frac{l^2}{205\,000} = \frac{1650l^2}{9.84 \times 10^6 \, d} = \frac{l^2}{5964 \, d}$$

but δ must not exceed $l/360$, hence

$$\frac{l}{360} = \frac{l^2}{5964 \, d}$$
$$360l = 5964 \, d$$
$$\frac{l}{d} = \frac{5964}{360} = 17 \quad \text{or} \quad l = 17 \, d$$

Similar rules can be derived in the same manner for rectangular timber sections. Table 11.2 gives values for the stresses and values of E taken from BS 5268: Part 2: 1991. These rules are only applicable to beams simply supported at each end and carrying a UDL over their full length.

Note that for each of the grades of timber mentioned in Table 11.2 two values of the elastic modulus are given. Clause 14.7 of BS 5268: Part 2: 1991 states: 'The deflections of solid timber members acting alone should be calculated using the minimum modulus of elasticity for the strength class or species and grade.' For load–sharing systems (floor and ceiling joists, rafters, etc.) the mean value should be used subject to limitations given in Clause 13 of the code.

Table 11.2 Span/depth ratios for timber beams with UDL over full span and $\delta \not> 0.003 \times$ span

Strength class	Bending stress (N/mm²)	E (N/mm²)		L_e/h
		Minimum	Mean	
SC2	4.1	5000		17.6
			8000	28.1
SC3	5.3	5800		15.8
			8800	23.9
SC4	7.5	6600		12.7
			9900	19.0
SC5	10.0	7100		10.2
			10 700	15.4

Deflection of reinforced concrete beams

The composite nature of reinforced concrete beams complicates the determination of their deflections and, although these may be calculated, the process is tedious and time-consuming. Clause 3.4.6.3 of BS 8110: Part 1: 1985 states that normally the beam will not deflect excessively provided the span/effective depth ratios are kept within the following limits:

Cantilever	7
Simply supported	20
Continuous	26

Summary

For beams which behave elastically (steel, timber) and for standard types of loading, the *actual deflection* is

$$\delta = c \times \frac{W}{E} \times \frac{l^3}{I}$$

where c = a numerical coefficient taking into account the load system (UDL, point load etc.) and the manner in which the beam is supported (fixed or simple supports); see Table 11.1

W = the total (unfactored) load on the beam

l = the span

E = modulus of elasticity of the material

I = moment of inertia (second moment of area) of the section

Limitations are imposed on the maximum amount of deflection by the British Standard specifications and codes of practice. In general:

For steel the limit is $(1/360) \times$ span

For timber the limit is $0.003 \times$ span

Reinforced concrete beams (and slabs) are deemed to satisfy the limitation if their specified span/depth ratio is not exceeded.

Exercises

Note: The value of E for steel is 205 000 N/mm^2. The values of E for timber are given in Table 11.2.

1 A 457 × 191 UB98 is simply supported at the ends of a span of 7.2 m. The beam carries an inclusive UDL of 350 kN. Calculate the maximum deflection.

2 A 406 × 178 UB60 is simply supported at the ends of a span of 6.0 m. The beam carries a point load of 140 kN at midspan. Calculate the deflection due to this load, ignoring the weight of the beam.

3 A 356 × 171 UB45 is simply supported at the ends of a span of 5.5 m. It carries an inclusive UDL of 12 kN/m

and a central point load of 75 kN. Calculate the maximum deflection.

4 Calculate the safe, inclusive UDL for a 356 × 171 UB67 simply supported at the ends of a span of 9 m. The permissible bending stress is 165 N/mm^2 and the deflection must not exceed $(1/360) \times$ span.

5 Calculate the maximum deflection of a 305 × 165 UB40 cantilevered 3 m beyond its fixed support and carrying an inclusive UDL of 30 kN.

6 A 203 × 133 UB30 is fixed at one end and cantilevered for a distance of 1.2 m. The beam supports a point load of 15 kN at its free end. Calculate the maximum deflection ignoring the weight of the beam.

7 A 254 × 146 UB43 is fixed at one end and cantilevered for a distance of 1.5 m. The beam carries a UDL of 12 kN/m and a point load of 10 kN at the free end. Calculate the maximum deflection.

8 A 75 mm wide and 150 mm deep beam in SC4 class timber carries an inclusive UDL over a simply supported span of 1.8 m. Calculate the maximum deflection for UDL of 10 kN.

9 A timber (SC2) beam 75 mm wide and 240 mm deep carries a point load of 5.5 kN at the centre of its simply supported span of 3 m. Calculate the maximum deflection due to this load.

10 A timber (SC3) beam, 100 mm × 300 mm, spans 4 m and carries a central point load of 5 kN in addition to an inclusive UDL of 10 kN. Calculate the maximum deflection.

11 A 75 mm × 225 mm SC4 timber beam spans 5 m on simple supports. Calculate the value of the safe UDL for the following conditions: (a) permissible bending stress of 7.5 N/mm^2 and deflection is not important; (b) the deflection must not exceed 0.003 × span.

12 A beam in SC3 timber carries a UDL of 7.5 kN on a simply supported span of 5 m. Assuming the beam to be 89 mm wide calculate its depth when (a) the permissible bending stress is 5.3 N/mm^2 and deflection is not important; (b) maximum deflection is limited to 0.003 × span.

13 Calculate the maximum deflection of a cantilever beam in SC4 timber 75 mm wide and 240 mm deep. The beam carries an inclusive UDL of 5 kN over a span of 2 m.

14 A timber (SC2) cantilever beam is 1.2 m long, 75 mm wide and 150 mm deep. It carries a UDL of 1.0 kN and a point load of 0.9 kN at its free end. Calculate the maximum deflection.

15 Floor joists in SC2 timber spaced at 360 mm c/c are to carry an inclusive load of 2.0 kN/m^2 over a simply

supported span of 4.0 m. Determine a suitable size for the joists if the bending stress is limited to 4.5 N/mm^2 and the deflection must not exceed 0.003 × span. (Use the mean value of E (see Table 11.2, page 201).)

16 Calculate the maximum span/depth ratio for a steel cantilever beam supporting a UDL so that deflection does not exceed (1/360) × span. The permissible bending stress is 155 N/mm^2.

17 Calculate the minimum depth for a steel beam simply supported at the ends of a 7.5 m span carrying a point load at midspan. The weight of the beam may be ignored. The deflection must be limited to the usual (1/360) × span and the permissible bending stress is 165 N/mm^2.

18 Determine the deflection of a 150 mm × 400 mm SC2 timber beam which is simply supported at the ends of its 5 m length. The beam is subjected to a maximum bending moment of 21 × 10^6 N mm due to an inclusive UDL.

19 A solid rectangular beam is simply supported at the ends of a 4.8 m span. The beam is subjected to a maximum bending moment of 90 × 10^6 N mm due to an inclusive UDL. Determine a suitable section for the beam given:

Modulus of elasticity of the material = 4000 N/mm^2
Maximum permissible bending stress = 7.5 N/mm^2
Deflection must be limited to 15 mm

20 Calculate the minimum depth for a simply supported SC4 timber beam, which is to carry a UDL over a span of 4.0 m, if the deflection must not exceed 0.003 × span and the permissible stress in bending is 7.5 N/mm^2. Assuming the beam to be 150 mm wide, determine the value of the UDL.

21 A timber beam 75 mm × 150 mm simply supported at the ends of a 2.0 m span deflects 5 mm under a 10 kN UDL. Without calculating the modulus of elasticity, determine the maximum deflection of a beam of similar timber 150 mm × 300 mm due to a UDL of 40 kN on a span of 4.0 m.

12 Axially loaded columns

In this chapter the factors affecting the column's load-carrying capacity are investigated. The connection between the slenderness of the column and its tendency to buckle is discussed. The influence of the 'fixity' of the ends of the column, and the shape of its section on that slenderness, is considered in relation to timber, steel and reinforced concrete columns.

When the line of action of the resultant load is coincident with the centre of gravity axis of the column (Fig. 12.1(a)), the column is said to be **axially loaded** and the stress produced in the material is said to be a *direct compressive stress*. This stress is uniform over the cross-section of the column. The term *concentric loading* is sometimes used instead of *axial loading*.

When the load is not axial, it is said to be eccentric (i.e. off-centre) and bending stress is induced in the column as well as a direct compressive stress (Fig. 12.1(b)). Eccentric loading is dealt with in Chapter 14.

Fig. 12.1 Loading of columns: (a) axial loading; (b) eccentric loading

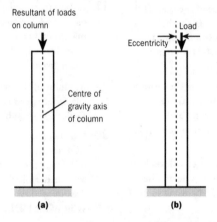

Resultant of loads on column

Load

Eccentricity

Centre of gravity axis of column

(a) (b)

Other words used to describe members which are subjected to compressive stress are *pillar, post, stanchion, strut*. There are no definite rules as to when any one of these words should be used, but the following convention is fairly general.

Column and pillar can usually be applied to any material, e.g. timber, stone, concrete, reinforced concrete, steel. Post is usually confined to timber. Stanchion is often used for rolled steel I-sections and channel sections. Strut has a more general significance than stanchion or post but normally it is not applied to a main supporting member of a building. The word is often used for compression members of roof trusses whether the material is timber or steel.

Design factors

The maximum axial load a column can be allowed to support depends on:

- the material of which the column is made
- the slenderness of the column

The slenderness involves not only the height or length of the column, but also the size and shape of its cross-section and the manner in which the two ends of the column are supported or fixed.

The majority of columns are designed by reference to tables of permissible stresses contained in British Standard specifications and codes of practice. These tables of permissible stresses (which are reproduced on pages 202 and 216) have been constructed from complex formulae which have been derived as the result of a great deal of research, mathematical and experimental, into the behaviour of columns under load. It is not possible in this book to deal with the mathematical theories of column design but an attempt will be made to give an explanation of the general principles.

A very short column will fail due to crushing of the material, but long columns are likely to fail by 'buckling', the failing load being much less than that which would cause failure in a short column of identical cross-sectional dimensions.

Consider Fig. 12.2(a), which represents a strip of pliable wood, say 6 mm × 54 mm in cross-section and 600 mm long. A small vertical force

Fig. 12.2 Design factors: (a) buckling; (b) three members of equal cross-sectional area

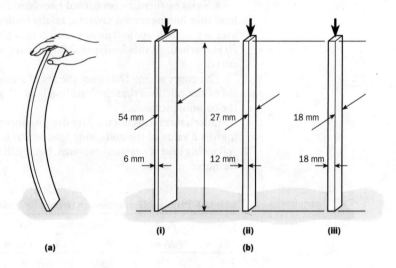

applied as shown will cause buckling. It should be obvious (but can be confirmed by experiment if necessary) that a larger force would be required to cause failure if the member were only 300 mm high. In other words, the 600 mm high member is more slender than the 300 mm high member of equal cross-sectional dimensions.

Now consider Fig. 12.2(b). All three members are of equal cross-sectional area (324 mm^2) and of equal height, yet member (ii) will require more load to cause buckling than member (i) and member (iii) is the strongest column.

Slenderness ratio   By reference to Fig. 12.2(b) it can be seen that the smaller cross-sectional dimension of the column is very important from the point of view of buckling and it appears that the **slenderness ratio** of a column can be defined as

$$\frac{\text{Effective length of column (in millimetres)}}{\text{Least width of column (in millimetres)}} = \left(\frac{l}{b}\right); \quad \left(\frac{L_e}{b}\right)$$

e.g. slenderness ratio of

Column (i) $= 600/6 \ = 100$

Column (ii) $= 600/12 = 50$

Column (iii) $= 600/18 = 33$ approx.

Since most timber posts and struts are rectangular in cross-section, it is reasonable to express the slenderness ratio in terms of the length and least width (i.e. least lateral dimension). This method is, in fact, adopted by BS5268 (see Table 12.1) but an alternative slenderness ratio is also given, i.e.

$$\frac{\text{Effective length}}{\text{Least radius of gyration}} = \left(\frac{l}{r}\right) \text{ or } \left(\frac{L_e}{i}\right)$$

and it is this slenderness ratio which must be used when the post is not of solid rectangular cross-section.

For explanation of the terms *least radius of gyration* and *effective length* see pages 208 and 211, respectively.

A Swiss mathematician named Leonhard Euler (1707–1783) showed that a long thin homogeneous column, axially loaded, suffers no deflection as the load is gradually applied until a critical load (the collapsing or buckling load P) is reached. At this load, instability occurs and the column buckles into a curve.

The curve of Fig. 12.3 is not the arc of a circle, and Euler found (with the aid of the calculus) that the buckling load P gets less as the slenderness of the column increases.

Euler's formula is not used for design, since (except for very long columns) it gives a value of the collapsing load which is much higher than the actual collapsing load of practical columns, but it still forms part of modern column formulae.

Table 12.1 Permissible compression stresses for timber struts

L_e/i (N/mm²)	L_e/b (N/mm²)	Strength class		L_e/i (N/mm²)	L_e/b (N/mm²)	Strength class	
		SC2	SC4			SC2	SC4
< 5	1.4	5.30	7.90	90	26.0	2.47	3.47
5	1.4	5.17	7.70	100	28.9	2.16	3.00
10	2.9	5.05	7.52	120	34.7	1.66	2.28
20	5.8	4.79	7.13	140	40.5	1.30	1.77
30	8.7	4.53	6.73	160	46.2	1.03	1.40
40	11.6	4.23	6.27	180	52.0	0.84	1.14
50	14.5	3.91	5.73	200	57.8	0.69	0.94
60	17.3	3.55	5.17	220	63.6	0.58	0.78
70	20.2	3.18	4.57	240	69.4	0.50	0.67
80	23.1	2.81	3.99	250	72.3	0.46	0.62

(*Source*: Adapted from Tables 9 and 22 BS 95268: Part 2: 1991)

The values of permissible compressive stresses for timber struts are the product of the *grade stress* and *modification factors* appropriate to given conditions of services. Those given in Table 12.1 have been compiled for use in the examples and exercises in this chapter only.

Fig. 12.3 Euler's formula

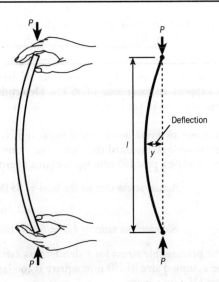

Deflection

l

y

P

P

P

P

Example 12.1 Calculate the permissible axial load for the following timber posts of SC2 class timber, all the posts having an effective length of 3.47 m:

- 150 mm × 150 mm
- 225 mm × 100 mm
- 300 mm × 75 mm

Note: All posts have a cross-sectional area of 22 500 mm^2.

Solution *150 mm × 150 mm*

Slenderness ratio = 3470/150 = 23.1

From Table 12.1

Permissible stress = 2.81 N/mm^2

Permissible axial load = 2.81 × 22 500 = 63.2 kN

225 mm × 100 mm

Slenderness ratio = 3470/100 = 34.7

By interpolation

Permissible stress = 1.66 N/mm^2

Permissible axial load = 1.66 × 22 500 = 37.3 kN

300 mm × 75 mm

Slenderness ratio = 3470/75 = 46

Again, by interpolation

Permissible stress = 1.03 N/mm^2

Permissible axial load = 1.03 × 22 500 = 23.2 kN

It is interesting to note that the 150 mm × 150 mm post can carry almost three times the load permitted for the 300 mm × 75 mm post.

Example 12.2   A post made from SC2 class timber of 3.45 m effective length is required to support an axial load of 35 kN. Determine suitable dimensions for the cross-section of the post.

Solution   Dimensions must be assumed because it is not possible to determine the permissible stress until the slenderness ratio is known. For example, let the first trial be a post 150 mm square (area of cross-section = 22 500 mm^2).

$$\text{Actual stress due to the load} = 35\,000/22\,500$$
$$= 1.56 \text{ N/mm}^2$$
$$\text{Slenderness ratio} = l/b = 3450/150 = 23$$

The permissible stress for a slenderness ratio of 23 is 2.82 N/mm^2 therefore the assumed size of 150 mm square is too large. As a second trial, assume a post 125 mm square.

$$\text{Actual stress} = 35\,000/15\,625 = 2.24 \text{ N/mm}^2$$
$$\text{Slenderness ratio} = l/b = 3450/125 = 27.6$$

From Table 12.1

$$\text{Permissible stress for } l/b \text{ of } 26 = 2.47 \text{ N/mm}^2$$
$$\text{for } l/b \text{ of } 28.9 = 2.16 \text{ N/mm}^2$$
$$\text{Difference in stress for } 2.9 = (2.47 - 2.16) = 0.31$$
$$\text{Difference in stress for } 1.3 = \frac{1.3}{2.9} \times 0.31 = 0.14$$
$$\text{Permissible stress for } l/b \text{ of } 27.6 = 2.16 + 0.14$$
$$= 2.30 \text{ N/mm}^2$$

Since the permissible stress is slightly more than the actual stress due to the load, a 125 mm square post is suitable.

Radius of gyration   It has already been stated that the shape of the cross-section has an important influence on the load-carrying capacity of a column. A square timber post can support more load than a post of rectangular cross-section of equal area and equal height. It has been shown also that in the case of rectangular cross-section columns it is reasonable to base permissible stresses on a slenderness ratio obtained by dividing the length of the column by its least width. It is not possible, however, to use the dimension of least width when designing columns which have other than solid rectangular sections. Steel columns, for example, are made in various shapes, some of which are shown in Fig. 12.4.

Fig. 12.4 Typical steel column cross-sections

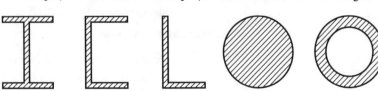

For such columns it is necessary to use some method of calculating slenderness ratios which can be applied to any shape of cross-section. A property which takes into account not only the size of the section (i.e. area) but also its shape (i.e. the arrangement of the material in the cross-section) is the **radius of gyration**. It is obtained by dividing the moment of inertia I of the section by its area A and then extracting the square root. The symbol r is commonly used to denote radius of gyration but g, i and k are sometimes used.

$$\text{Radius of gyration } r = \sqrt{\left(\frac{I}{A}\right)}$$

The use of the word *gyration* when applied to stationary columns in buildings may appear strange until it is realized that the word is also used in dynamics, the branch of mechanics dealing with bodies in motion. For example, consider Fig. 12.5(a) which represents a disc (such as a flywheel) rotating about

Fig. 12.5 Radius of gyration: (a) flywheel; (b) structural beam

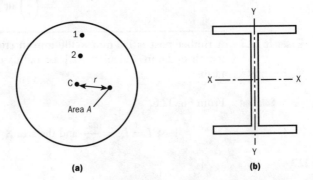

its centre C. In dynamics, it is usually the mass (weight) of the wheel which enters into calculations but in this instance the area A of the disc will be considered. Different particles of the disc travel at different velocities. For example, in one revolution of the wheel, particle 1 travels a greater distance than particle 2. In estimating the total energy of the disc, the term Σay^2 enters into the calculations. Σay^2 is the sum of the second moments of all the particles of area about the centre C of the disc, i.e. the moment of inertia I.

Imagine that all the area A of the disc is concentrated into *one* imaginary particle at a distance r from the centre of the disc. The moment of inertia about C of this particle is Ar^2 and, if the total energy of the disc is to remain unaltered, Ar^2 must equal the total moment of inertia I of the disc. Therefore

$$Ar^2 = I$$

and

$$r = \sqrt{\left(\frac{I}{A}\right)}$$

The radius of gyration is, in this connection, the distance from the centre C of the disc to the point at which the whole area of the wheel can be assumed to be concentrated so that the total energy remains unaltered.

In structural work, it is convenient to use the property $\sqrt{(I/A)}$ in conjunction with the length of the column for estimating slenderness ratios.

Least radius of gyration The structural engineer is not concerned with moment of inertia about a point as in the disc discussed above, but with moment of inertia with reference to a given axis. If a column of I-section buckles under its load, the bending will be about the weaker axis (axis Y–Y), as indicated in Fig. 12.5(b). Therefore, the radius of gyration must be calculated from I_{yy} which is the least moment of inertia.

$$\text{Least radius of gyration} = \sqrt{\left(\frac{\text{least moment of inertia}}{\text{area of cross-section}}\right)}$$

$$\text{i.e.} \quad r = \sqrt{\left(\frac{I_{yy}}{A}\right)}$$

$$\text{Slenderness ratio} = \frac{\text{effective length of column (mm)}}{\text{least radius of gyration (mm)}}$$

$$= \left(\frac{l}{r}\right) \text{ or } \left(\frac{L_e}{i}\right)$$

Example 12.3 A timber post is 150 mm × 100 mm in cross-section and has an effective length of 2.6 m. Calculate its least radius of gyration and the slenderness ratio.

Solution From Fig. 12.6,

$$\text{least } I = I_{yy} = \frac{db^3}{12} \text{ and the area } A = db$$

Fig. 12.6 Example 12.3

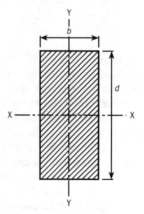

Therefore

$$\text{least } r = \sqrt{\left(\frac{I}{A}\right)} = \sqrt{\left(\frac{db^3}{12} \times \frac{1}{db}\right)} = \sqrt{\left(\frac{b^2}{12}\right)} = \frac{b}{\sqrt{12}} = 0.289b$$

$$r_{yy} = 0.289b$$

(Note that $r_{xx} = 0.289d$)

Least $r = 0.289b = 0.289 \times 100 = 28.9$ mm
Effective length $= 2600$ mm
Slenderness ratio $= l/r = 2600/28.9 = 90$

By reference to Table 12.1 it will be seen that $l/r = 90$ corresponds to $l/b = 26.0$; and in this case $l/b = 2600/100 = 26.0$. Thus, when designing timber posts of solid rectangular cross-section by reference to Table 12.1, it is immaterial whether the slenderness ratio is taken as effective height divided by least radius of gyration (l/r) or as effective height divided by least width (l/b).

No such dilemmas arise, however, in the design of steel columns. Because of their varied shapes the allowable compressive strength, p_c, shown in Table 12.2, is based on the l/r slenderness ratio.

Table 12.2 Compressive strength p_c for rolled section struts

λ^*	Compressive strength p_c (N/mm^2)							
	P_y axis of buckling X–X (a)				P_y axis of buckling Y–Y (a)			
	265	*275*	*340*	*355*	*265*	*275*	*340*	*355*
15	265	275	340	355	265	275	340	355
25	261	270	333	347	258	267	328	342
35	254	264	324	338	247	256	313	327
45	247	256	313	326	235	243	296	308
50	242	251	306	318	229	237	286	298
55	237	245	267	309	221	228	275	285
60	232	239	288	298	214	221	263	272
65	224	232	275	285	205	211	249	257
70	217	224	262	270	196	202	235	242
75	208	214	246	252	187	192	221	226
80	198	203	230	235	177	181	206	211
85	188	192	214	217	167	171	192	196
90	177	180	198	201	157	161	178	181
95	166	168	183	185	148	150	165	168
100	155	157	169	171	138	141	153	155
105	145	147	155	157	129	131	142	144
110	135	137	144	145	121	123	132	134
115	126	127	133	134	113	115	123	124
120	118	119	124	125	107	108	114	116
125	109	110	115	116	100	101	106	107
130	103	103	107	108	94	95	100	101
135	96	97	100	101	88	89	93	94
140	90	91	94	94	83	84	88	88
145	85	85	88	88	78	79	82	83
150	80	80	82	83	74	74	77	78
155	75	75	77	78	70	70	73	73
160	71	71	73	73	66	66	69	69
165	67	67	69	69	63	63	65	66
170	63	64	65	65	59	60	62	62
175	60	60	61	62	56	57	58	59
180	57	57	58	58	54	54	55	56
250	30	30	31	31	29	29	30	30
350	16	16	16	16	15	15	16	16

*λ = (effective length of strut)/(radius of gyration about relevant axis)
(*Source*: Adapted from Tables 27(a) and (b) BS 5950: Part 1: 1990)

Effective length of columns

It should be noted that in discussing the slenderness ratio the length of the column was qualified by the term 'effective'. This is in accordance with the provisions of the relevant BS codes of practice which state that, for the purpose of calculating the slenderness ratio of columns, an effective length should be assumed. This **effective length** can be defined as that length of the column which is subject to buckling.

The relevant codes of practice give guidance on the relationship of the actual length of the column between lateral supports, L, to its effective length. This is summarized in Table 12.3.

The reason why the effective length of a column may be less than or greater than its actual length in a building or structure is as follows. The safe compressive stress for a column depends not only on the actual length and cross-sectional dimensions of the column but also on the manner in which the ends of the column are restrained or fixed. Tables 12.1 and 12.2 have been derived for one condition of end-fixing (both ends pinned or hinged). To make allowance for other conditions of end-fixing, instead of constructing further tables of permissible stresses, adjustment is made by using a different length of column when calculating the slenderness ratio.

Before considering the end-fixing of columns, it may be instructive to study the behaviour of beams as indicated in Fig. 12.7(a).

In Fig. 12.7(a)(i) both ends of the beam are free to bend upwards when the load is applied; in other words, the ends of the beam are not restrained in direction, although they are held in position.

In Fig. 12.7(a)(ii) one end of the beam is firmly fixed so that the end is restrained in direction, and in Fig. 12.7(a)(iii) both ends of the beam are restrained in direction.

The load-carrying capacities of these three beams will be different (other conditions being equal) because of the manner in which the ends of the beams are held or supported.

Table 12.3 Effective length of columns

Type of 'fixity'	Effective length of column	
	BS 5950, BS 5268	BS 8110
1 Effectively held in position and restrained in direction at both ends.	0.7L	0.75L
2 Effectively held in position at both ends and restrained in direction at one end.	0.85L	0.75L– L
3 Effectively held in position at both ends, but not restrained in direction at either end.	L	0.75L– L
4 Effectively held in position and restrained in direction at one end, and at the other partially restrained in direction but not held in position	1.5L	L–2.0L
5 Effectively held in position and restrained in direction at one end, but not held in position nor restrained in direction at the other end.	2.0L	L– 2.0L

(*Source*: Adapted from Table 21 BS 5268, Table 24 BS 5950 and Tables 3.21 and 3.22 BS 8110)

Fig. 12.7 Behaviour of beams and columns: (a)(i) both ends of beam freely supported (bending exaggerated), (ii) one end fixed, (iii) both ends fixed; (b) all column ends are held in position with (i) top and bottom not restrained in direction, (ii) top not restrained, (iii) both ends restrained

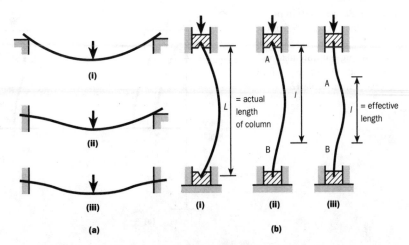

In a comparable manner, columns will buckle into differently shaped curves according to the way in which the ends of the columns are held.

The columns of Fig. 12.7(b) are all held in position, i.e. the ends of the columns are not free to move sideways or backwards or forwards. A greater load is required to cause column (ii) to fail than column (i), and column (iii) is the strongest. The length of curve marked AB in columns (ii) and (iii) is similar to the whole length of curve of column (i), and this length is called the effective length of the column.

Column (i) is similar to case 3 of Table 12.3 on page 213, column (ii) is similar to case 2, and column (iii) is similar to case 1.

Case 5 in Table 12.3 can be explained by reference to Fig. 12.8(a). The tops of the columns are not restrained either in position or direction and will tend to buckle as shown. The effective length is therefore taken as twice the actual length of the column.

Figure 12.8(b) represents columns which are held in position and restrained in direction at the bottom and only restrained in direction at the top. This case is a little better than case 4 of Table 12.3.

Practical interpretation

Appendix D to BS 5950: Part 1: 1990 gives typical examples of stanchions for single-storey buildings and the effective lengths which may be used in their design. Attention is also drawn to the fact that, although Tables 12.1 and 12.2 give values of permissible stresses for slenderness ratios up to 250 and 350, respectively, it is only in special cases that the slenderness ratio is allowed to exceed 180. Clauses 15.4 of BS 5268: Part 2: 1991 and 4.7.3.2 of BS 5950: Part 1: 1990 should be consulted for details.

Example 12.4

Calculate the compression resistance of a 203 × 203 UC60 stanchion which is 3.6 m high (between lateral supports). Both ends of the stanchion are held effectively in position but only one is also restrained in direction. (Grade 43 steel.)

Solution

The end fixity of the stanchion corresponds to case 2 of Table 12.3, so

Effective length of the column, $l = 0.85L = 3060$ mm

Fig. 12.8 Representation of 'fixity' types: (a) case 5 from Table 12.3; (b) restrained at top and bottom, but only held in position at the bottom – similar to case 4 from Table 12.3

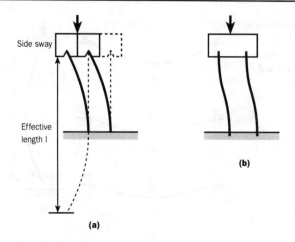

From Table 12.4 on pages 216–17

$$r_{yy} = 52.0 \text{ mm} \qquad A_g = 7640 \text{ mm}^2 \qquad t = 14.2 < 16 \text{ mm}$$

Therefore

$$l/r_{yy} = 3060/52.0 = 59$$

From Table 12.2 (by interpolation)

$$p_c \text{ for } l/r_{yy} \text{ of } 59 = 222 \text{ N/mm}^2$$

Hence, the compression resistance

$$P_c = p_c \times A_g$$
$$= 222 \times 7640 = 1696 \text{ kN}$$

Example 12.5   A stanchion of I-section is required to support a factored axial load of 2000 kN. The effective length of the stanchion is 4 m. Choose a suitable section in Grade 43 steel.

Solution   It is necessary to make a guess at the size of the stanchion. For example, assume a 203 × 203 UC60 stanchion then, working from first principles:

From Table 12.4

$$r_{yy} = 52.0 \text{ mm} \qquad A_g = 7640 \text{ mm}^2 \qquad t = 14.2 \text{ mm} < 16 \text{ m}$$

Therefore

$$l/r_{yy} = 4000/52.0 = 77$$

From Table 12.2, $p_c = 188 \text{ N/mm}^2$

The compression resistance, P_c, of the stanchion is

$$P_c = p_c \times A_g$$
$$= 188 \times 7640 = 1436 \text{ kN} < \text{applied factored load}$$

i.e. the assumed section is not adequate.

By trying a 254 × 254 UC73 in the same way it will be found that its compression resistance, P_c, is 2020 kN, which is satisfactory.

Reinforced concrete columns

The design of reinforced concrete columns by the elastic method was discontinued on the introduction of the 1957 edition of CP114, after test results had shown that steel and concrete do not behave elastically at failure.

The calculations for the permissible axial load, P_0, on a short reinforced concrete column, i.e. one where the ratio of effective height to the least lateral dimension does not exceed 15, were based on

$$P_0 = p_{cc}A_c + p_{sc}A_{sc}$$

where
p_{cc} = the permissible stress for the concrete in direct compression

A_c = the cross-sectional area of concrete excluding any finishing material and reinforcing steel

p_{sc} = the permissible compression stress for column bars

A_{sc} = the cross-sectional area of the longitudinal steel

For the purpose of the exercises in this chapter the values of p_{cc} may be taken as follows:

7.6 N/mm² for 1:1:2 mix (or C30 grade)

6.5 N/mm² for 1:1½:3 mix (or C25 grade)

5.3 N/mm² for 1:2:4 mix (or C20 grade)

The value of p_{sc} for hot rolled steel bars complying with BS4449 up to 40 mm dia. used as longitudinal reinforcement in columns was given as 125 N/mm².

BS 8110: Part 1: 1985 distinguishes between braced and unbraced as well as short and slender columns. A braced column is one which is part of a structure stabilized against lateral forces by walls, bracing or buttressing. The definition of a short braced (or unbraced) column is based on a ratio similar to that given above. (See Clause 3.8.1.3 for details.)

Clause 3.8.4.3 of BS 8110 states that the design ultimate axial load, N, supported by a short column which cannot be subjected to significant moments because of the nature of the structure, may be calculated from

$$N = 0.4f_{cu}A_c + 0.75A_{sc}f_y$$

where
f_{cu} = characteristic strength of concrete

f_y = characteristic strength of reinforcement

For short braced columns supporting an approximately symmetrical arrangement of beams, Clause 3.8.4.4 of BS 8110 permits the following version of the above equation to be used in the calculation of the design ultimate axial load:

$$N = 0.35f_{cu}A_c + 0.67A_{sc}f_y$$

provided that the beams are designed for uniformly distributed imposed loads and the beam spans do not differ by more than 15 per cent of the longest beam.

In both equations allowance is made for the partial safety factor for strength of material, γ_m.

Table 12.4 Dimensions and properties of universal columns

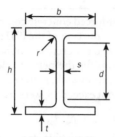

Designation	Area of section (cm^2)	Mass per metre (kg/m)	Dimensions (mm)						Ratios for local buckling	
			h Section depth	b Section width	s Web thickness	t Flange thickness	r Root radius	d Depth between fillets	b/2t Flange	d/s Web
356 × 406 × 634	808	633.9	474.6	424.0	47.6	77.0	15.2	290.2	2.75	6.10
356 × 406 × 551	702	551.0	455.6	418.5	42.1	67.5	15.2	290.2	3.10	6.89
356 × 406 × 467	595	467.0	436.6	412.2	35.8	58.0	15.2	290.2	3.55	8.11
356 × 406 × 393	501	393.0	419.0	407.0	30.6	49.2	15.2	290.2	4.14	9.48
356 × 406 × 340	433	339.9	406.4	403.0	26.6	42.9	15.2	290.2	4.70	10.9
356 × 406 × 287	366	287.1	393.6	399.0	22.6	36.5	15.2	290.2	5.47	12.8
356 × 406 × 235	299	235.1	381.0	394.8	18.4	30.2	15.2	290.2	6.54	15.8
356 × 368 × 202	257	201.9	374.6	374.7	16.5	27.0	15.2	290.2	6.94	17.6
356 × 368 × 177	226	177.0	368.2	372.6	14.4	23.8	15.2	290.2	7.83	20.2
356 × 368 × 153	195	152.9	362.0	370.5	12.3	20.7	15.2	290.2	8.95	23.6
356 × 368 × 129	164	129.0	355.6	368.6	10.4	17.5	15.2	290.2	10.5	27.9
305 × 305 × 283	360	282.9	365.3	322.2	26.8	44.1	15.2	246.7	3.65	9.21
305 × 305 × 240	306	240.0	352.5	318.4	23.0	37.7	15.2	246.7	4.22	10.7
305 × 305 × 198	252	198.1	339.9	314.5	19.1	31.4	15.2	246.7	5.01	12.9
305 × 305 × 158	201	158.1	327.1	311.2	15.8	25.0	15.2	246.7	6.22	15.6
305 × 305 × 137	174	136.9	320.5	309.2	13.8	21.7	15.2	246.7	7.12	17.9
305 × 305 × 118	150	117.9	314.5	307.4	12.0	18.7	15.2	246.7	8.22	20.6
305 × 305 × 97	123	96.9	307.9	305.3	9.9	15.4	15.2	246.7	9.91	24.9
254 × 254 × 167	213	167.1	289.1	265.2	19.2	31.7	12.7	200.3	4.18	10.4
254 × 254 × 132	168	132.0	276.3	261.3	15.3	25.3	12.7	200.3	5.16	13.1
254 × 254 × 107	136	107.1	266.7	258.8	12.8	20.5	12.7	200.3	6.31	15.6
254 × 254 × 89	113	88.9	260.3	256.3	10.3	17.3	12.7	200.3	7.41	19.4
254 × 254 × 73	93.1	73.1	254.1	254.6	8.6	14.2	12.7	200.3	8.96	23.3
203 × 203 × 86	110	86.1	222.2	209.1	12.7	20.5	10.2	160.8	5.10	12.7
203 × 203 × 71	90.4	71.0	215.8	206.4	10.0	17.3	10.2	160.8	5.97	16.1
203 × 203 × 60	76.4	60.0	209.6	205.8	9.4	14.2	10.2	160.8	7.25	17.1
203 × 203 × 52	66.3	52.0	206.2	204.3	7.9	12.5	10.2	160.8	8.17	20.4
203 × 203 × 46	58.7	46.1	203.2	203.6	7.2	11.0	10.2	160.8	9.25	22.3
152 × 152 × 37	47.1	37.0	161.8	154.4	8.0	11.5	7.6	123.6	6.71	15.4
152 × 152 × 30	38.3	30.0	157.6	152.9	6.5	9.4	7.6	123.6	8.13	19.0
152 × 152 × 23	29.2	23.0	152.4	152.2	5.8	6.8	7.6	123.6	11.2	21.3

(*Source*: Structural sections to BS 4: Part 1 and BS 4848: Part 4 British Steel)

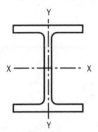

Second moment of area (cm⁴)		Radius of gyration (cm)		Elastic modulus (cm³)		Plastic modulus (cm³)		Parameters and constants				Designation
X–X axis	Y–Y axis	X–X axis	Y–Y axis	X–X axis	Y–Y axis	X–X axis	Y–Y axis	u Buckling parameter	x Torsional index	H Warping constant (dm⁶)	J Torsional constant (cm⁴)	
274800	98130	18.4	11.0	11580	4629	14240	7108	0.843	5.46	38.8	13720	356 × 406 × 634
226900	82670	18.0	10.9	9962	3951	12080	6058	0.841	6.05	31.1	9240	356 × 406 × 551
183000	67830	17.5	10.7	8383	3291	10000	5034	0.839	6.86	24.3	5809	356 × 406 × 467
146600	55370	17.1	10.5	6998	2721	8222	4154	0.837	7.86	18.9	3545	356 × 406 × 393
122500	46850	16.8	10.4	6031	2325	6999	3544	0.836	8.85	15.5	2343	356 × 406 × 340
99880	38680	16.5	10.3	5075	1939	5812	2949	0.835	10.2	12.3	1441	356 × 406 × 287
79080	30990	16.3	10.2	4151	1570	4687	2383	0.834	12.1	9.54	812	356 × 406 × 235
66260	23690	16.1	9.60	3538	1264	3972	1920	0.844	13.4	7.16	558	356 × 368 × 202
57120	20530	15.9	9.54	3103	1102	3455	1671	0.844	15.0	6.09	381	356 × 368 × 177
48590	17550	15.8	9.49	2684	948	2965	1435	0.844	17.0	5.11	251	356 × 368 × 153
40250	14610	15.6	9.43	2264	793	2479	1199	0.844	19.9	4.18	153	356 × 368 × 129
78870	24630	14.8	8.27	4318	1529	5105	2342	0.855	7.65	6.35	2034	305 × 305 × 283
64200	20310	14.5	8.15	3643	1276	4247	1951	0.854	8.74	5.03	1271	305 × 305 × 240
50900	16300	14.2	8.04	2995	1037	3440	1581	0.854	10.2	3.88	734	305 × 305 × 198
38750	12570	13.9	7.90	2369	808	2680	1230	0.851	12.5	2.87	378	305 × 305 × 158
32810	10700	13.7	7.83	2048	692	2297	1053	0.851	14.2	2.39	249	305 × 305 × 137
27670	9059	13.6	7.77	1760	589	1958	895	0.850	16.2	1.98	161	305 × 305 × 118
22250	7308	13.4	7.69	1445	479	1592	726	0.850	19.3	1.56	91.2	305 × 305 × 97
30000	9870	11.9	6.81	2075	744	2424	1137	0.851	8.49	1.63	626	254 × 254 × 167
22530	7531	11.6	6.69	1631	576	1869	878	0.850	10.3	1.19	319	254 × 254 × 132
17510	5928	11.3	6.59	1313	458	1484	697	0.848	12.4	0.898	172	254 × 254 × 107
14270	4857	11.2	6.55	1096	379	1224	575	0.850	14.5	0.717	102	254 × 254 × 89
11410	3908	11.1	6.48	898	307	992	465	0.849	17.3	0.562	57.6	254 × 254 × 73
9449	3127	9.28	5.34	850	299	977	456	0.850	10.2	0.318	137	203 × 203 × 86
7618	2537	9.18	5.30	706	246	799	374	0.853	11.9	0.250	80.2	203 × 203 × 71
6125	2065	8.96	5.20	584	201	656	305	0.846	14.1	0.197	47.2	203 × 203 × 60
5259	1778	8.91	5.18	510	174	567	264	0.848	15.8	0.167	31.8	203 × 203 × 52
4568	1548	8.82	5.13	450	152	497	231	0.847	17.7	0.143	22.2	203 × 203 × 46
2210	706	6.85	3.87	273	91.5	309	140	0.848	13.3	0.0399	19.2	152 × 152 × 37
1748	560	6.76	3.83	222	73.3	248	112	0.849	16.0	0.0308	10.5	152 × 152 × 30
1250	400	6.54	3.70	164	52.6	182	80.2	0.840	20.7	0.0212	4.63	152 × 152 × 23

Example 12.6   A 250 mm square reinforced concrete column with an effective length of 3 m contains four 25 mm dia. longitudinal bars. Calculate the safe axial load for the column if $p_{cc} = 5.3$ N/mm^2 and $p_{sc} = 125$ N/mm^2.

Solution
$$\text{Slenderness ratio} = \frac{\text{effective length}}{\text{least width}}$$
$$= 3000/250 = 12 < 15$$

Therefore the column is a short one.

$$\text{Gross area of concrete} = 250 \times 250 = 62\,500 \text{ mm}^2$$
$$A_{sc} = 4 \times 491 \text{ mm}^2 = 1964 \text{ mm}^2$$
$$A_c = A_{gross} - A_{sc} = 60\,536 \text{ mm}^2$$

Therefore
$$P_0 = p_{cc}A_c + p_{sc}A_{sc}$$
$$= 5.3 \times 60\,536 + 125 \times 1964$$
$$= 320.8 + 245.5$$
$$= 566.3 \text{ kN}$$

Example 12.7   A short reinforced concrete column is required to carry an axial load of 900 kN. Design a square column containing 0.8 per cent of steel and 8.0 per cent of steel.

Solution   Assume $p_{cc} = 5.3$ N/mm^2 and $p_{sc} = 125$ N/mm^2.
For 0.8 per cent steel:
$$A_{sc} = \frac{0.8}{100} \times A_g$$

where A_g is gross cross-sectional area.
Therefore
$$P_0 = p_{cc}A_c + p_{sc}A_{sc}$$
$$= 5.3(A_g - 0.008A_g) + 125 \times 0.008A_g$$
$$= 5.3A_g - 0.04A_g + 1.0A_g$$
$$= 6.26A_g$$

Hence
$$A_g = 900\,000/6.26 = 143\,770 \text{ mm}^2$$
$$\text{Length of side} = \sqrt{143\,770} = \text{(say) } 380 \text{ mm}$$
$$A_{sc} = 0.008 \times 0.144 \times 10^6 = 1155 \text{ mm}^2$$

Make column 380 mm square with four No. 20 mm dia. reinforcing bars (1260 mm^2).
Now for 8.0 per cent of steel:
$$A_{sc} = \frac{8.0}{100} \times A_g$$
$$P_0 = 5.3(A_g - 0.08A_g) + 125 \times 0.08A_g$$
$$= 14.9A_g$$

Hence

$$A_g = 900\,000/14.9 = 60\,400 \text{ mm}^2$$

$$\text{Length of side} = \sqrt{60\,400} = \text{(say) } 250 \text{ mm}$$

$$A_{sc} = 0.08 \times 60\,400 = 4832 \text{ mm}^2$$

Make column 250 mm square with four No. 32 mm dia. and four No. 25 mm dia. reinforcing bars (5180 mm^2).

Summary

Timber columns   Permissible stresses for various slenderness ratios are given in Table 12.1.

If the column is of solid rectangular cross-section, the slenderness ratio may be taken as effective length divided by the least lateral dimension. If not of rectangular cross-section, the slenderness ratio is effective length divided by least radius of gyration.

The effective length should be assumed in accordance with Table 12.3. The safe axial load for the column is obtained by multiplying the permissible stress by the area of the cross-section of the column.

Steel columns

$$\text{Slenderness ratio} = \frac{\text{effective length}}{\text{least radius of gyration}} = \frac{l}{r}$$

See Table 12.3 and Appendix D of BS 5950 for a guide to estimating effective lengths. Permissible stresses for various slenderness ratios are given in Table 12.2.

The safe axial load is obtained by multiplying the permissible stress by the area of the cross-section of the column.

$$\text{Least } r \text{ for solid rectangular section} = 0.289b$$

$$\text{Least } r \text{ for hollow square section} = \frac{1}{2}\sqrt{\left(\frac{B^2 + b^2}{3}\right)}$$

$$\text{Least } r \text{ for hollow circular section} = \tfrac{1}{4}\sqrt{(D^2 + d^2)}$$

Least r for rolled sections, e.g. I-sections, channels and angles can be obtained from Tables in BS 4: Part 1: 1993.

Reinforced concrete columns   A short column is one where the ratio of the effective column length to least lateral dimension does not exceed 15. A large number of columns in buildings are therefore 'short'.

For permissible stress design:

safe axial load $P_0 = p_{cc}A_c + p_{sc}A_{sc}$

where p_{cc} = permissible concrete stress

$\qquad A_c$ = area of cross-section of concrete = $A_g - A_{sc}$

$\qquad p_{sc}$ = permissible steel stress

$\qquad A_{sc}$ = area of steel

For limit state design see Clause 3.8 of BS 8110: Part 1: 1985, or Part 3 of that code which contains design charts for rectangular columns.

Exercises

1 Calculate the safe axial loads for the following posts all of SC2 Class timber: (a) 75 mm square, 2 m effective length; (b) 100 mm square, 2.5 m effective length.

2 Calculate the safe axial loads for the following posts of SC2 Class timber all of 3 m effective length:
(a) 150 mm × 50 mm; (b) 100 mm × 75 mm;
(c) 85 mm square.

3 A timber post of 3.6 m effective length is required to support an axial load of 270 kN. Determine the length of side of a square section post of SC4 Class timber.

4 A timber post of SC4 Class timber has an effective length of 3 m and is 300 mm diameter (solid circular cross-section). Calculate the value of the axial load.

5 A timber post of solid circular cross-section is of timber SC4 Class and 3 m effective length. It has to support an axial load of 300 kN. Determine the diameter of the post.

6 Two timber posts 200 mm × 74 mm in cross-section of SC2 Class timber are placed side by side without being connected together to form a post 200 mm × 150 mm. The effective length of the posts is 2.6 m. Calculate the safe axial load for the compound post and compare this load with the safe axial load of one solid post 200 mm × 150 mm in cross-section of the same effective length.

7 Calculate the compression resistance of a 254 × 254 UC89 stanchion for the following end fixings and effective lengths: (a) 12 m, both ends fully restrained; (b) 9 m, one end fully restrained the other held in position; (c) 6 m, both ends held in position only.

8 Determine the maximum permissible effective length of a 254 × 254 UC73 section. Calculate the compression resistance for such effective length.

9 Choose a suitable stanchion of I-section to support a factored axial load of 2000 kN, the effective length = 3 m.

10 Make calculations to determine which of the following stanchions has the greater compression resistance: (a) a stanchion consisting of two 152 × 152 UC23 sections placed side by side without being connected together; effective length = 2.6 m; (b) one 203 × 203 UC46 section of 2.6 m effective length.

11 Calculate the safe axial load for a 400 mm square 'short' reinforced concrete column containing eight 25 mm diameter bars. Use 1 : 2 : 4 mix of concrete.

12 Calculate the safe axial load for a 450 mm diameter 'short' circular cross-section reinforced concrete column containing six 32 mm diameter bars.

$$p_{cc} = 6 \text{ N/mm}^2 \qquad p_{sc} = 125 \text{ N/mm}^2$$

13 Design a square cross-section column to support a load of 2 MN assuming 4 per cent of steel and $1 : 1\frac{1}{2} : 3$ concrete mix.

14 Design a short column to carry 2.5 MN, assuming (a) $p_{cc} = 7.6 \text{ N/mm}^2$, $p_{sc} = 140 \text{ N/mm}^2$, 8 per cent steel; (b) $p_{cc} = 4.3 \text{ N/mm}^2$, $p_{sc} = 125 \text{ N/mm}^2$, 0.8 per cent steel.

15 Compare the load-carrying capacities of a timber and a reinforced concrete column both 3 m effective length and 300 mm square. The timber is SC4 Class and the reinforced concrete column contains eight 25 mm diameter bars.

$$p_{cc} = 6.0 \text{ N/mm}^2 \qquad p_{sc} = 125 \text{ N/mm}^2$$

16 One dimension of a 'short' reinforced concrete column must not exceed 250 mm. The column has to support an axial load of 730 kN. Assuming 4 per cent steel, design the column using 1 : 2 : 4 concrete mix.

17 A column is 400 mm square and 6 m effective length and it contains eight 25 mm diameter bars. Calculate the safe axial load assuming

$$p_{cc} = 5.7 \text{ N/mm}^2 \quad \text{and} \quad p_{sc} = 125 \text{ N/mm}^2$$

18 A 'short' column is 450 mm square and it has to support an axial load of 2 MN. Calculate the area of steel required.

$$p_{cc} = 6.0 \text{ N/mm}^2 \qquad p_{sc} = 125 \text{ N/mm}^2$$

19 A square concrete column with 5 per cent of reinforcement is to carry an axial load of 554 kN. Given that $p_{cc} = 8 \text{ N/mm}^2$ and $p_{sc} = 125 \text{ N/mm}^2$ determine (a) a suitable size for the column and the cross-sectional area of the reinforcement; (b) the maximum length of the column so that it may be treated as a 'short' column, assuming full restraint at both ends.

13 Connections

The treatment of connections in this chapter is confined to direct shear connections in steelwork. The behaviour of rivets and bolts is presented in some detail. The criterion value of rivets or bolts is explained and its use in design calculations is demonstrated. High strength friction grip bolts are dealt with briefly to give an introduction to this type of connection. Finally, welding is also discussed because of its undoubted versatility and the resulting popularity.

Note: All the loads used in the design of connections in this chapter are **factored loads** (i.e. specified loads multiplied by the relevant partial factor) in accordance with Clause 6.1.1 of BS5950: Part 1: 1990.

Riveting and bolting

A rivet or bolt may be considered simply as a *peg* inserted in holes drilled in two or more thicknesses of steel in order to prevent relative movement. For example, the two steel plates in Fig. 13.1(a) tend to slide over each other, but

Fig. 13.1 Riveting

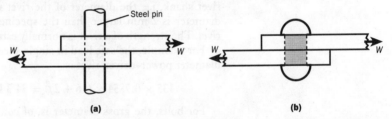

(a)   (b)

could be prevented from doing so by a suitable steel pin inserted in the holes in each plate, as shown. In order to prevent the *steel pin* from slipping out of holes, bolts with heads and nuts are used or rivet heads are formed, and these produce an effective connection (Fig. 13.1(b)).

The rivet heads (or bolt heads and nuts) do, in fact, strengthen the connection by pressing the two thicknesses of plate together, but this strength cannot be determined easily, and so the rivet or bolt strength is calculated on the assumption that its shank (shown shaded) only is used in building up its strength.

Single shear

If the loads W in Fig. 13.1(b) are large enough, the rivet or bolt could fail, as in Fig. 13.2, in *shear*, i.e. breaking by the sliding of its fibres along line A–A. This type of rivet or bolt failure is known as **failure in single shear**. The area of steel rivet resisting this failure is the circular area of the rivet shank, shown hatched in Fig. 13.2, i.e.

$$\tfrac{1}{4}\pi \times (\text{diameter of rivet})^2 \quad \text{or} \quad 0.7854d^2 = A$$

The shear strength, p_s, for rivets may be assumed to be as follows:

110 N/mm^2 for hand-driven rivets

135 N/mm^2 for power-driven rivets

Fig. 13.2 Failure in single shear

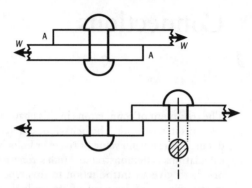

For ordinary bolts of grade 4.6 the value of p_s is given in Table 32 of BS 5950: Part 1: 1990 as 160 N/mm^2.

Power-driven rivets are usually driven by a special machine. The rivets and the rivet heads are formed more accurately than is possible in the case of hand-driven rivets and they are therefore permitted a higher stress. The holes are drilled 2 mm larger in diameter than the specified sizes of the rivets.

Since rivets are driven while hot and, therefore, their material fills the hole completely, it is necessary to distinguish between the nominal and the gross diameter of the rivets. The nominal diameter refers to the specified size of the rivet shank, i.e. the diameter of the rivet when it is cold, whilst the gross diameter is 2 mm larger than the specified (i.e. nominal) diameter of the rivet. The strength of a rivet is normally estimated on its *gross* diameter.

For example, the safe load in single shear (safe stress × area) of a 16 mm diameter power-driven rivet is

$$135 \times 0.7854 \times (16 + 2)^2 = 34.3 \text{ kN}$$

For bolts, the gross diameter is, of course, equal to the nominal diameter. Therefore the safe load in single shear, or single shear value (SSV) of a 16 mm diameter ordinary bolt of grade 4.6 is

$$160 \times 0.7854 \times 16^2 = 32 \text{ kN}$$

Double shear   In the type of connection shown, for example, in Fig. 13.3 (a double cover butt joint), the rivets or bolts on one side of the joint would have to shear across *two* planes, as shown. This is known as **failure in double shear**.

A rivet or bolt under these circumstances will need twice as much load to break it compared with a rivet or bolt in single shear, so the safe load on a

Fig. 13.3 Failure in double shear

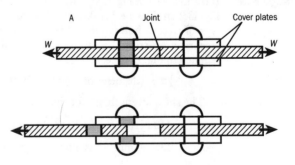

rivet in double shear is twice that on the same rivet in single shear. Thus, using the circular area of the rivet shank as before,

> Safe load in double shear
>
> = area × 2 × single shear safe stress

Therefore,

> Safe load for the above ordinary bolt
>
> = 2 × shear strength × area
>
> = 2 × 160 × 201
>
> = 64 kN

or, simply,

> 2 × SSV = 2 × 32 = 64 kN

Bearing   The two main ways in which the rivet or bolt itself may fail have been discussed. This type of failure assumes, however, fairly thick steel plates capable of generating sufficient stress to shear the rivet.

Consider Fig. 13.4(a). The heavy load of 120 kN taken by the 25 mm steel plates would certainly shear the 12 mm diameter rivet (single shear).

Fig. 13.4 Failure in bearing: (a) plates would shear rivet; (b) rivet would tear plates; (c) bearing force; (d) effective area (section A–A)

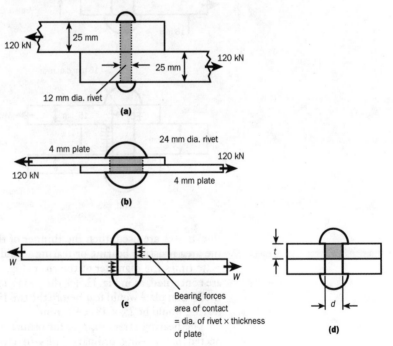

Now consider the opposite type of case, as in Fig. 13.4(b), where a thick steel rivet (24 mm diameter) is seen connecting two very thin steel plates. The steel plates in this case are much more likely to be torn by the rivet than the rivet to be sheared by the weaker steel plates.

This type is known as **failure in bearing** (or *tearing*), and note should again be taken of the area which is effective in resisting this type of stress

(Fig. 13.4(c)). The area of contact of the rivet with the plate on one side of it is actually semicylindrical, but since the bearing stress is not uniform, it is assumed that the area of contact is the plate thickness times the rivet diameter. This area is shown shaded in Fig. 13.4(d).

For bearing purposes, as for shear, the gross diameter of the rivet can be taken as the nominal diameter plus 2 mm.

When two plates of the same thickness are being connected, then *either* plate could tear, and the area resisting bearing would be the thickness of one plate times the diameter of the rivet (Fig. 13.5(a)). Where plates of different

Fig. 13.5 Area resisting bearing: (a) plates of equal thickness; (b) plates of unequal thickness; (c) three plates

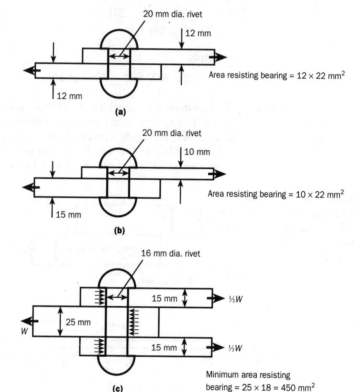

thicknesses are used, then the thinner of the two plates would tear first, so the area resisting bearing or tearing would be the thickness of the thinner plate times the diameter of the rivet (Fig. 13.5(b)). Where three thicknesses are concerned, as in Fig. 13.5(c), the two 15 mm plates are acting together and the 25 mm plate would tear before the two 15 mm plates, so the area resisting tearing would be $25 \times 18 = 450$ mm^2.

The bearing strength, p_{bb}, for ordinary bolts of grade 4.6 and for connected parts using ordinary bolts in clearance holes, p_{bs}, are given in BS 5950 as 460 N/mm^2 for Grade 43 steel.

The value of p_{bb} for hand-driven rivets may be taken as 350 N/mm^2 and for power-driven rivets as 405 N/mm^2.

Criterion value   It will be seen that rivets or bolts may be designed on the basis of their strength in shear, or their strength in bearing.

In actual design, the lesser of these two values will, of course, have to be used. This is called the **criterion value** of the rivet or bolt.

When designing this type of connection, the following questions should be asked:

- Is the connection in single or double shear?
- What is the safe appropriate shear load on one rivet or bolt?
- What is the safe bearing load on one rivet or bolt?

The criterion value, that is the lower of the values given by the last two questions, will then be used in determining the safe load on the connection.

Example 13.1   Calculate the safe load W on the lap joint shown in Fig. 13.6 in terms of the shear or bearing values.

Fig. 13.6 Example 13.1

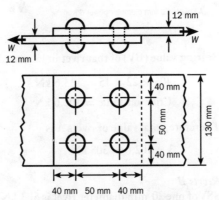

Four 20 mm diameter power-driven rivets

Solution   There are only two plates, and therefore the connection is in single shear.
Single shear value of one rivet is

$$135 \times 0.7854 \times 22^2 = 135 \times 380 = 51.3 \text{ kN}$$

Bearing value of the rivet in 12 mm plate is

$$405 \times 22 \times 12 = 405 \times 264 = 106.9 \text{ kN}$$

The shear value is, therefore, the criterion value in this case, and the safe load for the connection is

$$4 \times 51.3 = 205 \text{ kN}$$

Example 13.2   What is the safe load W in kN on the tie shown in Fig. 13.7 with respect to rivets A and B?

Solution   Rivets A are in double shear and rivets B are in single shear.

Rivets A
DSV of one 20 mm diameter rivet is

$$2 \times 135 \times 380 = 102.6 \text{ kN}$$

Fig. 13.7 Example 13.2

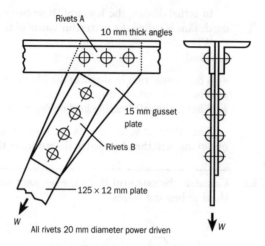

Rivets A

10 mm thick angles

15 mm gusset plate

Rivets B

125 × 12 mm plate

W

All rivets 20 mm diameter power driven

W

Bearing value (BV) of that rivet in 15 mm plate is

$$405 \times 22 \times 15 = 133.6 \text{ kN}$$
$$\text{Criterion value} = 102.6 \text{ kN}$$

Therefore, total value of rivets A is

$$3 \times 102.6 = 308 \text{ kN}$$

Rivets B

SSV of one 20 mm diameter rivet is 51.3 kN
Bearing value of that rivet in 12 mm plate is

$$405 \times 22 \times 12 = 106.9 \text{ kN}$$
$$\text{Criterion value} = 51.3 \text{ kN}$$

Therefore, total value of rivets B is

$$4 \times 51.3 = 205 \text{ kN}$$

It appears, therefore, that the safe load W is decided by rivets B and is 205 kN. The strength of the plate in tension, however, should be investigated before the 205 kN load is accepted as the safe load.

The 125 mm × 12 mm plate is weakened by one 22 mm diameter rivet hole. Hence the net cross-sectional area of the plate is

$$(125 - 22) \times 12 = 1236 \text{ mm}^2$$

and the permissible tensile stress for Grade 43 steel is 275 N/mm² for plates up to 16 mm thick.

Tension value of the plate is

$$275 \times 1236 = 340 \text{ kN}$$

In this case, therefore, the strength of the connection is decided by rivets B. It should be noted, however, that the angles have not been checked, although they are usually satisfactory.

Example 13.3 A compound bracket, shown in Fig. 13.8, is connected to the 13 mm thick web of a stanchion by six 16 mm diameter ordinary bolts of grade 4.6. The bracket carries a reaction of 225 kN from a beam. Is the connection strong enough in terms of the bolts?

Solution There are three thicknesses – the web, the angle and the cover plate – but the bolts are in single shear because the angle and the cover plate act as one.

SSV of one 16 mm diameter bolt is

$$160 \times 0.7854 \times 16^2 = 160 \times 201 = 32.2 \text{ kN}$$

Bearing value of the bolt in 13 mm plate is

$$460 \times 16 \times 13 = 460 \times 208 = 95.7 \text{ kN}$$

$$\text{Criterion value} = \text{shear value} = 32.2 \text{ kN}$$

$$\text{Safe load} = 6 \times 32.2 = 193.2 \text{ kN}$$

This is less than the applied load (reaction). Therefore, either the number of the 16 mm diameter bolts should be increased or larger diameter bolts will have to be used.

In practice the bolts in this type of connection would be also investigated for direct tension since, according to Clause 4.7.6 of BS 5950: Part 1: 1990, the reaction must be assumed to be applied at least 100 mm from line A–A (Fig. 13.8), thus creating an eccentricity of loading.

Fig. 13.8 Example 13.3

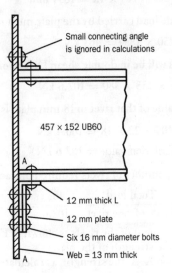

Small connecting angle is ignored in calculations

457 × 152 UB60

A

12 mm thick L

12 mm plate

Six 16 mm diameter bolts

Web = 13 mm thick

A

Double cover butt connections In designing butt and other similar types of connections, it should always be borne in mind that not only can failure occur through an insufficient number of rivets or bolts being provided, but that the member itself may fail in tension.

Consider, for example, Fig. 13.9 noting, in particular, the layout of the rivets in what is called a 'leading rivet' arrangement.

One possible chance of failure is that the plate being connected would fail by tearing across face A–A or B–B under a heavy load. Therefore, no matter

Fig. 13.9 Double cover butt connections

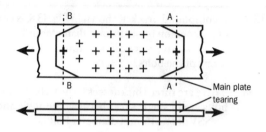

Main plate
tearing

how many rivets are employed, the safe strength in tension across this and other faces could never be exceeded.

The strength of the rivets must be approximately equal to the strength of the member in tension for the connection to be considered economical.

Example 13.4

A 125 mm × 18 mm steel plate used as a tension member in a structural frame has to be connected using a double cover butt connection with two 12 mm cover plates and 20 mm diameter power-driven rivets.

Design a suitable connection assuming that the permissible stress in tension for the steel plate is 250 N/mm^2.

Solution

However the rivets are arranged, the section will be weakened by having at least one rivet hole so the net cross-sectional area of the plate is

$$(125 - 22) \times 18 = 1854 \text{ mm}^2$$

and the safe load carried by the plate must not exceed

$$250 \times 1854 = 463 \text{ kN}$$

The rivets will be in double shear. DSV of one 20 mm diameter rivet is

$$2 \times 135 \times 380 = 102.6 \text{ kN}$$

Bearing value of that rivet in 18 mm plate is

$$405 \times 22 \times 18 = 160.4 \text{ kN}$$

Criterion value = 102.6 kN

Therefore, number of rivets required on each side of joint is

$$\frac{\text{Total load}}{\text{Value of one rivet}} = \frac{463}{102.6} = 4.5, \text{ say 5 rivets}$$

The arrangement of the rivets is shown in Fig. 13.10(a).

Check the strength of the plate:

At section A–A, the strength, as calculated, is 463 kN.

Fig. 13.10 Example 13.4: (a) five rivets on each side of the joint; (b) six rivets on each side

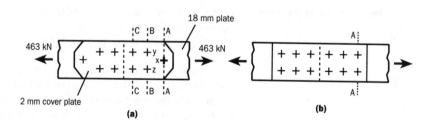

At Section B–B, the plate is weakened by two rivet holes, but, in the event of tearing of the plate across B–B, the connection would not fail until the rivet marked X also had failed.

Thus the strength across B–B is

$$250 \times (125 - 2 \times 22) \times 18 + 102\,600$$
$$= 250 \times 1458 + 102\,600 = 467 \text{ kN}$$

and the strength at C–C is

$$250 \times 1458 + 3 \times 102\,600 = 672 \text{ kN}$$

since in this case the rivets X, Y and Z have to be considered.

Finally, the cover plates have to be designed. The critical section here is section C–C. Assume 12 mm thick plates then the strength at C–C is

$$250 \times (125 - 2 \times 22) \times 24 = 486 \text{ kN}$$

Therefore the above connection would carry 463 kN.

It is sometimes useful to check the 'efficiency' of the connection. This is given by

$$\text{Efficiency} = \frac{\text{safe load for the connection}}{\text{original value of the undrilled plate}} \times 100$$

In the above case, the efficiency would be

$$\frac{463\,000}{250 \times 125 \times 18} \times 100 = 82.3\%$$

Increasing the number of rivets above that which is required may, in some cases, actually weaken the connection.

Consider the connection in Example 13.4. Had six rivets been used as in Fig. 13.10(b), instead of the required five, the value of the plate at section A–A would now be

$$250 \times (125 - 2 \times 22) \times 18 = 364 \text{ kN}$$

as against 463 kN for the leading rivet arrangement.

High strength friction grip bolts

The rivets and bolts, discussed so far in this chapter, relied on their shear and bearing strength to produce an effective connection capable of transmitting a load from one member to another, e.g. from beam to column.

The performance of high strength friction grip (HSFG) bolts is based on the principle that the transfer of the load may be effected by means of friction between the contact surfaces (interfaces) of the two members. To produce the necessary friction a sufficiently high clamping force must be developed, and this is achieved by tightening the bolts to a predetermined tension. In this way the bolts are subjected to a direct (axial) tensile force and do not rely on their shear and bearing strength.

The substantial forces needed to produce the necessary 'friction grip' require the bolts to be of high tensile strength and the interfaces to be meticulously prepared. Various methods are used to ensure that the bolts are tightened to the required tension in the shank. These include torque control by means of calibrated torque wrenches and special load-indicating devices.

Table 13.1 High strength friction grip bolts (general grade)

Nominal size and thread diameter	Proof load, (minimum shank tension) (kN)
M16	92.1
M20	144
M22	177
M24	207
M27	234
M30	286
M36	418

(*Source*: Adapted from BS 4395: Part 1)

HSFG bolts and their use are specified in BS 4395 and BS 4604, respectively, and further details may be obtained from manufacturers' literature.

Considering connections subject only to shear between the friction faces, the slip resistance, P_{SL}, may be determined from the following:

$$P_{SL} = 1.1 \times K_s \times \mu \times P_0 \quad \text{for parallel shank and}$$
$$P_{SL} = 0.9 \times K_s \times \mu \times P_0 \quad \text{for waisted shank fasteners}$$

The slip resistance is the limit of shear (i.e. the load) that can be applied before slip occurs. For parallel shank fasteners, however, it may be necessary to check the bearing capacity of the connection (see the note to Clause 6.4.1 of BS 5950: Part 1: 1990).

The value of the factor K_s is 1.0 for fasteners in clearance holes (as in most cases). It is less for oversized and slotted holes.

μ is the slip factor (i.e. the coefficient of friction between the surfaces) and may be taken as 0.45 for general grade fasteners. Its value is limited to 0.55.

P_0 is the minimum shank tension as given in Table 13.1.

Example 13.5

Consider the connection in Example 13.3 (Fig. 13.8). Assume that the six bolts are 16 mm diameter HSFG bolts (general grade). Is the connection strong enough now?

Solution

$$\text{Safe load} = 1.1 \times 1.0 \times 0.45 \times 92.1 \times 6 = 273 \text{ kN} > 225 \text{ kN}$$

i.e. the connection is now satisfactory.

It must be pointed out again that here, as in the case of Example 13.3, the bolts would also be subject to tension caused by the eccentricity of loading. This tension reduces the effective clamping action of the bolts and, therefore, the safe load would have to be suitably decreased. (See Clause 6.4.5 of BS 5950: Part 1: 1990.)

Welding

Welding for structural purposes is governed by the requirements of BS 5135, *Process of welding of carbon and carbon–manganese steels*, and the design of welds is covered by Clause 6.6 of BS 5950: Part 1: 1990.

The two types of weld used are butt welds and fillet welds.

Butt welds

These require the edges of the plates to be prepared by bevelling or gouging as shown in Fig. 13.11(a). This preparation and the need for careful alignment when welding make the butt weld generally the more expensive of the two.

For the purpose of strength calculations, butt welds are treated as the parent metal, i.e. the allowable stresses for the weld are the same as those for the connected plates.

Fillet welds

No special preparations are needed and the strength of the weld is calculated on the throat thickness (see Fig. 13.11(b)).

The design strength, p_w, depends on the grade of the steel of the connected parts and is 215 N/mm^2 for Grade 43 steel.

The size of the weld is specified by the minimum leg length of the weld, e.g. the strength of an 8 mm fillet weld for Grade 43 steel is

$$8 \times 0.7 \times 215 = 1204 \text{ N/mm}$$

Fig. 13.11 Welding: (a) butt welds; (b) fillet welds

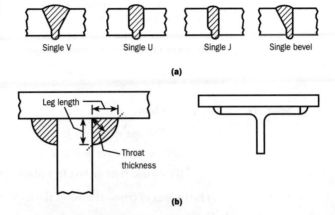

Single V   Single U   Single J   Single bevel

(a)

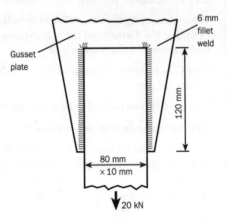

Leg length

Throat thickness

(b)

i.e. each millimetre length of this weld is capable of carrying a load of 1204 N.

When deciding on the size of a weld it is wise to consider that the amount of weld metal increases faster than the strength of the weld, e.g. compare 6 mm and 8 mm welds:

Increase in strength 33%

Increase in weld metal 78%

Example 13.6   A tension member in a framework consists of an 80 mm × 10 mm flat and is subject to a direct force of 206 kN. Design a suitable fillet weld connection using a gusset plate, as shown in Fig. 13.12.

Fig. 13.12 Example 13.6

Gusset plate

6 mm fillet weld

120 mm

80 mm × 10 mm

20 kN

Solution   Welding along the two edges of the flat requires a minimum length of weld of 80 mm on each side (Clause 6.6.2.3 of BS 5950), i.e. minimum length of weld is 160 mm.

Use 6 mm weld

$$\text{Required length} = \frac{206\,000}{6 \times 0.7 \times 215} = 228 \text{ mm}$$

The weld should be returned continuously around the corner for a distance not less than $2 \times$ weld size to comply with Clause 6.6.2.1 of BS 5950 and an allowance of one weld size should be made at the open end of the weld.

The overall length of the welds should be

$$\tfrac{1}{2} \times 228 + 1 \times 6 + \text{return end}$$

Summary

Rivets and bolts

SSV of one rivet or bolt $= A p_s$

DSV of one rivet or bolt $= 2 A p_s$

BV of one rivet or bolt in a plate of thickness t mm $= d t p_{bb}$

A is the area of cross-section of the rivet shank or bolt shank.

For rivets, A may be taken as the area of a circle 2 mm greater in diameter than the specified (nominal) diameter.

For bolts, A is the area calculated from the nominal diameter.

d is the diameter of the rivet or bolt.

For rivets, $d =$ nominal diameter plus 2 mm

For bolts, $d =$ nominal diameter

$p_s =$ shear strength

$p_{bb} =$ bearing strength

In certain problems, the strength of the plate in tension may have to be investigated. The permissible tension stress should not exceed $0.84 \times$ minimum ultimate tensile strength.

When deducting the areas of rivet or bolt holes to determine the strength of a plate, the diameter of the hole is taken as 2 mm greater than the nominal diameter of the rivet or bolt.

HSFG bolts rely on their tensile strength to induce friction between the connected parts.

A *butt weld* is considered to be as strong as the parent metal.

The strength of a *fillet weld* per millimetre of its length is calculated as

$$0.7 \times \text{size of weld} \times \text{design strength,} \ p_w$$

Exercises

Note: Permissible tension stress for Grade 43 steel may be assumed to be 250 N/mm². These examples may be adapted for HSFG bolts and fillet welds.

1 The size of each plate in a simple lap joint is 100 mm × 12 mm and there are six 20 mm diameter ordinary bolts in a single line. Calculate the safe load in tension.

2 In a double cover butt connection, the joined plate is 125 mm × 12 mm and the cover plates are 125 mm × 8 mm. There are two 20 mm diameter power-driven rivets each side of the joint (four rivets in all). Calculate the maximum safe tension for the plates.

3 A simple lap joint with five 24 mm diameter hand-driven rivets is shown in Fig. 13.Q3. Calculate the maximum safe pull, W.

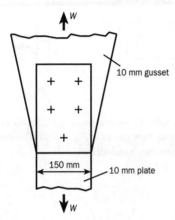

Fig. 13.Q3

4 Eight bolts, 24 mm diameter, connect a flat bar to a gusset plate as shown in Fig. 13.Q4. Calculate the safe load, W.

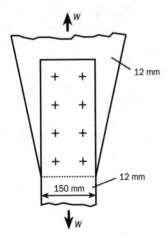

Fig. 13.Q4

5 Figure 13.Q5 gives two different bolted connections (a) and (b). In each case, calculate the safe load, W.

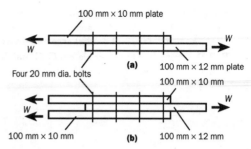

Fig. 13.Q5

6 A 100 mm wide plate is connected, by means of a 10 mm gusset plate to the flange of a 254×254 UC107 stanchion, as shown in Fig. 13.Q6. Calculate the required thickness of the plate and the number of 20 mm diameter bolts.

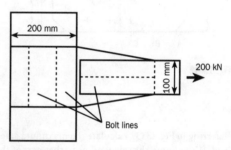

Fig. 13.Q6

7 A 10 mm thick tie member is to be connected to a 10 mm thick gusset plate, as indicated in Fig. 13.Q7. Calculate the necessary width, x, of the tie and calculate the number of 16 mm diameter bolts to connect the tie to the gusset.

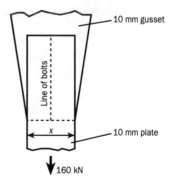

Fig. 13.Q7

8 Figure 13.Q8 shows a joint in a tension member. Determine the safe load, *W.* (Calculations are required for the strength of the middle plate at sections A–A and B–B; the strength of the cover plate at C–C, and the strengths of the rivets in shear and bearing.)

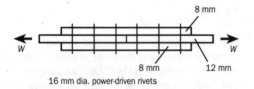

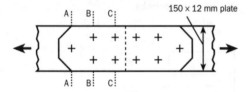

Fig. 13.Q8

9 Referring to Fig. 13.Q9, calculate the maximum safe load, *W,* confining the calculations to that part of the connection which lies to the right of line A–A. Then, using this load, find the required thickness of the 75 mm wide plate and the necessary number of 16 mm diameter hand–driven rivets.

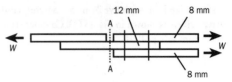

All rivets 16 mm dia. hand-driven rivets

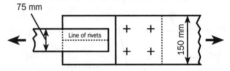

Fig. 13.Q9

10 Calculate the value of the maximum permissible load, *W,* for the connection shown in Fig. 13.Q10. The rivets are 20 mm diameter, hand–driven.

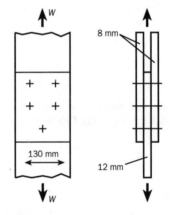

Fig. 13.Q10

14 Addition of direct and bending stress

Previous chapters have dealt largely with two main types of stress:

- *Direct or axial stress* This occurs when a load (tensile or compressive) is spread evenly across a section, as in the case of a short column axially loaded. Here, the unit stress or stress per mm^2 is found by dividing the total load by the sectional area of the member, i.e. direct stress $= f_d = W/A$.
- *Bending stress* In this case, the stress has been seen to vary – having different values at different distances from the neutral axis of the section. The stress at a distance y from the neutral axis is $f_b = M \times y/I$, or where the stress at the extreme fibres is required $f_{bmax} = M/Z$.

Cases frequently arise, however, where *both* these types of stress occur at the same time, and the combined stress due to the addition of the two stresses is to be determined.

Consider, for example, the short timber post shown in plan in Fig. 14.1. The load passes through the centroid of the section, and the resulting stress is pure axial stress

$$W/A = 90\,000/22\,500 = 4 \text{ N/mm}^2$$

The stress is the same at all points of the section, and this is shown by the rectangular shape of the 'stress distribution diagram'.

Fig. 14.1 Direct stress: (a) loading of a short timber post; (b) stress distribution

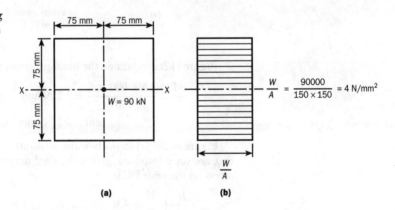

Figure 14.2(a)(i) shows the same section with a load of 90 kN as before, but this time the load, whilst still on the X–X axis, lies 20 mm away from the Y–Y axis.

As before, there will still be a direct stress everywhere on the section of $W/A = 4 \text{ N/mm}^2$, but the eccentricity of the load with regard to the Y–Y axis causes a moment of $90\,000 \times 20 = 1.8 \times 10^6$ N mm, which has the effect of increasing the compression on the area to the right of Y–Y (shaded) and de-creasing the compression on the portion to the left of Y–Y.

Figure 14.2(a)(ii) shows the direct stress W/A in compression.

Fig. 14.2 Direct stress plus bending stress: (a) Y–Y axis eccentricity $= 20$ mm; (b) Y–Y axis eccentricity $= 25$ mm; (i) loading diagram, (ii) direct stress, (iii) bending stress, (iv) combined stress distribution

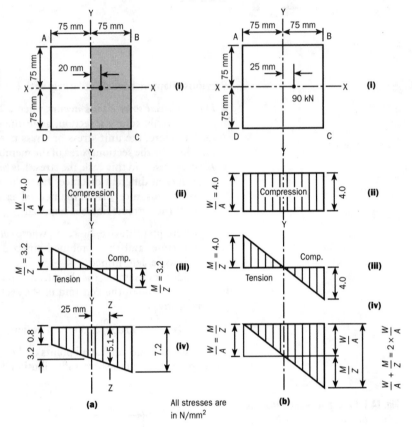

All stresses are in N/mm^2

Figure 14.2(a)(iii) shows the bending stress of

$$\frac{M}{Z} = \frac{1\,800\,000 \times 6}{150^3} = 3.2 \text{ N/mm}^2 \text{ compression at BC}$$

$$\text{and} \quad 3.2 \text{ N/mm}^2 \text{ tension at AD}$$

Figure 14.2(a)(iv) shows how these two stress distribution diagrams, (ii) and (iii), are superimposed, giving the final diagram as shown. The compressive stress on the edge BC is

$$\frac{W}{A} + \frac{M}{Z} = 4.0 + 3.2 = 7.2 \text{ N/mm}^2$$

whilst the stress on edge AD is

$$\frac{W}{A} - \frac{M}{Z} = 4.0 - 3.2 = 0.8 \text{ N/mm}^2$$

Note: The stress is still compressive at any point, but it varies from face AD to face BC.

The stress at section Z–Z (25 mm from Y–Y) is

$$\frac{W}{A} + \frac{My}{I_{yy}} \quad \text{where } y = 25 \text{ mm}$$

$$= 4.0 + \frac{1\,800\,000 \times 25 \times 12}{150^4}$$

$$= 4.0 + 1.1 = 5.1 \text{ N/mm}^2$$

Figure 14.2(b)(i) shows the same short column as in the previous example, but this time the eccentricity about the Y–Y axis has been increased to 25 mm. As before,

$$\text{Direct stress} = W/A = 90\,000/22\,500 = 4 \text{ N/mm}^2$$

but the M is now $90\,000 \times 25 = 2.25 \times 10^6$ N mm. Thus

$$\frac{M}{Z} = \frac{2.25 \times 10^6}{562\,500} = 4 \text{ N/mm}^2$$

Figure 14.2(b)(ii) and (iii) show the direct and bending stresses respectively, whilst in Fig. 14.2(b)(iv) the two diagrams have been superimposed, showing that:

- The stress on face BC is

$$\frac{W}{A} + \frac{M}{Z} = 4.0 + 4.0 = 8.0 \text{ N/mm}^2$$

- The stress on face AD is

$$\frac{W}{A} - \frac{M}{Z} = 4.0 - 4.0 = 0$$

The original compression of W/A on face AD has been cancelled out by the tension at AD caused by the eccentricity of the loading away from that face, and the stress intensity at AD $= 0$, whilst that on face BC $= 8.0$ N/mm^2.

This interesting example demonstrates a general rule that, in the case of square or rectangular sections, the load may be eccentric from one axis by no more than $\frac{1}{6}$th of the width unless tension on one face is permissible.

In the case of a steel, timber or reinforced concrete column tension would be allowable within reasonable limits, but, in a brick wall, tension would be undesirable and the resultant load or thrust in a masonry structure should not depart more than $d/6$ (or $b/6$) from the axis.

This is usually called the **law of the middle third**, and it states that if the load or the resultant thrust cuts a *rectangular* section at a point not further than $d/6$ from the centre line (i.e. within the middle third), then the stress everywhere across the section will be compressive.

Consider the masonry retaining wall shown in Fig. 14.3(a). The earth pressure P and the weight W of the wall combine to form a resultant R which cuts the base, as shown, at a distance e from the neutral axis (which, in this case, is also the centre line) of the base.

The vertical component V of the resultant causes a moment of $V \times e$, which adds compressive stress between the wall and the earth below at A, and reduces the pressure between the surfaces at B.

Fig. 14.3 The law of the middle third

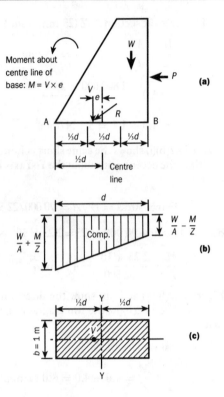

If V cuts the base within the middle third of its width d, then the stress everywhere between A and B will be compressive, as in Fig. 14.3(b).

Note: As the wall itself may be of any length, it is customary in design to consider only one metre length of wall, i.e. to estimate the weight of one metre of wall and the earth pressure on one metre of length also.

Thus, if the rectangle shown in Fig. 14.3(c) represents the area of wall resting on the base AB (as seen in a plan), then the eccentricity of the load V will tend to rotate this area about the axis Y–Y, so that, in calculating Z, the value $bd^2/6$ will be in effect $1 \times d^2/6 = \frac{1}{6}$th of d^2, and will normally be in m³.

Eccentricity about both axes

The short pier shown (Fig. 14.4) carries one load only, eccentric to both X–X and Y–Y axes. Consider the stresses at A, B, C and D.

At C, the stress is

$$\frac{W}{A} + \frac{M_{xx}}{Z_{xx}} + \frac{M_{yy}}{Z_{yy}}$$

because the load has moved towards C from both axes.

Stress at corner B will be

$$\frac{W}{A} - \frac{M_{xx}}{Z_{xx}} + \frac{M_{yy}}{Z_{yy}}$$

for the load is away from the Y–Y axis on the same side as B, but is away from the X–X axis on the side furthest from B.

Fig. 14.4 Eccentricity about both axes

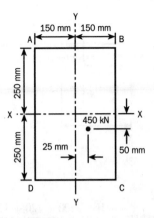

Similarly,

$$\text{Stress at corner A} = \frac{W}{A} - \frac{M_{xx}}{Z_{xx}} - \frac{M_{yy}}{Z_{yy}}$$

$$\text{Stress at corner D} = \frac{W}{A} + \frac{M_{xx}}{Z_{xx}} - \frac{M_{yy}}{Z_{yy}}$$

The solution is outlined below:

$$\frac{W}{A} = \frac{450\,000}{300 \times 500} = 3.0 \text{ N/mm}^2$$

$$\frac{M_{xx}}{Z_{xx}} = \frac{450\,000 \times 50 \times 6}{300 \times 500^2} = 1.8 \text{ N/mm}^2$$

$$\frac{M_{yy}}{Z_{yy}} = \frac{450\,000 \times 25 \times 6}{400 \times 300^2} = 1.5 \text{ N/mm}^2$$

Hence,

$$\text{Stress at corner A} = 3.0 - 1.8 - 1.5$$
$$= -0.3 \text{ N/mm}^2 \text{ tension}$$
$$\text{Stress at corner B} = 3.0 - 1.8 + 1.5$$
$$= +2.7 \text{ N/mm}^2 \text{ compression}$$
$$\text{Stress at corner C} = 3.0 + 1.8 + 1.5$$
$$= +6.3 \text{ N/mm}^2 \text{ compression}$$
$$\text{Stress at corner D} = 3.0 + 1.8 - 1.5$$
$$= +3.3 \text{ N/mm}^2 \text{ compression}$$

Example 14.1 A short stanchion carries three loads, as shown in Fig. 14.5(a). Calculate the intensity of stress at corners A, B, C and D.

$$\text{Area} = 7640 \text{ mm}^2 \qquad Z_{xx} = 584\,000 \text{ mm}^3 \qquad Z_{yy} = 201\,000 \text{ mm}$$

Fig. 14.5 Example 14.1: (a)
method 1; (b) method 2

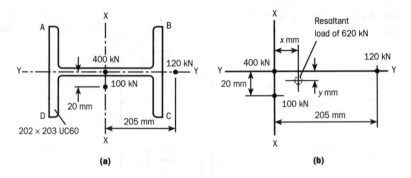

(a)   (b)

Solution   *Method 1*

$$\text{Direct stress } \frac{W}{A} = \frac{400 + 120 + 100}{7640} = 81.2 \text{ N/mm}^2$$

$$\text{Bending stress } \frac{M_{xx}}{Z_{xx}} = \frac{120\,000 \times 205}{584\,000} = 42.1 \text{ N/mm}^2$$

$$\text{Bending stress } \frac{M_{yy}}{Z_{yy}} = \frac{100\,000 \times 20}{201\,000} = 9.9 \text{ N/mm}^2$$

Therefore, the combined stresses are

$$\text{At A}\quad 81.2 - 42.1 - 9.9 = +29.2 \text{ N/mm}^2$$
$$\text{At B}\quad 81.2 + 42.1 - 9.9 = +113.4 \text{ N/mm}^2$$
$$\text{At C}\quad 81.2 + 42.1 + 9.9 = +133.2 \text{ N/mm}^2$$
$$\text{At D}\quad 81.2 - 42.1 + 9.9 = +49.0 \text{ N/mm}^2$$

Method 2

The value and position of the resultant of the three loads is calculated and the resultant load applied to the column in place of the three loads.

To find distance x (see Fig. 14.5(b)) take moments of loads about the X–X axis:

$$(400 + 120 + 100) \times x = 120 \times 205$$
$$x = 24\,600/620 = 39.7 \text{ mm}$$

For distance y take moments about the Y–Y axis:

$$620 \times y = 100 \times 20$$
$$y = 2000/620 = 3.22 \text{ mm}$$

Now, the direct stress W/A remains the same and the bending stresses are

$$\frac{M_{xx}}{Z_{xx}} = \frac{620\,000 \times 39.7}{584\,000} = 42.1 \text{ N/mm}^2$$

$$\frac{M_{yy}}{Z_{yy}} = \frac{620\,000 \times 3.22}{201\,000} = 9.9 \text{ N/mm}^2$$

Both are the same as in Method 1.

Application to prestressed concrete beams   The addition of direct and bending stresses is not confined to columns only. It is used extensively in the design calculations for prestressed concrete beams. The prestressing force is generally applied in the direction of the longitudinal axis of the beam, thus producing direct stresses in the beam. The bending stresses result mainly from the loads the beam has to support, but there is usually also bending caused by an eccentricity of the prestressing force.

Example 14.2   A plain concrete beam, spanning 6 m, is to be prestressed by a force of 960 kN applied axially to the 240 mm × 640 mm beam section.

      Calculate the stresses in the top and bottom fibres of the beam if the total inclusive UDL carried by the beam is 136 kN.

Solution   The direct stress caused by the prestressing force P is

$$\frac{P}{A} = \frac{960\,000}{240 \times 640} = 6.25 \text{ N/mm}^2$$

The bending stresses, on the other hand, are produced by the inclusive UDL and these are M/Z where

$$M = \tfrac{1}{8}Wl = \tfrac{1}{8} \times 136 \times 6 \times 10^6 = 102 \times 10^6 \text{ N mm}$$
$$Z = \tfrac{1}{6}bd^2 = \tfrac{1}{6} \times 240 \times 640^2 = 16\,384\,000 \text{ mm}^3$$

Therefore,

$$\frac{M}{Z} = \frac{102\,000\,000}{16\,384\,000} = 6.23 \text{ N/mm}^2$$

and the combined stresses are

$$\text{At the top} \quad 6.25 + 6.23 = +12.48 \text{ N/mm}^2$$
$$\text{At the bottom} \quad 6.25 - 6.23 = +0.02 \text{ N/mm}^2$$

as shown diagrammatically in Fig. 14.6(a).

In order to increase the beam's load bearing capacity it is usual to apply the prestressing force below the centre line of the section. This induces tensile bending stresses in the top fibres and compressive bending stresses in the bottom fibres of the beam. The applied loads produce bending stresses of the opposite sign, i.e. tension in bottom and compression in top fibres.

      The combination of the stresses is shown in Fig. 14.6(b).

Example 14.3   Consider the beam in Example 14.2 with the prestressing force applied on the vertical axis of the section but 120 mm below its intersection with the horizontal axis. Calculate the stresses in top and bottom fibres.

Solution   Direct stress

$$\frac{P}{A} = \text{(as before)} + 6.25 \text{ N/mm}^2$$

Bending stress due to prestressing force

$$\frac{P \times e}{Z} = \frac{960\,000 \times 120}{16\,384\,000} = \pm 7.03 \text{ N/mm}^2$$

Fig. 14.6 Example 14.2: (a) axial prestressing; (b) prestressing below the centre line; (i) beam details, (ii) stress diagrams

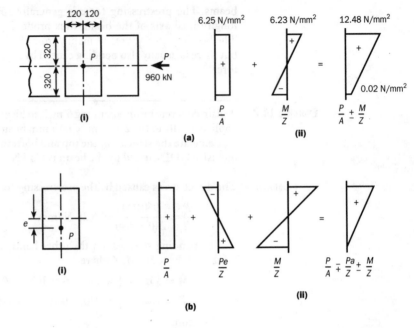

Bending stress due to applied loads

$$\frac{M}{Z} = \text{(as before)} \pm 6.23 \ \text{N/mm}^2$$

The combined stresses are

At the top $+ 6.25 - 7.03 + 6.23 = +5.45 \ \text{N/mm}^2$

At the bottom $+ 6.25 + 7.03 - 6.23 = +7.05 \ \text{N/mm}^2$

It may be concluded from this example that the UDL could be doubled without indicating tensile stresses in the concrete when the beam is fully loaded. There are, however, various problems which have to be taken into account when designing prestressed concrete beams and which are outside the scope of this textbook.

Summary

Maximum combined stress = direct stress + bending stress. When load W is eccentric to axis X–X only:

$$\text{Maximum stress} = \frac{W}{A} + \frac{We_x}{Z_{xx}}$$

When load W is eccentric to axis Y–Y only:

$$\text{Maximum stress} = \frac{W}{A} + \frac{We_y}{Z_{yy}}$$

When load W is eccentric to both axes:

$$\text{Maximum stress} = \frac{W}{A} + \frac{We_x}{Z_{xx}} + \frac{We_y}{Z_{yy}}$$

In prestressed concrete beams, where opposing bending stresses are produced by the prestressing force P and the applied loads W,

$$\text{Maximum stress} = \frac{P}{A} \mp \frac{P \times e}{Z} \pm \frac{M_w}{Z}$$

Exercises

1 A short timber post carries a load of 40 kN eccentric from one axis only as shown in Fig. 14.Q1. Calculate the intensity of stress (a) at face BC (f_{bc}), (b) at face AD (f_{ad}).

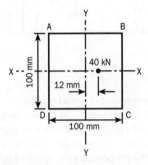

Fig. 14.Q1

2 Calculate the compression stress on the faces BC and AD of the short timber post, as shown in Fig. 14.Q2.

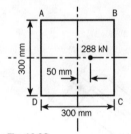

Fig. 14.Q2

3 Repeat Question 2, but increasing the eccentricity to 60 mm.

4 Calculate the stress at the corner C of the short timber post shown in Fig. 14.Q4.

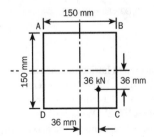

Fig. 14.Q4

5 A short steel stanchion consists of a 203 × 203 UC60 and carries loads as shown in Fig. 14.Q5. Calculate the stress at faces AD and BC.

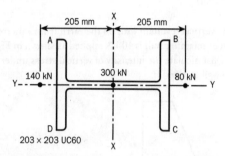

Fig. 14.Q5

6 A short steel post consisting of a 152 × 152 UC30 carries three loads, as shown in Fig. 14.Q6. Calculate the stress at each of the corners A, B, C and D.

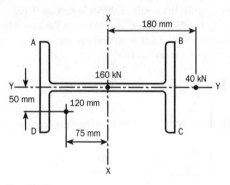

Fig. 14.Q6

7 The short steel post, as shown in Fig. 14.Q7, is a 152 × 152 UC37. There is a central point load of 120 kN and an eccentric load of W kN as shown. The maximum stress on the face BC is 80 N/mm². What is the value of the load W?

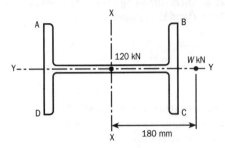

Fig. 14.Q7

8 The vertical resultant load on the earth from a metre run of retaining wall is 40 kN applied as shown in Fig. 14.Q8. Calculate the intensity of vertical stress under the wall at A and B.

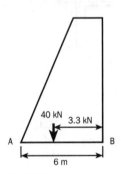

Fig. 14.Q8

9 A triangular mass wall, as shown in section in Fig. 14.Q9, weighs 20 kN/m³ and rests on a flat base AB. What is the intensity of vertical bearing stress at A and B per metre length of wall?

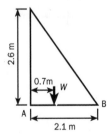

Fig. 14.Q9

10 The concrete beam shown in Fig. 14.Q10 is prestressed by the application of two loads of 300 kN applied at points 40 mm below the neutral axis of the section. (All dimensions are in mm.) (a) What are the stresses at the upper and lower faces of the beam? (b) At what distance below the neutral axis would the loads have to be applied for there to be no stress at all on the top face of the beam? (All dimensions are in mm.)

Note: Neglect self-weight of beam.

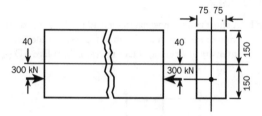

Fig. 14.Q10

11 A double angle rafter member in a steel truss is subjected to an axial compression of 60 kN and also to bending moment from a point load of 10 kN, as shown in Fig. 14.Q11. The member is composed of two angles, as shown, for which the properties (given for the double angle section) are

$$A = 3000 \text{ mm}^2 \qquad I_{xx} = 1.75 \times 10^6 \text{ mm}^4$$

Calculate the maximum compressive stress in the member.

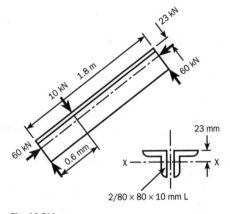

Fig. 14.Q11

12 A timber beam subjected to horizontal thrusts at each end, shown in Fig. 14.Q12, is loaded with a central concentrated load of 5 kN (all dimensions are in mm). The beam itself weighs 150 kN. What is (a) the maximum compressive stress and (b) the maximum tensile stress?

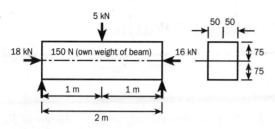

Fig. 14.Q12

13 A plain concrete beam, 240 mm × 600 mm, spans 6 m and weighs 3.2 kN/m. A horizontal compressive force of 900 kN is applied to the beam at 180 mm above its underside.

Calculate (a) the stresses in the top and bottom fibres of the beam; (b) the additional UDL the beam could carry without developing tension in its bottom fibres.

15 Gravity retaining walls

The previous chapter has been concerned mainly with the addition of direct and bending stress when these two types of stress occur within a material, e.g. the variation of stress across the face of a column section which arises due to eccentricity of loading.

The principles involved in those cases are used in much the same way on occasions where, for example, a wall, resting on soil or on a concrete footing and acted upon by horizontal forces, is transmitting to the soil or footing stresses which consist of:

- direct stress from the wall's weight
- stress due to the overturning moment

Before these resultant stresses are considered, it will be necessary to study the effect of the combined action of the vertical and horizontal (or inclined) forces on the overall behaviour of the wall.

As a result of that action the wall may fail in three ways:

- sliding
- overturning
- overstressing

But since the main purpose of a retaining wall is to provide resistance to the horizontal (or inclined) forces caused by the retained material, the nature of the pressures these forces exert on the wall will be investigated first.

Horizontal forces – wind pressure

This is the simplest case, because wind pressure is assumed to be uniform. Therefore, the total resultant force acts at the centre of the area over which the pressure is applied and is given by

$$P = p \times \text{area newtons}$$

Or since, in the case of a wall, 1 m length of wall is generally considered, P is given by

$$P = p \times 1 \times H = p \times H \text{ newtons}$$

where p = unit wind pressure in N/m^2

   H = the height of the part of the wall subject to that wind pressure (see Fig. 15.1(a)) in metres

The determination of the unit wind pressure p is detailed in CP3: Chapter V: Part 2: Wind loads. It is based on several factors, including not only the basic speed of the wind and the type of topography in the locality of the wall, but also a statistical factor which takes into account the probability of the basic wind speed being exceeded within the projected life span of the wall.

Fig. 15.1 Horizontal forces:
(a) wind pressure; (b) water
pressure (i) vertical retaining
wall, (ii) pressure variation;
(c) action of resultant force;
(d) action of resultant force
where the liquid does not reach
the top of the dam

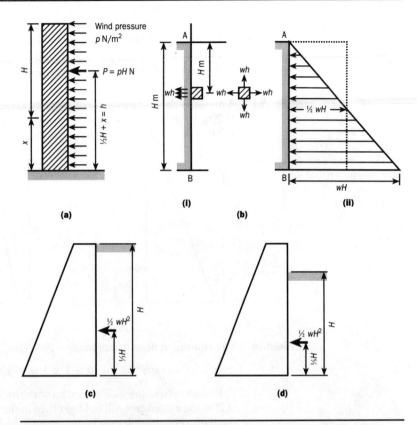

Horizontal forces – liquid pressure

Consider the vertical surface AB, shown in Fig. 15.1(b)(i), to be the face of a wall which is retaining a liquid. It can be shown that a cubic metre of liquid, situated at a depth h metres below the surface, exerts a pressure of $w \times h$ kN outwards on all its six side surfaces. w in this case is the equivalent density or the unit weight of the liquid in kN/m^3.

Thus the intensity of outward pressure varies directly with the depth and will have a maximum value of $w \times H$ kN/m^2 at H m, the maximum depth as indicated in Fig. 15.1(b)(ii).

At the surface of the liquid (where $h = 0$), the pressure will be zero. So, as the maximum is wH kN/m^2, the average pressure between A and B is $\frac{1}{2}wH$ kN/m^2.

In dealing with retaining walls generally, it is convenient, as was said earlier, to consider the forces acting on one metre length of wall, that is, an area of wall H m high and 1 m measured perpendicular to the plane of the diagrams (Fig. 15.1(b)). Thus, as the 'wetted area' concerned is H m^2, the total force caused by water pressure on a one metre strip of wall is

$$\text{'Wetted' area} \times \text{average rate of pressure} = H \times \tfrac{1}{2}wH$$

$$= \tfrac{1}{2}wH^2 \text{ kN}$$

This total resultant force on the wall's vertical surface from the liquid is (as will be seen from Fig. 15.1(b)(ii)) the resultant of a large number of forces, which range from zero at the top to wH at the base. The resultant will therefore act at a point $\frac{1}{3}H$ from the base, as shown in Fig. 15.1(c).

Note: If the liquid does not reach the top of the wall, as, for example, in Fig. 15.1(d), then the resultant force is calculated with H as the depth of the liquid and not as the height of the wall. The force is again $\frac{1}{2}wH^2$ and it acts at a point $\frac{1}{3}H$ (one-third the depth of the liquid) from the wall's base.

Example 15.1　A masonry dam retains water on its vertical face. The wall is, as shown in Fig. 15.2(a), 3.7 m, but the water level reaches only 0.7 m from the top of the wall. What is the resultant water pressure per metre run of wall?

Fig. 15.2 Example 5.1: (a) vertical dam; (b) non-vertical dam

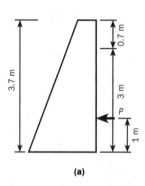
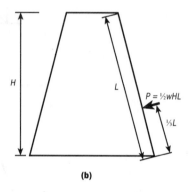

(a)　　　　　　　　　　(b)

Solution　The equivalent density w of water $= 10$ kN/m^3. Therefore

$$P = \tfrac{1}{2}wH^2 = \tfrac{1}{2} \times 10 \times 3 \times 3 = 45 \text{ kN acting at 1 m above base}$$

In cases where the wall in contact with the water is not vertical, as in Fig. 15.2(b), the wetted area will be larger than in the case of a vertical back, and the resultant pressure will thus be increased to $\frac{1}{2}wHL$ (i.e. the wetted area will be L m^2 instead of H m^2 considering one metre run of wall).

Horizontal forces – soil pressure

It is obvious that pressures on walls from retained soil or other granular materials cannot be determined with quite the same accuracy as with water. Soils vary in weight and character; they behave quite differently under varying conditions of moisture, etc., and, in general, the resultant pressures on vertical and non-vertical surfaces from soils are obtained from various soil pressure theories. Numerous theories exist for the calculation of soil pressures, and these theories vary in the assumptions which they make and the estimated pressures which they determine. A great deal of research is still being directed upon this subject, but most of it is beyond the scope of a volume of this type, and therefore only the well-tried Rankine theory will be dealt with in detail.

Rankine's theory of soil pressure

It has been seen that a cubic metre of liquid at a depth h below the surface presses outwards horizontally by an amount wh kN/m^2 (w being the equivalent density of liquid). In the case of soil weighing w kN/m^3, the outward pressure at a depth of h m below the surface will be less than wh kN/m^2, since some of the soil is 'self-supporting'.

Consider, for example, the soil retained by the vertical face AB in Fig. 15.3(a). If the retaining face AB were removed, then some of the soil would probably collapse at once, and in the course of time the soil would assume a line BC, as shown. The angle ϕ made between the horizontal and the line BC

Fig. 15.3 Rankine's theory of soil pressure: (a) angle of repose; (b) action of resultant force

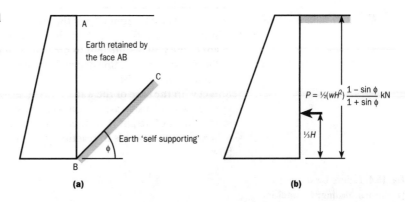

A

Earth retained by the face AB

C

Earth 'self supporting'

ϕ

B

(a)

$P = \frac{1}{2}(wH^2)\dfrac{1 - \sin \phi}{1 + \sin \phi}$ kN

$\frac{1}{3}H$

(b)

varies with different types of soil, and is called the **angle of repose** or *the angle of internal friction* of the soil. It can be said, therefore, that only part of the soil was in fact being retained by the wall and was exerting pressure on the wall. Thus, it follows that the amount of pressure on the wall from the soil depends upon the angle of repose for the type of soil concerned, and Rankine's theory states in general terms that the outward pressure per square metre at a depth of h m due to a level fill of soil is

$$wh\left(\frac{1 - \sin \phi}{1 + \sin \phi}\right) \text{ kN/m}^2$$

as compared with (wh) kN/m^2 in the case of liquids. Thus, by similar reasoning as used in the case of liquid pressure, the maximum pressure at the bottom of the wall is given by

$$\text{Maximum pressure} = wH\left(\frac{1 - \sin \phi}{1 + \sin \phi}\right) \text{ kN/m}^2$$

$$\text{Average pressure} = \tfrac{1}{2}wH\left(\frac{1 - \sin \phi}{1 + \sin \phi}\right) \text{ kN/m}^2$$

The soil acts at this average rate on an area of H m^2 of wall, so the total resultant force per metre run of wall is

$$P = \tfrac{1}{2}wH^2\left(\frac{1 - \sin \phi}{1 + \sin \phi}\right) \text{ kN}$$

and this acts, as shown in Fig. 15.3(b), at $\frac{1}{3}H$ above the base of the wall.

Example 15.2   Soil weighing 15 kN/m^3 and having an angle of repose $\phi = 30°$ exerts pressure on a 4.5 m high vertical face. What is the resultant horizontal force per metre run of wall?

Solution $\sin \phi = \sin 30° = 0.5$ and

$$P = \tfrac{1}{2}wH^2\left(\frac{1 - \sin \phi}{1 + \sin \phi}\right)$$

$$= \tfrac{1}{2} \times 15 \times 4.5 \times 4.5 \times \frac{1 - 0.5}{1 + 0.5}$$

$$= 151.875 \times \tfrac{1}{3} = 50.625 \text{ kN}$$

Modes of failure – sliding

It was stated at the beginning of the chapter that a retaining wall may fail by sliding. The possibility that the wall may slide along its base exists unless the weight of the wall is sufficient to prevent such movement.

The resistance to sliding depends upon the interaction of the weight of the wall and the friction between the material of the wall and the soil directly in contact with the base of the wall.

The basis of this interaction may be explained as follows. Consider a wooden block of weight W resting on a steel surface, as in Fig. 15.4(a). The steel surface presses upwards in reaction to the downward weight of the block, as shown.

Fig. 15.4 Failure by sliding: (a) normal loading; (b) addition of an angled force; (c) angle and coefficient of friction

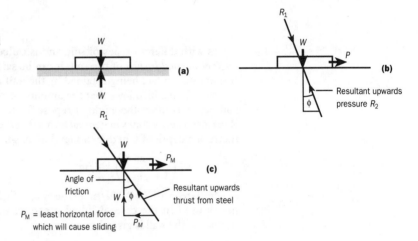

If, however, a small load P (not sufficient to move the block) is also placed on the block, as in Fig. 15.4(b), then the resultant of the forces W and P will be R_1, and the resultant upward pressure R_2 from the steel surface will also be inclined as shown. This resultant upward pressure will be inclined at an angle ϕ, as indicated.

If the force P is now gradually increased, the angle ϕ will also increase, until at a certain load (which depends on the nature of the two surfaces in contact and on the weight W) the block will move horizontally.

The angle ϕ which the resultant upward thrust makes with the vertical at the stage where the block starts to slide is known as the *angle of friction* between the two surfaces.

The tangent of ϕ is P_M/W (see Fig. 15.4(c)), so that

$$\tan \phi = \frac{\text{least force which will cause sliding}}{\text{weight of block}} \qquad (1)$$

and tan ϕ is known as the *coefficient of friction* for the two materials and is denoted by the letter μ (mu).

For most materials this coefficient of friction will vary between 0.4 and 0.7, and it will be appreciated from equation (1) that

$$P_M = \text{the least force that will cause sliding}$$

$$= W \times \text{coefficient of friction } \mu$$

In the case of retaining walls, P_M, the force which would cause sliding, can be calculated as $W \times$ coefficient of friction, but the horizontal force P of the retained material should not exceed approximately half of the force P_M.

Example 15.3   The masonry dam, shown in Fig. 15.5, retains water to the full depth, as shown. The coefficient of friction between the base of the wall and the earth underneath is 0.7. Check if the wall is safe against sliding.

Solution

$$P = \text{actual horizontal pressure on side of wall}$$
$$= \tfrac{1}{2}wH^2 = \tfrac{1}{2} \times 10 \times 4^2 = 80 \text{ kN}$$
$$P_M = \text{horizontal force which would just cause sliding}$$
$$= 0.7 \times W = 0.7 \times \tfrac{1}{2}(1+3) \times 4 \times 18$$
$$= 0.7(144) = 100.8 \text{ kN}$$

Fig. 15.5 Example 15.3

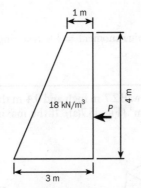

The actual pressure (80 kN) exceeds half the value of P_M, and the factor of safety against sliding is

$$\frac{\mu W}{P} = \frac{100.8}{80} = 1.26 < 2$$

which is undesirable.

Modes of failure – overturning   A retaining wall may have quite a satisfactory resistance to sliding, but the positive action of the horizontal forces may tend to overturn it about its toe (Fig. 15.6).

Fig. 15.6 Failure by overturning

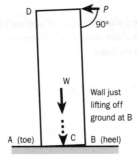

Equilibrium will be upset when the wall just rises off the ground at the heel, B, the turning point being the toe, A. (It is assumed that sliding of the wall will not occur.) When the wall is just lifting off the ground, the overturning moment due to the force P is just balanced by the restoring or balancing moment due to the weight of the wall, the wall being balanced on the edge A. Since the inclination of AB to the horizontal is very slight, distances can be taken assuming the wall is vertical.

Taking moments about A,

$$P \times \text{distance AD} = W \times \text{distance AC}$$

or

$$\text{Overturning moment} = \text{restoring (balancing) moment}$$

In practice it is usual to apply a factor of safety of 2, where

$$\text{Factor of safety against overturning} = \frac{\text{restoring (balancing) moment}}{\text{overturning moment}}$$

and, therefore,

$$\text{Overturning moment} = \tfrac{1}{2} \times \text{restoring (balancing) moment}$$

Example 15.4   A long boundary wall 2.7 m high and 0.4 m thick is constructed of brickwork weighing 18 kN/m^3 (Fig. 15.7(a)). If the maximum wind pressure uniformly

Fig. 15.7 Example 15.4

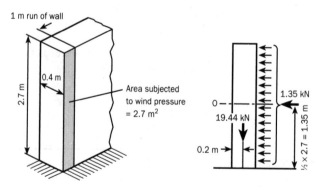

distributed over the whole height of the wall is 500 N/m^2, calculate the factor of safety against overturning, neglecting any small adhesive strength between the brickwork and its base.

Solution

$$\text{Weight of 1 m run of wall} = 2.7 \times 1.0 \times 0.4 \times 18$$
$$= 19.44 \text{ kN}$$
$$\text{Wind pressure on 1 m run of wall} = 2.7 \times 1.0 \times 0.5$$
$$= 1.35 \text{ kN}$$

and this can be taken as acting at the centre of height of the wall for purposes of taking moments about O (Fig. 15.7(b)).

$$\text{Restoring moment} = 19.44 \times \tfrac{1}{2} \times 0.4$$
$$= 3.888 \text{ kN m}$$
$$\text{Overturning moment} = 1.35 \times \tfrac{1}{2} \times 2.7$$
$$= 1.823 \text{ kN m}$$

Therefore

$$\text{FS against overturning} = 3.888/1.823 = 2.13$$

It will be found that a satisfactory factor of safety against overturning is achieved if the resultant of the horizontal and vertical forces crosses the base of the wall within its 'middle third', i.e. when no tensile stresses are allowed to develop in the wall.

This comes within the scope of the third mode of failure of retaining walls and is the topic of the following section of this chapter.

Modes of failure –
overstressing

When the weight of a wall per metre, W, and the resultant pressure from the soil or liquid, P, have been calculated, these two forces may be compounded to a resultant, as in Fig. 15.8(a). It will be shown that the position along the base at which this resultant cuts (i.e. at S) has an important bearing on the stability of the wall and on the pressures exerted by the wall upon the earth beneath.

Consider now the wall shown in Fig. 15.8(b)(i). The values in kN of W and P are as shown, and W acts through the centroid of the wall section.

Fig. 15.8 Failure by overstressing: (a) resultant of weight and pressure; (b)(i) loading diagram, (ii) resultant, (iii) stress or pressure distribution

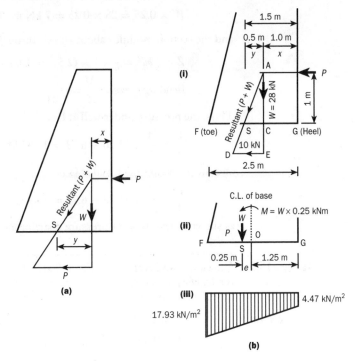

From the similar triangles ASC and ADE

$$\frac{y}{1.4} = \frac{10}{28} \quad \text{so} \quad y = \frac{10 \times 1.4}{28} = 0.5 \text{ m}$$

Thus the resultant force cuts the base at $1.0 + 0.5 = 1.5$ m from the point G, the heel of the wall.

It can be seen that the resultant force cuts the base at point S, as shown in Fig. 15.8(b)(i) and (ii).

In considering the effect of this resultant on the soil or concrete under the base FG, it is normally convenient to resolve this force into the vertical and horizontal components W and P from which this resultant was compounded. (Note that the vertical component of the resultant is equal to the weight W of the wall only when the soil pressure is horizontal.)

These two component forces, W and P, are shown again in Fig. 15.8(b)(ii). Component force P tends to *slide* the wall along the plane FG, so need not be considered when calculating the pressure on the soil beneath FG. Component force W, however, does exert pressure on the soil or concrete under FG, and as the point S is to the left of the centre line of the base the soil under the left-hand part of the base will have greater pressure exerted upon it than the soil under the right-hand half. Here

$$\text{Direct stress} = \frac{W}{A} = \frac{\text{weight of wall}}{\text{area of base}}$$

$$= \frac{28}{2.5 \times 1.0} = 11.2 \text{ kN/m}^2$$

The moment due to the eccentricity of the resultant is

$$W \times \text{distance of S from the centre of the base}$$

$$W \times 0.25 = 28 \times 0.25 = 7 \text{ kN m}$$

and the section modulus about an axis through the centre line of the base is

$$Z = \tfrac{1}{6}bd^2 = \tfrac{1}{6} \times 1 \times (2.5)^2 = 1.04 \text{ m}^3$$

$$\text{Bending stress} = \frac{M}{Z} = \frac{7}{1.04} = 6.73 \text{ kN/m}^2$$

Thus, the pressure under wall at F is

$$\frac{W}{A} + \frac{M}{Z} = 11.20 + 6.73 = 17.93 \text{ kN/m}^2$$

and the pressure under wall at G is

$$\frac{W}{A} - \frac{M}{Z} = 11.20 - 6.73 = 4.47 \text{ kN/m}^2$$

as shown in the stress or pressure distribution diagram of Fig. 15.8(b)(iii).

Example 15.5   A masonry wall is shown in Fig. 15.9(a) and weighs 20 kN/m^3. It retains on its vertical face water weighing 10 kN/m^3. The water reaches the top of the wall. Calculate the pressures under the wall at the heel and the toe.

Solution

$$W = \tfrac{1}{2}(1.0 + 3.0) \times 4.5 \times 20 = 180.0 \text{ kN}$$

$$P = \tfrac{1}{2}wH^2 = \tfrac{1}{2} \times 10 \times (4.5)^2 = 101.25 \text{ kN}$$

Fig. 15.9 Example 15.5: (a) loading
diagram; (b) the equilibrium
about the base's centroid

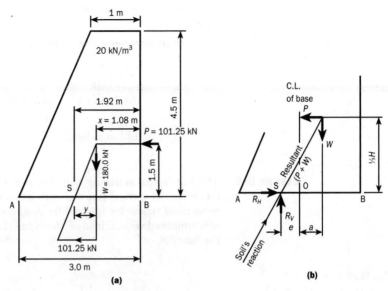

(a)

(b)

To determine the centroid of the wall,

$$x = \frac{1.0 \times 4.5 \times 0.5 + \frac{1}{2} \times 2.0 \times 4.5 \times 1.7}{1.0 \times 4.5 + \frac{1}{2} \times 2.0 \times 4.5}$$

$$= \frac{2.25 + 7.50}{4.50 + 4.50} = 1.08 \text{ m}$$

By similar triangles

$$\frac{y}{1.5} = \frac{101.25}{180.0}$$

$$y = \frac{101.25 \times 1.5}{180.0} = 0.84 \text{ m}$$

Thus, the resultant force cuts at $1.08 + 0.84 = 1.92$ m from B at S or 0.42 m to
the centre line of the base.

Hence, the pressure at A is

$$\frac{W}{A} + \frac{M}{Z} = \frac{180.0}{3.0 \times 1.0} + \frac{180.0 \times 0.42 \times 6}{1.0 \times 3.0 \times 3.0}$$

$$= 60.0 + 50.4 = +110.4 \text{ kN/m}^2$$

and the pressure at B is

$$\frac{W}{A} - \frac{M}{Z} = 60.0 - 50.4 = +9.6 \text{ kN/m}^2$$

The purpose of calculating distance y was to find the eccentricity of W, i.e.
distance e, so that $M = We$ could be determined. Distance e, however, can be
obtained directly by considering the equilibrium about the centroid of the
base of all the forces acting on the wall. This is illustrated in Fig. 15.9(b).

Let R_H and R_V be, respectively, the horizontal and vertical components of
the soil's reaction acting at the intersection of the resultant of P and W with
the base of the wall.

Then, for equilibrium, $R_H = P$, and $R_V = W$, both of which have been
calculated previously.

Hence, taking moments about O, the centre of area (or centroid) of the base:

$$W \times a - P \times \tfrac{1}{3}H + W \times e = 0$$

$$e = \frac{P \times \tfrac{1}{3}H - W \times a}{W}$$

$$= \frac{101.25 \times 1.5 - 180.0 \times (1.50 - 1.08)}{180.0}$$

$$= 0.42 \text{ m as before}$$

Example 15.6   The trapezoidal retaining wall shown in Fig. 15.10(a) weighs 22 kN/m^3 and retains on its vertical face soil with an equivalent density of 16 kN/m^3 and an angle of repose (or internal friction) ϕ of 30°. The retained soil carries a superimposed vertical load of 9.6 kN/m^2. Determine the pressure under the base AB.

Fig. 15.10 Example 15.6: (a) loading of a trapezoidal retaining wall; (b)(i) equivalent additional height of soil, (ii) determining the eccentricity, (iii) pressure distribution diagram

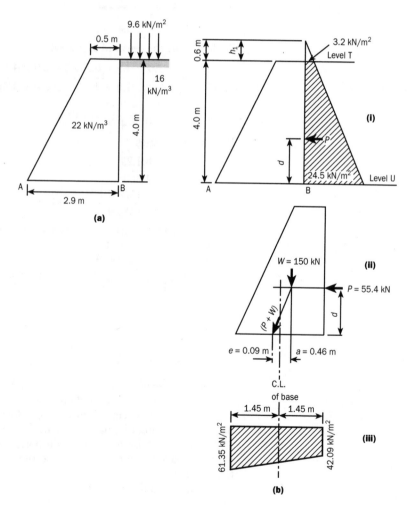

Solution The imposed load is converted to an equivalent additional height of soil as shown in Fig. 15.10(b)(i), where

$$h_1 = \frac{\text{intensity of imposed load kN/m}^2}{\text{equivalent density of soil kN/m}^3}$$

$$= \frac{9.6}{16.0} = 0.6 \text{ m}$$

Therefore, the pressure at the top of the wall, level T, is

$$w \times h_1 \times \frac{1 - \sin \phi}{1 + \sin \phi} = 16 \times 0.6 \times \tfrac{1}{3} = 3.2 \text{ kN/m}^2$$

and that on the underside of the wall, at level U, is

$$w \times (h_1 + H) \times \frac{1 - \sin \phi}{1 + \sin \phi} = 16 \times 4.6 \times \tfrac{1}{3} = 24.5 \text{ kN/m}^2$$

The force P, acting on the vertical face of the wall, is found from the shaded trapezoid of pressure and it acts at the centroid of the trapezoid, i.e.

$$P = \tfrac{1}{2}(3.2 + 24.5) \times 4 = 55.4 \text{ kN}$$

The distance d between the base of the wall and the line of action of force P is

$$d = \tfrac{1}{55.4}[3.2 \times 4 \times 2 + (24.5 - 3.2) \times \tfrac{4}{2} \times \tfrac{4}{3}]$$

$$= \frac{25.6 + 56.8}{55.4} = 1.49 \text{ m}$$

Now, the vertical force W due to the weight of the wall is

$$W = \tfrac{1}{2}(2.9 + 0.5) \times 4 \times 22 = 150 \text{ kN}$$

To find distance a between the line of action of force W and the centre line of the base, take moments about that centre line, then

$$a = \frac{0.5 \times 4 \times (1.45 - 0.25) + 2.4 \times \tfrac{1}{2} \times 4 \times (1.45 - \tfrac{1}{3} \times 2.4 - 0.5)}{0.5 \times 4 + 2.4 \times 2}$$

$$= \frac{2.4 + 0.72}{2.0 + 4.8} = \frac{3.12}{6.80}$$

$$= 0.46 \text{ m}$$

Finally, to determine the eccentricity, e, consider the moments of *all* the forces (i.e. the applied loads, P and W, and the reaction of the soil, R) about the centre line of the base, as demonstrated above. Therefore,

$$e = \frac{150 \times 0.46 - 55.4 \times 1.49}{150} = \frac{-13.5}{150}$$

$$= -0.09 \text{ m (i.e. to the left of centre line, Fig. 15.10(b)(ii)}$$

Thus, pressure at A is

$$\frac{W}{A} + \frac{We}{Z} = \frac{150}{1.0 \times 2.9} + \frac{150 \times 0.09 \times 6}{1.0 \times 2.9 \times 2.9}$$

$$= 51.72 + 9.63 = 61.35 \text{ kN/m}^2$$

and pressure at B is

$$51.72 - 9.63 = 42.09 \text{ kN/m}^2$$

The pressure distribution diagram is shown in Fig. 15.10(b)(iii).

The previous examples have been chosen so that the resultant force has cut the base within the middle third. Hence, although the pressures at B and A (the heel and toe of the base) have been of different amounts, the stresses have been compressive at both points. Consider, however, the case shown in Fig. 15.11(a)(i). The resultant here cuts outside the middle third and thus there will be a tendency to tension at B, i.e. M/Z will be greater than W/A.

Fig. 15.11 Overstressing where the resultant does not cut within the middle third: (a) resultant cuts outside middle third; (b) joint B cannot resist tension; (i) loading diagrams, (ii) pressure distribution diagrams

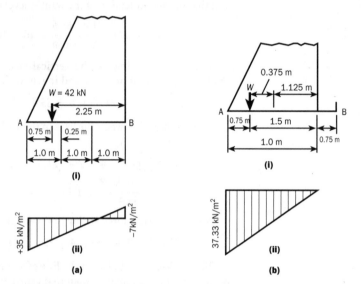

If the joint at B is capable of resisting tensile stress then the stress at A will be

$$\frac{W}{A} + \frac{M}{Z} = \frac{42}{1.0 \times 3.0} + \frac{42 \times (1.5 - 0.75) \times 6}{1.0 \times 3.0 \times 3.0}$$
$$= 14.0 + 21.0 = 35 \text{ kN/m}^2$$

and the stress at B will be

$$\frac{W}{A} - \frac{M}{Z} = 14.0 - 21.0$$
$$= -7.0 \text{ kN/m}^2 \quad \text{or} \quad 7.0 \text{ kN/m}^2 \text{ tension}$$

The diagram of pressure distribution under the base will be as in Fig. 15.11(a)(ii).

If the joint at B is not capable of resisting tension, then there will be a tendency for the base to lift where tension would occur. In this case, the point at which the resultant cuts the base should be considered the middle-third point, as in Fig. 15.11(b)(i), so that the effective width of the base becomes three times the distance AS, i.e.

$$3 \times 0.75 = 2.25 \text{ m}$$

The stress at A (the toe) then becomes

$$\frac{W}{A} + \frac{M}{Z} = \frac{42}{1.0 \times 2.25} + \frac{42 \times 0.375 \times 6}{1.0 \times 2.25 \times 2.25}$$
$$= 18.67 + 18.67 = 37.33 \text{ kN/m}^2$$

and the pressure distribution diagram is as shown in Fig. 15.11(b)(ii).

Example 15.7   For the retaining wall shown in Fig. 15.12 determine:

- The position of the intersection point of the resultant of forces P and W with the base AB when soil is level with top of wall.
- The reduced height of the soil which will cause the point of intersection to coincide with the edge of the middle third of the base.

Fig. 15.12 Example 15.7

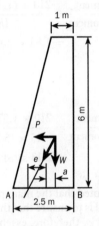

Assume:

$$w \text{ for wall} = 20 \text{ kN/m}^3$$
$$w \text{ for soil} = 15 \text{ kN/m}^3$$
$$\phi \text{ for soil} = 30°$$

Solution

$$W = \tfrac{1}{2}(1.0 + 2.5) \times 6 \times 20 = 210 \text{ kN}$$
$$P = \tfrac{1}{2}(15 \times 6^2) \times \tfrac{1}{3} = 90 \text{ kN}$$

To determine the values of a and e take moments about centre line of base.

$$a = \frac{1.0 \times 6.0 \times 0.75 - 1.5 \times \tfrac{1}{2} \times 6.0 \times 0.25}{\tfrac{1}{2}(1.0 + 2.5) \times 6.0}$$

$$= \frac{4.5 - 1.125}{10.5} = 0.32 \text{ m to right of centre}$$

$$e = \frac{90 \times \tfrac{1}{3} \times 6.0 - 210 \times 0.32}{210} = \frac{180 - 67.5}{210}$$

$$= 0.54 \text{ m} > 2.5/6 \text{ i.e. tension will develop}$$

Now to find the reduced height of the soil:

$$\tfrac{1}{2}(15 \times H^2) \times \tfrac{1}{3}H \times \tfrac{1}{3} - 210 \times (0.32 + \tfrac{1}{6} \times 2.5) = 0$$

$$H^3 = \frac{(67.5 + 87.5) \times 18}{15} = 186$$

$$H = 5.71 \text{ m}$$

The above example may be used quite conveniently to verify the final statement in the previous section on overturning. Check the stability of the above wall in both cases.

Moments are taken about the toe of the wall at A.

Soil level with the top

$$\frac{\text{Restoring moment}}{\text{Overturning moment}} = \frac{210 \times (1.25 + 0.32)}{180}$$

$$= 1.83 < 2 \text{ unsatisfactory}$$

Soil height reduced

$$\frac{\text{Restoring moment}}{\text{Overturning moment}} = \frac{210 \times 1.57 \times 18}{15 \times (5.71)^3}$$

$$= 2.13 \text{ satisfactory}$$

Finally, it should be noted that the main point of calculating the pressures under the base of the wall is to ensure that the soil on which the wall is founded is not overstressed. It is, therefore, essential to keep the actual pressures within the limits of the safe bearing capacities of the soil.

The safe bearing capacities vary over a very wide range of values from, say, 50 kN/m^2 for made-up ground to 650 kN/m^2 for compact, well-graded sands or stiff boulder clays. The values for rocks are, of course, higher.

Summary

The three *modes of failure* of a retaining wall are

- sliding
- overturning
- overstressing

The *horizontal force*, P, acting on 1 m length of the vertical face of a retaining wall is

- wind: $p \times H$
- liquid: $\tfrac{1}{2}wH^2$
- soil (or granular material), using Rankine's formula

$$\tfrac{1}{2}wH^2 \times \frac{1 - \sin\phi}{1 + \sin\phi}$$

where $\quad p$ = unit wind pressure in N/m^2

w = equivalent density (or unit weight) of retained material in kN/m^3

ϕ = angle of repose (or internal friction) of soil (or granular material)

H = height (or depth) of retained material (wind included) in metres

Resistance to *sliding*

$$\frac{W \times \mu}{P_M} = \text{factor of safety (usually 2)}$$

where $\quad W$ = weight of wall in kN

μ = coefficient of friction between wall and supporting soil (0.4 − 0.7)

P_M = the least force that will cause sliding, in kN

Resistance to *overturning*

$$\frac{\text{Restoring (balancing) moment}}{\text{Overturning moment}} = \text{factor of safety (usually 2)}$$

(Moments are taken about an outer edge — toe or heel — of the base of the wall.)

Resistance to *overstressing*

The maximum stress resulting from the combination of direct and bending stresses must be kept within limits of the safe bearing capacity of the soil supporting the wall.

For 'no tension', the resultant of the applied loads P and W must cross the base within its middle third or

$$e \not> \tfrac{1}{6} \times \text{width of base}$$

Exercises

1 Calculate the resultant horizontal force P acting on the wall shown in Fig. 15.Q1 for (a) a liquid of equivalent density $w = 10$ kN/m^3; (b) a soil of equivalent density $w = 16$ kN/m^3 and angle of repose $\phi = 30°$, given that $H =$ (i) 2 m, (ii) 3 m, (iii) 7 m.

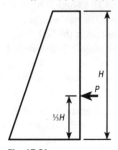

Fig. 15.Q1

2 The walls shown in Fig. 15.Q2(a) and (b) weigh 22 kN/m^3. Calculate the value of P in kN per m run of

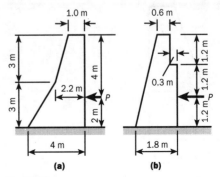

Fig. 15.Q2

wall, if there is to be a factor of safety of 2 against sliding; $\mu = 0.6$.

3 The horizontal forces on the walls shown in Fig. 15.Q3(a), (b) and (c) represent the forces on a 1 m run of wall. (All dimensions are in m.) All the walls weigh 22 kN/m³. Calculate the factor of safety against overturning of the walls. (Only the position of the vertical line in which the c.g. is situated need be calculated.)

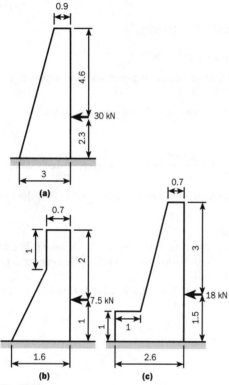

(a)

(b)   **(c)**

Fig. 15.Q3

4 The wall, shown in Fig. 15.Q4, has a resultant vertical force of 57 kN/m length, due to the weight of the wall

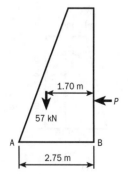

Fig. 15.Q4

W and the pressure *P*. Calculate the pressure at the toe (point A).

5 A masonry retaining wall is trapezoidal, being 4 m high, 1.4 m wide at the top and 2.0 m at the base. The masonry weighs 20 kN/m³ and the wall retains on its vertical face earth weighing 15 kN/m³ at an angle of repose of 40°. Calculate the maximum pressure under the wall.

6 A brick pier *H* m high is 450 mm × 300 mm in section and weighs 18 kN/m³. It has a uniform wind pressure of 720 N/m² on one side, as shown in Fig. 15.Q6 and rests on a concrete block to which it is not connected. How high is the wall if the resultant force cuts through the point A?

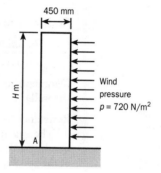

Fig. 15.Q6

7 A concrete wall of trapezoidal section, 4.8 m high, has a top width of 1.0 m and a base width of 2.5 m with one face vertical and a uniform slope on the other face. There is water pressure on the vertical face, with top water level 1 m below the top of the wall. If the water weighs 10 kN/m³ and the masonry 24 kN/m³, calculate the maximum and minimum pressures on a horizontal plane 2 m above the base of the wall.

8 A brick wall 450 m thick and 3.0 m high is built on a level solid concrete base slab. The wall weighs 20 kN/m³. Assuming no tension at the bed joint, calculate the intensity of the uniformly distributed horizontal wind force in N/m² over the full height of the wall that would just cause overturning of the wall.

9 A concrete retaining wall is as shown in Fig. 15.Q9. Using the Rankine formula, calculate the pressure under the wall at A and B. Assume the soil to weigh 16 kN/m³, and the natural angle of repose of the soil to be 30°. The concrete weighs 24 kN/m³.

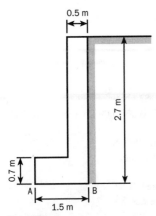

Fig. 15.Q9

10 Fig. 15.Q10 shows the section of a mass concrete retaining wall, weighing 24 kN/m³. The position and amount of the resultant horizontal force is as shown. Calculate the vertical pressures under the wall at A and B, assuming (a) tension stresses are permitted; (b) no tension stresses can develop.

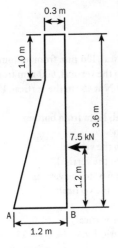

Fig. 15.Q10

11 A small mass concrete retaining wall, 2 m high, is trapezoidal in section, 0.3 m wide at the top and 1.2 m wide at the base. The wall weighs 24 kN/m³ and it retains on its vertical face soil weighing 14 kN/m³ at an angle of repose of 30°. Calculate the pressure under the base of the wall at the heel and at the toe.

12 A wall 0.23 m thick, weighing 19 kN/m³, rests on a solid foundation. It has a uniform horizontal pressure of 0.7 kN/m² on one face. If the resultant force at the base just cuts at the edge of the wall (i.e. if overturning is just about to take place), how high is the wall?

13 How high will be a wall, similar to that mentioned in Question 12, if the resultant pressure at the base cuts at the middle third, i.e. if tension is about to occur?

14 An L-shaped retaining wall is shown in Fig. 15.Q14. The masonry weighs 22 kN/m³. Calculate the maximum and minimum pressures under the wall in kN/m².

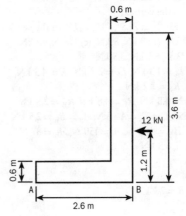

Fig. 15.Q14

Answers to exercises

Chapter 2

1 (a) 51.8 N, 73.2 N; (b) 183.0 N, 224.2 N; (c) 577.4 N, 288.7 N
2 14.5°
3 $L = 0.87$ kN, $R = 0.50$ kN, Reaction $= 1.36$ kN
4 $L = 2.04$ kN, $R = 0.75$ kN, $W = 2.88$ kN, $M = 1.06$ kN
5 19.9 kN at 31° to the vertical
6 7.3 kN at 25° to the vertical
7 9.6 kN at 81° to the vertical
8 $A = 32°$, resultant $= 17.2$ kN
9 7.3 kN at 70° to the vertical
10 559 N at 23° to the vertical
11 677 N at 17° to the vertical
12 774 N at 64°, 554 N at 22.5°, 793 at 37.5° to the vertical
13 (a) 35 kN, 40 kN; (b) 8 kN, 8 kN; (c) 17 kN, 10 kN
14 9.2 kN, $x = 22.5°$, 8.5 kN, 3.5 kN
15 $L = 2.5$ kN, $L_v = 1.3$ kN, $L_h = 2.2$ kN; $R = 4.3$ kN, $R_v = 3.7$ kN, $R_h = 2.2$ kN
16 5.5 kN, 6.7 kN, 4.8 kN, $R_L = 4.8$ kN, $R_R = 2.8$ kN
17 $T = S = 5$ kN, $A_v = B_v = 4.3$ kN, $A_h = B_h = 2.5$ kN
18 AB $= 173$ N, BC $= 100$ N, $A_v = 150$ N, $A_h = 87$ N
19 5.0 kN, 6.1 kN, 6.8 kN
20 82.5°, 6.1 kN
21 2.9 kN
22 $A = 9.0$ kN, $B = 7.3$ kN
23 0.29 kN
24 4.0 kN in inclined members, 2.3 kN in horizontal member and in cable, vertical reactions $= 2$ kN
25 BC $=$ CD $= 3.0$ kN, BD $= 3.5$ kN, AB $=$ AD $= 1.7$ kN, $R_A = 3.0$ kN
26 (a) $X = 64.6$ N down, $Y = 82.0$ N up; (b) $X = 4.6$ N down, $Y = 22.0$ N up; (c) $X = 1.6$ kN up, $Y = 3.2$ kN down; (d) $X = 0.4$ kN down, $Y = 1.2$ kN down; (e) 38.6 kN at 43° to vertical; (f) 41.4 kN at 40° to the vertical
27 13.7 kN, 11.8 kN.

Chapter 3

1 $A = 5.30$ kN, $B = 3.95$ kN
2 $R_A = 17.3$ kN, $R_B = 20.0$ kN, AC $= 20.0$ kN, BC $= 17.3$ kN, AB $= 10.0$ kN, $R_{Bv} = 10.0$ kN, $R_{Bh} = 17.3$ kN
3 Top $= 385$ N, bottom $= 631$ N (385 N horizontal, 500 N vertical)
4 Cable 1.9 kN, hinge 1.9 kN
5 Cable 1.4 kN, hinge 1.4 kN

6 $A = 0.50$ kN, $B = 0.87$ kN
7 $A = 0.67$ kN, $B = 1.20$ kN (1.00 kN vertical, 0.67 kN horizontal)
8 String 1000 N, hinge 900 N
9 Rope 6.1 kN, hinge 6.0 kN
10 500 N, 500 N
11 Cable 1.15 kN, hinge 1.53 kN
12 Rope 14.1 kN, hinge 22.4 kN
13 Rope 3.5 kN, hinge 7.9 kN
14 3.87 kN, 8.24 kN, 8.87 kN
15 $A = 7.64$ kN, $B = 5.77$ kN, AB $= 10.0$ kN, AC $= 2.75$ kN, BC $= 5.77$ kN
16 $x = 20°$
17 354 N, 250 N
18 $A = 334$ N, $B = 213$ N
19 $A = 315$ N, $B = 192$ N
20 $A = 0.37$ kN, $B = 0.78$ kN
21 433 N, 578 N, 807 N
22 4.72 kN, 4.25 kN, 2.51 kN
23 $A = 6.3$ kN, $B = 2.9$ kN
24 (a) 62.5 N at 70° to the vertical, 100 mm from bottom r.h. corner; (b) 60.0 N at 65° to the vertical, 600 mm from bottom l.h. corner; (c) 75.0 N at 48° to the vertical, 1.1 m from bottom l.h. corner
25 3.8 kN at 72° to the vertical, 1.7 m from bottom
26 6.0 kN, 6.0 m from bottom
27 38.5 kN, 41.0 kN at 32.5° to the vertical
28 60.0 kN, 118.6 kN at 37.5° to the vertical
29 $X = 11.4$ kN, $Y = 6.5$ kN at 6.6° to the vertical
30 50.0 kN, 22.4 kN at 63.4° to the vertical
31 55.0 kN, 65.0 kN
32 4.4 kN, 11.5 kN at 41° to the vertical
33 $R = 5.7$ kN at 41° to the vertical, $L = 9.4$ kN at 23° to the vertical
34 8.2 kN, 16.8 kN at 80° to the vertical
35 11.6 kN at 45° to the vertical, 8.7 kN at 71° to the vertical
36 43.3 kN, 49.8 kN at 20.5° to the vertical
37 $R = 44.2$ kN at 11.0° to the vertical, $L = 47.5$ kN at 10.5° to the vertical
38 13.2 kN, 15.4 kN at 27° to the vertical
39 $R = 40.5$ kN at 7° to the vertical, $L = 57.0$ kN at 5° to the vertical.

Chapter 4

1 110 N, 205 N
2 1.5 m

3   375 N, 225 N (upwards)
4   35 mm, 265 N
5   0.23 m
6   1.64 kN, 1.06 kN
7   4.23 kN, 1.53 kN (downwards)
8   $W = 22.5$ N, $A = 425.0$ N, $B = 237.5$ N
9   See Answers to Exercise 3
10 1.44 kN, 2.36 kN at 17.7° to the vertical
11 2.5 kN, 1.8 kN at 65.3° to the vertical
12 30.6 kN, 34.2 kN, 25.0 kN, 43.3 kN
13 6.0 kN, 5.34 kN downwards at 76.5° to the vertical
14 3.9 kN, 6.9 kN
15 (a) 0.41 m from hinge, 70 N, 176 N at 20° to the vertical;
    (b) 2.41 m from hinge, 4.12 kN, 6.10 kN at 42.5° to the
    vertical; (c) 2.27 m from hinge, 3.0 kN, 5.78 kN at 21.5° to
    the vertical
16 (a) 2.33 m from hinge, 3.5 kN, 5.7 kN at 38° to the verti-
    cal; (b) 1.8 m from the hinge, 30 kN, 39 kN at 40° to the
    vertical; (c) 3.17 m from the hinge, 15.8 kN, 21.8 kN at
    46.5° to the vertical; (d) 3.0 m from the hinge, 10.0 kN,
    22.4 kN at 27° to the vertical
17 15.0 kN, 9.92 kN at 41° to the vertical, 6.50 kN
18 300 mm, 45 000 N mm
19 (a) 50 kN, 70 kN; (b) 110 kN, 130 kN; (c) 120 kN, 10 kN;
    (d) 55 kN, 245 kN; (e) 70 kN, 150 kN
20 (a) 225 kN, 135 kN; (b) 199 kN, 249 kN; (c) 324 kN,
    108 kN; (d) 172.8 kN, 115.2 kN; (e) 221.5 kN, 60.5 kN;
    (f) 242.8 kN, 107.2 kN; (g) 523.1 kN, 231.9 kN
21 $A = 40$ kN, $B = 50$ kN, $C = 40$ kN, $D = 10$ kN,
    $E = 26.7$ kN, $F = 13.3$ kN
22 14.07 kN

Chapter 5

For solutions to Questions 1 to 31 see Figs A.1–A.31. All
    forces are in kN; '+' indicates compression and '–'
    indicates tension.

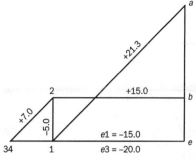

Fig. A.2

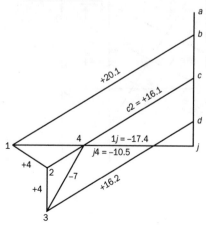

Fig. A.3

Fig. A.1

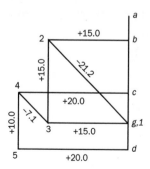

Fig. A.4

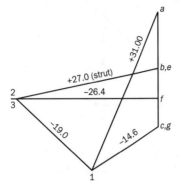

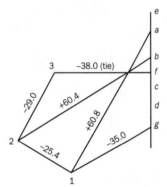

Fig. A.5

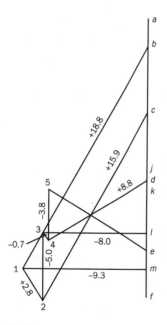

Fig. A.6

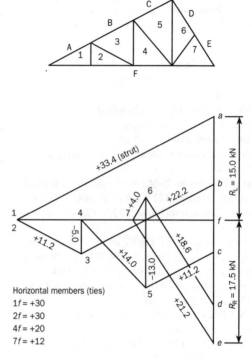

Horizontal members (ties)
1f = +30
2f = +30
4f = +20
7f = +12

Fig. A.8

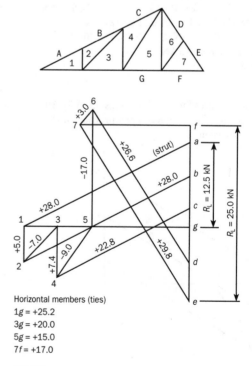

Horizontal members (ties)
1g = +25.2
3g = +20.0
5g = +15.0
7f = +17.0

Fig. A.9

Fig. A.7

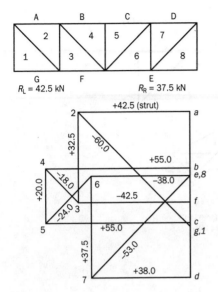

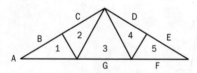

Fig. A.10

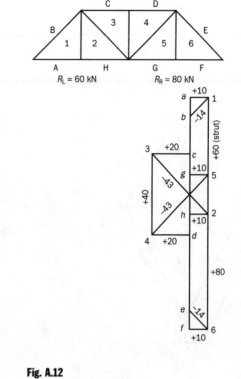

Fig. A.12

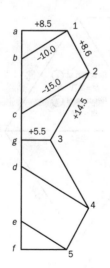

Fig. A.11

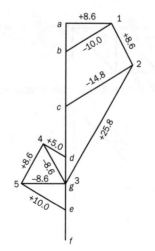

Fig. A.13

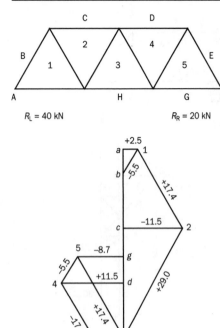

$R_L = 40$ kN $R_R = 20$ kN

Fig. A.14

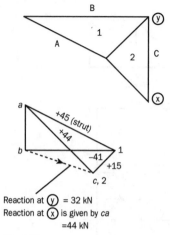

Reaction at y = 32 kN
Reaction at x is given by ca
=44 kN

Fig. A.16

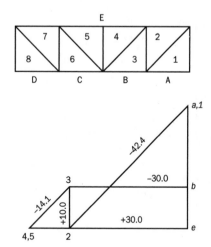

Fig. A.15

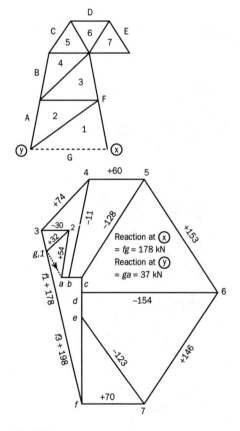

Reaction at x
= fg = 178 kN
Reaction at y
= ga = 37 kN

Fig. A.17

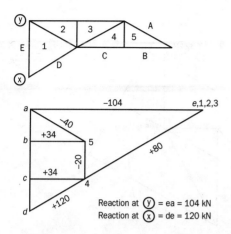

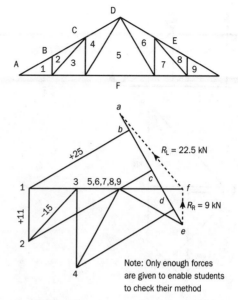

Reaction at (y) = ea = 104 kN
Reaction at (x) = de = 120 kN

Fig. A.18

R_L = 22.5 kN

R_R = 9 kN

Note: Only enough forces
are given to enable students
to check their method

Fig. A.20

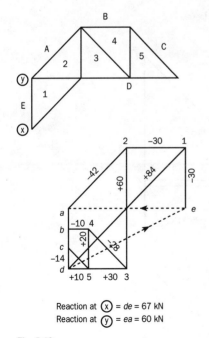

Reaction at (x) = de = 67 kN
Reaction at (y) = ea = 60 kN

Fig. A.19

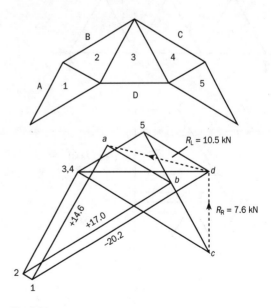

R_L = 10.5 kN

R_R = 7.6 kN

Fig. A.21

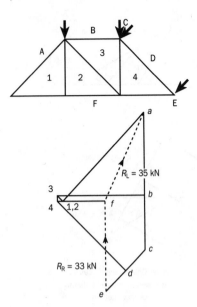

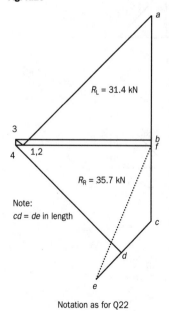

Fig. A.22

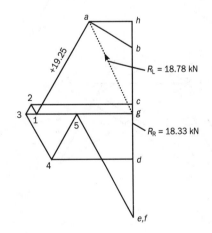

Fig. A.23

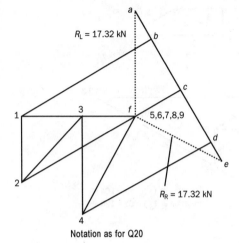

Notation as for Q20

Fig. A.24

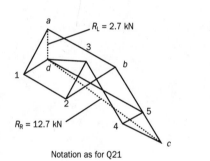

Notation as for Q21

Fig. A.25

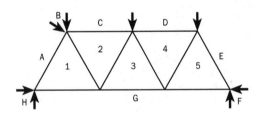

Note:
$cd = de$ in length

Notation as for Q22

Fig. A.26

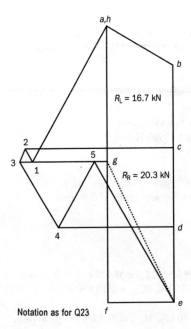

R_L = 16.7 kN

R_R = 20.3 kN

Notation as for Q23

Fig. A.27

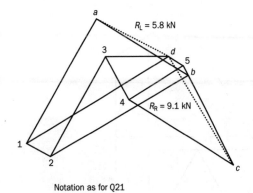

R_L = 5.8 kN

R_R = 9.1 kN

Notation as for Q21

Fig. A.29

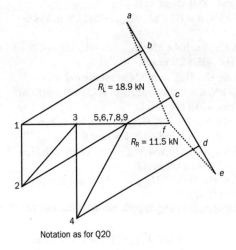

R_L = 18.9 kN

R_R = 11.5 kN

Notation as for Q20

Fig. A.28

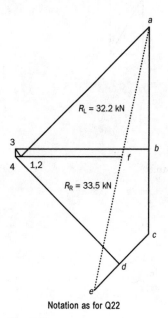

R_L = 32.2 kN

R_R = 33.5 kN

Notation as for Q22

Fig. A.30

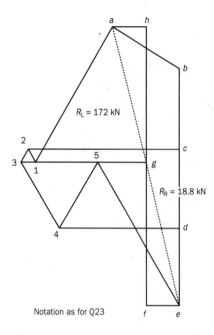

$R_L = 172$ kN

$R_R = 18.8$ kN

Notation as for Q23

Fig. A.31

32 AB = 2.39 kN ⎫
    AC = 0.98 kN ⎬ all struts
    AD = 2.25 kN ⎭
33 AB = 16.62 kN   strut
    AC = 9.82 kN   tie
    AD = 7.25 kN   tie
34 AB = AC = 0.7 kN   ties
    AD = AF = 2.43 kN   struts
         AE = 3.33 kN   tie
         BC = 0.5 kN   strut
    BF = DC = zero

Chapter 6

1   135 N/mm²
2   69.75 kN
3   8.9 mm
4   122.2 N/mm²
5   157.8 kN
6   18.5 mm
7   80 kN, 26 mm
8   55 mm, 99 kN
9   9.3 mm
10 460 mm
11 5

12 442 mm
13 (a) 68 N/mm², (b) 516 mm, (c) 2.45 m
14 (a) 251 N/mm², (b) 404 N/mm², (c) 176 839 N/mm²
15 1.5 mm
16 141 N/mm², 2 mm
17 8727 N/mm², 1.6 mm
18 146.25 kN, 1.6 mm
19 Copper 0.27 mm, brass 0.28 mm
20 15.5 N/mm², 0.001, 15 500 N/mm²
21 202.5 kN, 2 mm
22 2 × 6.6 mm
23 Timber 5.1 N/mm², steel 129.8 N/mm², 1.9 mm
24 69.4 kN, steel 106.6 N/mm², brass 41.6 N/mm².

Chapter 7

1   $L = 91$ kN, $R = 59$ kN, $M_{max} = 58$ kN m 1.03 m from L
2   $L = 145$ kN, $R = 135$ kN, $M_{max} = 151.9$ kN m 1.75 m from L
3   $L = 100$ kN, $R = 60$ kN, $M_{max} = 83.3$ kN m 1.67 m from L
4   $L = 24$ kN, $R = 36$ kN, $M_{max} = 57.6$ kN m 2.8 m from L
5   $L = 31.5$ kN, $R = 73.5$ kN, $M_{max} = +16.5$ kN m 1.05 m from L and -15.0 kN m at R
6   $L = 140.4$ kN, $R = 114.6$ kN, $M_{max} = 219.1$ kN m 3.18 m from L
7   $L = 99.6$ kN, $R = 104.4$ kN, $M_{max} = +14.3$ kN m 1.79 m from L and -35.0 kN m at R
8   $L = R = 60$ kN, $M_{max} = 96$ kN m at centre of span
9   $L = 112.1$ kN, $R = 27.9$ kN, $M_{max} = +7.38$ kN m 0.87 m from L and -30.0 kN m at L
10 $L = 120$ kN, $R = 200$ kN, $M_{max} = +120$ kN m between 1 m and 5 m from L and -120 kN m at R
11 $L = 97.5$ kN, $R = 187.5$ kN, $M_{max} = +35.1$ kN m 0.25 m from L and -120.0 kN m at R
12 $L = 75$ kN, $R = 175$ kN, $M_{max} = +5$ kN m 1 m from L and -140 kN m at R
13 $L = 169$ kN, $R = 201$ kN, $M_{max} = 282.95$ kN m 2.04 m from L
14 $L = 230$ kN, $R = 140$ kN, $M_{max} = 512.88$ kN m 4.5 m from L
15 $M_{max} = +28.9$ kN m and -80.0 kN m
16 37.63 kN m
17 2.83 m ($\sqrt{8}$ m)
18 (a) $L = 150$ kN, $R = 145$ kN; (b) $+158.75$ kN m 4.5 m from L, -70 kN at L
19 (a) $L = 142$ kN, $R = 148$ kN; (b) 181.4 kN m 2.4 m from L
20 $R = 10$ kN, $M_{max} = 30$ kN m at support
21 Max. SF $= 146.7$ kN at R, $M_{max} = 136.67$ kN m 2.5 m from L

22 (a) $L = 215.5$ kN, $R = 235.5$ kN; (b) $M_{max} = 708.25$ kN m
   5.5 m from L
23 (a) 15.0 kN m; (b) 245.0 kN m; (c) 357.5 kN m
24 400 kN m
25 (a) $R_L = 20$ kN, $R_R = 40$ kN; (b) $M_{max} = 46.20$ kN m; (c)
   $M_C = 31.11$ kN m
26 (a) $+20$ kN m; (b) -10 kN m; (c) $+5$ kN m
27 120 kN
28 6 m
29 60 kN
30 (a) $\frac{1}{2}Wl$; (b) $\frac{3}{8}Wl$.

Chapter 8
1 444 mm east and 444 mm south of A
2 (a) 22.3 mm; (b) 55.9 mm; (c) 82.1 mm; (d) 104.2 mm
3 (a) $\bar{x} = 83$ mm, $\bar{y} = 325$ mm; (b) $\bar{x} = 134.7$ mm,
   $\bar{x} = 62.9$ mm; (c) $\bar{x} = 76.3$ mm, $\bar{y} = 49.6$ mm; (d)
   $\bar{x} = 83.8$ mm, $\bar{y} = 148.5$ mm; (e) $\bar{x} = 55.0$ mm,
   $\bar{y} = 41.7$ mm
4 $I_{xx} = 381.24 \times 10^6$ mm^4, $I_{yy} = 47.50 \times 10^6$ mm^4
5 $I_{xx} = 247.30 \times 10^6$ mm^4, $I_{yy} = 126.25 \times 10^6$ mm^4
6 $a = 29.4$ mm, $I_{xx} = 4.20 \times 10^6$ mm^4,
   $I_{yy} = 9.54 \times 10^6$ mm^4
7 $a = 53.8$ mm, $I_{xx} = 11.6 \times 10^6$ mm^4,
   $I_{yy} = 2.15 \times 10^6$ mm^4
8 $a = 49.5$ mm, $b = 24.5$ mm, $I_{xx} = 6.56 \times 10^6$ mm^4,
   $I_{yy} = 2.37 \times 10^6$ mm^4
9 $a = 398$ mm, $I_{xx} = 998.79 \times 10^6$ mm^4,
   $I_{yy} = 76.00 \times 10^6$ mm^4
10 $a = 71$ mm
11 $I_{xx} = 12.13 \times 10^9$ mm^4, $I_{yy} = 3.19 \times 10^9$ mm^4
12 $x = 127.7$ mm, $I_{xx} = 199.34 \times 10^6$ mm^4
13 $x = 90.5$ mm, $I_{xx} = 117.51 \times 10^6$ mm^4
14 $x = 49.5$ mm, $I_{xx} = 6.56 \times 10^6$ mm^4
15 $I_{xx} = 37.83 \times 10^6$ mm^4, $I_{yy} = 9.02 \times 10^6$ mm^4
16 $I_{xx} = I_{yy} = 3.54 \times 10^6$ mm^4
17 $x = 37$ mm, $I_{xx} = 161.97 \times 10^6$ mm^4,
   $I_{yy} = 74.73 \times 10^6$ mm^4
18 $I_{xx} = 159.31 \times 10^6$ mm^4, $I_{yy} = 44.95 \times 10^6$ mm^4
19 24 mm
20 160 mm
21 24 mm
22 221 mm
23 $I_{xx} = 2.69 \times 10^9$ mm^4.

Chapter 9
1 25 kN
2 258 mm
3 409 kN
4 169.5 kN
5 150 N/mm^2

6 96 kN
7 6.8 kN
8 8.8 kN
9 8.88 kN/mm^2
10 (a) 4.86 N/m^2; (b) 0.156 $\times 10^6$ mm^3
11 (a) 137 mm; (b) $I_{xx} = 280.9 \times 10^6$ mm^4,
   $Z_{xxC} = 1.26 \times 10^6$ mm^3, $Z_{xxT} = 2.05 \times 10^6$ mm^3;
   (c) 25.3 kN/m
12 453 mm
13 (a) 305×165 UB40; (b) and (c) 305×102 UB83;
   (d) 610×229 UB101; (e) 457×191 UB82;
   (f) 254×146 UB37
14 (a) 51.0 N/mm^2; (b) 17.8 N/mm^2
15 152×89 UB16
16 Joists: 142 mm $\times$ 75 mm, 133 mm $\times$ 48 mm;
   A $= 305 \times 102$ UB28, B $= 254 \times 102$ UB25
17 $l = 6.4$ m, beam $= 356 \times 127$ UB39
18 22 m
19 406×178 UB74
20 305×102 UB33
21 10.5 kN.

Chapter 10
1 (a) Steel: 120 N/mm^2, timber: 8 N/mm^2;
   (b) 81.6 $\times 10^6$ N mm
2 99.5 kN
3 74.8 $\times 10^6$ N mm
4 (a) 44.9 kN; (b) 20 mm
5 592 mm
6 $d_1 = 127$ mm, $A_{st} = 475$ mm^2
7 (a) 590 mm; (b) 1520 mm^2
8 $d_1 = 289$ mm, $A_{st} = 776$ mm^2
9 $d_1 = 380$ mm, $A_{st} = 1246$ mm^2, $W = 12.5$ kN
10 $d_1 = 318$ mm, $A_{st} = 856$ mm^2
11 $d_1 = 200$ mm, $A_{st} = 844$ mm^2
12 Beam A: $d_1 = 323$ mm, $A_{st} = 865$ mm^2;
   beam B: $d_1 = 167$ mm, $A_{st} = 224$ mm^2
13 $d_1 = 430$ mm, $A_{st} = 1879$ mm^2.

Chapter 11
1 18 mm
2 14 mm
3 16 mm
4 105 kN
5 5.8 mm
6 1.5 mm
7 1.4 mm
8 5.5 mm
9 7.2 mm
10 11.5 mm
11 (a) 7.6 kN; (b) 4.3 kN

12 (a) 205 mm; (b) 247 mm
13 8.8 mm
14 7 mm
15 75 × 160 mm
16 7.5
17 127 mm
18 13.7 mm
19 200 mm × 600 mm
20 316 mm, 37.4 kN
21 10 mm.

Chapter 12

1 (a) 13.5 kN; (b) 25.9 kN
2 (a) 4.8 kN; (b) 9.9 kN; (c) 11.7 kN
3 225 mm
4 443 kN
5 255 mm
6 49.8 kN; 106.5 kN
7 (a) 1096 kN; (b) 1266 kN; (c) 1797 kN
8 11.6 m, 502 kN
9 203 × 203 UC71
10 (a) 1180 kN; (b) 1385 kN
11 1.3 MN
12 1.6 MN
13 422 mm square, 7117 mm^2
14 (a) 370 mm square, 10 994 mm^2; (b) 690 mm square, 3800 mm^2
15 587.1 kN, 1007.2 kN
16 250 mm × 290 mm, 2900 mm^2
17 980 kN
18 6597 mm^2
19 (a) 200 mm square, 2000 mm^2; (b) 4.0 m.

Chapter 13

1 264 kN (plate)
2 205 kN (rivets)
3 292 kN (rivets)
4 323 kN (plate)
5 (a) and (b) 201 kN (bolts)
6 9.3 mm, 4 bolts each in plate and column

7 $x = 76$ mm, 5 bolts
8 AA, BB 408 kN, CC 488 kN, rivets 208 kN
9 224 kN, 16 mm, 8 rivets
10 324 kN (plate).

Chapter 14

1 (a) $+6.88$ N/mm^2; (b) $+1.12$ N/mm^2
2 $f_{bc} = 6.4$ N/m^2, $f_{ad} = 0$
3 $f_{bc} = +7.04$ N/mm^2, $f_{ad} = -0.64$ N/mm^2
4 7.81 N/mm^2
5 AD $= +89$ N/mm^2, BC $= +47$ N/mm^2
6 $A = +8.8$ N/mm^2, $B = -7.4$ N/mm^2, $C = +158.4$ N/mm^2, $D = +174.6$ N/mm^2
7 160.6 kN
8 $A = +8.67$ kN/m^2, $B = +4.67$ kN/m^2
9 $A = 52.0$ kN/m^2, $B = 0$
10 (a) $+12$ N/mm^2 (bottom), $+1.33$ N/mm^2 (top); (b) 50 mm
11 72 N/mm^2
12 (a) $+8.0$ N/mm^2; (b) -5.6 N/mm^2
13 (a) -0.25 N/mm^2 (top), $+12.75$ N/mm^2 (bottom); (b) $W = 244.8$ kN.

Chapter 15

1 (a) (i) 20 kN/m^2, (ii) 45 kN/m^2, (iii) 245 kN/m^2; (b) (i) 10.7 kN/m^2, (ii) 24 kN/m^2, (iii) 131 kN/m^2
2 (a) 69.3 kN; (b) 29.7 kN
3 (a) 8.3; (b) 9.3; (c) 8.2
4 35.42 kN/m^2
5 80 kN/m^2
6 5 m
7 66.1 kN/m^2, 36.9 kN/m^2
8 1350 N/m^2
9 $A = 47.5$ kN/m^2, $B = 18.1$ kN/m^2
10 (a) 101.5 kN/m^2, 11.5 kN/m^2; (b) 102.9 kN/m^2
11 31.1 kN/m^2, 28.9 kN/m^2
12 1.44 m
13 0.48 m
14 37 kN/m^2, 20 kN/m^2.

Index

MITCHELL'S BUILDING SERIES

External Components

MICHAEL McEVOY

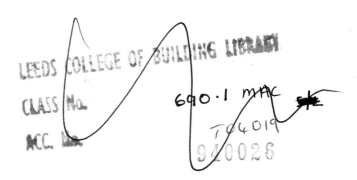

Longman
Scientific &
Technical

Longman Scientific & Technical
Longman Group UK Limited
Longman House, Burnt Mill, Harlow,
Essex, CM20 2JE, England
and Associated Companies throughout the world

First published 1994

ISBN 0-582-21255-3

British Library Cataloguing in Publication Data
A CIP record for this book is available from the British Library

Set by 4 in Compugraphic Times and Melior
Produced by Longman Singapore Publishers (Pte) Ltd
Printed in Singapore

Contents

vi   Contents

Acknowledgements

Midway through this latest revision of Mitchell's *Components* the series changed its publisher, consequently an uncommonly large number of people have been involved in its production. My apologies for omissions from this inevitably abbreviated list but acknowledgements are especially due to Thelma Nye of Batsfords for her very able editorial direction over many years, to Jean Marshall who was responsible for the majority of detailed drawings that have been retained here, to Basil Wilby, Longman's consulting editor, and to the in-house team at Harlow. Many manufacturers have been approached for up-to-date information, and acknowledgement of their assistance is given alongside the details and photographs they have provided.

Many thanks also to Alan Blanc, author of this book's companion volume *Internal Components*, for invaluable suggestions gleaned from his years spent in practice with Sylvia Blanc and his extensive teaching, on both sides of the Atlantic. The following individuals and organizations helped with advice and have proof-read the text of *External Components*; I am also grateful for their permission to include items from their own publications:

A. Smith: Prefabricated Building Manufacturers' Association, M. McLellan: Terrapin Ltd, D. Kimmins: CLASP Development Group, J. Weir and F. Cooper: Glass and Glazing Federation, S. Margolis: British Woodworking Federation, M. Malone: Aluminium Window Association, M. Schlotel: Crittall Windows Ltd, D. McNaughton: British Plastics Federation, B. Martin: Association of Rooflight Manufacturers, R. Simmons: Patent Glazing Contractors' Association, J. Hobday and C. Stewart: Pilkington Glass Ltd, P. Newman: British Flat Roofing Council, B. Jenkins: Flat Roofing Contractors Advisory Board, J. Blowers: Mastic Asphalt Council and Employers' Federation, Dr A. Strong: Single Ply Roofing Association, F. Coote: Lead Sheet Association, C. Atkins: Copper Development Association, R. Thilthorpe: Zinc Development Association, C. Jones: British Steel Strip Products, and M. Rich: Eternit Ltd.

Extracts from British Standards are reproduced with the permission of the BSI. Complete copies can be obtained by post from: BSI Sales, Linford Wood, Milton Keynes, MK14 6LE.

My particular thanks go to Steve Chapman and Roy Townend for their painstaking efforts in producing the large number of new illustrations that have been required, and to my wife and family who have for too long suffered the sounds of a word processor in action.

Michael McEvoy

Preface

The decision to divide the updating of the Mitchell's volume entitled *Components* into two parts follows the tragic and untimely death of the previous edition's author Derek Osbourn. *Components* was last revised by him in 1989 but some of the style, and pertinent illustrations, were retained from earlier Mitchell's (the 1971 and 1963 editions, by Harold King and Denzil Neal respectively). The future form of the series was under discussion at that time, notwithstanding the necessity for the dozen or so Mitchell's titles to retain an overall identity. Throughout the subsequent period, invaluable advice and continuity have been provided by Yvonne Dean, Derek's colleague at the University of North London.

Increasingly, the trend in construction is away from the crafting of details on site to designer's drawings. Instead, the procurement of buildings is becoming a matter of selection between pre-engineered components. Methods of specification and contractual relationships are changing to reflect this trend. The components market is becoming international to the extent that sub-components produced in different countries may be brought together elsewhere for assembly into the final manufactured product. The variety of choice that this implies is better able to meet varying requirements of performance and cost, as the definition of requirements is itself becoming more precise (e.g. the National Building Specification).

For all these reasons the subject of 'components' has outgrown its previous containment within one volume of the Mitchell's Building Series. The two new books, *Internal Components* and *External Components*, continue the direction suggested by the 1989 edition; in each chapter a discussion of 'reasons why' precedes Mitchell's familiar 'hands on' approach to construction. The logic of the new division generally holds good although inevitably some items (such as certain types of door) fall less clearly into one category or the other. ironmongery has been included in *Internal Components* simply because most ironmongery is operated from within buildings.

It is greatly to be welcomed that Longman, Mitchell's new publishers, are fully committed to developing the series to the requirements of today.

1994   Michael McEvoy

1 Introduction

1.1 The ecology of components

There are considerable repercussions for the global environment in the process of design and specification of building components. Energy is consumed at every stage of their fabrication and throughout their working lives. In the first instance there is the energy cost of obtaining the necessary raw materials, then manufacture, incorporation into a building and finally in terms of the efficiency of their performance in use. The relative ecological importance of these factors can be gauged from the size of the construction industry in Western economies (in Europe, it is the second largest industry after agriculture).

This environmental impact manifests itself in various ways. For example, in the short to medium term, the depletion of the world's store of oil can be expected to profoundly affect patterns of energy consumption. This will be felt in the way buildings are occupied and constructed and in the pricing of the wide range of oil-based products used in construction. Of particular and immediate concern is the extent to which tropical rain forests are being felled to supply hardwoods to the building sites of the northern hemisphere.

Alternatives to the use of non-renewable tropical hardwoods are described in Friends of the Earth's *The Good Wood Manual*. Because materials that are 'closer to nature' are more environmentally benign, most of their preferred alternatives are timbers from managed forests in Europe and North America. Some synthetic substitutes, as well as the treatments used to preserve low-durability softwoods, are associated with other problems such as pollution, high energy consumption in manufacture and health hazards. Building construction is still a major user of hardwoods (see chapters 4 and 7). Friends of the Earth advise that unless a source of supply is known to be managed and sustainable, tropical hardwoods should not be specified. The guide discusses alternatives for the detailing of timber components, such as doors and windows, to achieve durability without the use of tropical hardwoods.

Since the 1970s, concern about resource depletion has resulted in more stringent measures to reduce energy consumption being incorporated into the *Building Regulations*. Attention has recently been focused on environmental pollution and the greenhouse effect. The emission into the atmosphere of carbon dioxide, an inevitable product of the combustion of fossil fuels, is a major contributory factor. Ironically, the foamed plastics insulation materials being used in building to conserve energy, require the use in their manufacture of CFC gases that have a highly deleterious effect on the ozone layer and are also an agent responsible for global warming. Alternative insulants are polystyrene (which is CFC free), cork or mineral fibre (see section 8.2.3). Manufacturers are now producing foamed plastics that employ blowing agents other than CFCs but some of the alternative gases are also contributors to the greenhouse effect. When deciding which insulation to specify, full technical descriptions should be sought.

The internal environment within buildings is altered by the chemical content of the materials used in their construction. Little is known about the health effects of exposure to the long-term release of gases from organic compounds such as wood-based boards and plastics, although this is a current cause of concern. Some chemicals and fibres used in building materials produce an allergic reaction in some people but only universally dangerous materials such as asbestos, formaldehyde, radon and lead are subject to regulation. Manufacturers are required to display on their products the presence of hazardous materials in order to conform with *Health and Safety at Work* and *Control of Substances Hazardous to Health Regulations*.

In response to the need to make construction more environmentally responsible, the Building Research

Establishment (BRE) have introduced a method of rating building performance — *BRE environmental assessment method* (BREEAM). At present this is applicable only to office blocks but versions for other building types are being developed. The method evaluates aspects of design for which there is good evidence of the environmental problems they can cause, and credits are given under three headings:

1. *Global issues*: global warming (carbon dioxide emission per year), ozone depletion (absence of CFCs in insulation materials and refrigerating systems, halons in fire-fighting equipment), rain forest destruction (timber obtained from sustainable sources) and resource depletion (use of recyclable materials).
2. *Neighbourhood issues*: Legionnaires' disease (from cooling towers), local wind effects, reuse of existing sites.
3. *Indoor issues*: Legionnaires' disease (from water supplies), lighting, indoor air quality, hazardous materials (absence of lead in paint, asbestos products, urea formaldehyde insulation).

These topics obviously have considerable implications for the design of the external envelope of buildings. The BRE have also designed a domestic energy model (BREDEM) which is the basis of energy classification schemes such as the *national home energy rating* (NHER), developed by the National Energy Foundation.

Building components and materials not only vary in terms of their energy performance in use but also have different levels of *embodied energy* (the amount consumed in extracting raw materials, fabricating components and incorporating them into buildings). Several studies have been made of the energy used to produce building materials, demonstrating a ratio of 1 : 23 between the lowest and highest levels of embodied energy. Aluminium, for example, uses particularly high levels of energy in its extraction whilst employing only 10 per cent of the excavated material. On the other hand, to a large extent, the countries producing aluminium do so using hydroelectricity rather than energy derived from fossil fuel. Conversely, although timber's embodied energy level is much lower than that of metals, an increased use of quick-growing softwood for structural purposes would require its treatment with toxic chemicals. Growing more trees would however, as a result of increased photosynthesis, positively reduce the concentration of atmospheric carbon dioxide. The discussion concerning embodied energy and building components is a complex and contradictory one which is as yet unresolved. A recent reference is *The Green Construction Handbook* by Ove Arup and Partners, published by JT Design and Build.

A comprehensive standard has been introduced for environmental management systems, BS 7750:1990. This describes how companies can quantify the environmental impact of their activites to arrive at a total environmental performance rating and so reduce waste, pollution and the consumption of raw materials and fuel. The design of products should also minimize the environmental results of their manufacture, use and eventual disposal. An overall environmental strategy is established and independently assessed; the company is then entitled to use the British Standard kitemark (see *MBS: Internal Components*).

1.2 Durability

A related issue is the assessment of the total lifetime cost of building materials to arrive at a specification suitable for a building's use and anticipated lifespan. The factors included in this calculation are initial purchase and construction, fuel consumption and running costs, maintenance costs and the cost of eventual removal and disposal.

Life-cycle costing is recommended in BS 7543:1992 *Guide to durability of buildings and building elements, products and components* to determine at what stage it is no longer economically viable to retain and repair rather than replace components or whole buildings. BS 7543 was necessitated by the expanding range of materials used in building and the increasingly sophisticated methods available to understand the agents that cause deterioration. Durability as defined within the standard as 'the ability of a building and its parts to perform its required functions over a period of time and under the influence of agents'.

The various types of agent are:

- *Weathering agents* such as changes in ambient temperature, water, atmospheric gases, air contaminants such as pollution or sea spray, freeze/thaw cycles and the wind.
- *Biological agents* including micro-organisms, insects and plants.
- *Stress agents* whether causing sustained stress such as structural creep or intermittent stresses such as earthquakes.
- *Chemical* and *physical agents* such as incompatible materials or soil contaminants.
- *Use agents*, i.e. normal wear and tear due to everyday use.

The British Standard outlines a procedure to predict the anticipated *service life* of products, components and buildings by reference to previous experience. Also, by measuring rates of deterioration during accelerated tests and making projections from the results (qualified of course by the expectation of proper maintenance and adequate detailing). 'Service life' is the actual period during which no excessive expenditure is required on the operation,

Designer's statement to client giving information on the proposed durability of the building and its parts

General details:
Site: 32 High Road
Date: 1.1.94
Drwgs: 100 Sk — 1312

Design life of Building:
Category: normal life (minimum 60 years)
Maintenance provision:
lifetime scheduled maintenance & repair

Basic site data:
Basic wind speed: 46 m/s
Driving rain index: moderate exposure
Other factors: 1 in 50 year flood risk

Construction elements	Materials/type	Category of design life			Maintenance, repair and required replacements (anticipated)	Means of maintenance/ repair	Category of failure effect (a) danger of collapse (d) costly repair (e) costly because repeated (f) interrupts use of building
		replaceable	maintainable	lifelong			
Substructure:							
Foundations	strip			✓	—		a, d, f
ground floor slab	*in situ*/solid			✓	—		d, f
Walls:							
exterior to 1st floor	brick/cavity		✓		repoint—every 50 years	scaffold	e
above 1st floor to eaves	boarding/conc. block walls		✓		re-stain—every 5 years		e
internal walls	block/plastered		✓		redecorate—every 5 years	trestles	e
Superstructure:							
upper floor: timber joist	timber joists/ chipboard panel			✓	—		a, d, f
Roof:							
structure	timber joists			✓			d
system	felt/warm roof	✓			felt replaced — 20 years	ladder access — temp. scaffold rail required	
Finishes:							
floors	carpet	✓			replacement — 15 years	ladder/trestles	e
plastered walls	emulsion		✓		repaint — every 5 years	ladder/trestles	e
plasterboard ceilings	emulsion		✓		repaint — every 5 years	ladder/trestles	e
woodwork	oil/alkyd paint		✓		repaint — every 5 years	ladder/trestles	e

Services: An example of a *Design life data sheet* as described in BS 7543:1992 *Guide to durability*…

maintenance or repair of a component or construction.

Also included in the British Standard is a format for *Design life data sheets* to be completed by the designer in order to alert the client to the implications, in terms of durability and requirements for maintenance, contingent on the choice of a particular component. The data sheet requires the material or type of construction to be categorized according to:

- *Design life of components* (the period of use intended by the designer):
  1. *Replaceable*
     Shorter life than the building's life — e.g. floor finishes, services installation
  2. *Maintainable*
     To last with periodic attention during the building's life — External cladding, doors and windows
  3. *Lifelong*
     To last for the life of the building — Foundations and main structure

- *Required levels of maintenance*:
  1. *Repair only*
     Maintenance restricted to restoring items to their original function after a failure — e.g. reglazing windows
  2. *Scheduled maintenance and repair*
     Scheduled work carried out at a predetermined interval of time — e.g. painting external joinery
  3. *Condition-based maintenance and repair*
     Carried out as a result of knowledge of an item's condition — e.g. 5-year inspection of historic churches

- *Categories of failure effects*:

	Effect	Example
1.	Danger to life	Sudden collapse of structure
2.	Risk of injury	Loose stair nosing
3.	Danger to health	Serious damp penetration
4.	Costly repair	Extensive scaffolding required
5.	Costly because repeated	Window fastening replacement
6.	Interrupts building's use	Heating failure
7.	Security compromised	Broken door latch
8.	No exceptional problems	Replacement of light fittings

Concern about the precise durability of components has been particularly marked in the case of house construction. The durability and maintenance requirements of components employed by many housing associations are now governed by the the Housing Association Property Mutual (HAPM) *Component Life Manual*. Using life-cycle costing techniques, components have been graded to achieve insurance cover between 'grade A' (35 years) and 'grade U' (uninsurable).

General rules in achieving durability are to use materials that have proved their longevity in the past, to employ forms of building that protect more vulnerable components (e.g. eaves of roofs projecting over and protecting walls), and constructions appropriate to local climatic conditions, also to establish that components are predicted to be durable by Agrément certificates (see *MBS: Internal Components*) or by conformity with British Standards.

1.3 Note on European standards

See also *MBS: Internal Components* section 1.4 *Component testing and quality assurance*.

Throughout this book reference is made to British Standards which define national product standards in the UK and *British Standard Codes of Practice* for design and workmanship. As a result of the Single European Act which came into force in 1987, new technical standards consistent throughout the EC are soon to be introduced.

CEN standards   The *CE mark* will mean that a product complies with a harmonized European standard written by CEN (the European standards organization), or that it has obtained a *European Technical Approval*, described below, or that it complies with a national standard that has achieved recognition at community level.

European Technical Approval (ETA) offers an alternative way, other than compliance with CEN standards, for manufacturers to have their products CE marked. The nearest UK current equivalent is the British Board of Agrément certificate which is given following an independent assessment of goods for which there is no British Standard. As in the case of Agrément certificates, the cost of assessment and testing will be borne by the manufacturer. To obtain an ETA, the product will have to conform with the 'essential requirements' of the *EC's Construction Products Directive* (i.e. mechanical resistance and stability, fire safety, hygiene and health, safety in use, protection against noise and energy efficiency). As a result, if a component has a CE mark it must be accepted under the building control regulations of each member state.

Among the first building products for which CEN standards have been introduced are profiled steel panels for cladding (see section 8.4.3). The Approved Document to support Regulation 7 of the *1991 Revision to the Building Regulations* recognizes the forthcoming necessity for materials and workmanship to comply with European harmonized standards or European Technical Approvals. A complete interpretation of the subject is contained in *The Building Regulations Explained and Illustrated*, 9th edition by Vincent Powell-Smith and M.J. Billington.

The general requirements for the manufacture and dimensional coordination of components is described in *MBS: Internal Components* chapter 1.

2 Prefabricated building components

2.1 Introduction

Since the beginnings of the modern movement in architecture, the *industrialization of building* has been advocated and explored. These attempts to remove the production of components from site to factory have at their most ambitious resulted in the prefabrication of complete buildings (see photos 2.1 and 2.2) including composite wall or roof panels with exterior and interior pre-finished surfaces, intended to achieve all the requirements of the external environmental envelope.

The commonly perceived *advantages* of industrialization for the construction industry are:

- One-half of the on-site *cost* of conventional construction is manual labour; this can be significantly reduced by prefabrication.
- Removal from the building programme's critical path of labour-intensive activities results in *faster construction* and earlier completion.
- The greater the amount of work that can be prefabricated the greater will be the benefit of the superior equipment and organization of the factory. An increase in *quality* results from the use of better production tools, careful choice of materials bought in bulk and better quality control.

In several significant ways construction differs from manufacturing industry, resulting in the *difficulties* that have beset the industrialized production of buildings:

- Buildings are expected to have a long lifespan and their pattern of use may change requiring them to be flexible, in both respects contrary to the usual characteristics of consumer products.
- A building has its own distinctive requirements and location that mitigates against standardization. Increasingly, catalogue ranges of individual factory-produced components are standardized whilst still allowing design freedom when determining the overall scheme.
- The building site is often cold, wet and uncomfortable. In terms of working environment there are considerable advantages in moving as many activities as possible to the factory.
- Buildings have many functional requirements and consequently a large number of skills are involved in making them, between 20 and 30 different trades being quite usual. The success of industrialization depends on the extent to which comprehensive assemblies can be produced that perform a number of functions subject to the preconditions outlined above.
- Manufacturing production requires a single management responsible for design and production. Buildings have traditionally been the result of the independent inputs and divided interests of the client, designer and contractor being brought to bear at various stages of the design and its construction.

2.2 Economic and social factors

For industrial processes to be viable there are several underlying requirements:

- To make the best use of equipment, production is centralized at a factory, sized and located to be best able to achieve *economies of scale*.
- The greater the production volume the lower the *unit cost*. The investment in plant is then spread over the large number of items produced.
- The output is to as large an extent as possible *standardized*. Large-volume production of a standardized product allows specialization of the staff employed and yet greater productivity as a result. The building products industry is, however, largely organized in favour of batch production rather than manufacturing large, standardized volumes.

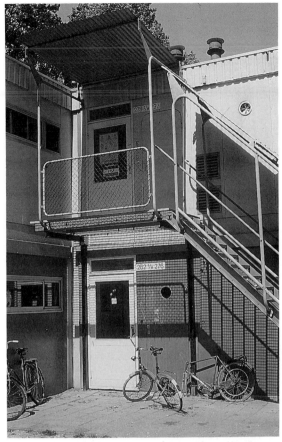

Photos 2.1 and 2.2 Prefabricated 'semi-permanent' student housing at the Delft University of Technology, Holland. Architects: 'Gimmie Shelter', 1978–81.

The 224 units were constucted from stacked lightweight steel framed boxes.

- A sophisticated organization is required to plan the design, production and marketing of the product which must be co-ordinated by, and in the charge of, a single management.

In recent years there has been little enthusiasm for the large-scale production of completely factory prefabricated multi-storey flats either in this or other Western countries. Most of these systems were designed for high-volume public housing programmes which have greatly reduced in number. The subsequent private sector dominated housing market has been too volatile to justify the investment risk. Also, the previous generation of industrialized construction produced repetitive and monotonous buildings that proved unpopular and of poor technical performance. In Japan, however, the majority of new housing now comes to site in the form of modular units that are capable of being very quickly assembled.

2.3 Open and closed systems

The terms *open system* and *closed system* are used to describe different approaches to industrialized building. An open system of building is where components made by one manufacturer may be combined with those produced by a different company. The purpose of this is to allow the designer a choice of a range of components, which are interchangeable over a wide selection of building types.

However, the success of an open system depends on the extent to which dimensional co-ordination and universal methods of jointing are accepted. It is also necessary for manufacturers to work to a series of standard sizes, including thicknesses, for specific groups of components. Open systems based on nation-wide standards have been adopted in some Scandinavian countries but aesthetic objections and the costs associated with establishing the system have precluded their use elsewhere. Open systems offer the possibility of great economies of scale but also with some wastage of material; standardization implies over-specification as a result of restricted choice.

In a closed system of building the components are not interchangeable with any other system, the building is assembled from components specifically designed for and applicable to a particular building type, for example schools or housing. Although a closed building system is carefully developed to meet the requirements of its users, the designer's choice of components is limited by the variations allowed within the particular system. The notion of client participation at the design stage (as in local authority school

For office loadings including partitions

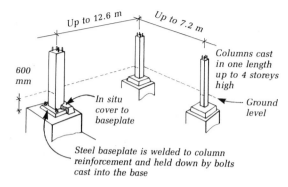

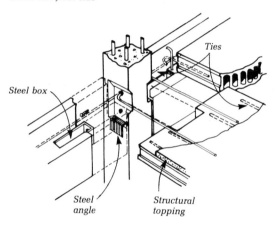

Junction between edge beam and floor slab

Junction between edge beam and spine beam

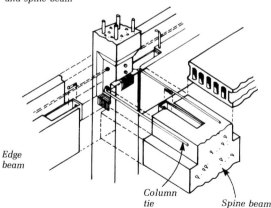

Fig. 2.1 Details from Bison's precast concrete frame system (Bison Frames Ltd)

programmes) and close co-operation with component manufacturers is to produce buildings which perform their function within precise economic limits.

The restrictions imposed by closed systems and the difficulty of establishing completely open systems have led to the development of *flexible (open/closed) systems*. These comprise a fixed series of components that are combined in a prescribed way but with considerable freedom in layout. Some of these are structural systems only, the cladding and finishes being applied by conventional on-site application

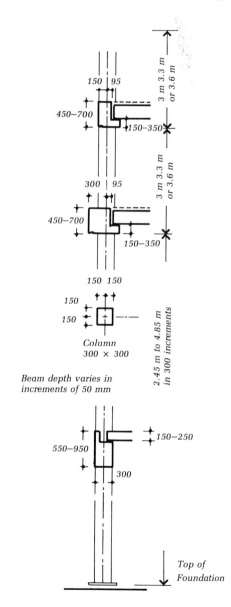

Fig. 2.2 Range of frame dimensions within Bison's precast system (Bison Frames Ltd)

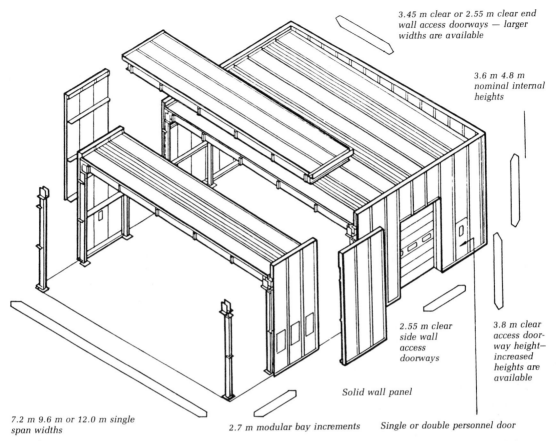

*3.45 m clear or 2.55 m clear end
wall access doorways — larger
widths are available*

*3.6 m 4.8 m
nominal internal
heights*

*2.55 m clear
side wall
access
doorways*

*3.8 m clear
access door-
way height—
increased
heights are
available*

Solid wall panel

*7.2 m 9.6 m or 12.0 m single
span widths*

2.7 m modular bay increments   *Single or double personnel door*

(a) Terrapin Matrex Light Industrial Building System: mass produced modules which can be transported in 'packs' and then erected and linked on site. Modules of lightweight galvanized cold-rolled steel sections, with cladding and roof panels of rigid polyurethane or isocyanurate foam insulation between steel facing sheets (U = 0.6 W/m²K) finished with PVC 'Plastisol'. Window and door frames are of extruded aluminium alloy sections, satin anodized finish. Each roof unit can incorporate a 900 mm wide translucent double-skinned glass fibre panel to provide daylighting

Fig. 2.3 Terrapin Matrex light industrial building system (Terrapin Ltd)

(see the example shown in figs 2.1 and 2.2). Traditional forms of enclosure such as brickwork can be employed, although this undermines the potential efficiency of industrialized methods. Many systems of building which were initially closed became more flexible, mainly by combining the supply of components and fittings across various systems.

2.4 Contractor-sponsored and client-sponsored systems

Systems of building have been developed and sponsored by both contractors and client organizations. In the case of contractor sponsored systems, sometimes referred to as *proprietary systems*, the contractor provides a building of the type in which they specialize, including delivery and

installation on site. Examples of these 'modular' constructions are timber-framed housing systems; dry assembled industrial/commercial systems specifically for agricultural, office and educational buildings; and smaller-scale prefabricated and/or temporary 'volumetric' or 'pod' systems used for toilets and site offices, etc. Also included in the latter category are air-supported and tent structures. Some examples are illustrated and described in figs 2.3−2.6.

'Modular' systems have a limited range of plan dimensions and structural spans and can be clad externally and finished internally in a prescribed number of alternative materials. Shown in fig. 2.3(a) is a system originally intended for use in industry as unit factories. This has been further developed (fig. 2.3(b)) into a low-rise steel-framed assembly principally employed as offices.

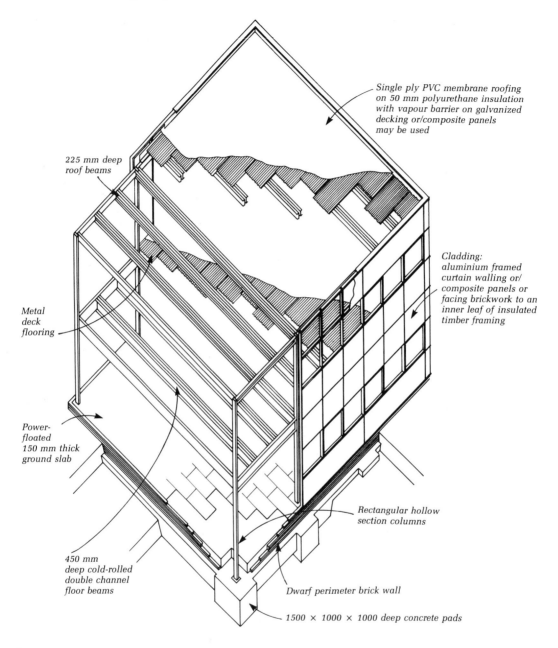

Single ply PVC membrane roofing on 50 mm polyurethane insulation with vapour barrier on galvanized decking or/composite panels may be used

225 mm deep roof beams

Cladding: aluminium framed curtain walling or/ composite panels or facing brickwork to an inner leaf of insulated timber framing

Metal deck flooring

Power- floated 150 mm thick ground slab

Rectangular hollow section columns

450 mm deep cold-rolled double channel floor beams

Dwarf perimeter brick wall

1500 × 1000 × 1000 deep concrete pads

(b) Terrapin r.b. series steel-framed building system. A dry- assembly building system using cold-formed steel beams combined with hot-rolled hollow section steel columns, principally used for offices

Fig. 2.3 *continued*

The structure comprises rectangular hollow section steel columns and double beams made from cold-formed, galvanized channel sections. Since the buildings are light in weight, their foundations are mass concrete pads connected by a trench-fill foundation to a perimeter wall that rises to floor level and forms a shutter to the ground-floor slab. The superstructure is 'dry' construction. The upper floors are of steel decking fixed to the beams, the floor surface being carpet on chipboard that is bonded to rigid polystyrene insulation with a backing sheet of plasterboard (to reduce sound transmission). The roof can be waterproofed either with a single ply PVC membrane or steel-faced insulated composite panels. The elevations can be clad in either curtain walling, composite panels or brickwork. This type of lightweight modular system using factory prefabricated elements dry-assembled on site is well suited to contemporary 'fast-track' construction (see below), and is responsible for a considerable volume of building in the UK.

In the case of client-sponsored systems, because of the necessity for a large and continuing building programme and in order to make a particular approach viable, various local authorities (which included new towns as well as government departments) created associations to develop

Fig. 2.4 Glasdon Modular Building System: small scale factory made module being off-loaded to site

systems principally for building schools and housing. The last remaining of these client-sponsored systems is CLASP (see below).

2.5 Consortium of Local Authorities Special Programme (CLASP)

In 1957 CLASP was invented in the Nottinghamshire county architect's office when eight local authorities agreed to develop a prefabricated system of school building exploiting the potential of industrialized methods to achieve economy, quality and speed. It was also required to solve a specific technical problem, how to design school buildings for sites which were subject to mining subsidence.

The light steel frame structure of CLASP comprises lattice floor and roof beams. The Mark 1 to 4 versions included a prefabricated timber roof deck and floors of timber construction either prefabricated or made *in situ*. In common with the later versions of the system, the cladding materials included precast concrete slabs with exposed aggregate facing, tile hanging, metal sheeting or timber boarding on timber cladding frames. The window frames which were factory glazed later incorporated gasket glazing; opening lights were in metal frames. In 1971 CLASP introduced the Mark 5 version with a number of modifications, including a steel deck diaphragm roof instead of timber panels and much enhanced precautions against the spread of fire.

Fig. 2.5 Glasdon Modular Building System: available in a range of sizes from 1.2 × 1.2 up to 18 × 5.4 m composed of GRP panels (see figure 2.6). Suitable for a range of building types, including offices, gatehouses, security posts, car park kiosks and sheds

In 1982 CLASP launched Mark 6 (see fig. 2.7 and photo 2.3) in response to changing patterns of building procurement; it was agreed that the revised system should have:

- The maximum benefits of a standard closed system but with more design options for the external appearance of buildings.

- Simplified technical design, administration and documentation.
- Reduction in the number of site activities and shorter component delivery times leading to reductions in site labour and contract periods.
- Improved maintenance costs and costs in use including those of energy.
- Commercial exploitation potential for use both at home and for export.

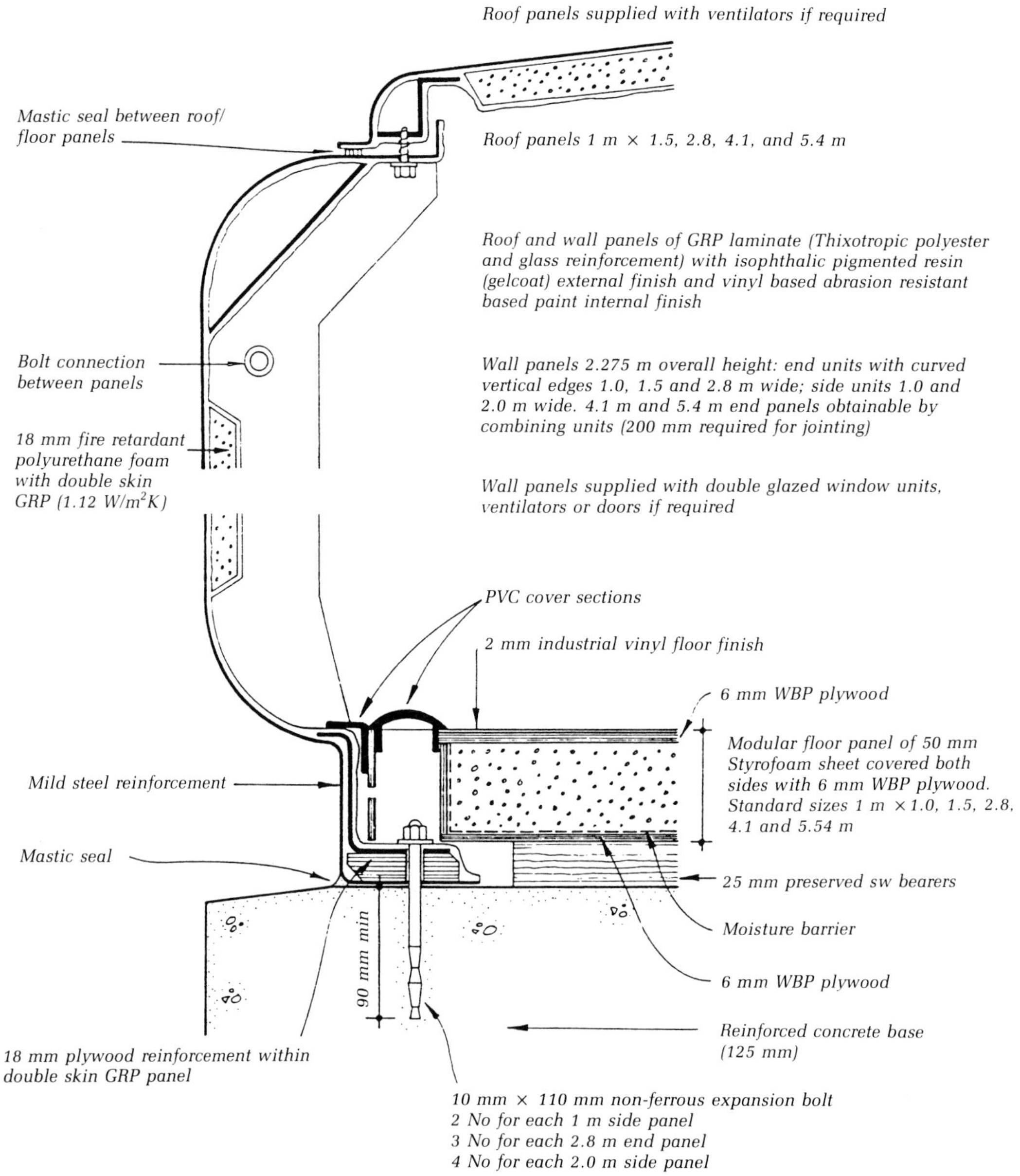

Roof panels supplied with ventilators if required

Mastic seal between roof/ floor panels

Roof panels 1 m × 1.5, 2.8, 4.1, and 5.4 m

Roof and wall panels of GRP laminate (Thixotropic polyester and glass reinforcement) with isophthalic pigmented resin (gelcoat) external finish and vinyl based abrasion resistant based paint internal finish

Bolt connection between panels

Wall panels 2.275 m overall height: end units with curved vertical edges 1.0, 1.5 and 2.8 m wide; side units 1.0 and 2.0 m wide. 4.1 m and 5.4 m end panels obtainable by combining units (200 mm required for jointing)

18 mm fire retardant polyurethane foam with double skin GRP (1.12 W/m²K)

Wall panels supplied with double glazed window units, ventilators or doors if required

PVC cover sections

2 mm industrial vinyl floor finish

6 mm WBP plywood

Modular floor panel of 50 mm Styrofoam sheet covered both sides with 6 mm WBP plywood. Standard sizes 1 m × 1.0, 1.5, 2.8, 4.1 and 5.54 m

Mild steel reinforcement

Mastic seal

90 mm min

25 mm preserved sw bearers

Moisture barrier

6 mm WBP plywood

Reinforced concrete base (125 mm)

18 mm plywood reinforcement within double skin GRP panel

10 mm × 110 mm non-ferrous expansion bolt
2 No for each 1 m side panel
3 No for each 2.8 m end panel
4 No for each 2.0 m side panel

Fig. 2.6 Glasdon modular building systems: construction details

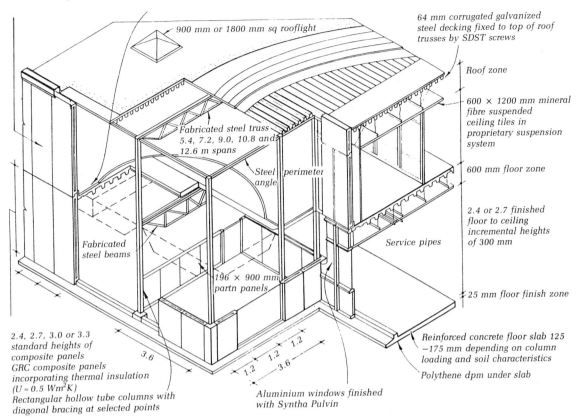

External wall cladding panels of metal siding in situ masonry with tiles or boards or composite panels

900 mm wide and 900 mm long (heavy deck capacity) or 1.8 m long (normal deck capacity) precast concrete decks supported on steel floor beams
25 mm screed and floor finish zone

900 mm or 1800 mm sq rooflight

64 mm corrugated galvanized steel decking fixed to top of roof trusses by SDST screws

Roof zone

Fabricated steel truss 5.4, 7.2, 9.0, 10.8 and 12.6 m spans

600 × 1200 mm mineral fibre suspended ceiling tiles in proprietary suspension system

Steel perimeter angle

600 mm floor zone

Fabricated steel beams

2.4 or 2.7 finished floor to ceiling incremental heights of 300 mm

Service pipes

196 × 900 mm partn panels

25 mm floor finish zone

2.4, 2.7, 3.0 or 3.3 standard heights of composite panels GRC composite panels incorporating thermal insulation (U = 0.5 Wm²K)
Rectangular hollow tube columns with diagonal bracing at selected points

3.6

1.2 *1.2* *1.2* *1.2* *3.6*

Aluminium windows finished with Syntha Pulvin

Reinforced concrete floor slab 125 −175 mm depending on column loading and soil characteristics
Polythene dpm under slab

Fig. 2.7 CLASP: Consortium of Local Authorities Special Programme, Mark 6. A comprehensive description of this latest system can be found in 'CLASP: An Introduction to the System'.

2.5.1 Dimensional co-ordination

Horizontal co-ordinating planes As indicated in fig. 2.7, the basic module used by CLASP is 100 mm (= M the internationally agreed basic modular dimension for building, see *MBS: Internal Components* section 1.8 *Modular co-ordination*), the planning grid 300 × 300 mm (3M × 3M) and the structural grid 1.8 × 1.8 m (18M × 18M) — see fig. 2.8.

Columns These are normally located with centre-lines on intersections of the structural grid. The external wall may change direction at any 1.8 m (18M) grid position compatible with the perimeter column spacings of 1.8 and 3.6 m (18M and 36M), with a maximum of 5.4 m (54M) if supporting pitched roof trusses (see figs 2.9 and 2.14).

The maximum length of continuous structure is 45 m (450M), after which expansion joints are required.

Partitions These are 200 mm (2M) thick centred on the planning grid or 100 mm (1M) thick with one face on the planning grid. Openings in partitions for door frames or screens have 100 mm (1M) flexibility of position, but the relationship of partitions to internal linings is such that the junction normally occurs at a 300 mm (3M) interval.

Floor beams These have spans of from 1.8 to 9 m (18M to 90M,) in 1.8 m (18M) increments and within a 600 mm (6M) deep floor zone (fig. 2.9).

Roof zone Within a 600 mm (6M) roof zone, the maximum span for flat roofs is 9 m (90M) when using

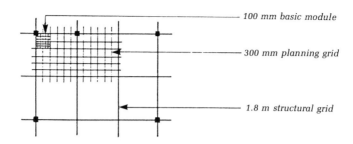

REFER TO STEEL FRAME AND PITCH ROOF FOR BEAM SPANS

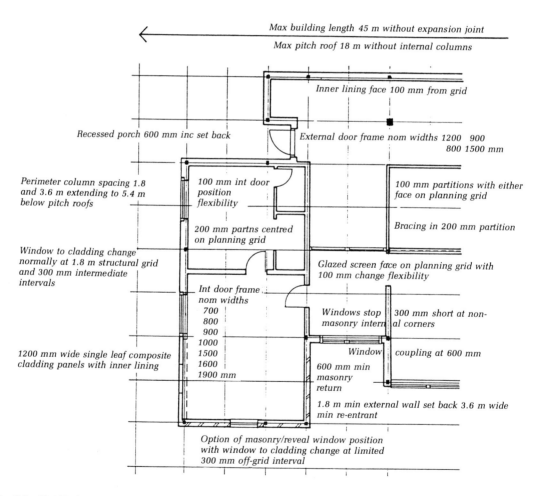

Fig. 2.8   CLASP: horizontal co-ordinating planes using basic module of 100 mm

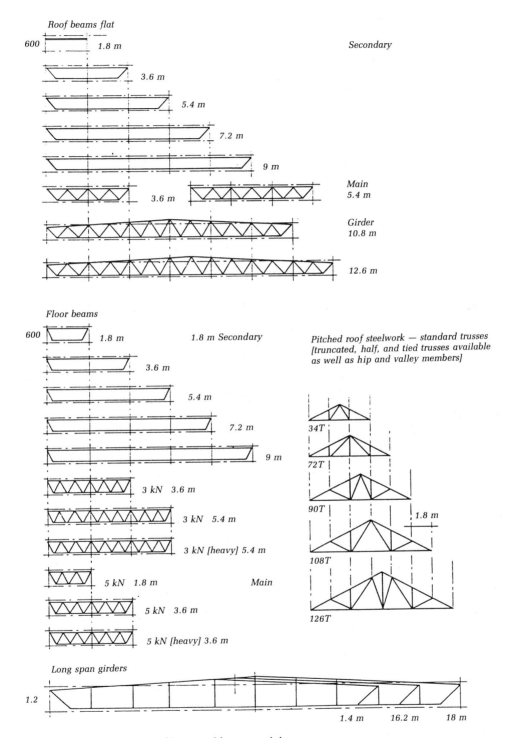

Fig. 2.9 CLASP: range of floor and roof beams and long-span girders

Photo 2.3 CLASP Mark 6: the completed frame. Photo courtesy of the CLASP Development Group.

beams, or 12.6 m (126M) using lattice girders (fig. 2.10). Roof beams of 3.6 m (36M) and below are flat, and above are generally cambered. Sports or assembly halls may require the use of the long-span girders which fit within a 1.2 m (12M) roof zone. Pitch roof spans are from 5.4 to 12.6 m (54M to 126M), in 1.8 m (18M) increments.

Vertical co-ordinating dimensions   The basic module is 100 mm (1M), and the range of vertical flexibility is indicated in fig. 2.11. The floor zone and the roof zone are both 600 mm (6M), except in the case of long-span girders (12M) and pitched roofs. The normal maximum building height is three storeys, subject to ground conditions. Buildings of four storeys are possible but may involve modifications and limitations on the components to be used for external walls. Changes of ground level are standardized at 600 mm (6M) and 1.2 m (12M), although larger changes on a 300 mm (3M) increment are possible.

2.5.2 Construction

This may vary with site conditions and/or project.

Ground slab and foundations   Normally a 125 mm (sometimes a 175 mm for heavier loading) *in situ* reinforced concrete slab is laid on a blinded sub-base and polythene sheet. The polythene sheet is not designed as a damp-proof membrane but acts as a slip joint between the slab and the ground if a site is subject to mining subsidence, and generally prevents the migration of fines from the concrete into the sub-base. The square-section steel columns are usually fixed at the slab by a steel base pin resin-grouted into a hole drilled in the concrete (see fig. 2.12), although separate column bases may be provided when loading conditions are high. In situations where ground movement created by subsidence may occur, floor finishes are chosen which are not affected by damp (e.g. hot applied bitumen compounds or asphalt). The floor finish zone is 25 mm which is insufficient for the use of screed.

Steel frame   The frame consists of galvanized tubular columns with fabricated horizontal steel beams, trusses and ties. All these components and their connections have been designed to accommodate ground movements, including those resulting from mining subsidence. Although the

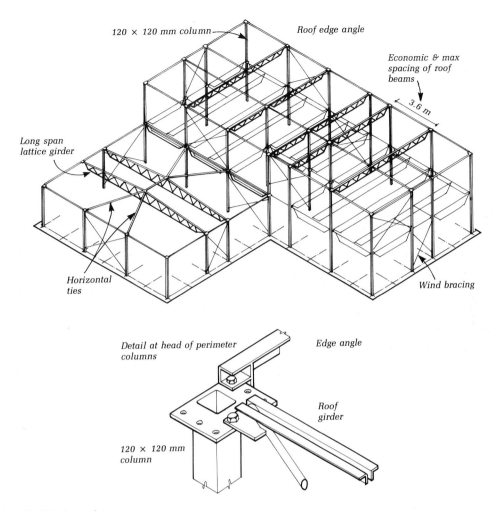

120 × 120 mm column

Roof edge angle

Economic & max spacing of roof beams

3.6 m

Long span lattice girder

Horizontal ties

Wind bracing

Detail at head of perimeter columns

Edge angle

Roof girder

120 × 120 mm column

Fig. 2.10 CLASP: flat roof structure

structure has this flexibility, substantial movements can of course cause major disruption to the rest of the building's fabric (in this regard it is interesting to compare the relative ability of traditional wooden frame buildings to settle over time without catastrophic damage being incurred).

For the frame to be stable, horizontal bracing or diaphragms are required and these are provided by the ground-floor slab as well as the horizontal floor and roof constructions. Together with other frame components, they ensure that the building maintains its plan shape by transmitting wind forces to the vertical bracing, which normally consist of diagonal braces between columns at selected points. These braces should form a continuous line to the perimeter of the building on a structural grid. As they are usually contained within external walls and/or internal partitions, their precise location should be established during the design stage of the project.

External walls The choice of materials for external walls may include metal sidings, brickwork and small unit cladding (tiles or boards) supported on cladding frames. The cladding forms the outer leaf of a two-layer wall and an internal lining of plasterboard, or plastics-faced metal sheeting is used (see fig. 2.13.)

Intermediate floors These are constructed as ribbed precast reinforced concrete decks, nominally 50 mm thick between ribs, supported on steel floor beams. The ribs taper at their ends so the decks can be seated on the top chord of the floor beams which have welded shear studs grouted between the edges of planks. The floor then performs as a structural diaphragm. At the perimeter, the decks extend beyond the grid line to support the external wall. Floor decks are 900 × 1800 mm with beams at 1800 mm centres. Standard floor loadings are 3 or 5 kN/m², the higher

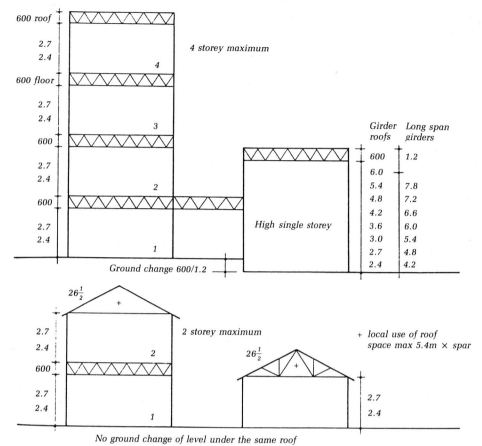

Fig. 2.11 CLASP: vertical co-ordinating dimensions using basic module of 100 mm. Floor and roof zones are both 600 mm, except in the case of long-span girders (12 m) and pitched roofs

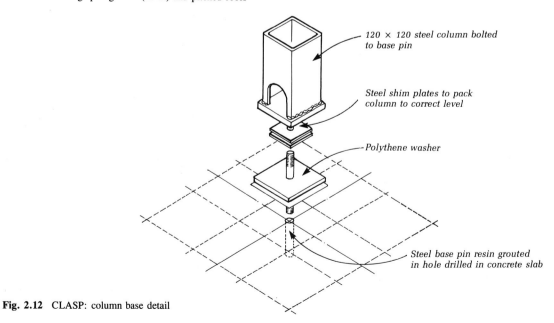

Fig. 2.12 CLASP: column base detail

loading is achieved by by the use of stronger floor beams. The decks, which provide half-hour fire resistance, are located below the upper floor co-ordinating plane by 25 mm to allow for a floor finish, and can have 'knock-out' holes to permit penetration of service pipes. Larger holes through the floor may be provided by using beams or galvanized mild steel trimming angles.

Flat roof When a flat roof is required, it is provided by corrugated galvanized steel decking which is fixed to the top of roof beams. The decking follows the camber of the roof beams by warping with the deck, and constitutes a structural diaphragm to assist in keeping the frame stable. The decking supports a vapour barrier, insulation, built-up felt finish and solar reflective chippings.

A pitched roof structure This is an integral part of the steel frame, consisting of galvanized mild steel trusses (fig. 2.14) hips and valley members supported on perimeter columns linked by a beam at the eaves. Timber rafters span between cold formed galvanized mild steel 'Z' purlins supported by the trusses, hips and valley members.

The roof has a pitch of 26 degrees and is suitable for roof coverings weighing up to 50 kg/m^2. Horizontal wind bracing is normally located at ceiling level, but may be raised locally into the inclined plane of the roof. This bracing enables the roof to span between vertical wind braces provided to the perimeter columns at the ends of the building.

2.6 Fast-track construction

The commercial developments of recent years achieved considerable overall cost savings by being constructed at speed, the method being termed *fast-track construction*. Implied in this description may be the necessity to reduce

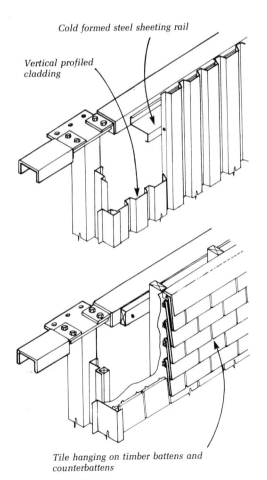

Cold formed steel sheeting rail

Vertical profiled cladding

Tile hanging on timber battens and counterbattens

Fig. 2.13 CLASP: wall cladding options

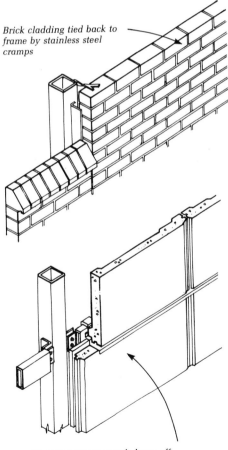

Brick cladding tied back to frame by stainless steel cramps

Precast concrete panels hung off horizontal RHS

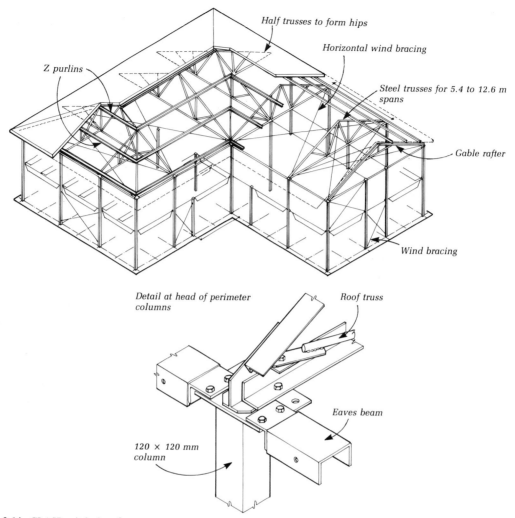

Half trusses to form hips

Horizontal wind bracing

Z purlins

Steel trusses for 5.4 to 12.6 m spans

Gable rafter

Wind bracing

Detail at head of perimeter columns

Roof truss

Eaves beam

120 × 120 mm column

Fig. 2.14 CLASP: pitched roof structure

the total project time from inception of the design to handover of the completed building. Certainly, the less time that is spent on site, the lower the interest costs associated with financing the project and the greater the developer's return. These reductions in programme time are achieved by overlapping tasks (rather than the traditional linear progression of design office procedures and trades on site). Fast-track construction has required new methods of working, organization and assembling components.

To achieve speed on site much use is made of factory prefabrication, following the same principles that governed the building systems of the 1960s.

This has several implications:

- As many items as possible that can be taken off the

critical path for the project, the better the chances of the programme being met. As a result site construction becomes a relatively simple assembly operation, enabling finishing trades to get started earlier and to work in the dry.

- The controlled environment of the factory enables components to be manufactured more quickly, to a better standard (since *quality control* can be more easily applied) and consequently to tighter tolerances.

- The specialist knowledge of fabrication companies can be brought to bear on the detailed design of components. These items may have been put out to tender described only by outline drawings accompanied by a *performance specification*. The design solution put forward by the specialist supplier can as a result make best use of that

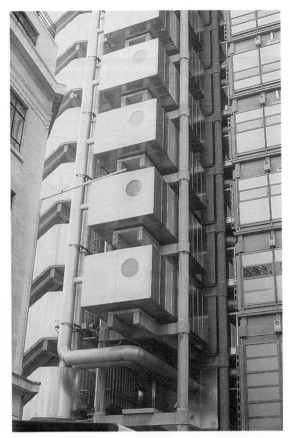

Photo 2.4 Habitat, Montreal. Architect: Moshe Safdie, 1966.

The concrete pre-fabricated units were lifted into position complete, except for their roofs which were cast *in situ* to reduce weight. Even so, they weighed up to 80 tons each. The boxes are stacked twelve high, the lowest level taking the accumulated weight of the upper units, as a result the structural design of the walls is not standardized.

Although this very large unit concrete prefabrication has remained a rarity, smaller scale concrete 'box' construction is currently used, for example, to prefabricate factory-finished bathroom units for hotels and hostels.

Photo 2.5 Prefabricated lavatory modules at the Lloyds of London Building. Architects: Sir Richard Rogers and Partners, 1986.

The pods were made in Bristol, transported to the site and craned into position at night. The units sit on steel plates fixed to the precast concrete beams; the pods were not slid into place, instead they and the supporting structure were built up together. They each contain a concrete floor slab which enables them to cantilever beyond the line of the beams. The external stainless steel panels were supplied by the cladding manufacturer and form a complete enclosure including the roof and soffit. The pods, as a consequence, have a large exposed area — a negative factor in terms of heat loss and future maintenance requirements.

firm's system and processes.
* Sub-contracts are let as *packages* of work and because there are considerable advantages in reducing the number of responsible parties, the packages tend to be large and inclusive of a number of elements. Prefabricated curtain walling, for example, is likely to come to site in storey-height 'unitized' panels including stone cladding and factory-glazed windows.

One of the principal technical problems of 1960s system building was the lack of resolution at the joints. The 'Consortia Schools' systems (like CLASP) overcame this problem as a result of extensive research and development, but the problem of tolerances that was intrinsic in the jointing together of large factory-made components led to widespread problems of weather-proofing particularly in concrete panel housing systems. Particular problems in fast-track construction can be encountered at the junction of two or more packages. Establishing the sequence of construction and the 'buildability' of the finished assembly is a major part of the designer's role.

Labour-intensive components, such as the lavatories in office buildings (which involve approximately 25 different trades in their manufacture), have increasingly been constructed off-site as *modular pods* (see photos 2.4 and 2.5), installed complete and requiring only to be connected to the building's services. Modular 'pods' are particularly associated with fast-track construction. Realization that European pre-fabrication methods are lagging behind those of the U.S. and Japan led in 1991 to the establishment of the *European Initiative for the Development of Building Module Technology*. The same constraints apply as for other prefabricated components:

- Their overall size is subject to transport limitations.
- They are only suited to use on large projects, the greater the number of pods and the fewer number of types the lower the unit cost. Ideally the pods should be repetitive from floor to floor; to have numerous different layouts in one building would be very expensive.
- The weight of the pods is another important consideration because they have to be lifted into position by crane; this may influence the choice of materials and form of construction.

Similarly lift shafts, in these types of building, may also be prefabricated. The *modular shafts* (see photo 2.6) are delivered complete in storey-height lengths and are erected with the main structure. The modules are stacked to very fine tolerances and the electricity supply is connected, all that is then required is for the installation to be tested and handed over (instead of having to wait for a watertight shaft and scaffolding before the guides, etc. are fixed, as is usual in lift construction). As a simple comparison, modular lift shafts are considerably more expensive than using conventional methods but the savings in construction time offset the additional costs. The shaft walls are thinner than in monolithic construction and yet the sound attenuation has been found to be better (because of the greater accuracy with which the guide rails can be assembled in the factory and because their brackets are isolated from the main structure of the building).

Fast-track construction requires repetitive and simple details that reduce the number of operations required. The greatest saving is achieved not by reducing the amount of time spent on an individual task but by reducing the number of tasks. Similarly it is necessary, to as great an extent as possible, to eliminate wet trades. So particular components are associated with this form of building. For example, drywall partitions fixed to galvanized steel studs are likely to be used rather than block walls, raised floors on power-floated concrete slabs are employed in preference to screed. Great vigilance is required on the part of the designer and contractor. Drywall partitions, for example, require extra pieces to be framed into the metal studs if the partition is going to receive fixings for shelves, sinks, etc. Whilst plasterwork could be varied in thickness to obscure out-

Photo 2.6   Prefabricated lift shaft. Photo courtesy of Schindler Ltd.

The lift shaft, including the lift itself, the pit and the machine room are made to heights to suit the building's floor to floor dimensions. The module is lifted into the building, through pre-formed rectangular holes in the floor slabs.

In conventional lift shaft construction, the main contractor builds-in pockets at prescribed locations so the lift company can fit the guide rails and door entrances (a dry, weatherproof shaft and machine room are required before assembling the equipment).

Using this modular system, the steel box pit section, containing the buffers and other equipment, is first lowered into place. Then, each storey-height steel-framed box, complete with doors and guide rails, is carefully placed into position. As each section is slotted into the next and plumbed, it is secured against the floor edges by brackets. Neoprene gaskets are used between the brackets and the shaft to reduce the transfer of vibration from the lifts to the rest of the building.

The assembled car and counterweight are finally installed, followed by the machine room containing its pre-tested equipment. Connection to main services, testing and commissioning, are the only tasks left before completion. Overall, lift installation time can be reduced by more than 60 per cent.

of-tolerance masonry, dry-assembled components are less forgiving of mistakes and accuracy may be difficult to reconcile with speed of construction.

2.7 Trinity modular microflat

The technology of lavatory pods and prefabricated lift shafts

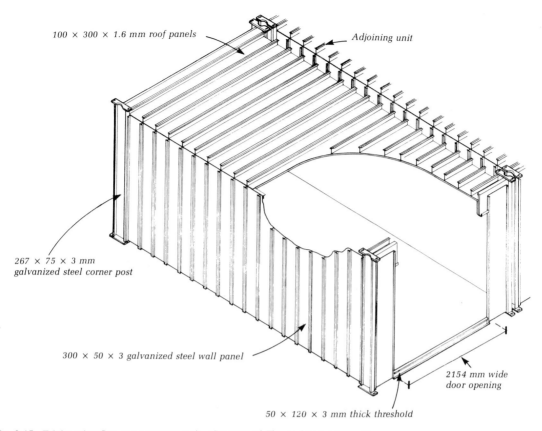

100 × 300 × 1.6 mm roof panels

Adjoining unit

267 × 75 × 3 mm
galvanized steel corner post

300 × 50 × 3 galvanized steel wall panel

2154 mm wide
door opening

50 × 120 × 3 mm thick threshold

Fig. 2.15 Trinity microflat: one apartment unit. (Courtesy of *The Architects' Journal*)

has been brought to bear on the design of the *Trinity modular microflat*. The microflat is a prefabricated one- or two-person dwelling intended to provide an economical form of housing in central cities, hostel or workers' accommodation for industrial applications (fig. 2.15).

The module size is basically that of a standard industrial container 8.13 m by 3.25 m, the finished size is slightly larger but it is transported in the same way. Each unit is constructed from a welded and galvanized cold-formed steel frame with 100 mm deep profiled steel decking spanning the width. The walls and ceiling are made from 300 mm wide steel panels joined at their flanged steel edges which are 50 mm deep for the walls and 100 mm for the roof. They are craned into position, stacking one above the other. The corner posts support separate 165 × 132 mm universal beams which span the width of the corridor between banks of units; these carry precast concrete planks forming the corridor floor (fig. 2.16).

The structure of a block of microflats is a mass concrete box frame made by pouring concrete between the parallel walls of two adjoining flats. The walls are made in two

separate pours to prevent bulging of the steel panels which form permanent shuttering to the concrete cross walls. The floors are formed similarly by infilling the depth of the flanged ceiling panels with concrete. This slab then has sufficient strength to support the weight of the suspended floor above it (should it for some reason collapse). No reinforcement is required in the walls until they are 5 storeys high and the system is suitable for buildings up to 15 storeys. The 200 mm thick composite steel and concrete walls achieve a fire separation between flats of 6 hours and sound insulation of 55 dB. Tolerances in the assembly are taken up in the somewhat variable thickness of the concrete walls. The floor construction is a fifth of the weight of an equivalent concrete slab but, together with the suspended decking floor above, it gives better sound insulation.

The front end of each unit has a framed opening for a patio door giving on to a balcony. The balcony floor and flank walls hinge back to the face of the unit for transportation, and the flank walls provide privacy between units and the necessary fire separation. All the services for the flat are installed before delivery and are ready for

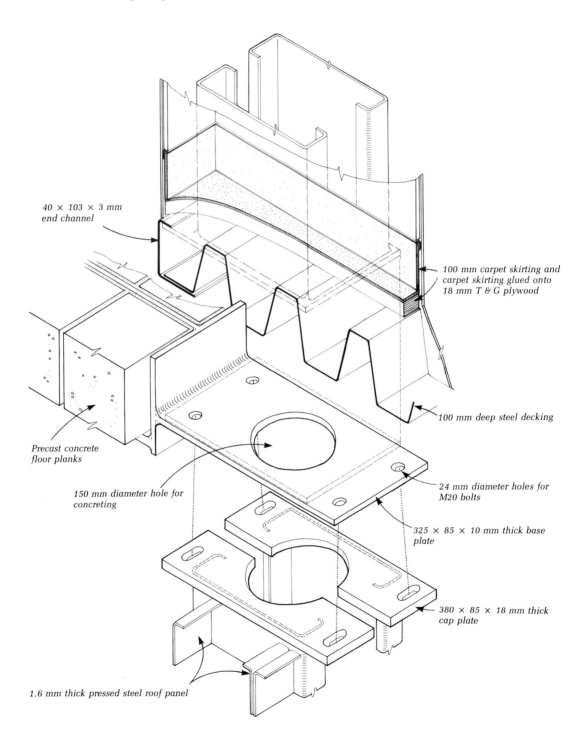

40 × 103 × 3 mm
end channel

100 mm carpet skirting and
carpet skirting glued onto
18 mm T & G plywood

100 mm deep steel decking

Precast concrete
floor planks

150 mm diameter hole for
concreting

24 mm diameter holes for
M20 bolts

325 × 85 × 10 mm thick base
plate

380 × 85 × 18 mm thick
cap plate

1.6 mm thick pressed steel roof panel

Fig. 2.16   Trinity microflat: detail showing the corner junction of two units and the central corridor

connection at the other (corridor) end of the unit without access being required into the unit. A siphonic drainage system is used for flexibility in the organization of the plumbing within the building. The flats are waterproof, mastic tape is used between the riveted flanges of the galvanized deck panels, so the modules can be left in the open prior to construction. Under these circumstances there is a danger of condensation so the units are delivered with an internal dehumidifier running which is disconnected at handover; a warning indicator shows if the humidifier ceases to function.

For this large unit prefabrication to be economically viable, the modules have to be delivered with as much added value as possible, so the flats are constructed to be ready for occupation, decorated with fabric-covered dry-assembled lining panels, fully equipped and furnished.

2.8 Component design: accuracy and tolerances

Much contemporary construction consists of the assembly on site of components that have been prefabricated elsewhere. In many cases these elements have factory-applied finishes and consequently are not amenable to alteration on site. Because all construction is to some extent inaccurate (and to a lesser extent so are manufacturing processes), the accommodation of these errors has to be designed into the joints between components.

BS 5606: 1990 *Guide to accuracy in building* specifies permissible tolerances for building and the method by which they should be calculated. The previous (1978) edition of the British Standard allowed inaccuracies to be accumulated one on another without the work being deemed unsatisfactory. This sanctioned the possibility of quite unacceptable discrepancies and the British Standard was largely unused by specification writers who inserted their own clauses instead.

The reasons for inaccuracy in building are divided into two categories: *Induced deviations* are an inevitable result of the construction process due to, for example, errors in setting out. *Inherent deviations* are inevitable inaccuracies due to the physical properties of materials, for example thermal expansion or contraction.

The new standard adopts the same methodology as the old. A survey of the extent of inaccuracy on building sites was carried out from which a middle range of tolerances was selected to represent a permissible degree of accuracy called the *characteristic accuracy* of different tasks. This range has now been reduced and the method of calculation has been changed to refer measurements back to reference points at regular intervals throughout the height of the building so tolerances cannot accumulate unreasonably. To

achieve these more stringent standards, correspondingly more stringent methods of site checking will also be required. For the first time the British Standard also defines the accuracy to be expected of components made in a factory, based once again on survey information. Accuracy of fabrication is however dealt with in several British Standards for individual elements and BS 5606 is cross-referenced to them. Alternatively, information should be sought from specialist manufacturers. Although factory-made components will have a higher order of accuracy than on-site construction, manufacturing tolerances are important in the context of repeated elements in large buildings.

BS 5606 recommends particular procedures for the determination of the tolerances required between elements, how they should be accommodated and procedures for monitoring dimensions on site. Firstly, tolerances should be specified which reflect the overall requirements of the design with regard to aesthetic, legal and other criteria and then the specific tolerances required between particular components. The normal standards of accuracy that can be expected are the basis of Table 1 (on-site construction) and Table 2 (manufactured components) of BS 5606. For particular situations requiring special accuracy the sizes with their associated tolerances should be shown on the drawings and incorporated within the specification which should also indicate the methods of checking and measurement to be employed. In some cases tests and mock-ups may be required to determine the feasible level of accuracy.

Sources of induced deviations include *setting out* inaccuracies due to human inability to measure with absolute accuracy, the variability of locating grids and inevitable inaccuracies in site surveys and of measuring equipment. In *manufacturing* and construction, errors arise due to differences in the size and shape of manufactured items which are intended to be identical. Moulds and jigs should be checked before and during fabrication and the sizes of finished components should be checked as early as possible during the production run. Finally, during *assembly*, inaccuracies are inevitable due to the impossibility of exactly locating components relative to setting out lines and if fixing methods are used which have insufficient capability for adjustment (fig. 2.17).

An example is given (fig. 2.18) for the calculation of the tolerance required to locate a window (with a target size of 2390 mm) within a structural opening (target size 2430 mm). From table 1 of BS 5606, the space between the columns, if constructed from *in situ* concrete, will be subject to variation by plus or minus 18 mm. The window, from Table 2, will have a tolerance of plus or minus 4 mm. At the junction between the window and the concrete reveal the tolerance will be equal to the square root of the sum of the squares of each individual deviation which in this

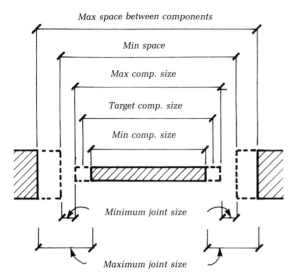

Fig. 2.17 Variation that can be anticipated between the dimensions of components and the structure to receive them: from BS 5606: 1990

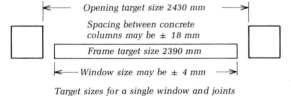

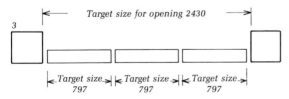

Fig. 2.18 Sizes of joints and methods of accommodating tolerances when fixing components into prepared openings: from BS 5606: 1990

case is 18.4 mm. Consequently the anticipated maximum joint size will be 29.2 mm and the minimum 10 mm and the jointing technique used should be able to accommodate this variability (say 30 mm to 10 mm). If the extent of the tolerance is too great, other measures can be employed, for example changing the design so the windows are fixed at the face of the columns, detailing the columns with rebates to join the windows, using a cover strip over the joints or using more than one window to create more joints able to absorb the tolerance.

Particular situations, such as the necessity to locate a building very accurately in relation to a boundary, may require a particularly specific degree of accuracy. Under other circumstances a high degree of inaccuracy may be inevitable. Such is the case in fast-track construction. Building at speed is inevitably inaccurate and wide tolerances have to be accommodated in the method used to joint components. Typically the main frame of the building, whether steel or to an even greater extent concrete, is subject to considerable deviations so it is afterwards obscured within a casing of metal, drywall or other sheet material that can be more accurately assembled and which finally represents the form of the structure. Although the additional cost involved in erecting the structure to finer tolerances may be only a small percentage increase in the overall cost of the project, it may be impossible to accommodate the additional time required.

In fast-track construction there is a particularly wide disparity between the accuracy of assembly of prefabricated elements such as the cladding system and the relatively slipshod standard of the structure. A usual method to accommodate these tolerances is to establish a *buffer zone* within the grid layout of the plan which can be filled by a flexible or variable element that can be formed on site typically as a 'flash gap' made in, for example, pressed metal. This principle can be used at junctions between all factory-made and other relatively inaccurate elements of the building. In fast-track construction these tolerances may have to be that much greater because of the telescoped nature of the design and construction programme; the elevations, for instance, are likely to have been finalized prior to the detailing of the structure. Similarly the foundations will have been detailed before the final design of the superstructure.

The method of tendering sub-contracts as separate packages entails that tolerances are allowed between them. The junction between several packages, as frequently happens at single grid intersections presents particular problems of fit and weathering. The necessity to stay on programme and the fact that the prime agent in these contracts is likely to be a project manager rather than a designer may preclude the possibility of unsatisfactory work being done again. Because much of the detailed design is now carried out by specialist sub-contractors, it is particularly important that performance specifications embody a predetermined approach to the achievement of the accuracy required and describe how tolerances between packages are to be accommodated.

Some recent types of component such as rain-screen cladding are both slender in section and relatively quick to assemble on site. They require either an accurately constructed background structure to which they are attached or methods of attachment that are able to accommodate the tolerances to be expected.

Fast-track construction requires the question of accuracy in construction to be considered at the earliest stages of a project. Rather than the logical sequence of design and construction that was expected within traditional contracts, fast-track design is governed by lead-in times and the availability of components and materials. This requires that issues of buildability, fit and accuracy form part of the initial design discussions on fast-track contracts.

3 External glazing

3.1 Introduction

Of all the advances in technology that have made their mark on 20th century building, perhaps none has been as significant as in the manufacture of glass. The aesthetic capabilities of the material have been increased by the variety of types that are now produced. Reference should be made to literature and samples available from the major manufacturers, and the publications of the *Glass and Glazing Federation*.

BS 6262: 1982 *Glazing for buildings* defines *glazing* as 'The securing of glass or plastics in prepared openings in, for example, windows, door panels, screens and partitions'. A description that holds good although to some extent the requirements of BS 6262 have been superseded by BS 8000: 1990 *Workmanship in building* Part 7: *Code of practice for glazing*.

External glazing is described as *inside glazing* when the glass or plastic is inserted from inside the building; and as *outside glazing* when the glass or plastic is inserted from outside. *Internal glazing* is where neither face of the glass or plastic is exposed outside the building.

MBS: Materials chapter 12 gives details of the properties associated with each type of glass used in the construction of buildings. Most glass specified is soda/lime/silica-based with the addition of a small percentage of magnesium and aluminium oxides and colouring agents, conforming with BS 952: *Glass for glazing* Part 1: 1978 *Classification*, and Part 2: 1980 *Terminology for work on glass*. A number of different high-temperature processes can be used to form glass to shape, it is then cooled slowly (*annealed*) to relieve the in-built stresses which would otherwise result. The term *annealed glass* is used to distinguish it from *toughened glass* which is chemically or heat treated (cooled rapidly) during manufacture to put its external surfaces into compression and interior into tension. This balanced prestressing makes it more resistant to breakage than ordinary glass and when shattered it divides into small blunt-edged fragments. The available types of glass for building are as summarized below.

Annealed flat glass can be classified as follows:

- *Clear float* for general use. After heating the components in a furnace, the molten material is poured in a flat sheet over a bath of molten tin. The float process is now responsible for the vast majority of glass that is produced.
- *Solar control glass* tinted for solar absorption and reflection. The tint can be achieved either by colouring the whole thickness of the glass or by coating the surface with a thin layer of metal oxide.
- *Roughcast glass* with an obscured surface on one side. This used to be manufactured by casting the glass on to a bed of sand, imparting its characteristic appearance; it is now made by passing the molten glass through rollers but the terminology remains.
- *Opal glass* forms a brilliant surface when backlit.
- *Patterned glass* which has a pattern rolled into the surface to make it obscure.
- *Wired glass* is produced by rolling; it can have a cast finish or be polished. A very limited number of sizes of mesh are available but since wired glass is employed for reasons of both fire separation and as a safety glass, different wire thicknesses are now available according to the anticipated use. Should the glass be broken, the wire holds the pieces safely together; it also gives the material fire resistance.

Those based on flat glass, annealed and/or toughened are:

- *Toughened glass*, strengthened for impact resistance and resistance to thermal fracture. Because it cannot be cut or worked after it has been made, toughened glass has to be used in the sizes obtained from the manufacturer. It is available in a variety of body-tinted and surface-coated finishes.

Photo 3.1   Power Centre for the Performing Arts, University of Michigan, Ann Arbor, US. Architects: Kevin Roche, John Dinkeloo and Associates, 1965.

The building stands within a wooded park, the modelled facade of the foyer exploiting the reflectivity of its mirrored glass.

Mirror-effect solar control glazing is made by coating clear glass with metal oxide either by pyrolysis (application of the oxide during glass manufacture while the glass is still warm) or by electro-magnetic coating. The layer can reflect between 10 and 30 per cent of the sun's energy which is 2 to 3 times more effective in this regard than normal clear glass. When laminated with a polyvinyl-butyl interlayer (i.e. both coatings being sandwiched between the two layers of glass), virtually all UV radiation is stopped from passing through.

- *Laminated glass* reinforced with sandwiched plastic sheet to make it resistant to impact. Although it will crack, the cracking is limited and the plastic interlayer helps hold the glass together and prevent splintering.
- *Laminated solar control glass* with a tinted interlayer for solar absorption or incorporating tinted or coated glass for solar absorption and reflection.
- *Laminated sound control glass* — the sound-damping properties of the plastic interlayer in combination with the thicknesses of glass give some degree of sound resistance.

- *Laminated ultraviolet (UV) light control glass* which contains an interlayer able to reflect 98 per cent of ultraviolet radiation (see photo 3.1).
- *Anti-bandit laminated glass* to BS 5544: 1978 (1985) that is designed to resist breakage for a short length of time.
- *Bullet-resistant laminated glass* to BS 5051: Part 1: 1988 that is tested to avoid penetration of the glass under gunfire.
- *Blast-resistant laminated glass* combines safety and anti-bandit capabilities and bullet resistance, it is custom manufactured.
- *Insulating glass units.* Two or more panes of glass with hermetic seals at their edges and containing dehydrated air or inert gas (usually argon for thermal insulation or sulphur hexafluoride for acoustic separation) between them. They can be manufactured from any one of the available flat glasses and may also be tinted or incorporate low-emissivity or other special glasses. *Low-emissivity glass* has a transparent coating applied to one side that reflects long-wave length energy, generated by plant and lighting, back into the building whilst still permitting short-wave solar energy to enter. The effective *U*-value of the glazing is consequently improved.
- *Fire-resisting.* As well as wired glass, in the past copper light glazing was used for this purpose, now specially formulated fire-resisting glasses are available including laminated glass containing an intumescent interlayer (see p. 46).
- *Lead X-ray glass* containing lead oxide for X-ray resistance.

Those having applied surface treatment, annealed, are:

- *Obscuring.* Obscuration by sandblasting, grinding or acid embossing.
- *Brilliant.* Surface cutting and polishing.
- *Engraving.* Cutting surface with small wheel.
- *Enamelling.* Staining or painting, and coating with fusible pigment then firing.
- *Stoving.* As above but fired at lower temperature.
- *Gilding.* Application of metal leaf to surface.
- *Silvering.* Deposition of silver on the glass surface (for mirrors).
- *Striped silvering* or '*Venetian*' stripe. Alternate silver and clear bands ('one-way glass').
- Applying *metallic film* resulting in reduction in light transmission/increase in reflectance.

With the important exception of toughened glass, all other types of single thickness glass sheets can be worked after manufacture, including cutting and shaping, hole cutting, notching, edge finishing, and bending for curved windows and domes. In addition to those listed above, hand-made

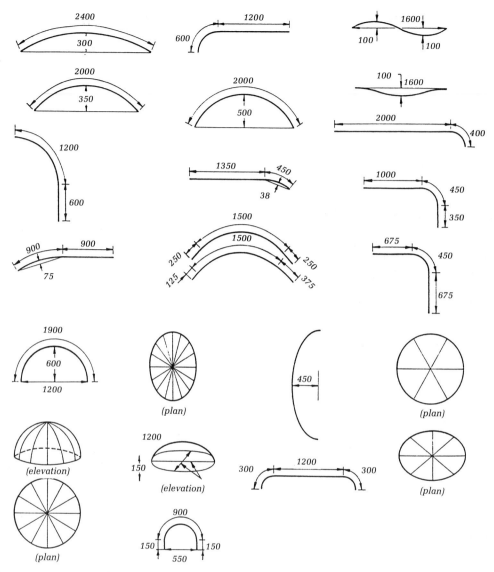

Fig. 3.1   Standard curves for bent glass in BS 952: Part 2: 1980

antique, bullion (bull's-eye), glass blocks, channel section, stained and leaded light glass are available.

Bent and curved glass can be produced from most types of flat annealed glass by kiln heating until it is soft enough to allow formation over a mould. Although largely superseded by the use of plastics for rooflights, curved glass is still employed for shop windows and glazed façades and for mirrors (convex and concave). Figure 3.1 shows standard curves for bent glass as given in BS 952: Part 2. Various publications of the Glass and Glazing Federation (GGF) are useful, particularly their *Glazing Manual*.

MBS: Materials chapter 13 gives details of the properties associated with each type of *plastic* used in the construction

of buildings. Plastics materials commonly used for glazing include: polycarbonate (PC), polymethyl methacrylate (PMMA) — commonly referred to as 'acrylic', unplasticized polyvinyl chloride (PVC-U), glass-reinforced plastics (GRP), hollow section extruded profiles in PMMA and PVC-U, cellulose acetate butyrate (CAB), and polystyrene (internal applications only). Many are preferred for situations where safety and security are important because of their relative high strength and impact resistance. When compared with thick glass, the significantly reduced weight allows economy in the extent of framing required. Plastics for glazing are available wired, laminated, tinted, textured and with solar control. Composites made from both glass and plastics are also available, as are anti-bandit and bullet-resistant materials, and single-, double- and triple-skinned extruded polycarbonate sheet. All plastics are combustible but this characteristic can be modified by the addition of retardants during manufacture so that some used for glazing are self-extinguishing (see p. 121). A useful publication by the GGF is their data sheet 4.5 *Glazing with plastics*.

3.2 Performance requirements

Glazing technology has reached a high level of complexity; section 15 of BS 6262 places great emphasis on the need for designers to make their requirements clear at each stage of the design process. However, successful performance specification of glazing obviously depends to a very large extent on characteristics of the glass or plastics used, including the selection of the appropriate type to fulfil specific functions, and the correct thickness relative to size of opening. These criteria are covered in outline here except where a more detailed explanation is necessary to understand methods of securing glass or plastics within its surround. This section on performance specification for glazing should be read in conjunction with section 4.2.

3.2.1 Appearance

The various capabilities of glass to form a transparent surface that is almost invisible or a mirror reflective of trees and sky, plus its transformation at night, make it one of the most evocative materials available to the designer.

The relationship to the surrounding wall of window and door openings and their shapes, materials and the techniques used for construction, contribute greatly to the character of a building. Window frames can be expressed boldly by the use of large sections, or have minimal influence when thin frames are employed. The amount of external light passing through or reflected back by the glazing plays an influential part in the appearance of a building, as can the reverse situation at night, when rooms and their occupants may either be concealed from or exposed to view from outside.

A performance specification should recognize the range of relevant issues. For instance, the psychological benefit of providing an awareness of the outside and factors arising from the interaction of physiological and psychological aspects such as the perception of comfort, security, privacy, light, glare and colour rendering within the building.

3.2.2 Strength

Glass Annealed glass, including wired glass, in the thicknesses which are commonly used, is easily broken and

Table 3.1 Design wind pressures from BS 6262

Basic wind speed (from fig. 3.6)	Height to eaves	Design wind pressure (N/m^2) for a ground roughness category (see table 3.2) of:			
(m/s)	(m)	cat. 1	cat. 2	cat. 3	cat. 4
52	3	1600	1200	950	750
	5	1800	1450	1150	850
	10	2350	2000	1450	1050
50	3	1500	1150	900	700
	5	1700	1350	1050	800
	10	2150	1850	1300	1000
48	3	1400	1050	850	650
	5	1550	1250	1000	750
	10	2000	1750	1200	900
46	3	1250	950	750	600
	5	1400	1150	900	650
	10	1850	1600	1100	850
44	3	1150	900	700	550*
	5	1300	1050	850	600
	10	1200	1450	1050	750
42	3	1050	800	650	500*
	5	1200	950	750	550*
	10	1550	1350	950	700
40	3	950	750	600	450*
	5	1100	850	700	500*
	10	1400	1200	850	650
38	3	850	650	550*	400*
	5	1000	800	600	450*
	10	1250	1100	750	550*

* In order to avoid excessive deflection, it is recommended that a minimum design wind pressure of 600 N/m^2 is used here.

Table 3.2 Ground roughness categories from BS 6262

Ground description	Category
Long fetches of open, level or nearly level country and all coastal situations	1
Open country with scattered windbreaks	2
Country with many windbreaks; small towns; outskirts of large cities	3
Surfaces with large and frequent obstructions, e.g. city centres	4

the fragments can be dangerous: wired glass holds the broken glass pieces in position and fragments of laminated glass are retained in position by the interlayer of plastic. Toughened glass is much stronger than annealed glass and when shattered is relatively harmless.

Glass must not only be sufficiently thick to resist stresses

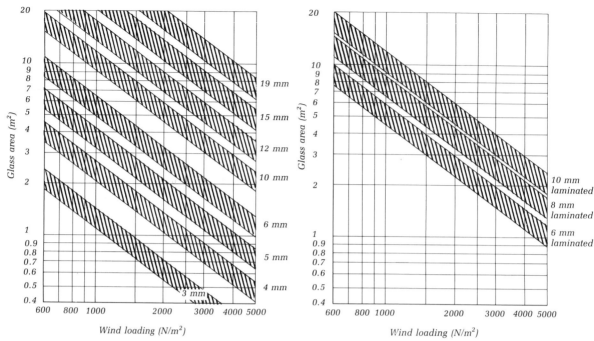

Fig. 3.2　Wind loading graph for annealed float glass (Pilkington Glass Ltd)

Fig. 3.3　Wind loading graph for laminated float glass (Pilkington Glass Ltd)

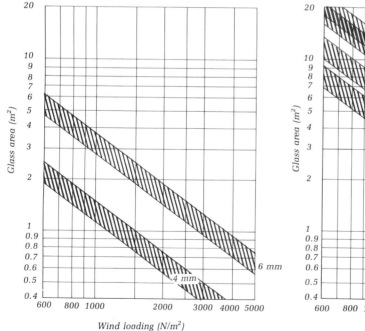

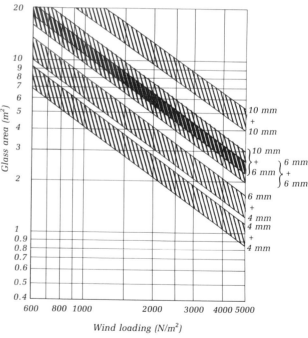

Fig. 3.4　Wind loading graph for rough cast, patterned and wired glass (Pilkington Glass Ltd)

Fig. 3.5　Wind loading graph for 'Insulight' float double glazing units (Pilkington Glass Ltd)

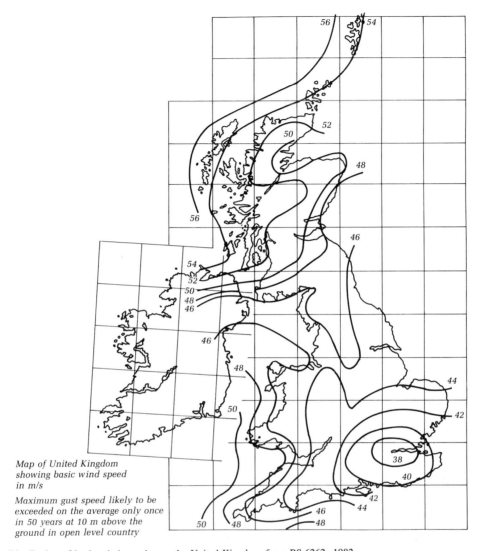

Map of United Kingdom
showing basic wind speed
in m/s

Maximum gust speed likely to be
exceeded on the average only once
in 50 years at 10 m above the
ground in open level country

Fig. 3.6 Distribution of basic wind speed over the United Kingdom from BS 6262: 1982

due to minor impact but also pressures exerted by wind, the severity of which will vary with exposure conditions, the type of glass and method of fixing. Figures 3.2 to 3.5, tables 3.1 and 3.2 and also fig. 6.2 provide a guide to the minimum safe thicknesses relative to wind loadings for the more common types of glass, glazed vertically in rectangular panes and supported at four edges in windows of a stated area. The lower lines are for square panes and the top of each shaded band corresponds to a length/width ratio of 3:1. The bands may be divided proportionally to determine intermediate ratios between 1:1 and 3:1. The design wind loading for pressure and suction should be found from CP3 *Code of basic data for the design of buildings* chapter V Part 2: 1972 *Wind loads*, or for low-

rise buildings only, an abbreviated method of calculation is contained in BS 6262. Using the BS 6262 method of calculation, the basic wind that can be anticipated is obtained from a wind speed contour map of the UK (see fig. 3.6). The least basic wind speeds are in the vicinity of London (38 m/s) and are greatest off the north-west coast of Scotland (56 m/s). According to the Meteorological Office, these are likely to be exceeded on the average only once in 50 years at 10 m above the ground in open level country.

To obtain the design wind speed, the basic wind speed is multiplied by a correction factor which accounts for the topography of the surrounding landscape and the height of the window above ground level. The correction factors range from 0.56 for windows 3 m or less above ground

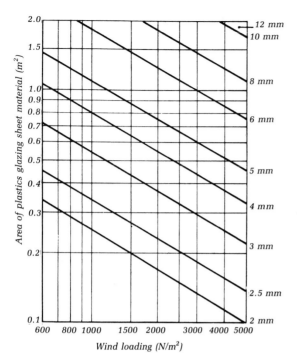

Fig. 3.7 Wind loading graph for plastics glazing sheet material, three second mean wind loading, 15 mm edge cover and aspect ratio of 1 : 1 to 1.5 : 1 (BS 6262:1982)

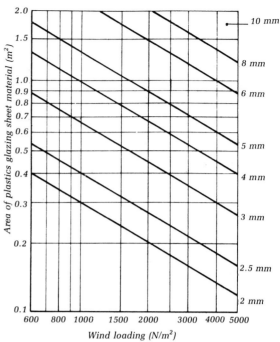

Fig. 3.8 Wind loading graph for plastics glazing sheet materials, three second mean wind loading, 15 mm edge cover and aspect ratio greater than 1.5 : 1 up to and including 2.5 : 1 (BS 6262:1982)

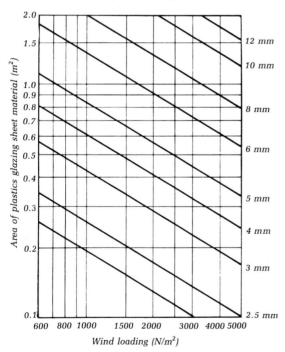

Fig. 3.9 Wind loading graph for plastics glazing sheet materials, three second mean wind loading, 5 mm edge cover and aspect ratio of 1 : 1 to 1.5 : 1 (BS 6262:1982)

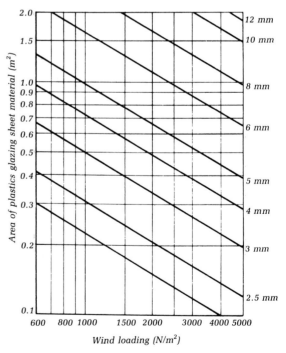

Fig. 3.10 Wind loading graph for plastics glazing materials, three second mean wind loading, 5 mm edge cover and aspect ratio greater than 1.5 : 1 up to and including 2.5 : 1 (BS 6262:1982)

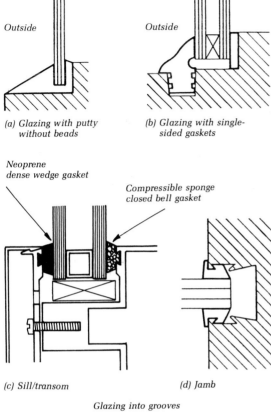

(a) Glazing with putty without beads

(b) Glazing with single-sided gaskets

Neoprene dense wedge gasket

Compressible sponge closed bell gasket

(c) Sill/transom

(d) Jamb

Glazing into grooves

Zipper strip

Drainage

(e) Glazing into grooves with channel gaskets

(f) Glazing into grooves with structural gaskets

Fig. 3.11   Rebate and grooved glazing ((c) Tremco Ltd)

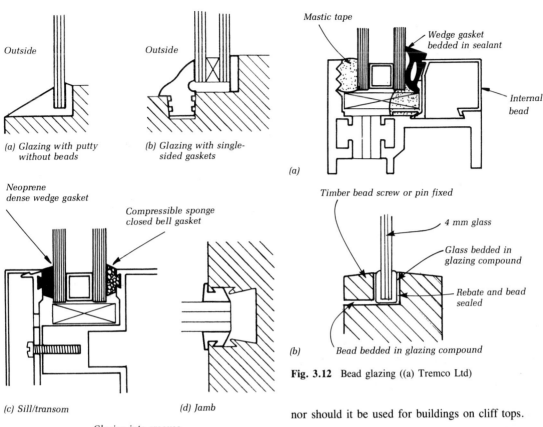

Mastic tape

Wedge gasket bedded in sealant

Internal bead

(a)

Timber bead screw or pin fixed

4 mm glass

Glass bedded in glazing compound

Rebate and bead sealed

(b)   Bead bedded in glazing compound

Fig. 3.12   Bead glazing ((a) Tremco Ltd)

level on a city centre site to 1.0 for windows 10 m above ground level in open country or coastal areas. The table published in BS 6262 has the correction factors and a conversion to wind speed in N/m² built in for simplicity of calculation. The BS 6262 abbreviated method is not suitable for any buildings higher than 10 m from ground level or where the basic wind speed is higher than 52 m/s,

nor should it be used for buildings on cliff tops.

Plastics   Figures 3.7–3.10 similarly provide a guide to minimum safe thickness and pane proportion/size for plastics. The wind loading should be determined using one of the methods described above. If the pane is situated where it may be subject to accidental breakage or is required to be particularly secure or vandal resistant, the thickness should be increased or the specification modified accordingly.

3.2.3 Glazing techniques

In order to take full advantage of their strength characteristics it is important that all edges of glass and plastics panes are adequately supported, normally within or against a frame of timber, metal or plastics. Glazing *compounds* or *gaskets* are used not only to form a joint which will hold the pane firmly in position but also to space the glazing away from the frame and so allow for differential movement between the two components. Glazing can be held by a supporting frame within a rebate, the glass or plastic being held by the glazing materials alone (figs 3.11(a) and (b)) or, alternatively, a bead is provided for this purpose and the glazing material only fulfils the other functions of weathersealing and allowing for

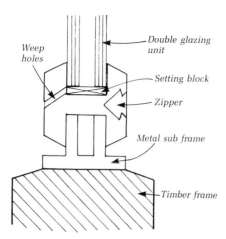

Fig. 3.13 Gasket glazing over nib (metal sub-frame)

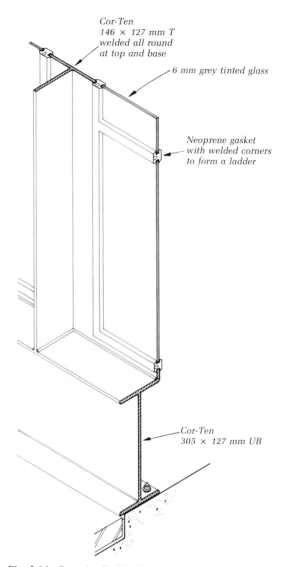

Fig. 3.14 Cummins Engine Factory, Darlington ladder gasket detail. (Architects: Kevin Roche, John Dinkeloo and Associates. Courtesy of *The Architects' Journal*) When constructed in 1965, this glazing system was very innovative, the single glass panels are large and at transoms they are joined only by the neoprene ladder gaskets. These were found to deflect too much in high winds so restraining steel sections were welded behind the horizontal gaskets but the principle of the design holds good.

differential movements (fig. 3.12). In either case the functional requirements are stringent and may often cease to be met unless the installation is well maintained.

The recent tendency has been for much more use to be made of dry-glazing techniques, particularly where windows are to be factory glazed. Gaskets can be held within grooves in a frame or glazed over a projecting nib (fig. 3.13) or the gaskets can be mitred and welded at corners to form a ladder gasket into which panes are fixed to form a window wall (see fig 3.14). Structural gaskets generally are made from rigid polychloroprene (neoprene), and the locking strip or 'zipper' made from a yet harder grade synthetic rubber, is pushed into a groove in the side of the main gasket after the glazing has been assembled. The zipper forces the inner sides of the main gasket into contact with the glass, a form of construction that requires assembly to close tolerances.

Alternatively, the gasket can be held in compression by an external pressure plate as in curtain walling or aluminium windows (see section 4.5). A soft gasket is used on one side of the glass (fig. 3.11(c)) and a more rigid, wedge-shaped gasket on the other. Some manufacturers' systems combine sealants, tapes and gaskets (fig. 3.12(a)); this optimizes the characteristics of the materials, gaskets can be cheaper to install and sealants provide positive adhesion.

3.2.4 Glazing compounds, sealants and gaskets

Materials used for glazing are described in detail in BS 6262, and briefly, the classifications are given under the headings below.

Putties
• *Linseed oil putty* (BS 544: 1969 *Specification for linseed*

oil putty for use in wooden frames). The initial hardening of putty is by absorption of oil into the timber of the frame, consequently its use is for glazing into softwood and absorbent hardwood windows. The surface of the putty should be painted as soon as it has established a hard outer layer (allow not less than 7 days) but the finishing coat must be applied within 28 days otherwise the putty will start to deteriorate.

- *Metal casement putty.* This is formulated to set hard and adhere to non-porous surfaces. It also should be painted after 7 and within 28 days of application.

The characteristics of the two types are now often combined into what is available as *general purpose putty*.

Flexible compounds After a period of time putty completely hardens so it has no capability to flex with movement of the glass and is not suitable for large windows (see p. 50). Where glass is subject to structural or thermal movement which cannot be accommodated by putties, *non-setting glazing compounds* may be used in conjunction with beads or in grooves (although there are now alternatives with better performance characteristics, as described below). Only the outer surface of these compounds hardens within 5−7 days. Because their non-setting nature is dependent on the volume used, their minimum recommended joint thickness is 3 mm, and 5 mm is the maximum.

Two-part rubberized compounds, comprise two components that mixed together form a rubbery material. For application by hand or gun, they are flexible and used with beads in exposed situations and for frames subject to movement or distortion. Because once cured the material is not easily displaced, it can be used for factory glazing and transported to site without damage. The minimum joint thickness is 1.5 mm; special primers have to be used if these materials are to be painted.

Sealants

- *General sealants* are gun applied and used for capping and bedding applications; they often contain hazardous or toxic constituents.
- *One-part sealant: curing types.* These have a polysulphide, silicone or urethane base which undergoes a chemical reaction to form a firm resilient seal. These are able to tolerate large movements over a range of temperatures but vary in the length of time they take to cure; silicones and urethanes cure quickly, polysulphides more slowly. It is important not to choose a slow-curing type if the joint is soon going to be subject to stress.
- *One-part sealant: solvent release type.* These have butyl or acrylic bases which do not undergo chemical reaction and remain soft and pliable but are subject to shrinkage due to loss of solvent. Butyl-based sealants adhere to most surfaces and primers are not required. Some grades are sensitive to UV light and so their principal use is for bedding beads and glass. Acrylic types are similarly compatible with most surfaces and are UV resistant but

some types degrade in the presence of water and should not be used in permanently wet conditions.

- *Two-part sealant: curing types.* These are supplied as two components, each either polysulphide or polyurethane based, that react and cure within 14 days to form rubbery material. They have to be mixed at the correct speed and for the correct time; the curing time and their durability are also affected by temperature and humidity. Polysulphide sealants are particularly inert and adhere to most surfaces although primers are required on masonry. Some polyurethane types are less resistant to UV light and primers are required on glass for best adhesion.

Preformed strip materials These are often used as a convenient alternative to non-setting compounds; they have similar properties but greater hardness. Separate strips are used on each side of the glass and they have to be held by screw-tightened pressure beads to maintain a seal. On the outside, the weather edge of the joint may be finished with non-setting compound or sealant. Types are:

- *Preformed mastic tapes*, butyl or polyisobutene based, are available in a variety of widths and thicknesses.
- *Extruded solid sections*, PVC or synthetic rubber based.
- *Cellular strips*, made of self-adhesive synthetic rubber, when used on the outside of the glass they usually have to be capped with a strip of sealant other than in sheltered situations.

Preformed compression-type gaskets

- *Structural gaskets*, made from vulcanized polychloroprene rubber compounds to BS 4255: Part 1: 1986 *Specification for non-cellular gaskets*. Integral locking strips or 'zipper' type insertions produce a compression grip on the frame structure and the glass.
- *Non-structural gaskets*, of synthetic rubber or plastics, are supplied in solid, tubular or dense sponge form. They should comply with BS 4255: Part 1: 1986, and are used as primary seals when maintained under positive pressure, or need to be mechanically retained in rebates and top sealed with one- or two-part sealant.

Table 3.3 provides general guidance on the selection of suitable glazing materials for particular types of glass. Linseed oil putty, metal casement putty, extruded solid sections with internal trims, and gaskets, are suitable for frames having a rebate, groove or nib for glazing without the use of beads — see p. 48. Non-setting compounds, two-part rubberized compounds, capping sealants, mastic tapes and compressible cellular strips are suitable when the frames incorporate glazing beads.

The glazing materials used for glazing into rebates

Table 3.3 Glazing materials for external applications from BS 6262

Type	Basic components	Typical life expectancy*	Form	Uses	Wind pressure resistance†	Comments
0. Linseed oil putty	Linseed oil	See comments columns	Bulk	Bedding and fronting	Up to 900 N/m²	Paint protection essential. Life dependent on paint maintenance. Inspect annually
1. Metal casement putty	Drying oils	See comments columns	Bulk	Bedding and fronting	Up to 900 N/m²	Paint protection essential. Life dependent on paint maintenance. Inspect annually
2a. Non-setting compound	Oils, plasticizers, polymers	See comments column	Bulk	Bedding glass and beads	Up to 1500 N/m²	Paint protection not essential, but painting prolongs life. Inspect every year and maintain as necessary. Sealer required for porous surround
2b. Preformed mastic tape (non-load bearing)	Oils, plasticizers, polymers	Over 15 years	Reels	Bedding	Up to 1500 N/m²	Limited load-bearing capability. Use with distance pieces
3. Gun grade sealant, solvent release type	Synthetic rubber or acrylic polymers and plasticizers	Over 15 years	Bulk or gun packs	Heel bead or bedding beads	Up to 1500 N/m²	
4a. Preformed mastic tape, load bearing	Synthetic rubber and plasticizers	See comments column	Reels	Bedding glass	Exceeding 1500 N/m²	Drainage is essential

	Material	Life expectancy	Form	Application	Wind pressure	Comments
4b. Solid sections	Vulcanized synthetic rubber or PVC	See comments column	Coils	Bedding glass	Varies according to design and materials	Drainage is essential
4c. Expanded sections	Cross-linked synthetic polymer or PVC	Over 10 years	Rolls	Bedding glass	Up to 1000 N/m^2	Drainage should be provided. Resistance to wind pressure can be improved by capping or heel beading with sealant
5a. Curing gun grade sealant, one-part	Solvent release acrylic sealants, polysulphide, silicone, polyurethane	Over 20 years	Gun packs or bulk	Heal bead and capping	Exceeding 1500 N/m^2	Cure may be slow owing to concealed position of sealant
5b. Curing gun grade sealant, two-part	Polyurethane, polysulphide, flexibilized epoxide	Over 20 years	Proportioned packs for on-site mixing	Heel bead and capping	Exceeding 1500 N/m^2	Cure is unaffected by being in concealed positions
6. Two-part rubberizing compound	Synthetic rubber, plasticizers	Over 20 years	Bulk	Bedding glass and beads	Exceeding 1500 N/m^2	Compound is load bearing when cured
7. Reinforced mastic tape	Foil or fabric, coated or impregnated with mastic compound	2–10 years	Rolls	Covering joints	Up to 1000 N/m^2	Generally used as a repair system, lapped over glass and frame
8a. Extruded sections structural gaskets	Vulcanized synthetic rubber	Over 20 years	Factory moulded	Bedding glass	Exceeding 1500 N/m^2	Drainage essential
8b. Extruded sections, non-structural gaskets	Vulcanized synthetic rubber or PVC	Over 20 years	Rolls or factory moulded	Bedding glass	Exceeding 1500 N/m^2	Drainage essential. Joints should be sealed

* Typical life expectancy is defined as the number of years during which the compound or sealant can be expected to perform its intended function satisfactorily under normal conditions of weather and exposure when used in accordance with the manufacturer's instructions.

† These figures should be taken as a comparative guide only, and may be exceeded or attained according to the pane size, the particular material or the glazing system used.

without beads should also be employed when using plastics materials in similar frames. If glazing beads are to be employed, they should be made from metal or wood combined with the superior properties of silicone and polysulphide mastics if edges are in excess of 500 mm; use silicone sealant only for sides over 2 m in length.

3.2.5 Thermal control

The type and extent of glazing can play a very important role in regulating the total annual energy consumption of a building. Windows are responsible for a large proportion of heat lost during the winter. Single-glazed windows in a typical semi-detached house, will account for about 20 per cent of the total heat loss; this can be reduced to 13 per cent with the introduction of double glazing. The implications of both heat losses and thermal gains should be considered in relation to the overall thermal balance of the building and the glazed areas designed accordingly. This process involves analysis of all the functional requirements of windows in terms of natural lighting particularly in relation to the use of special thermal control glasses, solar gains and the effects of ambient heat gains from artificial lighting, the buildings' occupants and its mechanical plant. The provision of natural and/or artificial ventilation is the other crucial factor. Further reference should be made to section 4.2.5 and *MBS: Environment and Services*.

Double glazing with clear glass reduces outward thermal transmission, but does not significantly alter the extent of solar gain. In vertical glazing, a cavity width of about 20 mm halves the amount of heat lost by a single pane, a 3 mm gap achieves about a 30 per cent reduction. The thermal performance of air gaps over 20 mm is practically constant and this optimum performance is almost maintained as the gap is reduced down to 12 mm (the reduction in effectiveness of a 12 mm space is approximately 2 per cent); below this width heat transmission becomes progressively greater. The thickness of glass has no practical effect on thermal insulation. Double glazing methods include *hermetically sealed glazing units, coupled windows, converted single glazing* and *insulated panels* (see photos 3.2 and 3.3).

Photo 3.2 B2 Building, Stockley Park, Heathrow. Architects: Troughton McAslan, 1989.

The translucent glazed panels are made from 'Kalwall'. They consist of two skins of GRP bonded either side of an I-section aluminium frame made up into a grid. The void is filled with fibre glass insulation which achieves a U-value of 0.4 W/m^{2}K with 30 per cent light transmission. The sun shades are tensioned PVC-coated polyester fabric containing a sheet of black PVC to reduce light transmittance.

Photo 3.3 B4 Building, Stockley Park, Heathrow. Architects: Troughton McAslan, 1991.

On this later phase of the Apple Computer UK Headquarters, a translucent panel with a larger-scale framing grid was used. They consist of a sandwich of two panes of clear glass either side of a slab of 'Ozalux' — a translucent insulant made from extruded acrylic filaments. The material is produced in thicknesses up to 40 mm which can achieve a U-value of 1.14 W/m^{2}K. Even fairly thick panels have quite high light transmittance because the filaments are at right angles to the glass and act as optical fibres.

(1) Single seal unit with hollow spacer frame

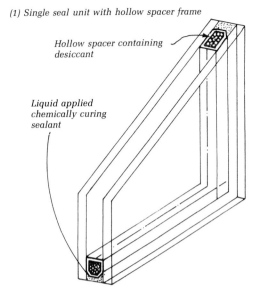

Hollow spacer containing
desiccant

Liquid applied
chemically curing
sealant

*(2) Single seal unit with pre-extruded
reinforced tape*

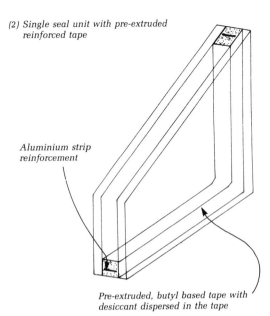

Aluminium strip
reinforcement

Pre-extruded, butyl based tape with
desiccant dispersed in the tape

(3) Dual seal with hollow spacer frame

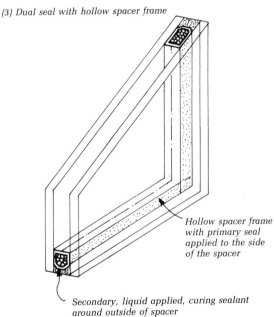

Hollow spacer frame
with primary seal
applied to the side
of the spacer

Secondary, liquid applied, curing sealant
around outside of spacer

Fig. 3.15 Types of double glazed units (GGF)

Hermetically sealed multiple glazing units These consist of multiple (usually two or three) parallel panes sealed together so that there is a gap between each pane. The spacer may be a tube which contains a desiccant to absorb any moisture between the panes sealed at the edges or a butyl-based tape with an embedded aluminium strip reinforcement. Various types are illustrated in fig. 3.15 and they are categorized according to whether they are singly

or double sealed. Double glazing units are factory produced and can be fabricated with many types of glasses used in combination — annealed, toughened, laminated, solar control etc. A range of sizes for the units is given in table 3.4, and table 3.5 and fig. 3.16 show glazing details. A *low-emissivity coating* to one of the surfaces of a double-glazed unit with a 12 mm air space can increase the insulation from about 2.8 to 2.0 W/m²K in a sheltered area. Argon gas filled double glazing has a *U*-value of approximately 2.5 W/m²K and with the addition of a low-emissivity coating this can be lowered to 1.5 W/m²K. Manufacturers will produce purpose-made hermetically sealed triple-glazed units. The handling and fixing of double glazed units call for considerable care and skill. Figure 3.17 shows a stepped unit which is used for double glazing existing windows with small rebates. Otherwise, timber frames in the UK will generally accept 14 mm thick units (two 4 mm panes and a 6 mm gap) or up to 20 mm (4−12−4). PVC-U and aluminium windows are able to accommodate wider types such as 4−16−4 or 4−20−4.

The GGF data sheet *Glazing techniques for insulating glass units* sets a higher standard than the recommendations in BS 6262 and it is to be expected that the standard will be revised accordingly. The GGF describe two alternative approaches to the glazing of insulating glass units. The edge seals of double glazing must not be in continuous contact with trapped water or the seal will deteriorate. The first method seeks to overcome this problem by building a void behind the edge seals that is connected to the outside by a series of drain and vent holes. The second is to solid bed

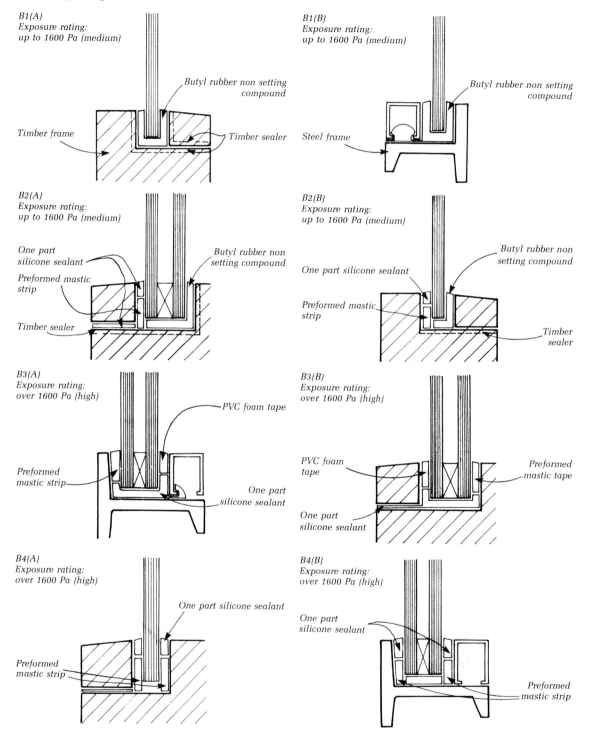

Fig. 3.16 Alternative details and specifications for external glazing (Hodgson's sealants)

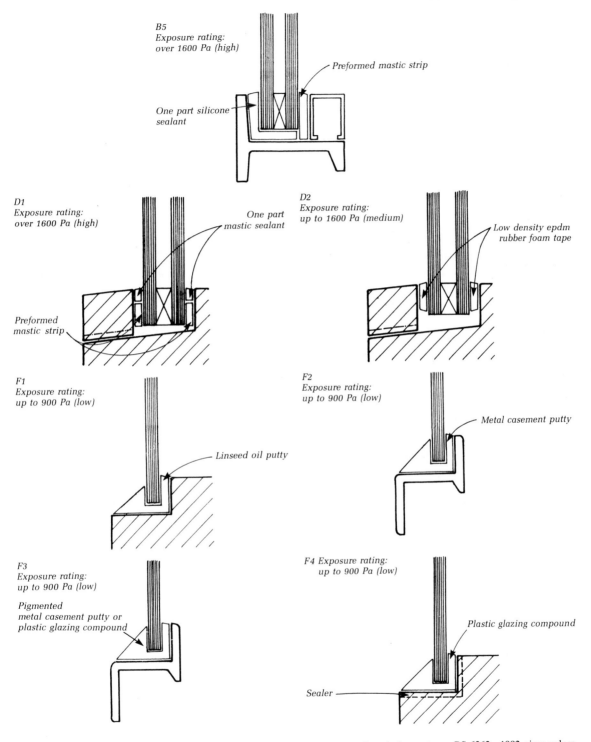

B5
Exposure rating:
over 1600 Pa (high)

Preformed mastic strip

One part silicone
sealant

D1
Exposure rating:
over 1600 Pa (high)

One part
mastic sealant

D2
Exposure rating:
up to 1600 Pa (medium)

Low density epdm
rubber foam tape

Preformed
mastic strip

F1
Exposure rating:
up to 900 Pa (low)

Linseed oil putty

F2
Exposure rating:
up to 900 Pa (low)

Metal casement putty

F3
Exposure rating:
up to 900 Pa (low)

Pigmented
metal casement putty or
plastic glazing compound

F4 Exposure rating:
up to 900 Pa (low)

Plastic glazing compound

Sealer

Note: These are the manufacturers' guidances — there is no BS test for non-gasket glazing systems. BS 6262 : 1982 gives values in N/m². The conversion to the now generally used SI unit (Pascal) is 1 Pa = 1 N/m².

the glazing using a silicone sealant which is impermeable to water but permeable to water vapour. As a consequence, any water that might become trapped within the joint can evaporate out through the body of the sealant.

Figure 4.53 illustrates the application of the vented method, that depending on the design of the PVC-U frame and gaskets is suitable for low, medium and high exposure ratings. The glass is held by PVC-U or synthetic rubber extruded sections profiled to fit into grooves in the frame and the glazing bead. The gap formed beneath the edge seal of the unit is drained/ventilated through slots cut into the underside of the rail at the bottom corners and bottom centre of the frame and ventilated through openings made at the top corners of the sash. Ventilation is necessary to dry out the cavities and to prevent any moisture being trapped by surface tension. Figure 3.16 D1 and D2 show the same principle applied to a beaded timber window; the underside of the beads has drainage ways grooved into it. The solid bedding method is shown in fig. 3.16 B2(A), B3, B4 and B5.

Adherence to recommended minimum rebate dimensions is essential to the successful installation of insulated glass units; reference should be made to the GGF's data sheet *4.2 Glazing techniques for insulating glass units*. Clearances and depths of rebate for bead fixing have been discussed under those headings. More detailed information should be sought from manufacturers.

Coupled windows   These are made as separate sashes which are coupled so they open together, each could be single glazed or high performance versions have one single pane and the other double glazed (see fig. 4.30). A catch enables the sashes to be released for cleaning. The air space between the frames is vented to the outside to prevent condensation, this has little effect on the overall thermal performance (see p. 88 and BRE digest 140 April 1972 *Double glazing and double windows*).

Converted single glazing   There are many proprietary systems which convert existing windows to 'double glazing', either to enhance thermal performance or to reduce sound transmission (see 'double windows' in section 3.2.7). Most of them attach a second pane of glass or plastics, in aluminium alloy or plastics channels, by clips and/or screws to the inside of the primary window frame or to a subframe within the reveals of the window opening. If the secondary glazing is to reduce heat loss, there should be an airtight seal around the perimeter of the subframe or outer frame to reduce air penetration; a tray of desiccant crystals can be placed between the two panes in order to help prevent 'misting'. Refer to the GGF's data sheet *Design, installation and performance of secondary sashes made of aluminium, PVC-U or wood*, June 1987.

Table 3.4 Maximum manufactured sizes for symmetrical, hermetically sealed double glazing units

Nominal total thickness (mm)	Thickness of each pane (mm)	Air space (mm)	Square units (mm)	Rectangular units (mm)
11	3	5	1270	1780 × 1270
12		6	1270	1780 × 1270
14		8	1270	1780 × 1270
15		9	1270	1730 × 1270
16		10	1270	1780 × 1270
18		12	1270	1780 × 1270
12		6	1270	1780 × 1270
13	4	5	1300	2130 × 1300
14		6	1300	2130 × 1300
16		8	1300	2420 × 1300
17		9	1300	2300 × 1300
28		10	1300	2400 × 1300
15	5	5	1700	2600 × 1300
16		6	1700	3050 × 1300
18		8	1830	2600 × 1400
19		9	1930	3050 × 1450
20		10	2100	4270 × 1600
22		12	2130	3050 × 1600
17	6	5	2000	4270 × 1370
18		6	2000	4270 × 1370
20		8	2130	5000 × 1700
21		9	2240	4270 × 1680
22		10	2440	4270 × 2000
24		12	2440	4270 × 2000
32	10	12	3000	5000 × 3000
36	12	12	3250	5000 × 3180

Source: Architects' Journal, May 1985.

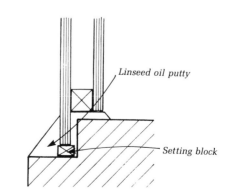

Fig. 3.17   Stepped double glazed unit

Linseed oil putty

Setting block

Table 3.5 Specification key to alternative glazing methods (Hodgson's sealants)

External Windows and doors			Glass							Plastics	
			Single control	Double Glazing Unit			Solar control	Safety		Single	
			Clear float	Flush Edge			Single Glass	Toughened	Laminated	Polycarbonate	PVC polyacrylate
Frame material	Frame finish	Glazing method	Clear float	Clear float	Laminated	Solar control					
Wood	Site painted	rebated beaded	F1 B1	B3	B3	B3	B3	F1 B1	B1, B3	B1, B4	B1, B4
	Site stained	rebated beaded drained	F4 B1 to B4 D1, D2	B2, B3, B4, B6 D1, D2	B3, B4, B6 D1, D2	B3, B4, B6 D1, D2	B3, B4, D1, D2	B4 B1 to B4 D1, D2	B3, B4 D1, D2	B4 D1, D2	B4 D1, D2
Steel	Site painted	rebated beaded	F2 B1	B3	B3	B3	B3	F2 B1	B1, B3	B1, B4	B1, B4
	Factory painted	rebated beaded drained	F3 B1, B2, B4 D1, D2	B4, B5 D1, D2	B4, B5 D1, D2	B4, B5 D1, D2	B4, B5 D1, D2	F3 B1, B2, B4 D1, D2	B2, B4 D1, D2	B4 D1, D2	B4 D1, D2
Aluminium	Anodized or factory painted	rebated drained	B1, B2, B4 D1, D2	B4, B5 D1, D2	B4, B5 D1, D2	B4, B5 D1, D2	B4, B5 D1, D2	B1, B2, B4 D1, D2	B2, B4 D1, D2	B4 D1, D2	B4 D1, D2

Note: Alternative specifications imply different levels of maintenance requirements and performance (see exposure ratings shown on figures). For details see *A Guide to Good Glazing Practice* by Peter Hodgson and Company.

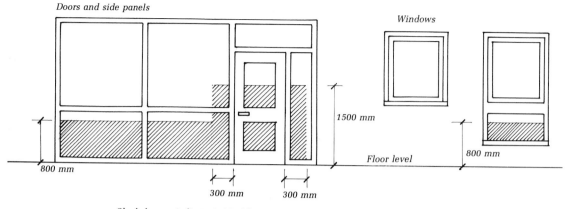

Fig. 3.18 Safety glazing requirements from the *Building Regulations 1991*

It is common practice in the US to fit winter 'storm windows', an additional pane that is fixed to the outside of the window by screws or clips. Also available in the UK are fly screens, mesh panels used to exclude insects in the summer.

3.2.6 Fire precautions

Windows of any type or construction are regarded within section B of the *Building Regulations 1991* as *unprotected areas* within an external wall and are therefore limited in extent where adjacent to a boundary (as discussed in section 4.2.6). Sometimes, however, it is necessary for external glazing to be of fire-resistant construction to satisfy BS 476 *Fire tests on building materials and structures* Part 22: 1987 *Methods for determination of the fire resistance of non-loadbearing elements of construction*. For example it may be necessary for an external escape staircase to be shielded from the possibility of an outbreak of fire within the building or for an unprotected external steel structure to be similarly shielded. Window frames made from steel hot-rolled sections (see 'Steel windows' p. 88) are able to achieve a 1-hour integrity rating under test with a variety of fire-resisting glazing materials of maximum pane size 1.3 m² (see the Steel Window Association's fact sheet no. 6).

Annealed glass cracks in fires, but wired glass and plastics interlayers in laminated glass hold broken particles in place for a time. Laminated fire-resistant glass is manufactured with an intumescent interlayer that expands with heat to form an insulating core.

Plain fire-resisting glass is available in three forms:

1. *Prestressed borosilicate glass* which softens at high temperatures without cracking and, although it can only be cut at the works, is manufactured in sheet sizes up to 2×1.2 m.
2. *Laminated with a special intumescent gel interlayer* which foams under heat to give the glass integrity, stability and insulation from radiant heat.
3. *Toughened calcium/silica flat glass*, a relatively new product produced by a reheat and cooling process which enables the finished product to satisfy the requirement of BS 476: Parts 20 and 22.

Plastics are combustible and PMMA (acrylic) materials are not self-extinguishing. Unplasticized PVC has inherent self-extinguishing properties. Most plastics used for glazing, although often having good surface spread of flame characteristics when tested to BS 476: Part 7 1987, are considered combustible and have little fire resistance. For fire regulations with regard to plastic rooflights see p. 121.

3.2.7 Sound control

Several detailed design aspects of glazing in windows can help reduce the transmission of sound, see section 4.2.7. The basic principles to be observed are:

- Gaps around the edges of frames should be eliminated by the use of weatherstripping; even small gaps can impair sound control.
- The thicker the glass the better; increase in mass is the simplest way of increasing sound insulation. For relatively lightweight components such as windows, cavity construction can considerably improve performance. Double glazing may increase sound reduction by 10−15 dB particularly if the panes differ in thickness from each other by at least 30 per cent. The

cavity can be filled with sulphur hexafluoride gas which absorbs sound in the air space.

- A sizeable space between two panes and absorbent linings to reveals will substantially increase sound attenuation (see p. 65). The cavity needs to be much wider than is required for thermal insulation; a minimum gap of 100 mm is required and preferably 200–300 mm.
- Laminated glass can increase insulation by up to 3 dB compared with solid glass of the same thickness; the composition of laminated glass enables its mass to be assisted by the dampening properties of the interlayer.

Double windows These comprise two single glazed windows, fitted separately into the structural opening. This construction is suitable where sound reduction is a main consideration. The panes should be appropriately spaced apart and of different thicknesses (which improves sound reduction particularly at higher frequencies). The frames must be sealed at their edges as tightly as possible and a sound-absorbent lining fixed to the reveals. Venetian or other blinds can be installed within the space between the windows, controlled by cords or rods. One leaf of the window should preferably be a fixed light, the other being openable so the cavity space can be cleaned. See Building Research Establishment digest 140 *Double glazing and double windows*.

3.2.8 Security

Security glazing systems are available for protection against manual attack, the use of firearms, and the effect of explosions — see *MBS: Introduction to Building* chapter 12. Materials suitable for security glazing include toughened glass, anti-bandit laminated glass, bullet-resistant laminated glass and blast-resistant laminated glass, as well as certain single plastics sheets and laminated plastics. Laminated glass and plastics can incorporate an alarm-sounding interlayer which activates when attacked. The selection of an appropriate material depends on the precise level of protection required. The proposed application and performance should always be discussed with the client and the manufacturer.

Further reference on this subject should be made to BS 5051 *Security glazing* Part 1: 1988. Recommendations for installing anti-bandit framed glazing, and bullet-resistant framed or unframed glazing for internal use are given in BS 5357: 1976 (1985) *Code of practice for the installation of security glazing*. This recommends appropriate forms of construction for the surrounding building fabric, for the fixing of the glass given the anticipated direction of attack and the specification of glazing compounds and framing.

3.2.9 Safety

Taking into account intended use, BS 6262 puts considerable emphasis on the need for glass and plastics in glazing to be selected relative to suitable type, thickness and size so as to provide an appropriate degree of safety. The standard identified particular 'risk areas', a concept which has now been incorporated in the revision of Part N of the *Building Regulations 1991* (see fig. 3.18). The regulations' requirements significantly exceed those of the British Standard; in Part N1 the following critical locations are identified:

- Glazed doors and side panels, and windows and glazing within 300 mm either side of door openings and to a height of 1500 mm above ground-floor level.
- Low-level glazing, glass or plastics within 800 mm of floor level (at which level children are particularly vulnerable).

Glass in these situations has either to be safety glass (laminated or toughened), annealed glass of sufficient thickness to be sufficiently robust or used in small panes, or be permanently protected by a screen with openings less than 75 mm and without horizontal rails which children would be able to climb.

Performance requirements and the method of testing the energy absorption (impact) of flat safety glass and plastics which are intended to reduce the risk of cutting and piercing injuries are given in BS 6206: 1981 *Specification for impact performance requirements for flat safety glass and safety plastics for use in buildings*. This British Standard specifies permanent markings that are required on all safety glazing and are to be located in a readily visible position. These must indicate the BS number, the class of glazing listed in its clauses (A, B or C), the type of material (i.e. 'T' for toughened), and the trade mark or name of the company who last cut the glass if the original manufacturer's mark has been removed. The GGF has compiled a register of marks to assist identification.

Requirement N2 of the *Building Regulations 1991* is applicable to all building types other than dwellings. It prescribes particular patterns of warning marks that have to be attached to glass in the critical locations described in Part N1. If the presence of the glass is made evident by a substantial frame or by being divided by mullions or transoms, they also satisfy requirement N2.

3.2.10 Durability and maintenance

Account must be taken of the relative durability of glass, plastic and glazing materials when selecting a glazing system. The glazing will form part of the environmental envelope of the building and its overall performance will depend on the various interactions that take place between

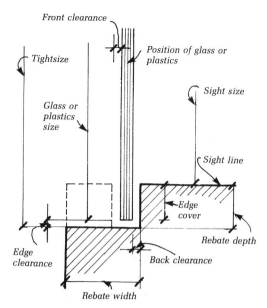

Fig. 3.19 Glazing terminology (from BS 6262)

the glazing and adjacent surfaces and components.

Before installation, glass and plastics must be adequately protected on site and during building work may require special protection (both externally and internally) from the effects of plastering, welding spatter, adhesives, alkaline paint removers, stone cleaning chemicals, etc. The durability, scratch and abrasion resistance of plastics will depend on their chemical nature, added constituents or surface coating.

Once completed, the building owner should be given precise guidance as to appropriate methods of inspection, maintenance, specialist work and replacement instructions. For comments on maintenance, including cleaning, see chapter 4 under each framing material and general points given in section 4.2.8. Refer also to BS 8000: 1990 *Workmanship in building* Part 7: *Code of practice for glazing*, also CP 153 Part 1:1969 *Cleaning and maintenance* and Part 2: 1970 *Durability and maintenance*.

3.3 Glazing of windows

3.3.1 Rebated surrounds

Rebates are made in a window or door frame to form a joint with the glass or plastics sufficient to withstand wind loads and to be airtight and watertight. The glass must not come into contact with the frame but be spaced away from it so the two materials are free to move independently of one another. Figure 3.19 shows the terminology used to describe the dimensions of a rebated glazing joint.

BS 8000 Part 7 lays down precise recommendations for the procedures to be followed when glazing with glass or

Table 3.6 Minimum frame edge clearance for glasses and plastics

Glass types	Edge clearance for a length or breadth:	
	Up to 2 m	Over 2 m
Float, sheet, cast, patterned and wired, up to 12 mm thickness	3 mm	5 mm
Toughened, up to 12 mm thickness	3 mm	5 mm
Laminated, up to 12 mm o/a thickness	3 mm	5 mm
Float, sheet, cast, patterned and wired, over 12 mm thickness	5 mm	5 mm
Toughened, over 12 mm thickness	5 mm	5 mm
Laminated, over 12 mm but under 30 mm o/a thickness	5 mm	5 mm
Laminated, over 18 mm o/a thickness	5 mm	5 mm
Insulating glass units, over 18 mm o/a thickness	5 mm	5 mm
Laminated, exceeding 30 mm o/a thickness	10 mm	10 mm

Based on nominal glass cutting sizes. Edge clearance may need to be greater for some glazing systems, e.g. some gaskets, drained glazing.

Plastics: length of side	Reduction on tight rebate size to allow for thermal movements
Up to 1000 mm	3 mm
1000–2000 mm	5 mm
2000–3000 mm	7 mm

For cutting at 18–20°C and use in ambient temperatures up to 35°C.

Source: BS 6262: 1982, tables 20 and 23.

plastics. The minimum width of rebate required depends on the thickness of glass or plastic and the type of fixing (glazing compound or gasket) which varies according to the framing material (wood, metal or plastics). The minimum rebate depth for sheet plastics depends on the allowances required for their thermal expansion, which is greater than if using glass, as well as the anticipated severity

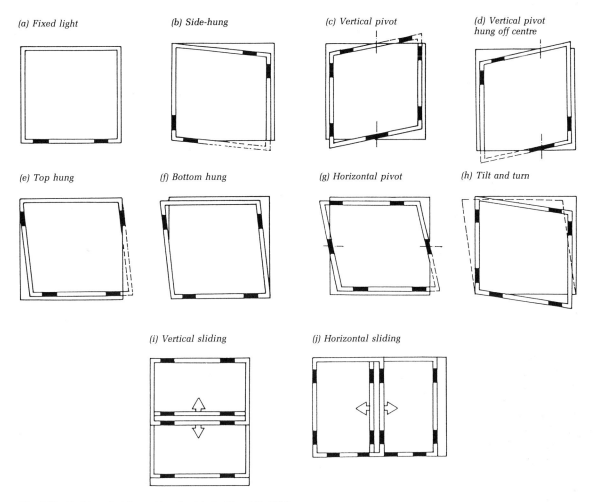

(a) Fixed light   *(b) Side-hung*   *(c) Vertical pivot*   *(d) Vertical pivot hung off centre*

(e) Top hung   *(f) Bottom hung*   *(g) Horizontal pivot*   *(h) Tilt and turn*

(i) Vertical sliding   *(j) Horizontal sliding*

Fig. 3.20   Position of setting and location blocks (from BS 6262)

of exposure. The advice of manufacturers should be sought when making these decisions.

In all cases, it is necessary to provide clearance between the edge and back of the glass or plastic and the frame so that whilst retaining an adequate seal against the weather, differential thermal movement can be accommodated. The bottom edge of panes over, say 0.2 m², must be located on spacers called *setting blocks* that establish this clearance. The resilient materials used traditionally, lead and sealed hardwood, have been superseded by rigid nylon or unplasticized PVC (though this should not be employed with heat absorbing glasses, double glazing units or large heavy panes of ordinary glass). The setting blocks need to be thicker than the pane(s) and generally from 25 to 75 mm long. The exceptions are vertical pivot windows which require a single block 150 mm long centred at the pivot point and side-hung casements which have a single block

close to the hinged side so the weight of the glass is borne directly by the hinges.

Similarly, clearance is provided at the sides and top of the frame by 25 mm long *location blocks* that also ensure that the glazing is centred on the frame. They should be of plasticized PVC or an equivalent softer material than used for setting blocks. The positions of setting and location blocks are shown in fig. 3.20 and table 3.6 indicates the recommended minimum edge clearances for glass and plastics. More generous clearances may however be needed for dark-coloured glass, solar control glass and spandrel panels that have a dark background. Body-tinted glasses that absorb the sun's heat are subject to breakage as a result of thermal stress due to differential expansion of exposed areas of the glass relative to the edges which are obscured. Other factors such as orientation and the internal temperature of the building are important and any defects at the edges of the glass can become points of failure. When using these glasses it is important to seek the advice of the manufacturer.

In addition to setting and location blocks, distance pieces are used to give extra restraint to flexible glazing compounds that may be displaced by wind pressure. They should be placed between the surfaces of the glass and the rebate and bead as shown in fig. 3.21. Distance pieces are usually 25 mm long and of a height that will allow a 3 mm cover of mastic. They should be under slight compression when beads are fixed. Distance pieces must not coincide with setting blocks or location blocks and they should be of softer material than setting blocks, usually plasticized PVC.

Putty in wood frames with rebates Putty is a commonly used method of fixing small panes of glass (fig. 3.11) in ordinary quality external painted joinery; it is rarely employed internally. Putty becomes brittle with age, it must be regularly painted and it is difficult to remove without damaging the surrounding wood. It is too rigid to accommodate large differential movements and is only suitable if wind loads are not expected to exceed 2300 N/m^2. For wind loads upto 1900 N/m^2 the maximum length plus height should not exceed 2700 mm, reducing to 2300 mm for wind loads up to 2300 N/m^2.

Linseed oil putty to BS 544: 1969 (1987) *Specification for linseed oil putty for use in wooden frames* sets partly by absorption of the oil in wood surrounds, and partly by oxidation. For this reason, linseed oil putty should not be used on dense, non-absorbent hardwood, such as teak, because initial setting will be considerably delayed. To prevent excessive and premature absorption of oil into more absorbent woods, it is necessary to prime the rebates. The primer should be one coat to conform with BS 2523: 1966 (1983) *Specification for lead-based priming paints*, or be an aluminium based wood primer to BS 4756: 1971 (1983) *Specification for ready mixed aluminium priming paints for woodwork*. External putty needs to be protected from the weather by a coating of impermeable paint. For this reason putty is not suited to windows that are are to be finished with highly microporous coatings such as exterior wood stains. Also, the greater degree of movement of timber frames finished with microporous stain leads to breakdown in adhesion between the putty and the glass and frame.

Setting blocks are placed in position for the glass and the rebates are puttied with bedding putty. The glass is then pressed into position and secured with glaziers' sprigs spaced at about 450 mm apart around the perimeter of the frame. On pressing in the glass, the remaining *back putty* should be not less than 2 mm thick between the glass and the rebate. Some putty will be squeezed out and this should be cut back inside at an angle, to prevent shrinkage causing

a groove in which condensation and dirt can accumulate.

The glass is then *front puttied* and formed with a putty knife at about 40 degrees to throw off water at this vulnerable point. The putty should be stopped about 2 mm from the sight line of the rebate so that paint can be applied over the glass up to the sight line to seal the edge of the putty to the glass without being seen from inside. The putty must be protected with paint as soon as possible after the initial hardening of the surface, to prevent long-term shrinkage and cracking.

As an alternative to linseed oil putty, polymer-based flexible glazing compounds can be used. They are formulated to be compatible with microporous paints and stains and have to be applied under carefully controlled conditions. An alternative method is to bed the glass in mastic tape and finish the joint in the same way as for putty but instead with curing gun grade sealant either one part (type 5a) or two part (type 5b), although it is difficult to achieve the smoothness and neat lines associated with putty glazing. See the GGF's *Manual* Section 5 'Glazing techniques for timber windows with microporous stain finishes'. Bead glazing techniques are preferred by the GGF and their recommended methods include both drained and solid bedded methods for glazing insulated glass units on microporous stain finished frames.

Microporous stains permit changes in the moisture content of the timber, and the resulting movement in the timber has to be accommodated at joints. The junction between the frame and surrounding wall at the jambs should be not less than 6 mm wide and at least 3 mm of compound should separate the glass from the beads and rebate upstand.

Putty in metal frames with rebates Metal casement putty is able to set hard and adhere to non-porous surfaces. The putty needs to be painted as soon as it has become firm otherwise loss of adhesion or cracking can occur.

Rebates in galvanized steel windows, unless suitably treated or painted 'at works', should be primed with calcium plumbate or self-etching primers.

This method is similar to that described for wood frames, but as no absorption of the oil in the putty can take place, a volatile solvent is added that gives pliability for working and then evaporates, acting as a hardening agent. The putty should be left for 7–14 days to firm sufficiently for painting to commence. A wire clip, used to retain the glass before the front putty is applied, is shown in fig. 3.21. Synthetic compounds are now available for site glazing into polyester powder coated frames, no further painting being required; their standard colour is white. Single glazing and a thin double glazing unit in metal frames are shown in fig. 3.16.

Beads in wood frames with rebates Beads are neater than putty fillets, more easily removed and they are necessary

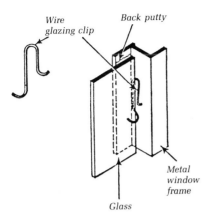

Fig. 3.21 Wire glazing clip

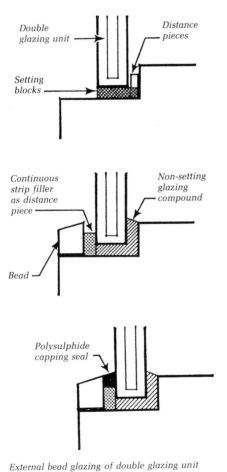

External bead glazing of double glazing unit

Fig. 3.22 Frame section showing setting and location blocks and distance pieces

to retain large panes. However, as replacement beads cannot always be guaranteed to be waterproof and are less secure, beads may be fixed on the inside of the glass. Broken glass can then be replaced from within the building, a considerable aid to maintenance for most traditional types of window in taller buildings. On the other hand, if internal beads are employed, care has to be taken that water does not penetrate at the outside junction of glass and rebate. External beads can be splayed in section so at their lower edge they form a tight joint with the glass, and the mastic filler is squeezed to give a watertight seal. Screws for attaching the beads should be at approximately 300 mm centres. The sloping top surface of the bead not only sheds water but angles the screw fixings into the sash which tightens the beads against the glass. In addition to setting and location blocks, where non-setting compounds may be displaced by wind pressure, distance pieces should be placed between the back edge of glass and the rebate (see fig. 3.22).

For lower-priced work which is to be painted, softwood beads are often fixed by panel pins, with some risk of the beads being damaged if they have to be removed later. For better quality windows, powder-coated aluminium beads or clear finished hardwood beads are fixed with screws, preferably in cups. In windows, both the glass and beads are set in glazing compound or sealant; timber beads may have to be primed on the back depending on the glazing material used. Beads which provide fire resistance are described in *MBS: Internal Components* chapter 6.

Figure 3.12(b) shows an example of single glazing with external beads. The rebate and beads are first painted with a proprietary sealer. Once this is dry, non-setting compound is applied to the rebate, setting and location blocks and distance pieces are inserted. The glass is then centred within the opening, seated on the setting blocks and pressed into the non-setting compound to engage the distance pieces. Preformed mastic strip is applied round the outer edges of the pane with its top 6 mm below the sight line, and it is capped with a bead of sealant finished with a slope away from the glass. The beads screwed or pinned in position are bedded to the frame on a fillet of sealant, usually applied by gun. For internal bead glazing, the bedding for the bead can be omitted. Figures 3.16 B and D show alternative specifications and examples of bead glazing of double-glazed units using both solid bedded and drained methods.

Beads in metal frames with rebates Pressed metal beads are either fixed by being screwed into threaded holes or they are clipped over studs; solid steel beads improve the fire resistance of metal frames. The use of non-setting compound, preformed mastic strip plus silicone sealant or gaskets is associated with bead glazing in metal frames. Typical applications are shown in fig. 3.16 (B1, B4 and B5).

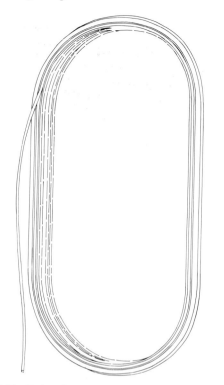

Fig. 3.23 Gasket glazing of windows in composite panels

3.3.2 Grooved surrounds

Dry glazing with gaskets is much used, particularly for the factory glazing of windows. The weather-seal works by the gasket being tightly restrained between the glass and the frame which often has a groove at the head and jambs and a bead at the sill — see fig. 3.11(c)–(f). The grooves need to be of sufficient depth to allow the initial positioning of the pane, and extruded strip gaskets or channel gaskets are used to retain the glazing. Both types are likely to have external fins to help shed water and are internally finned to be more easily pushed into the narrow space between glass and frame. A large variety of gasket sections are made in different grades of hardness according to their application, and it is important that the correct grade is specified as a too flexible gasket may leak. Strip gaskets are placed in position after the glass is seated on its setting and location blocks. Channel-gasketted glazing has the frame built around it so beads are not required but the frame has to be disassembled if the glass is to be replaced (see aluminium sliding windows, p. 105). Distance pieces are not necessary when using channel gaskets, setting and location blocks may also not be required or be replaced with a continuous strip of neoprene between the gasket and the frame.

3.3.3 Nibbed surrounds

Instead of having a rebate or groove, a frame can incorporate a projecting nib for glazing. In this case a structural gasket is also used — see fig. 3.13. When the glass or plastic is located clear of the frame, structural

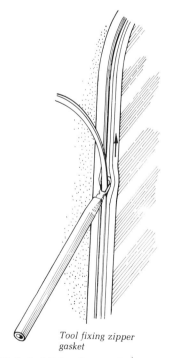

Tool fixing zipper gasket

Fig. 3.24 Method of fixing zipper gasket

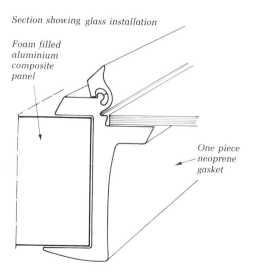

Section showing glass installation

Foam filled aluminium composite panel

One piece neoprene gasket

Photo 3.4 S.C. Johnson and Son Office Building, Racine, Wisconsin. Architect: Frank Lloyd Wright, 1939.

Even half a century later, the use of glass at Johnson Wax seems remarkable. Pyrex tubing and silicone sealant were both newly invented, this being a forerunner of the structural glazing applications described in chapter 6. The windows of the building have two skins of glass tube, the rooflights only one, but in both instances the tubing was held in place by cast aluminium racks. The butt joints between the tubes were continuously sealed with a bead of sealant. After some years of persistent leaks, due to ponding and thermal movement of the glass, the rooflights were covered over by external, conventional rooflights.

gaskets are used to connect the two, rather like a window in a car — see fig. 3.23. Structural gaskets not only have to provide an effective seal against the weather but also have to support the glass. The sealing pressure on the periphery of the glass and on the surround is the result of pushing in a 'zipper' strip made from a harder synthetic rubber than that used for the main gasket. The strip is inserted and removed (for reglazing) by the use of a special tool (fig. 3.24).

The corners of frames are vulnerable to water penetration unless either gaskets are mitred and welded or the windows are curved with gaskets made as a continuous length. Being resilient, structural gaskets located over nibs are well suited to curved openings and are much used for windows within steel, aluminium, GRP or glass fibre reinforced cement (GRC) composite panels. Alternatively to form angled corners, injection moulded L-pieces can be welded between lengths of extruded gasket to make a jointless assembly.

4 Windows

4.1 Introduction

The term *barrier* is a familiar description for many building elements. For example, roofs are obviously required to form a barrier to rainwater. Windows however, are a component of the external envelope that perform the complex task of *filtering*.

They have to admit air for ventilation whilst at other times excluding draughts, they admit light but may need to block direct sunlight and provide a view out whilst obstructing views in (see photo 4.1). The following list gives, in broad outline, the main points and questions to be considered in respect of window design:

- The disposition and proportion of window openings and frames have the most profound effect on a building's appearance. Arguably, once having established its overall shape and massing, this decision is the principal determinant of its architectural character. The choice of materials to be used is similarly subject to both aesthetic and functional considerations.
- The *Building Regulations 1991* require particular standards of thermal insulation and fire protection for external walls including window openings; and, if they are to open, require that they supply an appropriate level of ventilation (to habitable rooms, kitchens, bathrooms, common spaces and sanitary accommodation).
- The size and shape of window openings will determine the distribution and intensity of daylighting within the spaces they serve.
- Choice of the correct type and weight of glass.
- The extent, location and construction of opening parts of the window have to be decided, in terms of the proportion of total opening and the position and type of opening light or ventilator to admit enough fresh air for adequate ventilation.
- Determining whether or not the window openings will

have any special requirements for sound or heat insulation.
- The materials, method of manufacture and jointing must satisfactorily withstand the elements for an agreed period of time.
- The junction between the fixed and moving parts of the frame and the window opening must be designed to remain weathertight throughout their expected life (which should be defined).
- Ensuring that the combined components will fulfil all the current applicable statutory regulations and requirements.
- Establishing the permissible extent of openings within a wall relative to site boundaries (because of the need to guard against the possibility of a fire spreading from one building to another).
- Doing all this within an agreed cost limit.

Consequently a large number of technical and aesthetic decisions must be taken to arrive at a satisfactory design for a façade and the windows within it, and to control the comfort conditions in the rooms which they light and ventilate. The more important design points are summarized in the following paragraphs.

4.2 Performance requirements

4.2.1 Appearance

Visually important aspects of a fenestration pattern are the subdivision of the window, the proportion of the panes and the relationship between the glass and frame and the face of the wall. The location of transoms and mullions needs to be carefully considered so as not to obstruct lines of sight, when either standing or sitting, relative to views whether focused or panoramic.

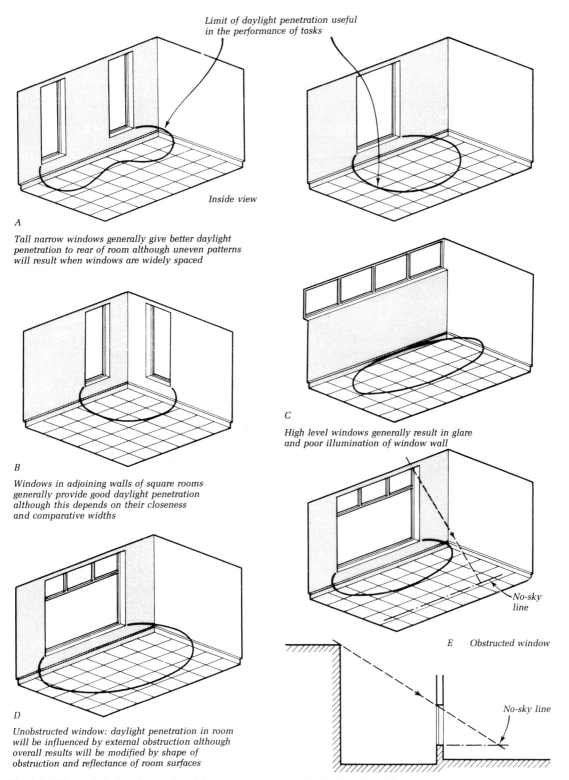

Limit of daylight penetration useful
in the performance of tasks

Inside view

A

*Tall narrow windows generally give better daylight
penetration to rear of room although uneven patterns
will result when windows are widely spaced*

B

*Windows in adjoining walls of square rooms
generally provide good daylight penetration
although this depends on their closeness
and comparative widths*

C

*High level windows generally result in glare
and poor illumination of window wall*

D

*Unobstructed window: daylight penetration in room
will be influenced by external obstruction although
overall results will be modified by shape of
obstruction and reflectance of room surfaces*

No-sky
line

E   *Obstructed window*

No-sky line

Fig. 4.1 Effects of window shape and position on penetration and distribution of daylight

4.2.2 Lighting

The extent of lighting, both natural daylight and artificial lighting, is a fundamental design consideration. Analysis should be carried out in greater detail at each stage of the design so that the lighting pattern has been fully evaluated before production drawings are made.

The window must light a room efficiently by providing the right amount of daylight in the right place with regard to the use of the room. Where the level of illumination is a critical factor, the amount of light falling on working surfaces may be calculated. This is a complex procedure; students are referred to *MBS: Environment and Services* chapter 4 *Daylighting*.

In addition to the number and size of windows, their shape and position and the amount of obstruction caused by the mullions and transoms, also determine the distribution of light in a room (see fig. 4.1).

In terms of comfort, the extent of contrast between the light area of the window and the relatively dark area of the surrounding wall is an important factor. Too sharp a contrast creates glare. This was avoided in Georgian houses, for example, by the use of tall windows with wide-splayed internal reveals painted a light colour; these reflect the light and lessen the contrast between the window and wall. The windows commonly start near the floor, distributing light over a large area of floor and reflecting light deeper into the room. These tall windows continue up to near the ceiling, producing the same effect at high level. Rather than concentrating windows into one wide opening, a number of openings were made, spaced apart to avoid areas of deep shade being formed in the corners of rooms. Their glazing bars are of narrow section and splayed, so that the inclined surfaces are illuminated and only a thin edge is left dark.

A good level of illumination results, without requiring excessive window area.

4.2.3 Ventilation

Window openings should be arranged to give an amount of ventilation most suited to the use of the room or space they serve. This may require a large amount of opening, to give a very rapid change of air or alternatively a small opening providing slow, regulated and controlled ventilation (see fig. 4.2). The mechanics of air movement is considered in more detail in *MBS: Environment and Services* chapter 3. It may be desirable to provide one or more modes of ventilation in the same window. Figure 4.3 shows common alternative ways of providing opening lights, or casements, in a window (see photos 4.2 and 4.3).

To describe the *handing* of a side-hung casement window the following convention is followed: the opening casement is right or left hand according to the side on which the casement occurs looking from the outside. The casement is always hinged on the outside of the frame. Most working drawings show an external view of the casement as shown in fig. 4.4. Note that the apex of the triangle drawn in elevation always indicates the hinged side of the opening light.

Georgian *vertical sliding sash windows* also provide effective ventilation. Warm and stale air within a room can be expelled by opening the upper sash. The size of opening is variable and the head of the window is normally protected from the ingress of rainwater by an arch or lintel. The sash can also be opened at the bottom for letting in fresh air in summer, the opening being up to half the total area of the window. For the most efficient natural ventilation and to serve more than just a few metres at the perimeter,

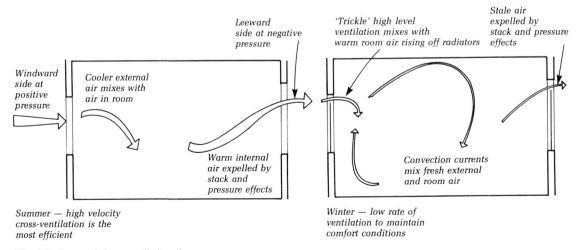

Fig. 4.2   Summer/winter ventilation diagrams

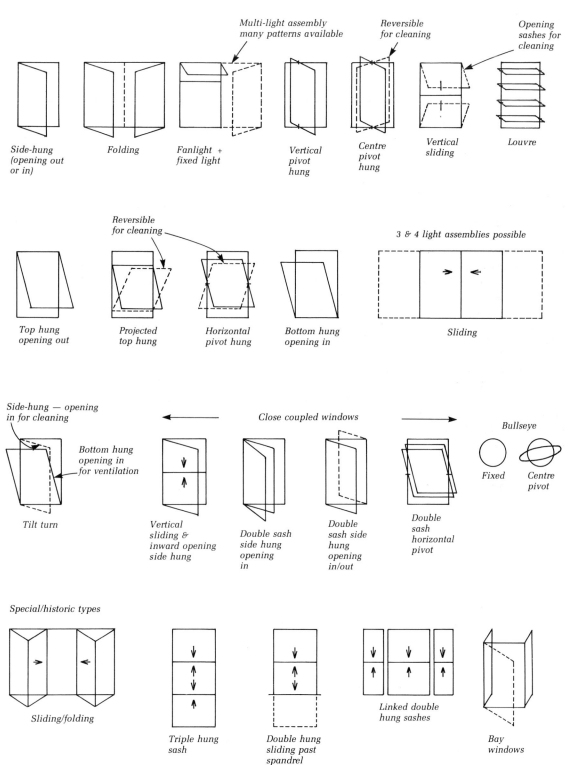

Multi-light assembly
many patterns available

Reversible
for cleaning

Opening
sashes for
cleaning

Side-hung
(opening out
or in)

Folding

Fanlight +
fixed light

Vertical
pivot
hung

Centre
pivot
hung

Vertical
sliding

Louvre

Reversible
for cleaning

3 & 4 light assemblies possible

Top hung
opening out

Projected
top hung

Horizontal
pivot hung

Bottom hung
opening in

Sliding

Side-hung — opening
in for cleaning

Close coupled windows

Bullseye

Bottom hung
opening in
for ventilation

Fixed

Centre
pivot

Tilt turn

Vertical
sliding &
inward opening
side hung

Double sash
side hung
opening
in

Double
sash side
hung
opening
in/out

Double
sash
horizontal
pivot

Special/historic types

Sliding/folding

Triple hung
sash

Double hung
sliding past
spandrel

Linked double
hung sashes

Bay
windows

Fig. 4.3 Types of window casements

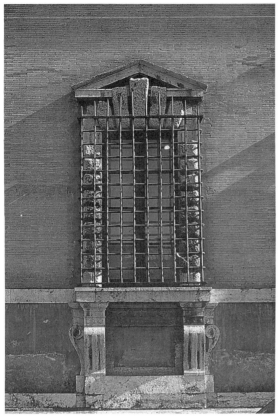

Photo 4.1 Ground floor window of the Villa Giulia (1551–5). Architect: Jacopo Vignola.

The windows of Roman Palazzi are a perfect fusion of functional — security, lighting and ventilation — and aesthetic requirements. This type of 'kneeling' window, with a protruding sill supported by corbels, was often used to frame a smaller window lighting a basement.

Photo 4.2 Shuttered window in Holland.

Although traditional practice in Europe, it is also becoming more usual in the UK for windows to be fitted with shutters — either side-hung jalousies or vertical roller shutters, as in this example.

through ventilation is required. Sash windows were sometimes fitted with draught boards that obstructed the flow the flow of air when the lower sash was opened but allowed relatively draught free ventilation through the gap formed between the meeting rails (see *MBS: Environment and Services* chapter 3). They did however, have the disadvantages of rattling in the wind, being less than airtight and being dangerous to clean outside, other than by standing on the window sill. Safety hooks for window clearners are now required for buildings with this type of window.

Bottom hung, opening-in and *top hung, opening-out* windows also provide an efficient means of ventilation; however, the *side-hung* opening casement does not allow draught-proof air flow since the vertical gap formed when the window is open results in a concentrated air stream. Often in conjunction with side-hung casements a small top

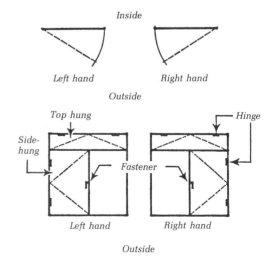

Fig. 4.4 Handing of casements

Photo 4.3 Bay window, Clare College Library, Cambridge. Architects, engineers and quantity surveyors: Arup Associates, 1985. The form of a bay window is capable of a wide variety of design interpretation, in this case forming a complete subsidiary space with its own furniture and environmental control.

hung opening is provided for slow or night ventilation.

The *Building Regulations 1991* are concerned only with ventilation to dwellings; buildings containing dwellings; rooms containing sanitary conveniences; and bathrooms. Approved document F of the *Regulations* requires either a mechanical means of ventilation for these spaces, or openable windows of prescribed size and positioning (table 4.1), louvres, air-bricks or progressively openable ventilators. Two rooms or spaces may be treated as a single room for ventilation purposes if there is an area of permanent opening between them equal to a minimum of one-twentieth the combined floor areas. Special requirements are stipulated when a ventilating window opens on to a courtyard.

The 1990 revision to Part F of the *Regulations* was introduced to counter the problem of condensation, particularly in kitchens and bathrooms. All bathrooms (not just internal ones as previously), are now required to be intermittently mechanically ventilated. Kitchens must have an extract fan or cooker hood. In addition a permanent trickle vent of minimum free area 4000 mm^2 should be incorporated in all habitable rooms and in kitchen windows. Internal kitchens are instead required to have a fan capable of extracting one air change per hour. Alternative approaches to the provision of mechanical ventilation to those given in the *Regulations* are provided in BS 5720: 1979 *Code of practice for mechanical ventilation and air conditioning in buildings* and BS 5250: 1989 *Code of practice: the control of condensation in dwellings*.

Apart from empirical data quoted in the *Building Regulations 1991* it is not easy to say in precise terms how much opening should be provided to give a certain ventilation rate because its extent is dependent on wind pressure and stack effect. Stack effect occurs mainly in winter, when warm air escapes from the upper part of the room and is replaced by cold, and therefore heavier air from outside. The large number of variables involved makes calculation difficult other than for the simplest cases, although computer programes are available (e.g. BRE's BREEZE).

4.2.4 Weather resistance

An important function of an external wall is its ability to provide weather resistance which in turn will help maintain the intended thermal control and ensure durability. Openings in the wall, such as those formed at windows, need special consideration as they require the combination of different materials and a complex set of performance requirements to be fulfilled. Wind and rain penetration often occurs at window openings as a result of poor detailing or faulty installation of the window. Careful analysis of construction methods which achieve a reasonable degree of weather resistance, an example of which is given in fig. 4.5, will indicate that the window itself — including its cross-sectional profiles — plays an important part in achieving protection to the inside of a building.

Standard tests for assessing the weather resistance performance of windows are defined by BS 5368 *Methods of testing windows*: Part 1: 1976: *Air permeability test* Part 2: 1980: *Watertightness test under static pressure*, Part 3: 1978: *Wind resistance tests* and Part 4: 1978: *Form of test report*; BS 6375 *Performance of windows* Part 1: 1989: *Classification for weathertightness* and Part 2: 1987: *Specification for operation and strength characteristics*.

BS 6375 lays down the criteria against which the results obtained from BS 5368 are classified, and includes methods to facilitate the selection and specification of windows in terms of exposure categories based on design wind pressure.

Table 4.1 Natural ventilation provided by openable windows

Room or space	Natural ventilation (opening areas)	Mechanical ventilation (extraction rates)
1 In dwellings		
(a) Habitable rooms	Rapid ventilation: ventilation opening equal to at least $\frac{1}{20}$ th room floor area. Some part at least 1.75 m above floor. Background ventilation: Ventilation opening equal to at least 4000 mm^2	*
(b) Kitchens		Rapid ventilation: 60 litres/s or 30 litres/s if in cooker hood. These rates may be intermittent (i.e. operated during cooking)
	Background ventilation: *Either*: Ventilation opening equal to at least 4000 mm^2	*Or*: Continuous operation at one air change per hour
2 In building containing dwellings		
(a) Common spaces	*Either*: Ventilation opening equal to at least $\frac{1}{50}$ th floor area if common space or communicating common spaces	*Or*: One air change per hour if common space is wholly internal
3 In any building		
(a) Sanitary accommodation	Rapid ventilation: *Either*: Ventilation opening equal to $\frac{1}{20}$ th room floor area. Some part at least 1.75 m above floor	*Or*: Three air changes per hour operated intermittently with 15 minute overrun
(b) Bathrooms	No recommendation given in AD F1	15 litres/s operated intermittently

* No recommendation given in AD F1 but see BS 5720: 1979 *Code of practice for mechanical ventilation and air conditioning in buildings.*
Source: *Building Regulations 1991*

The test apparatus comprises a chamber with an opening into which a window specimen can be fitted and incorporates a means of controlling changes of differential air pressure within defined limits, a system of spraying water at a specific flow rate across the face of the window, and a means of measuring water flow, differential pressure and deflection. In essence, the apparatus tests and window for *air permeability, watertightness* and *wind resistance*, but it is left for specifiers to choose their own acceptable level of performance in these areas relative to the degree of exposure to be experienced by a proposed building.

Windows are generally one of the weak points within the thermal envelope of a building not only due to their relative lack of insulation but also due to uncontrolled air leakage resulting in loss of heat. The design of windows to provide weather resistance has received much attention as a result of the increasing need to conserve energy within buildings (fig. 4.6). A *single rebated timber casement* is designed to form an '*open*' joint between the opening sash and frame which gives a flush appearance. A throating is run to the edge of the profile to discourage capillary action under

driving rain. Open joints are difficult to make water and airtight, to allow for moisture movement the gap between the frame and casement has to be at least 3 mm.

The double rebated window originated on the Continent, examples can be found in France and Switzerland dating back to the Middle Ages. Since the 1940s, this design has become usual for the UK manufacture of standard windows. Both the casement and the frame are rebated, this *double rebated (lipped) casement* results in a '*covered*' joint; the extra rebate was introduced here for draught-proofing. In this case the joint is protected, allowing a generous (often tapered) gap to improve capillary resistance. The window casement must however be fixed to give as efficient and tight a fit as possible against the edge of the rebates on the frame to prevent water penetration and draughts. Because this is difficult to ensure since timber is subject to shrinkage movement, standard windows have been further modified to make them more weather resistant.

Increasingly, even low-performance windows are being manufactured to include weatherstrip seals in the form of compressible strips of chloroprene rubber, EPDM (ethylene

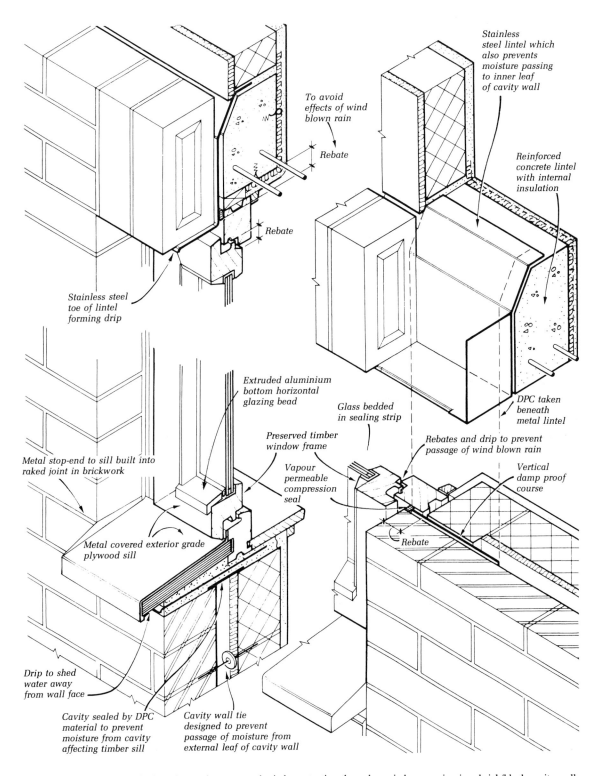

To avoid
effects of wind
blown rain

Rebate

Rebate

Stainless steel
toe of lintel
forming drip

Stainless
steel lintel which
also prevents
moisture passing
to inner leaf
of cavity wall

Reinforced
concrete lintel
with internal
insulation

DPC taken
beneath
metal lintel

Extruded aluminium
bottom horizontal
glazing bead

Glass bedded
in sealing strip

Rebates and drip to prevent
passage of wind blown rain

Vertical
damp proof
course

Metal stop-end to sill built into
raked joint in brickwork

Preserved timber
window frame

Vapour
permeable
compression
seal

Rebate

Metal covered exterior grade
plywood sill

Drip to shed
water away
from wall face

Cavity sealed by DPC
material to prevent
moisture from cavity
affecting timber sill

Cavity wall tie
designed to prevent
passage of moisture from
external leaf of cavity wall

Fig. 4.5 Construction method used to resist water and wind penetration through a window opening in a brick/block cavity wall

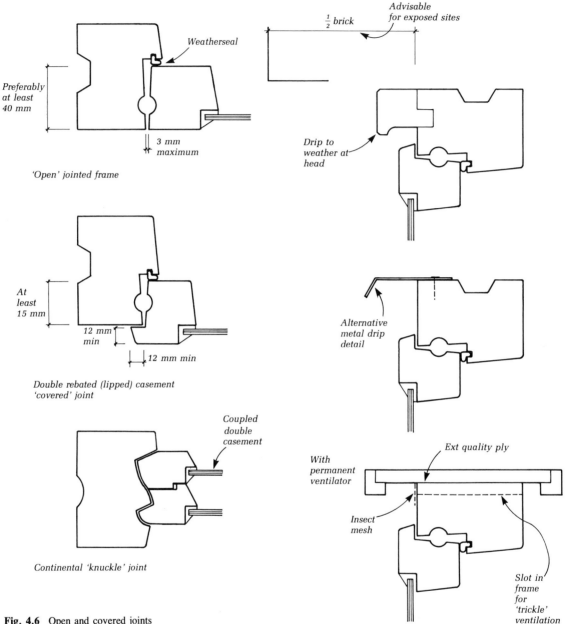

Fig. 4.6 Open and covered joints

polypropylene rubber), or plasticized PVC preferably fitted into grooves, not stapled and with welded corners (fig. 4.7); spring strips of metal alloy have been used in timber windows but mostly for the refurbishment of draughty old windows rather than for use in new ones. For the seal to succeed in resisting wind, air infiltration and water penetration it must be placed well back from the exposed

front face of the window where it will stay dry. If allowed to get wet, wind pressure will be able to 'pump' further rain through the seal. The most difficult window to weatherseal is an inward-opening casement; an example of a weatherstripping technique using neoprene is shown used in conjunction with an inward-opening metal casement in fig. 4.8.

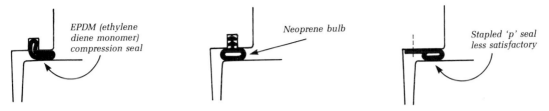

Fig. 4.7 Details of weatherstripping for casements

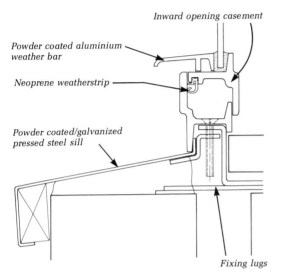

Inward opening casement

Powder coated aluminium weather bar

Neoprene weatherstrip

Powder coated/galvanized pressed steel sill

Fixing lugs

Fig. 4.8 Detail of inward-opening casement at sill (Crittall Windows Ltd)

4.2.5 Thermal control

Increasing requirements for energy conservation have resulted in greater consideration of the effect windows have on the overall heat loss from an external wall. A large area of glass, depending on its orientation, can also result in excessive solar heat gain which causes discomfort and is difficult to rectify speedily by opening windows or even the most sophisticated of artificial ventilation systems.

The *Building Regulations 1991*: Approved Document L1: 'Conservation of fuel and power' includes alternative procedures for meeting thermal control requirements in (a) dwellings, (b) other residential buildings, offices, shops and assembly buildings and (c) industrial, storage and other buildings.

The first method called the *elemental* approach requires certain *U*-values to be met for exposed walls, floors and ground floors in all building types: 0.25 for roofs of dwellings and 0.45 in other buildings and 0.6 for all semi-exposed walls and floors, i.e. elements that separate heated and unheated spaces. For each of these procedures, maximum areas of single-glazed windows and rooflights

Table 4.2 Maximum single glazed areas relative to purpose groups

Purpose group	Window(s) as % area of exposed wall	Rooflight(s) as % area of roof
RESIDENTIAL GROUP		
Dwelling house	Windows and rooflights together 15% of perimeter wall area	
Other residential	25%	20%
NON-RESIDENTIAL GROUP		
Places of assembly, offices and shops*	35%	20%
Industrial and storage	15%	20%

Areas which are double-glazed may have up to TWICE the single-glazed area.
Areas which are double-glazed, with a low-emissivity coating, or triple-glazed, may have up to THREE TIMES the permitted single-glazed area.

* Percentages shown not applicable to display windows in shops.
Note: Applies when following Procedures 1 and 2 of the *Building Regulations 1991* AD L1. For definition of purpose groups see *MBS: Internal Components* chapter 2 *Demountable partitions*.

are specified and these are indicated in table 4.2. Areas which are double glazed may have up to twice the single-glazed area; those which are triple glazed, or are double glazed and incorporate a low-emissivity coating, may have up to three times the permitted single-glazed area. When calculating these percentages it may be necessary to include the surface area occupied by lintels, jambs and sills (as these often fall below the thermal insulation standards of the body of the wall) but their *U*-value should not exceed 1.2 W/m²K. Also, an external door with 2 m² or more of glazed area should be included as part of the percentage allowed for windows and rooflights. The *U*-values required for construction are obtainable from the CIBSE (Chartered Institute of Building Services Engineers' Guide, Section A3: 1980: 'Thermal properties of building structures').

Alternative *Procedure 1* and *Procedure 2* provide calculation methods for designing the exposed-building fabric without limiting the areas of windows and rooflights to maximum percentages. When calculating the rate of heat loss that would occur, the *U*-value of windows and rooflights is taken as 5.7 W/m²K if single glazed, 2.8 for double glazing, and 2.0 for triple glazing or for double glazing incorporating a low-emissivity coating.

Procedure 1 allows a 'trade-off' between the area of windows and rooflights and the insulation value of walls, roofs and floors. The calculation has to show that the rate of heat loss is not greater than a building of the same size and shape designed in accordance with the elemental method. Although larger glazed areas are permitted, no allowance is made for solar gains. Procedure 2 also allows 'trade-off' between glazed and solid areas, but account can be taken of useful heat gains as well, including solar gains and those resulting from artificial light, industrial processes, etc. Acceptance of this 'energy target' method of calculation relies on proof that sufficient heating controls can be provided in respect of the contribution of useful heat gains.

Condensation   As thermal insulation standards increase so does the risk of condensation. It is a problem obviously associated with windows, caused by the interaction of a number of factors — ventilation rate, temperature and the form of construction. Reference should be made to the BRE report *Thermal insulation: avoiding risks*. The provision within the *Building Regulations* for trade-offs between areas of glazing and the *U*-value of other elements is likely to increase the extent of double glazing in new dwellings. Risks associated with this include the increased likelihood of condensation forming at any remaining cold surfaces, for example at windows left single glazed and on metal window frames without thermal breaks. If single and double glazing are both used in the same building it is particularly important that rooms are adequately heated and ventilated. Since moisture can cause the deterioration of the edge seals in double-glazed units, it is as well that double-glazed aluminium windows should be thermally broken. Conventionally, at the perimeter of window openings, the inner leaf or cavity construction joins the outer forming a cold bridge. To avoid this, insulated cavity closers are now widely employed to maintain an unbroken blanket of insulation up to the frame of the window (see fig. 4.9).

Even window frames with good thermal performance, for example PVC-U or timber frames glazed with sealed units, can still suffer from condensation. Condensation channels should be provided in the window sections used in those areas of buildings where condensation commonly occurs, such as bathrooms, shower and washing rooms, kitchen, utility and laundry rooms. Where condensation cannot be controlled it may be wiser to use single glazing and to collect the condensation in drainage channels before directing it to the outside. Increasing use is being made of sealed glazing units filled with argon gas and incorporating low-emissivity glass. These have a *U*-value as low as

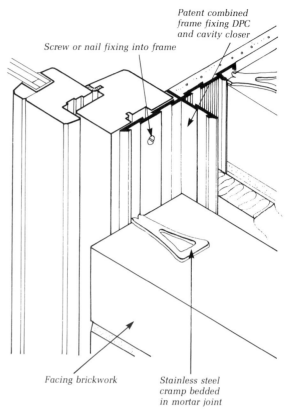

Patent combined frame fixing DPC and cavity closer

Screw or nail fixing into frame

Facing brickwork

Stainless steel cramp bedded in mortar joint

Fig. 4.9   Cavity closer used to avoid cold-bridging at jambs (Radway Plastics Ltd)

1.5 W/m²K, virtually eliminating the risk of condensation. Further information about thermal control and window design is given in section 3.2.5.

4.2.6 Fire precautions

Another limitation set upon the size and number of windows is the need to ensure that fire will not readily spread from one building to another. In the *Building Regulations 1991*, both doors in external walls and window openings (as well as other non-fire resisting components) are classified as *unprotected areas* since a fire within the building could rapidly spread through them to adjoining buildings. Some unprotected areas can however be discounted since they pose little threat in terms of fire spread, for example small openings some way distant from others in the same wall.

There are two methods of calculating the permissible extent of unprotected areas relative to their distance from a *site boundary*, or alternatively from a *notional boundary* (an imaginary boundary between buildings sharing common land). The first is suited to small residential buildings (i.e. a dwelling not more than three storeys high and not more than 24 m wide); a maximum of 5.6 m^2 of unprotected area is allowed a minimum of 1 m away from a boundary, increasing to 30 m^2 at a distance 5 m away, with no restriction thereafter. The second method is suited to all types of buildings. Maximum total percentages of unprotected area ranging from 1 to 100 per cent are permitted depending on the purpose group of the building and the distance from a relevant boundary. The *Building Regulations 1991* also enables the other methods described in the BRE report *External fire spread: building separation and boundary distances*, 1991: Part 1. These are the *enclosing rectangle* and *aggregate notional area* calculations that were first contained in the 1985 revision of the *Regulations*. Designers can use whichever of these methods give the most favourable result, which can be enhanced if a sprinkler system complying with BS 5306: Part 2 is to be installed in the building.

The *Regulations* now stipulate that windows that may be used as means of escape from fire should have a sill between 800 and 1100 mm above floor level and a minimum unobstructed opening 850 mm high and 500 mm wide. Section 7.2.5 and *MBS: Internal Components* chapter 6 provide more information about fire control in buildings.

4.2.7 Sound control

A window may have to achieve particular sound resistance requirements, these generally relate to those aspects of glazing design described on p. 45. See also comments on sound-resistant doors in *MBS: Internal components*. Considerations which can help to reduce transference of noise include:

- Elimination of air gaps at edges of opening lights by the use of draught excluders.
- Using thicker glass — doubling glass thickness can provide an additional 4 dB sound reduction.
- Providing double or multiple glazing in isolated frames — minimum gap of 100 mm (preferably 200 mm) between glass, and jambs lined with sound-absorbent material.
- Ensuring glass panels of sealed double-glazed units (normally only 6 mm gap) are of different thickness to avoid sympathetic resonance between panels.
- Specifying sealed glazed units with gap(s) filled with heavy inert gas.
- Providing laminated glass — the plastic interlayer will give a reduction of sound at different frequencies from glass.

CP 153: *Windows and rooflights* Part 3: 1972 *Sound insulation* deals with sound transmission through windows, rooflights and glazed curtain walling. It is an extension of the more basic information on the subject given in BS 8233:

1987 *Sound insulation and noise reduction for buildings* and provides guidance on likely external noise levels and degrees of sound transmission afforded by various forms of glazing (see chapter 3, also BRE Digests 96, 128 and 129 and *MBS: Environment and Service* chapter 6).

4.2.8 Durability and maintenance

The durability of a window depends largely on the characteristics of the different materials used for manufacture, how they are prepared for the rigours encountered during their life, the construction techniques used for installation in a building, and subsequent maintenance.

CP 153: *Windows and rooflights* Part 2: 1970 *Durability and maintenance* deals with the durability and maintenance of wood and metal windows, including protection on site, preconditioning, type of glues, selection of species and preservative treatment for wood windows, environmental suitability, frequency of washing and other notes on the behaviour of individual metals. Factors relating to the durability of windows arising from use of specific materials are covered later under each type of window.

The maintenance of windows can only be carried out with ease when consideration has been given during their initial design stage to all factors involved. Cleaning is an important consideration requiring careful thought; every effort should be made to ensure that building owners or maintenance personnel can carry out window cleaning in safety. For example, above the second or third storey of a building (and also where access for ladders is not convenient), windows should be designed so they can be cleaned and reglazed from inside the building. For this reason new forms of window such as tilt/turn have been developed. This type has two modes of operation — one for ventilation that is safe, being virtually impossible for children to fall through; is resistant to burglary and can be adjusted without disturbing plants and other paraphernalia on window sills and other horizontal surfaces. Also, by virtue of sophisticated ironmongery, tilt/turn windows can be opened inwards for cleaning.

By making the size and proportion of accessible glazing suitable to human reach (fig. 4.10) and by the use of special hinges, cleaning can be carried out without the aid of an external ladder. Special hinges have been developed for most window types (fig. 4.11). Timber windows can be side hung from pivots rather than hinges so they can be cleaned by reaching out from inside; similarly steel windows can incorporate 'stand-off' hinges (fig. 4.12).

Even vertical sliding sashes that are particularly dangerous to clean from within a room, can incorporate gear which enables the sashes to be released and opened inwards for cleaning. Some types of centre pivot window

Dangerous practices in cleaning the outside of windows from inside building

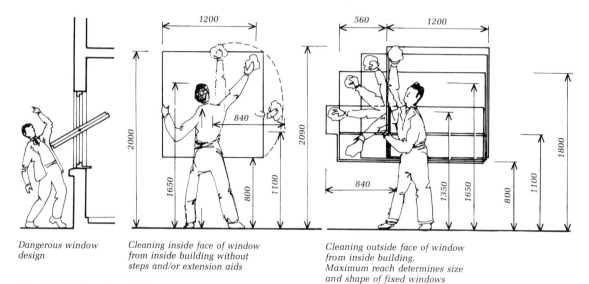

Dangerous window design

Cleaning inside face of window from inside building without steps and/or extension aids

Cleaning outside face of window from inside building. Maximum reach determines size and shape of fixed windows

Fig. 4.10 Design factors influencing safety of window design

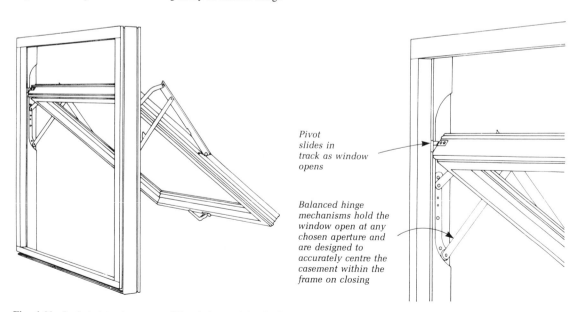

Pivot slides in track as window opens

Balanced hinge mechanisms hold the window open at any chosen aperture and are designed to accurately centre the casement within the frame on closing

Fig. 4.11 Projected top hung reversible window and detail of hinges (Centrum Windows)

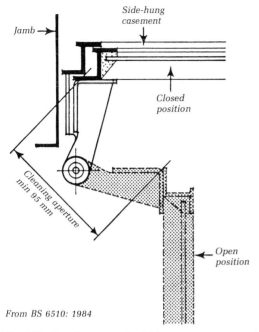

From BS 6510: 1984

Fig. 4.12 Extending (easy-clean) hinge, space requirements

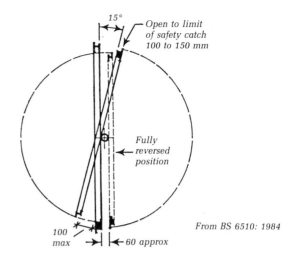

From BS 6510: 1984

Fig. 4.13 Reversible window space requirements

(as illustrated in fig. 4.13) are fitted with hinges that allow the window to turn through 180 degrees before locking in a reversed position. However, consideration must then be given to the effect of this swing on the design of the interior spaces of the building, such as the positioning of the curtains. When not wanting to use these types of window, the designer must consider alternative arrangements for cleaning such as access from balconies or the use of mechanical devices.

The cleaning of windows in tall buildings can be solved by the installation of specialist equipment such as cantilevered gantries, or specially profiled curtain wall mullions to allow safety clips to be inserted or for their use as wheel guides for external cradles. The appearance of the building will be affected by this decision; it is important that the designer and the future window cleaning contractor work together during early design stages.

CP 153: *Windows and rooflights* Part 1: 1969 *Cleaning and safety* deals with cleaning of glazing relative to problems of safety. See also section 3.2.10.

4.3 Timber windows

Timber is readily worked to produce profiled sections for both purpose-made and mass-produced windows but its tendency to move with variations in moisture content and to rot if not correctly detailed and specified have to be taken into consideration — see previous section 'Durability' and

Materials chapter 2. These problems are due to the inherent characteristics of timber; however, timber windows are relatively cheap and can have a life of 50 or 60 years given everyday maintenance skills. The lifespan achieved depends on the timber that is used. Oak can last almost indefinitely, examples of external oak joinery from the Renaissance still function today. Detailing which reduces the degree of exposure of the timber will also help a window to be longer lasting (the practice in Georgian England was for sash boxes to be masked from the weather by a rebate in the masonry forming the window opening). The timber specified should be suitable to the task performed by each individual component and the quality and cost which is envisaged. For lower priced work, preservative treated Douglas fir or larch might be employed instead of hardwood for sills and glazing bars.

Durability is a function of good design, skilled manufacture, correct installation and sufficient maintenance. The problem of decay has reduced with the now almost universal use of preservative treatment on softwood windows by immersion or pressure impregnation with organic solvent preservatives (which is often applied using double vacuum processes) in accordance with BS 5589: 1989 *Code of practice for preservation of timber*. Dry rot can occur in softwoods where the moisture content exceeds 22 per cent due to defective paint work or water penetrating at joints beneath stained finishes, so external softwoods should always be preservative treated. Care must, however, be taken during transportation, storage, installation and finishing of the windows on site. Hardwood windows should also be treated with preservative unless durable species are used and sapwood is precluded in the specification. Although hardwood is still widely used in the manufacture of windows in this country, the ecological implications of timber specification, as discussed in chapter 1, should be borne in mind.

Windows with a primer coat of paint or stain should, as soon as possible after installation, be given an opaque finish (paint) or a semi-transparent finish (stain) to approved specification and application methods — see *MBS: Finishes.* The newer types of paints and decorative stains can considerably reduce maintenance requirements for timber windows. Repainting should occur every 3–5 years, and after about 20 years it may be necessary to remove the thick layers and start again. Stain finishes normally require more frequent maintenance than good exterior quality paints, but the amount of maintenance can be reduced by careful preparation.

Like other elements within external walls, windows throughout a large part of the year are subject to different conditions externally than internally. In winter, warm and humid air within the building results in a difference in vapour pressure across the construction and the migration of humidity through permeable materials, in this case the timber of the window frames. Consequently it is desirable for the internal finish to act as a *vapour barrier*, oil alkyd gloss paint for example, and for the outer surface to be *permeable* to water vapour from the inside whilst being water repellent. Many decorative stains have this capability although they may still allow too great a fluctuation in the moisture content of the timber (and are not suitable for use over window putty); the alternative is to use microporous paint.

An example of a typical standard timber window and the names of the various components of the frame are shown in *MBS: Internal Components* chapter 5. The parts of a window most subject to the deteriorating effects of exposure to the weather, having the greatest flow of rainwater across them, are the sill and other horizontal members at the bottom of the window. The use of tropical hardwoods has become usual for these parts of windows but can be avoided by incorporating metal drips and flashings into the assembly, as has been continental practice. It is increasingly common for the glazing bead at the bottom rail of casements and fixed lights to be made from anodized or polyester-coated aluminium. This provides a more durable surface in this locally onerous situation. Ideally the bead should be face fixed to avoid cold-bridging the glazing assembly (see fig. 4.14). Fig. 4.20 (p. 75) shows the recommendations of the Norwegian BRE to limit water infiltrating the lower parts of the window. These include:

- Bottom rebate: sloped 1:8 and protected, e.g. with PVC tape carried up the sides 40 mm at each end.
- Bottom bead: drainage notches in underside at least 6 mm square at 300 mm centres; generous drip to front edge; concealed surfaces protected, e.g. polyurethane varnish; upper surface sloped 1:6; ends cut square; or better still use an aluminium bead as shown in fig. 4.14.
- Side beads: ends cut to leave gaps at least 6 mm above the top of the bottom bead to prevent water entering the end grain.

4.3.1 Standard and high performance

Standard timber windows in the UK are derived from the sections designed some 50 years ago by the English Joinery Manufacturers Association (thus EJMA sections) at a time when timber was scarce. This pattern was enshrined in earlier editions of the British Standard for timber windows BS 644. In 1976, the British Woodwork Manufacturers Association and the Joinery and Timber Construction Association amalgamated and the British Woodworking Federation (BWF) was formed. Since then the BWF have worked to improve EJMA sections in a number of ways, starting with the need to meet higher performance standards to embrace problems of wind and water penetration; also, the EJMA sections were by then less cheap to manufacture because of discrepancies in dimensions resulting from the conversion from imperial to metric. Timber must still be used efficiently however — in terms of both the material and the steps required for its conversion from tree to window section. The high costs involved in this process require that the finished window must provide a low cost-in-use ratio in competition with other materials, such as metals and plastics. Many manufacturers have obtained

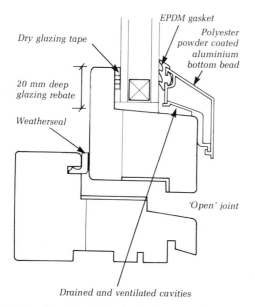

Dry glazing tape

EPDM gasket

Polyester powder coated aluminium bottom bead

20 mm deep glazing rebate

Weatherseal

'Open' joint

Drained and ventilated cavities

Fig. 4.14 Aluminium glazing bead at bottom rail of opening sash

British Board of Agrément test certificates and participate in the BSI Quality Assurance Scheme recommended in BS 5750 (see MBS: Internal Components chapter 1). BS 644: Wood windows Part 1: 1989 Specification for factory assembled windows of various types has replaced the previous prescriptive standard for timber windows with a performance standard. This approach had already been adopted for BS 4873 governing aluminium windows. The revised BS covers most window types with the exception of bay and oriel windows (although their component parts can be specified similarly). Included are definitions of terms and types, materials, workmanship, tolerances, weathertightness, sizes and finishes. It provides a method describing performance requirements against defined criteria for strength and weather resistance and is cross referenced to the other standards that cover these aspects of window design.

Weathertightness is determined by selection of an appropriate exposure category from BS 6375: Performance of windows Part 1: 1983 Classification for weathertightness. The exposure categories are established from test pressure levels for air permeability, watertightness and wind resistance. The test methods are as described in BS 5368: Methods of testing windows Part 1: 1976 Air permeability test Part 2: 1980 Weathertightness test under static pressure Part 3: 1978 Wind resistance tests Part 4: 1978 Form of test report. The appendix to BS 6375 shows how to find the exposure rating of any site in terms of design wind pressure; if using standard windows they should have been tested to achieve the rating required.

The strength of the frames must comply with BS 6375 Part 2: 1987 Specification for operation and strength characteristics (see also Weather resistance p. 59). The strength of the glass and its installation (whether in the factory or on site) and maintenance must satisfy BS 6262: 1982 Code of Practice for glazing of buildings and BS 8000: 1990 Workmanship in Building Part 7: Code of practice for glazing. BS 6262 sets out requirements for thickness of glass and describes the need for safety glass at locations where there is a particular risk (a stipulation which has now been incorporated into the Building Regulations). BS 644 windows that are factory glazed, will comply with BS 6262 but the requirement for safety glass must be specified. Critical locations are described in Part N1 of the Building Regulations 1991, they include windows within 800 mm of ground level and window panels adjacent to ground floor external doors up to a height of 1500 mm above ground level (see fig. 3.18).

Materials and workmanship are covered by BS 1186: Timber for and workmanship in joinery Part 1: 1991 Specification for timber and Part 2: 1988 Specification for workmanship. BS 644 windows are manufactured from timber as good as, or exceeding, the classifications in BS 1186 Part 1: Frames: class 3, Casements, sashes and beads: class 2 (except small section beads which shall be class 1, the highest grade, or class CSH which is for joinery made from clear grades of softwood and hardwood). Although most standard windows are made from a European softwood (such as Scots Pine — Pinus Sylvestris), hardwoods have been increasingly used by most manufacturers. Large sections of the higher classes of BS 1186 timber are not easily obtainable in European Redwood; as a result, window frames are increasingly being constructed from small timber sections glue-laminated together. The laminations have to be detailed to occur at locations sheltered from the weather and ingress of water. An alternative timber, such as Douglas Fir, would be suitable if a clear finish is required.

Figure 4.16 illustrates a typical casement window detail from a manufacturer's metric range. The sill is throated to catch water running down the face and blown under the sill, and to allow it to drip clear of the face of the wall. The sill and sash are also weathered, that is to say, the sloping surface is angled away from the building to take the water which runs down the face of the glass. This slope should be 1:8, or a minimum of 1:10. The internal surfaces also slope to discharge any condensation away from the

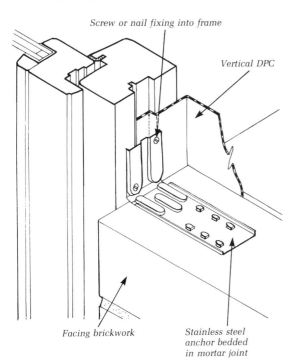

Screw or nail fixing into frame

Vertical DPC

Facing brickwork

Stainless steel anchor bedded in mortar joint

Fig. 4.15 Frame built-in as work proceeds. This type of 'through joint' greatly relies upon the adequacy of the seal made at the wall. It is preferable, particularly in exposed locations, to recess the frame behind the outer leaf as shown in 4.6

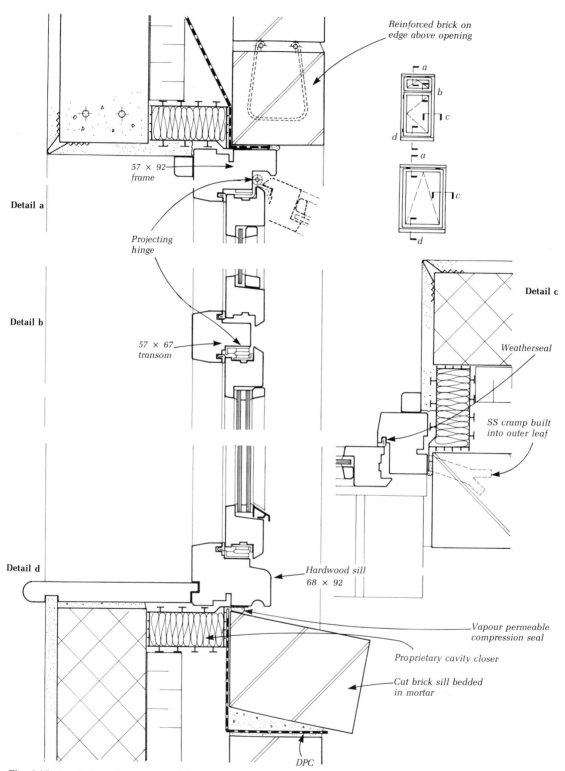

Fig. 4.16 Standard metric casement window

vulnerable junction of sash and glass. Some designs of metal or plastic bead incorporate a drainage route to the outside for condensation. The head of the opening fanlight is also protected by means of a weather-mould which projects over the opening.

Both the casement and the fixed frame are rebated so that a double draught and weather check is provided. The *capillary groove* between head and casement prevents water gaining access by capillary action and wetting the weatherseal. The jambs are detailed similarly — the vertical capillary grooves link with those in the top rail allowing water to be collected and run out onto the projected sill.

It is possible for standard windows to meet the less severe exposure ratings defined by BS 6375: Part 1: 1983 without windows being weathersealed. As was previously the case, the section of the double rebated frame is the only barrier to weather penetration and air infiltration. However, due to moisture movement in timber, even initially well fitting windows can develop sizeable gaps over time. It has become more usual for even low cost standard windows to be weatherstripped — they can then withstand the more severe exposure conditions laid down by the BS. The seals are best fitted in the factory to ensure that they are jointed properly and fit well. The seal must be located at the inside frame rebate. The system of grooves and drips between it and the outside will then ensure that it remains dry; if it is allowed to get wet, subsequent gusts of wind will pump rainwater past the weatherseal. The intention is then, to seal against air movement across the building envelope (which takes rainwater with it) — this is a physical process which attempts to equalize differential air pressure between the inside and outside of the building. This is the most significant factor in sealing against the weather since air seepage is the major cause of leaks in glazing systems. The length of joint between the sash and frame in front of the weatherseal forms an air buffer, slowing down the wind blowing into the joint. Consequently an 'air cushion' is formed that causes wind-borne rain to drop into the capillary groove and so be transmitted back to the outside. Weatherstripping is often fitted within a slot rebated into the frame and is always jointed at the corners. It must be compatible with all types of finishes to the window frame including preservatives, stains and paints, and be easily wiped clean.

Rebate widths for glazing, or *platform widths*, should be of an adequate size to receive at least a sealed double glazed unit (14 mm) plus timber beads — see Chapter 3 for detailed comments. Beads should be of preserved durable wood or metal, scribed or mitred at the corners and can be pinned in with brass pins or screwed with brass screws, in cups if required. Beads can be used externally provided they are appropriately profiled to ensure water is shed to the outside. Whether inside or out, the beads must be appropriately bedded (see section 3.3 *Glazing of windows*) and the glazing itself should have 'back and front' bedding of compatible materials to allow for even support and adequate seal. Putty can be used externally but generally not with sealed double-glazed units — it is cheaper and provides a more effective seal than glazing with beads but its use is limited to smaller sized windows (according to the area and thickness of glass). To fit the glazing, the putty is formed to a triangular bead with a putty knife.

A typical range of sizes for standard timber windows is shown below. Work sizes are 5 mm less than the co-ordinating sizes indicated. It is important to note that because of the flexibility of timber, manufacturers are able to make purpose made versions or special windows at little extra cost.

- Metric dimensionally co-ordinated range
  This is the most used range and encompasses the sizes suitable to most forms of construction, coupled with consideration of the production costs of a highly engineered mass-produced building component.
  Heights in increments of 150 mm to enable alignment with brick courses of the outer skin of the building: 450, 600, 750, 900, 1050, 1200, 1350, 1500 and 2100 mm.
  Widths based on multiples of standard casement sizes: 488, 631, 915, 1200, 1769 and 2338 mm.
- Modular co-ordinated range
  This range is based on a simple matrix of sizes based on heights of 150 mm increments and widths of 300 mm.
  Heights:
  600, 900, 1050, 1200, 1350, 1500, 1800 and 2100 mm.
  Widths:
  300, 600, 900, 1200, 1800 and 2400 mm.
- Imperial range
  This range has been in existence for over 30 years and is still being made.
  Heights by conversion:
  768, 921, 1073, 1226, 1378, 1530 and 2394 mm.
  Widths by conversion:
  438, 641, 921, 1226, 1810 and 2394 mm.

High performance timber windows   Whilst the standard range of windows in the UK is derived from EJMA profiles designed some 50 years ago and embodied in the previous British Standard, most high performance windows owe more to the continental, particularly Scandinavian, tradition and may be designed to Scandinavian standards (see fig. 4.17). These are designed to accept thicker types of double-glazing or triple-glazing and to meet the most severe exposure catagories of BS 6375: Part 1: *Classification for weathertightness*. These are related to test pressure levels for air permeability, watertightness and wind resistance as

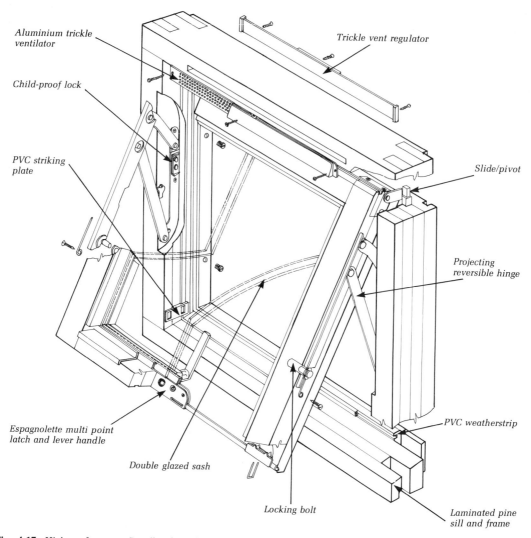

Aluminium trickle ventilator

Trickle vent regulator

Child-proof lock

PVC striking plate

Slide/pivot

Projecting reversible hinge

Espagnolette multi point latch and lever handle

PVC weatherstrip

Double glazed sash

Locking bolt

Laminated pine sill and frame

Fig. 4.17 High performance Scandinavian window (Reventa Products UK Ltd)

determined by the tests described in BS 5368. In terms of thermal insulation and air and water penetration, Scandinavian standards are considerably higher than in the UK.

These windows, some of which are imported, are constructed from a variety of timber species and pressure or vacuum impregnated with water soluble copper/chrome or organic solvent based preservatives. Water based processes increase the moisture content of the timber which has to be dried out before use; alternatively if organic solvents are employed they must be compatible with the glues used during manufacture.

High performance windows, which achieve low air permeability, are suitable for use in air conditioned buildings. The most sophisticated of these maintain an efficient weatherseal as well as good sound and thermal insulation by employing coupled opening lights with a still air gap of 70 to 80 mm between the outer single glazed leaf and the inner double glazed leaf. The two casements are locked together by catches that are designed to allow air circulation between them so as to avoid condensation forming. The outer window is single-glazed with thicker glass for sound insulation, and conventionally double-glazed on the inside for good heat conservation. These linked casement windows can achieve a U-value of 1.2 W/m^{2}K (if the sealed units to the inner leaf incorporate low-E glazing and the unit is gas filled), and a 52 dB sound reduction.

Photo 4.4 Conservatory window at Standen, East Grinstead, Sussex. Architect: Philip Webb, 1891–94.

These timber-framed windows were made up in the contractors joinery works and painted on site. This traditional method and material was easily adaptable to the tolerances required in the construction of the arched structural openings. Contemporary factory-finished windows lack this flexibility (see section 4.3.2).

4.3.2 Fixing timber windows

Figure 4.5 indicates construction details appropriate for a timber window fixed into an already constructed opening within a conventional cavity wall (see photo 4.4). When components are fitted into a pre-formed element of construction, they are called *second fixings*, for windows this is becoming more usual as factory-glazing becomes commonplace. Details suitable for windows to be built in as the walling progresses may not be suitable for second fix installation.

Whenever possible, wood windows should be fixed to the dry leaf of cavity walling. The use of a rebated reveal in the wall and a flexible damp proof course will provide a suitable seal against the penetration of rainwater and moisture across the jamb to the interior. This DPC also serves as a water barrier between the wet and the dry leaves of the wall. It should lap with the DPC over the lintel spanning the opening and with a DPC placed beneath the window sill to prevent deterioration due to rising moisture. Cold bridging at this junction can be avoided by not returning the inner blockwork leaf to close the cavity, but instead using an insulated PVC-U cavity closer, to which the frames can be fixed and which will also function as the DPC (see fig 4.9). Alternatively the blockwork closing the cavity around the sides and head/sill of the window should be made from lightweight aerated concrete blockwork that is itself insulative. For alternative approaches reference should be made the the BRE report *Condensation: avoiding risks*.

The sill must be *weathered* at a sufficient slope to throw off the water and *throated* with a groove near the front of its underside so that the water will drip off and not run back to the bed of the sill. In the past, timber sills have been built into the brickwork and pointed in mastic at the vulnerable junction between the end grain of the timber and the brickwork jambs of the opening; only the most durable hardwood will not rot under these circumstances. It is preferable for a metal sill to be used, as is common practice on the continent (see fig 4.5). Alternatively, sub-sills may be made from stone, canted brick or concrete, perhaps incorporating a water bar to locate the timber sill horizontally and act as an additional check against the penetration of moisture (see fig 4.18). A sub-sill will anyway be required if the window is set far back into the opening, resulting in a timber section that would be too large. The sub-sill must be provided with a throating or drip in order to shed water as far away as practicable from the wall face below. Traditionally, in the best work, projecting sills of stone or hardwood had stooled ends to direct water away from the junction with the wall (where staining below the sill is otherwise commonplace); drainage grooves can be cut into sill, just short of the ends, to achieve the same result (see fig. 4.23(b)). Whatever the material used for their construction, sills are very susceptible to defects caused by dampness. Either the weathering is not sufficiently steep to shed the water, or water can get back to the bed joint of the sill because the drip does not function, or because the slope of the concrete, stone or brick sub-sill allows the water to run back instead of outwards. Paint must not be relied upon to protect timber at this critical point.

If timber windows are to be fitted into a masonry wall as it is being built, care must be taken to prevent them being damaged or any load being imposed on them during the 'building-in' process. Although this is not easy to ensure, this method of installation enables a good fit for the window

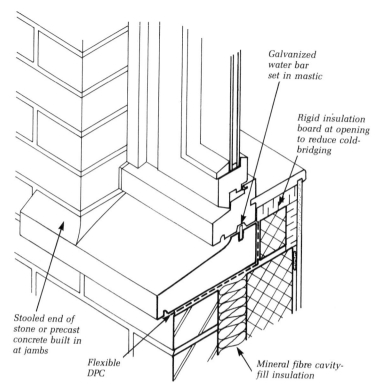

*Galvanized
water bar
set in mastic*

*Rigid insulation
board at opening
to reduce cold-
bridging*

*Stooled end of
stone or precast
concrete built in
at jambs*

*Flexible
DPC*

*Mineral fibre cavity-
fill insulation*

Fig. 4.18   Stone/concrete sub-sill to timber window

and makes possible an efficient seal between the wall and the frame. This is effectively accomplished by fixing a flexible DPC in a groove formed in the outside of the timber frame and then building the DPC into the wall (see fig. 4.15).

The head is not usually as troublesome with regard to water penetration, but the DPC in the cavity above must catch water running down the inside surface of the outer leaf and conduct it properly to the outside of the head. Note that fig. 4.5 shows internal insulation to the concrete lintel to alleviate cold-bridging at this point.

Window frames should be secured to walls by means of metal lugs (see fig. 4.15), fixed in the dry leaf of the cavity wall, or alternatively, screwed in place into plastic or similar plugs located at drilled holes within the surrounding masonry. Only non-ferrous metal or stainless steel should be used for these fixings, and in general, timber windows should be secured at the jambs only. However, large windows must also be fixed at the head and the sill (to locate but not secure the window). The frames are bedded in cement mortar. In the past the external joint at the perimeter of a window was usually sealed with mastic. Since this has the undesirable result of forming a vapour check on the outside, stopping internal humidity from escaping, what

is now recommended is that a vapour permeable compression seal should be used or a timber cover-piece acting as a rainscreen. The latter alternative is shown in fig. 4.19, the joint being mastic sealed at the inside only as is common Scandinavian practice.

High performance windows are likely to be factory glazed and finished and therefore not suited to being built in as work proceeds, as has been usual in the UK. It is similarly appropriate for the more sophisticated windows with complex ironmongery to be second-fix components. Windows that are to have a clear finish should also be fixed later as mortar or water stains will be visible through the finish. Where the window is placed in position after the opening is formed there may be difficulties in ensuring close tolerances that are equal and parallel round the frame and in providing a good connection with the jamb DPC, but there is likely to be less damage to the frame. A suitable template can be used to assist the process — one manufacturer supplies a PVC-U former that serves to line the opening as well as providing a cavity closer and obviating the need for a DPC to separate the inner and outer layers of the cavity (see fig. 4.21(a) and (b)). It is recommended that a maximum manufacturing tolerance on the size of the frame of second-fix windows should be 3 mm

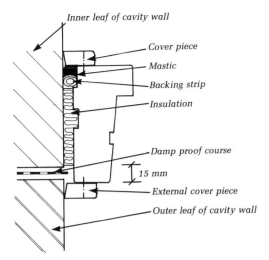

Fig. 4.19 Jointing between window frames and cavity walls

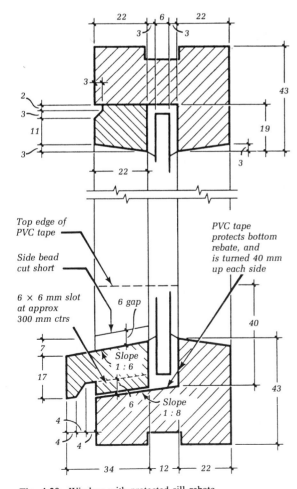

Fig. 4.20 Window with protected sill rebate

on all edges.

Factory glazed windows are installed by inserting folding wedges between the frame and the structural opening at the final fixing points. The window should be tested and the folding wedges adjusted to level and plumb as required, this is particularly necessary for tilt/turn and pivoting windows which have to be balanced to ensure proper functioning. Final fixing can be with tie-back straps into the inner skin of the wall or fixed with non-ferrous or stainless steel screws or expanding bolts. The holes drilled through the frame should be painted with preservative and plugged.

Unless the rebated jamb detail shown in fig. 4.5 is adopted for a window installed after the opening has been formed, the window must be carefully moved into the opening from the interior of the building so that the projecting DPC is collected by the groove in the side of the frame. Having accomplished this, the frame is fixed and either a vapour permeable compression seal is inserted at the the outside, or a rainscreen cover-piece is fixed to close the joint between the frame and the wall. The DPC beneath the sill can then be bent up behind the sill and the wall construction completed.

In terms of insulation and forming a seal, the joint to the wall should achieve the same performance as the outside wall. The frame-to-wall junction has to be sealed to prevent draughts. In conditions of the most severe weather exposure, the joint will need to be sealed with mastic on the outside and care must be taken to ensure that the relationship between the seal and the DPC does not trap moisture between the wall and the frame. External mastic seals are subject to direct exposure to the wind and rain, they age rapidly and will eventually allow rainwater to leak in under wind pressure. Consequently, in most situations, it is preferable to use either a vapour permeable compression seal at the outside, or for the joint to be mastic sealed at the inside only with an external cover strip scribed to fit the opening (although small cover pieces may not prove durable on very exposed sites). The mastic seal then prevents water vapour from within the building infiltrating the insulation. It also forms an airtight weatherseal that traps a cushion of air and reduces the velocity of air movement into the joint from outside, in a similar fashion to the weathersealing of the casements. The gap between the frame and structural opening should be packed with insulation to leave a drainage channel behind the external bead of 15 mm. Water vapour diffusing from within the construction and condensing within the drainage channel must be allowed to drain away by extending the sill beneath the channel at either side (see fig. 4.19).

Figure 4.22(a)–(d) illustrates various methods of locating timber windows in openings in the types of construction described under the headings below.

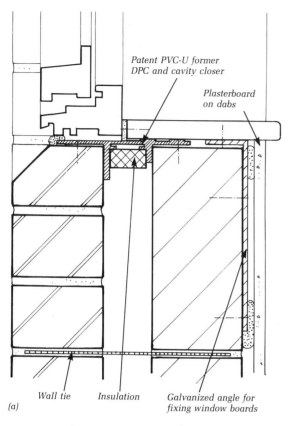

Patent PVC-U former
DPC and cavity closer

Plasterboard
on dabs

(a)

Wall tie   Insulation   Galvanized angle for
fixing window boards

Blockwork cavity wall construction Figure 4.22(a) shows a window opening formed within a cavity wall of fair-faced external concrete blockwork and an insulating block, plaster-finished inner skin. The head of the opening is supported by a reinforced concrete lintel that closes the cavity over the frame. It is made from lightweight concrete with prestressed reinforcement, making it easier to manhandle into position than traditional concrete lintels. The outer skin of concrete block is supported by a reinforced lintel block that spans the opening. A cavity tray separates the inner and outer layers of the construction and laps over the head of the window. Cavity trays (or cavity flashings) are typically made from a thin layer of metal (lead or copper) sandwiched both sides with bitumen-impregnated felt. The inner leaf of the wall is constructed from 100 mm insulating concrete blocks that with the addition of 40 mm mineral fibre partial cavity slabs can achieve a U-value of 0.45 W/m²K.

The timber sill is in non-tropical hardwood from a managed and sustainable source projected over the plinth block sub-sill below. This projection should be a minimum 35 mm and preferably 50 mm in order to shed water away. A cavity tray is built into the wall below the plinth brick sill. The inner leaf of blocks closing the top of the cavity at the sill are wider blocks from the same manufacturer's range. Insulation is contained between the inner and outer blockwork to complete the thermal envelope up to the window frame.

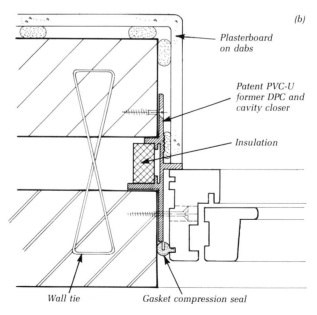

(b)

Plasterboard
on dabs

Patent PVC-U
former DPC and
cavity closer

Insulation

Wall tie   Gasket compression seal

Fig. 4.21 Proprietary former used as sub-frame to factory glazed windows: (a) detail at still; (b) detail at jambs (ESPE Window Systems Ltd)

At the jamb, the DPC is turned in to meet the side of the timber frame. In this instance, the frame is built into the wall by stainless steel cramps fixed to the frame and bedded in the horizontal masonry joints, as work proceeds. Note that the DPC projects into the cavity by about 40 mm to avoid any 'short circuiting' of water at the junction of the inner and outer leaves of the wall. At the jamb, the cavity is closed by return of the inner leaf blockwork against the vertical DPC. This DPC will be the same material used as the horizontal damp-proof course.

Externally insulated 'warm wall' construction   Figure 4.22(b) shows an increasingly used method of construction. An outer impervious layer of polymer-modified render is smoothed on to a stainless steel mesh fixed through rigid insulation to concrete block walling. The blockwork is fair-faced internally and painted. The head of the standard hardwood frame is fixed to a tanalized timber subframe of the same thickness as the insulation and render that also lines the reveals of the window opening. A reinforced concrete lintel supports the blockwork above and is finished to match.

The sub-sill is made from slate bedded in mortar on wedge-shaped block pads, and stainless steel straps are fixed at the wedges and hook around the front edge of the sill. Note the drainage groove cut at the end of the sill. The insulation continues under the sill to abut the subframe as at the head of the window thus maintaining a continuous envelope of insulation across the wall. The junction between the underside of the timber sill and the slate is pointed in mastic. Arguably, it is preferable in this externally insulated construction for the windows to be located closer to the outside of the wall and in line with the insulation, so as to avoid difficult details at the reveals and the sill (see fig. 4.35).

Internally insulated 'cold wall' construction   Although cavity walling has become the predominant method of construction in the UK, vernacular building employed a variety of methods that not only reflected the availability of particular materials but also the variety of climatic and exposure conditions in different parts of the country. The current *Building Regulations* permit a wide variety of approaches; solid brick walling is still possible in more sheltered areas and this approach is enjoying a revival. Figure 4.22(c) shows a window opening within a one brick thick wall insulated on the inside. Cold walls require care in the detailing and construction of vapour checks and the possibility of summertime condensation has led some designers to employ DPCs draining internal cavities and/or air-bricks to ventilate them, as shown here (see photo 4.5).

An internal precast concrete lintel to which is bolted a stainless steel angle supports the thickness of the wall above

Photo 4.5 Housing in Millman Street, London WC1. Architects: Hunt Thompson Associates; Structural Engineers: Alan Baxter and Associates, 1990.

A solid masonry, internally insulated construction (see fig. 4.22 (c)). The use of single skin cavity walls would have been structurally problematic for this six storey building, not only in terms of loading but also because of contemporary requirements for structures to resist collapse as a result of a gas or other explosion.

the opening. The lintel is faced by a half-brick self-supporting gauged arch which obscures the head of the window frame, and the joint between them is draught sealed with mastic at the bottom edge of the cavity tray. The 30 mm cavity within the external wall is formed by pressed metal channel framing that spans from floor to ceiling, the plasterboard is screw-fixed to it and the insulation contained within it. To reduce cold-bridging at the junction of the floor slab and walls, the timber floor and plasterboard ceiling are insulated for a minimum of 450 mm back into the building. The timber lining to the internal window reveals encases the thickness of construction required to achieve this detail and provides a fixing for curtains and blinds.

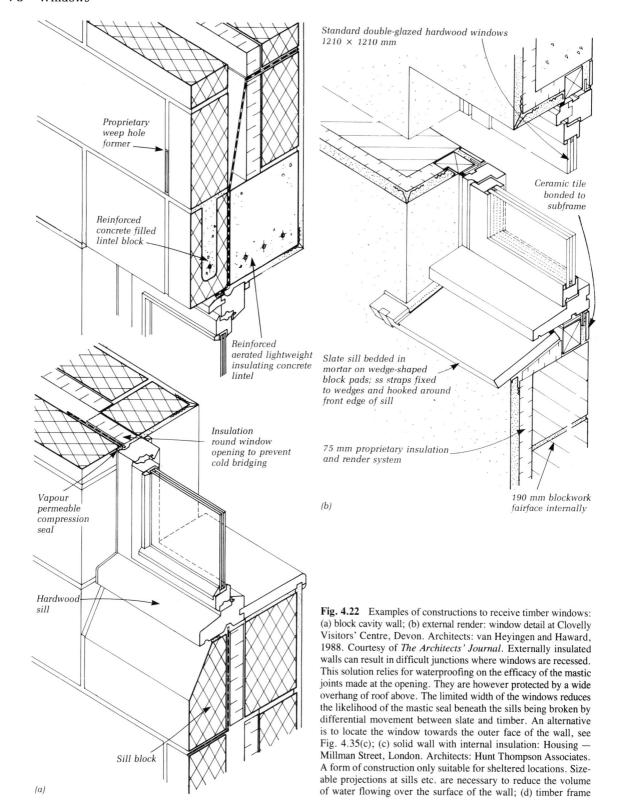

Proprietary
weep hole
former

Reinforced
concrete filled
lintel block

Reinforced
aerated lightweight
insulating concrete
lintel

Insulation
round window
opening to prevent
cold bridging

Vapour
permeable
compression
seal

Hardwood
sill

Sill block

(a)

Standard double-glazed hardwood windows
1210 × 1210 mm

Ceramic tile
bonded to
subframe

Slate sill bedded in
mortar on wedge-shaped
block pads; ss straps fixed
to wedges and hooked around
front edge of sill

75 mm proprietary insulation
and render system

190 mm blockwork
fairface internally

(b)

Fig. 4.22 Examples of constructions to receive timber windows:
(a) block cavity wall; (b) external render: window detail at Clovelly
Visitors' Centre, Devon. Architects: van Heyingen and Haward,
1988. Courtesy of *The Architects' Journal*. Externally insulated
walls can result in difficult junctions where windows are recessed.
This solution relies for waterproofing on the efficacy of the mastic
joints made at the opening. They are however protected by a wide
overhang of roof above. The limited width of the windows reduces
the likelihood of the mastic seal beneath the sills being broken by
differential movement between slate and timber. An alternative
is to locate the window towards the outer face of the wall, see
Fig. 4.35(c); (c) solid wall with internal insulation: Housing —
Millman Street, London. Architects: Hunt Thompson Associates.
A form of construction only suitable for sheltered locations. Sizeable projections at sills etc. are necessary to reduce the volume
of water flowing over the surface of the wall; (d) timber frame

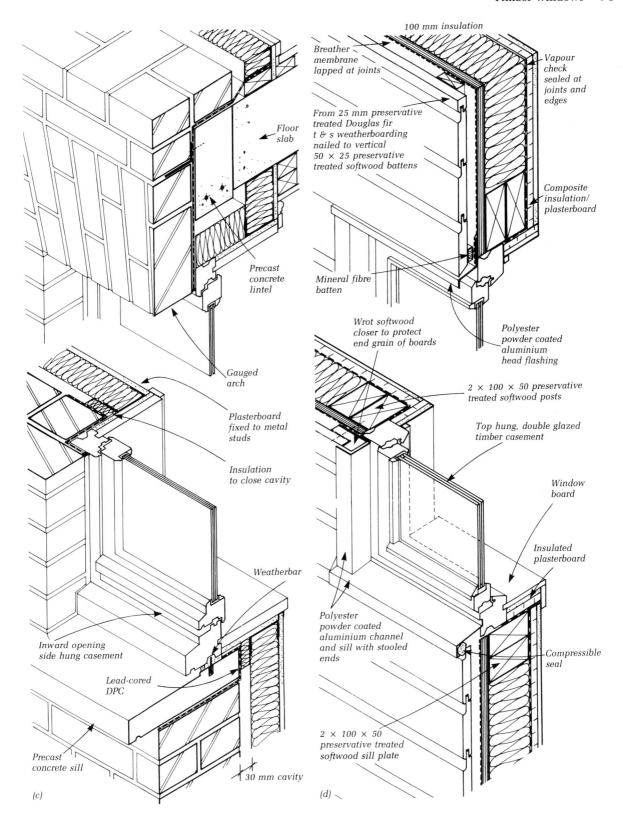

100 mm insulation

Breather membrane lapped at joints

Vapour check sealed at joints and edges

From 25 mm preservative treated Douglas fir t & s weatherboarding nailed to vertical 50 × 25 preservative treated softwood battens

Composite insulation/plasterboard

Floor slab

Precast concrete lintel

Mineral fibre batten

Wrot softwood closer to protect end grain of boards

Polyester powder coated aluminium head flashing

2 × 100 × 50 preservative treated softwood posts

Top hung, double glazed timber casement

Gauged arch

Plasterboard fixed to metal studs

Insulation to close cavity

Window board

Insulated plasterboard

Weatherbar

Polyester powder coated aluminium channel and sill with stooled ends

Compressible seal

Inward opening side hung casement

Lead-cored DPC

Precast concrete sill

30 mm cavity

2 × 100 × 50 preservative treated softwood sill plate

(c)

(d)

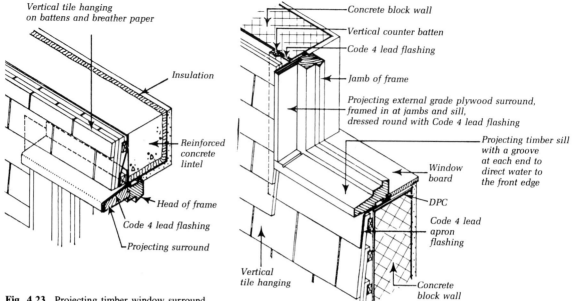

Fig. 4.23 Projecting timber window surround

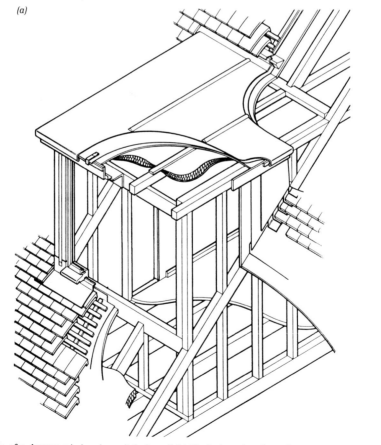

Fig. 4.24 (a) Example of a dormer window in a pitched roof (b) Typical section through a dormer window

(b)

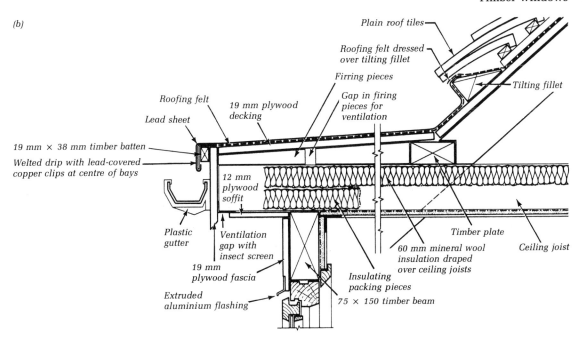

Plain roof tiles

Roofing felt dressed over tilting fillet

Firring pieces

Tilting fillet

Gap in firing pieces for ventilation

Roofing felt

19 mm plywood decking

Lead sheet

19 mm × 38 mm timber batten

Welted drip with lead-covered copper clips at centre of bays

12 mm plywood soffit

Timber plate

Ceiling joist

60 mm mineral wool insulation draped over ceiling joists

Plastic gutter

Ventilation gap with insect screen

19 mm plywood fascia

Extruded aluminium flashing

Insulating packing pieces

75 × 150 timber beam

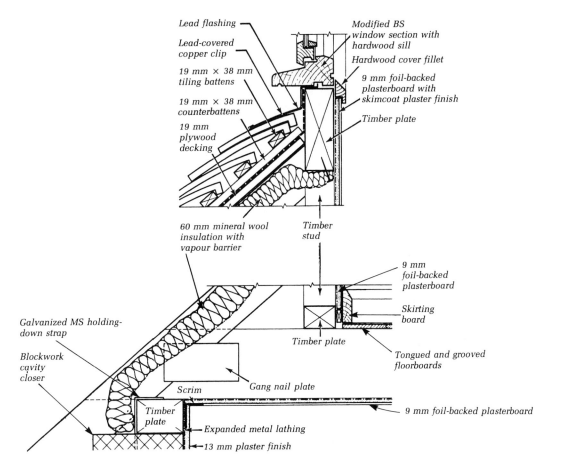

Lead flashing

Modified BS window section with hardwood sill

Lead-covered copper clip

Hardwood cover fillet

19 mm × 38 mm tiling battens

9 mm foil-backed plasterboard with skimcoat plaster finish

19 mm × 38 mm counterbattens

Timber plate

19 mm plywood decking

60 mm mineral wool insulation with vapour barrier

Timber stud

9 mm foil-backed plasterboard

Skirting board

Galvanized MS holding-down strap

Timber plate

Blockwork cavity closer

Tongued and grooved floorboards

Scrim

Gang nail plate

Timber plate

9 mm foil-backed plasterboard

Expanded metal lathing

13 mm plaster finish

Timber-framed construction   A timber window is most easily screwed into timber-framed construction as illustrated in fig. 4.22(d). At the head of the window is a pressed aluminium flashing over which is lapped the bottom of the breather paper (that provides a waterproof surface behind the horizontal weatherboarding whilst allowing internal moisture to escape). The aluminium flashing throws water at the outer face of the boarding away from the window below but a gap is left at the bottom edge of the boarding allowing ventilation of the void formed by the thickness of the vertical battens to which the boards are attached. At the jambs, an aluminium U-section masks the vulnerable end grain of the boards, fits under the head flashing and drains on to the pressed metal sill. Figure 4.23(a) and (b) shows the location of timber windows with projecting surrounds set in a *tile-hung* opening.

Dormer windows in pitched roofs   Figure 4.24 shows an example of the detailed design of a dormer window which is simply framed out from the timber roof structure. It is shown waterproofed by felt on plywood decking but could alternatively be roofed with metal sheeting that might also be used to clad the side cheeks of the dormer (see section 8.4).

Although dormers are not currently common in new dwellings in the UK, the need to economize on building costs and space heating could result in their increased adoption. The practice of providing the upper-floor accommodation within a roof space is common today in Sweden. By making the roof space a usable habitable space, the conventional upper-floor construction can be combined with the roof construction — particularly now that prefabricated trussed rafter configurations are manufactured which easily allow this to happen. Proprietary GRP dormer windows are also available. For further details of dormer window construction see *MBS: Structure and Fabric Part 1* chapter 7.

4.3.3 Horizontal pivot windows

Most manufacturers produce a 'standard' range for delivery from stock for this type of window. Other types of timber windows, e.g. *outward opening casements* and *multi-light frames* are also made to manufacturers' standard ranges.

Figure 4.25 shows a horizontal pivot-hung window — the word horizontal refers to the placing of the hinges, which are opposed horizontally. The hinges of a vertical pivot-hung window would be one above the other in a vertical line. The success of a pivot window depends upon the friction action of the hinge which should be strong enough to hold the window firmly in any open position. The pivot-hung window is much used because it gives a neat appearance to the façade of a building and provides

good control of ventilation. There may be some difficulty in hanging curtains or blinds on the inside of the window if the frame is set back into the window reveal. The example shown is for reasons of appearance fixed as far as possible towards the front of the brickwork opening (in which location the window is however more exposed to the ravages of the weather), and secured by stainless steel lugs. These lugs are screwed to the back of the frame and 'built-in' as the brickwork is carried up when the opening is formed, as shown on the plan in fig. 4.25. The opening casement is secured in this example by multi-point locking by an espagnolette bolt which holds the casement to the frame at four points.

The normal maximum casement width is 1200 mm, the maximum height 1350 mm. Some types almost fully rotate to allow cleaning from the inside but with some disturbance to curtains, plants on sills, etc.

4.3.4 Multi-light assemblies

Because of the difficulty of providing draught-free ventilation when using side-hung casements alone, it has been commonplace to incorporate a combination of top-hung and side-hung casements with fixed lights in one assembly. This is subject to the feasible sizes of particular window types; the maximum size of a side-hung casement for example is approximately 600 mm in width and 1350 mm in height.

The flexibility of timber window construction makes possible similar combinations of other window types with glazed or opaque panels or ventilators. Whole wall panels or storey-height panels can be made up in this way. In the example shown (fig 4.26), simple timber frames were designed to take insulated plywood spandrel panels, fixed light windows and double-glazed projected top-hung reversible casements. The vertical hardwood framing has internal Douglas fir stiffeners to span floor to floor where it is bolted to steel angles anchored to the concrete structure, the sill and horizontal drips are made from durable hardwood. Composite spandrel panels are manufactured for use in this type of construction. Typically, rigid insulation is bonded to a liner of steel or resin bonded timber board and an outer weather-face or plastisol coated or vitreous enamelled steel.

4.3.5 Vertical sliding sash

The manufacture of traditional double-hung sash windows continues and in recent years new companies have been established specifically to manufacture them, particularly for conservation projects. The frames were traditionally 'cased' with vertical hollow box frames containing weights supported by cords (or preferably chains) carried over

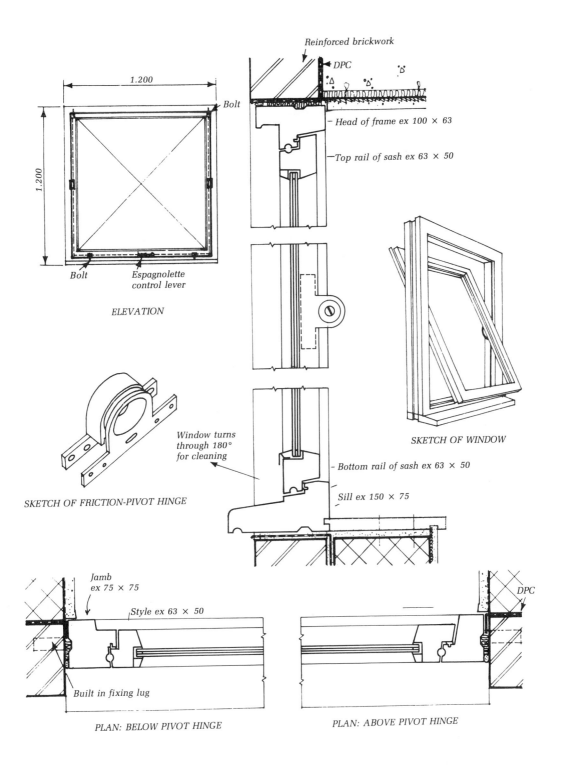

Fig. 4.25 Horizontal pivot window — single sash

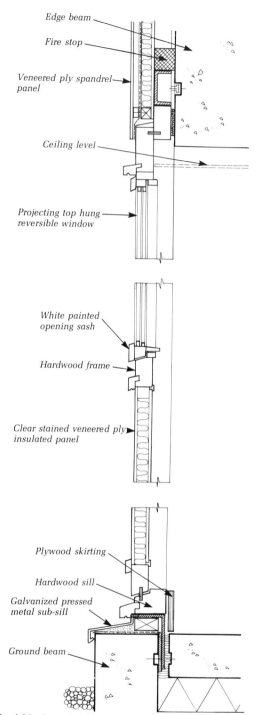

Edge beam

Fire stop

Veneered ply spandrel panel

Ceiling level

Projecting top hung reversible window

White painted opening sash

Hardwood frame

Clear stained veneered ply insulated panel

Plywood skirting

Hardwood sill

Galvanized pressed metal sub-sill

Ground beam

Fig. 4.26 Storey height multi-light assembly. Chilworth Science Park, Southampton. Architects: Edward Cullinan, 1989. Courtesy of *The Architects' Journal*. Depending on the degree of exposure, it may be preferable for drips to be made from metal rather than timber, as in the case of bottom glazing beads to timber windows (see fig. 4.16)

pulleys to counterbalance the sashes and hold them open in any required position.

Although still available with cast iron counterweights, for economy most sash windows now employ spiral spring balances, concealed within a circular groove within the sash, the frames being timber sections rather than sash boxes. The type has also been adapted to contemporary requirements; deeper section glazing beads are made to accommodate 14 mm double glazing units and timber parting beads have in some cases been substituted with uPVC sections incorporating a pile strip weatherseal. Similarly, meeting rails are fitted with compression seals. In this way double-hung sash windows can comply with the most severe exposure ratings of BS 6375 Part 1, the internal staff beads can also be weatherstripped to increase weather resistance yet further (see fig. 4.27).

To make it possible for windows to be cleaned from the inside and for glass to be replaced, the sashes can be fixed with *sliding/tilting* gear which enables the sashes to be released and opened inwards. The frames can be fitted with slide restricting bolts and other security locks.

4.3.6 Horizontal sliding windows

These are most commonly used as patio doors but the distinction is not important (see fig. 4.28). Several alternative track arrangements are possible, the simplest is where a single sliding window passes in front of a fixed light of the same size. This reduces the amount of track and simplifies the weather-proofing between the front and back doors. In standard units these are available as twin light units, one fixed and one opening in widths from 1600 to 2000 mm and three or four panel units, one or two glazed/ two fixed in widths from 2400 to 6000 mm. In timber-framed construction, a load-bearing timber mullion could be fitted between door units to couple them together and support a timber beam above the opening; in masonry, a steel beam or concrete lintel is likely to be used to frame the structural opening above the doors. Patio doors are made to fit within a conventional door opening 2100 mm in height but as in other timber windows, special sizes are readily available up to 2400 mm in height. Hermetically sealed double glazed units 18 or 20 mm in thickness are usual, when used in patio doors; these need to be of tempered safety glass to comply with section N1 of the *Building Regulations 1991* and BS 6206. The fixed panel is rebated into the frame, the opening panel is weathersealed by polypropylene brushes within an extruded aluminium channel attached within grooves in the head of the frame and the threshold. The thickness of the glass and the size of the patio doors make them relatively heavy. The bottom rail is likely to be around 230 mm in depth and up to 65 mm in width to accommodate the glass and the synthetic fibre

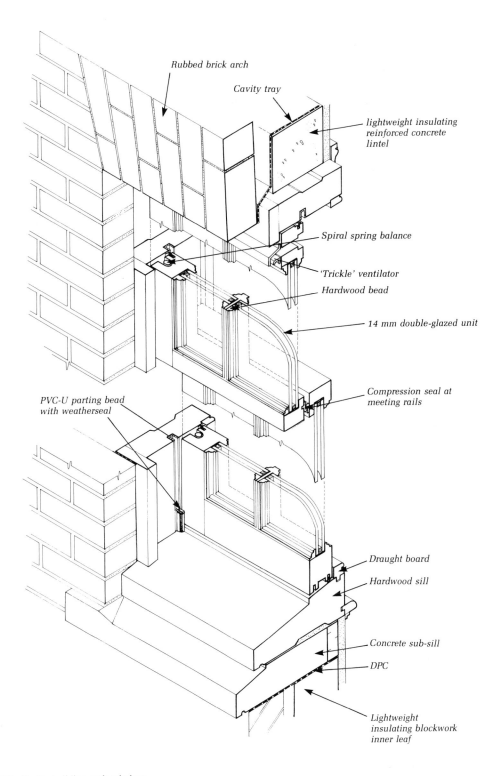

Rubbed brick arch

Cavity tray

lightweight insulating
reinforced concrete
lintel

Spiral spring balance

'Trickle' ventilator

Hardwood bead

14 mm double-glazed unit

Compression seal at
meeting rails

PVC-U parting bead
with weatherseal

Draught board

Hardwood sill

Concrete sub-sill

DPC

Lightweight
insulating blockwork
inner leaf

Fig. 4.27 Vertical sliding sash window

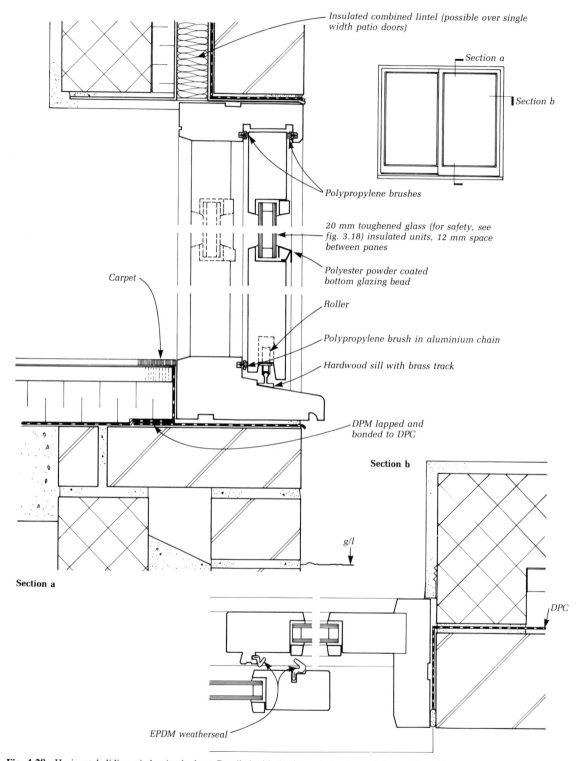

Insulated combined lintel (possible over single width patio doors)

Section a

Section b

Polypropylene brushes

20 mm toughened glass (for safety, see fig. 3.18) insulated units, 12 mm space between panes

Polyester powder coated bottom glazing bead

Carpet

Roller

Polypropylene brush in aluminium chain

Hardwood sill with brass track

DPM lapped and bonded to DPC

Section b

g/l

Section a

DPC

EPDM weatherseal

Fig. 4.28 Horizontal sliding window/patio door. Detailed with the frame to the front of the opening, this is suited only to sheltered locations or if protected by an overhang above

roller wheels rebated into the underside of the bottom rail which run on a brass track fixed into the threshold. Special 'lift and slide' ironmongery has been developed for patio doors; the locking handle turns through 180 degrees to 'lift' the sash on to the wheels for easy sliding. The reverse action 'drops' the doors down and compresses the seals for draught-proofing.

4.3.7 Tilt and turn windows

This type of window which originated in Germany, consists of a window incorporating sophisticated ironmongery, which is *bottom hung*, tilting inwards for draught-free ventilation and *side-hung* turning inwards for cleaning (see fig. 4.29). The combination of safety, security and the ability to maintain the windows without external access has resulted in their increasingly widespread use. The normal maximum casement width is 1200 mm and the maximum height 1800 mm. Advances in tilt and turn fittings have improved safety, including the provision of security stays, narrow gap restricted ventilation in either the tilt or turn mode, and the introduction of switch barriers to ensure the windows cannot be put into both modes at the same time. Insect screens or alternatively roller shutters can be fitted in combination with this design of window, the latter giving both added security and thermal insulation.

4.3.8 Projecting top-hung casements

Of the various types of reversible window invented in recent years, projecting top-hung casements have gained the widest acceptance (see figs 4.11 and 4.17). Projecting side-hung windows also exist but are less frequently used. The normal maximum casement width is 1200 mm and the maximum height is approximately 1350 mm. The principle of operation is similar to a fully rotating horizontal pivot window except that the special cantilevering hinges are designed to rotate the sash through 170 or 175 degrees outside of the opening; this is possible because the sash pivot is free to slide in a channel recessed within the frame. The window can consequently be cleaned without disturbing curtains, etc. inside the room. Although arguably not as safe as tilt/turn windows, a child-proof safety catch can be incorporated restricting the extent of the sash opening. Advantageously, this type when fully open affords virtually 100 per cent free area for ventilation.

4.3.9 Coupled sash windows

The horizontally pivoted double sash has for a long time been developed by window specialists on the continent and is now produced by several manufacturers in this country. Details of such a window are shown in fig. 4.30. The outer

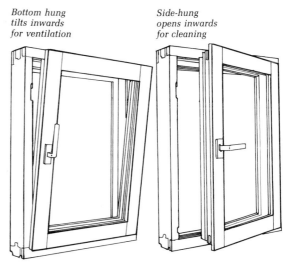

Bottom hung tilts inwards for ventilation   *Side-hung opens inwards for cleaning*

Fig. 4.29 Tilt/turn modes of operation

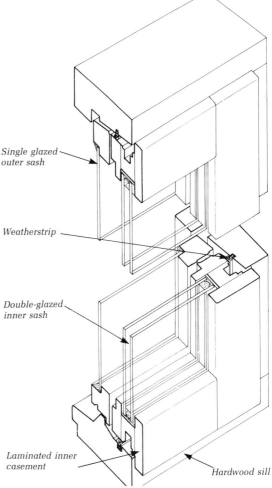

Single glazed outer sash

Weatherstrip

Double-glazed inner sash

Laminated inner casement

Hardwood sill

Fig. 4.30 Coupled sash window (Tema Fönster)

sash is secured to the inner sash and hinged so that the space between can be cleaned. The joint between the two sashes is not airtight, in fact the locking device is also a spacer to ensure that external air can circulate between the inner and outer panes. The air circulation should be enough to evaporate any condensation within the space but not sufficient to have any serious overall cooling effect. To separate the two sashes they have to be rotated through 180 degrees. Accordion pleated, or venetian blinds can be fixed in the space between the casements and operated from the side with cords. These windows, being balanced, can be made to a large size, a limiting factor being the distance the top swings into the room. The pivot mechanism holds the sash in any desired open position by friction. An espagnolette bolt, used to secure the window at all four corners, is recessed in a groove on three sides of the window and operated by one handle. This ironmongery also helps to stiffen the relatively slender timber window sections and prevents them from warping.

Close coupled windows are made in other combinations for example tilt/turn: the joint assembly can tilt inwards for ventilation but the windows are hinged separately to be side-hung when opening inwards for cleaning (see fig. 4.3). These windows are usually employed for their superior acoustic and thermal performance as described on p. 72 ('High performance windows'). Double windows can be made by combining other traditional types although the sashes are not coupled together, making operation less satisfactory, e.g. *vertical sliding sashes* with a side-hung inner window (opening inwards) also *side-hung casements*, one opening in, the other opening out (a form of secondary glazing but 'built-in' from the outset).

4.4 Steel windows

Steel and aluminium are the most common metals used in the manufacture of window sections although their characteristics and capabilities and the resulting forms of window made from the two materials are very different. If lifetime cost is considered, in terms of permanency and freedom from maintenance, bronze and gunmetal become possible alternatives. Stainless steel has potential for use as window frames particularly given its rapid development as a building material for more general applications.

Steel windows have been manufactured in the UK since the beginning of this century in basically the same form and by the same method used today. The system has long been admired for the slenderness of its glazing sections and the ability to combine windows into highly transparent window walls (see photos 4.6 and 4.7). This technology has left its mark on the history of modern design. Now, however, steel windows only constitute around 10 per cent of the window market in this country. Although technical

Photos 4.6 and 4.7   University of East Anglia. Architect: Rick Mather, 1985. (Photos courtesy of Crittall Windows Ltd.)
Steel framed windows can be coupled together to form complete window walls; in terms of design and geometry, the system is very flexible.

advances have been made, for example to incorporate weatherstripping and double glazing, the range of steel window types is still relatively restricted (see fig. 4.31).

Recommendations for steel windows are given in BS 6510: 1984 *Specification for steel windows, sills, window boards and doors.* Most manufacturers of steel windows have their product tested in accordance with the

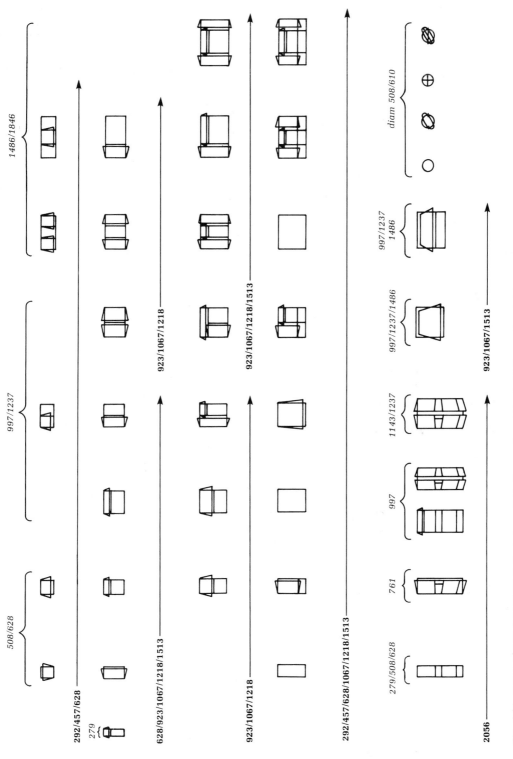

Fig. 4.31 Range of type F standard steel windows. Italic figures give widths, bold figures show heights

recommendations made in BS 5368 *Methods of testing windows* in order to meet the criteria laid down in BS 6375 *Performance of windows* (see section 4.2.4). Some have also obtained British Board of Agrément test certificates and participate in the BSI Quality Assurance Scheme recommended in BS 5750 (see *MBS: Internal Components* section 1.8 *Component testing and quality assurance*). There is no British Standard for stainless steel windows or windows made from pressed sections.

Steel window frames readily conduct heat (having 1200 times the conductivity of timber) and condensation can occur on their internal surfaces in rooms where humid conditions exist without proper heating and ventilation. Timber or plastic subframes (see p. 93) will help prevent staining of the adjacent wall surfaces but steel windows conforming with BS 6510 cannot be thermally broken. Double windows, separated by 100–200 mm, can give very good sound reduction up to 49 dB, especially if the reveals are lined with absorbent materials.

4.4.1 Standard

Steel windows are made in two types, *standard (type F)* and *W20*. Type F steel windows are made from lighter sections and are intended principally for domestic applications. Both types are made from steel bar that is hot-rolled into solid sections with specified profiles. The composition of the steel is specified in BS 6510. White hot steel ingots are passed through rollers to form a billet of steel about 50 mm square, 1200 mm long. The billet is then reheated and 'rerolled' through a further series of rollers under very heavy pressure which produces the correct section profile from which the window frame is formed. The frame section is cut to length and mitred, corners are welded and internal sub-divisions are either welded or tenoned and riveted.

The completed frames are then hot-dip galvanized to BS 729: 1971 *Hot-dip galvanized coating on iron and steel articles*. Undecorated galvanized steel windows have a life of more than 15 years in an urban environment, but *in situ* painting every 7 to 10 will increase this period to over 40 years. Increasingly, galvanized window frames are factory finished with a polyester powder coating (an organic stoved finish available in a fairly wide colour range) to BS 6497: 1984 *Powder organic coatings for application and stoving to hot-dip galvanized hot-rolled steel sections and preformed steel sheet for windows and associated external architectural purposes, and for the finish on galvanized steel sections and preformed sheet coating with powder organic coatings*. Regular washing of paint or polyester powder coating will further prolong the life of the steel window frame.

BS 6510 does not contain preferred modular sizes for steel windows. Manufacturers offer standard size ranges for type F windows but only as an indication of what is available, since many windows are made to order.

Most steel windows are now made for replacement/ refurbishment work so a metric modular range is no longer produced as standard, though windows to a 100 mm module can be made as a special in both F (standard) range windows and W20 sections. The Steel Window Association's range of non-modular sizes for various types of fixed and opening standard windows is as manufactured by its member companies (see fig. 4.31):

- *Heights*: 292, 457, 628, 923, 1067, 1218, 1513 and 2056 mm.
- *Widths*: 279, 508, 628, 997, 1237, 1486, 1846 and 1994 mm.
- *Circular*: 508 and 610 mm diameter.

Timber subframes were in the past used to co-ordinate the sizes of steel windows with the dimensions of brickwork openings and also for visual reasons, e.g. to provide a bolder outline to the window than is provided by the steel frame alone. Timber subframes are described in BS 1285: 1980 *Specification for wood surrounds for steel windows and doors* but they have mostly been superseded by the use of cellular PVC-U subframes. See p. 93 for further comments on subframes for metal windows.

4.4.2 Composite standard

Typical details illustrating the standard steel sections and coupling arrangements used to form composite windows are shown in fig. 4.32. The standard range includes *fixed lights, side-hung casements opening outwards, horizontally pivoted reversible casements* and *top-hung casements* opening outwards, with a selection of *casement doors* opening outwards. Windows and doors may be *coupled* together by the use of vertical coupling bars (mullions) and horizontal bars (transoms), and by the use of infill panels to form composite assemblies. There may be limits to the permissible overall size of an assembly, due to the difficulty of achieving manufacturing tolerances and the anticipated wind loadings. Advice should be obtained from window manufacturers when these questions arise. The windows are manufactured from test guaranteed steel. The main frames of the windows are constructed from bars, cut to length and mitred, with all the corners welded solid. Intermediate bars are tenoned and riveted to the outer frames, and to each other. The windows are hot-dip galvanized after manufacture. A manufacturing tolerance of 1.5 mm above or below (±) the standard dimensions and a maximum difference in measurements across the diagonals is allowed. Similarly a fitting tolerance of 3.0 mm maximum all round the window is anticipated between the

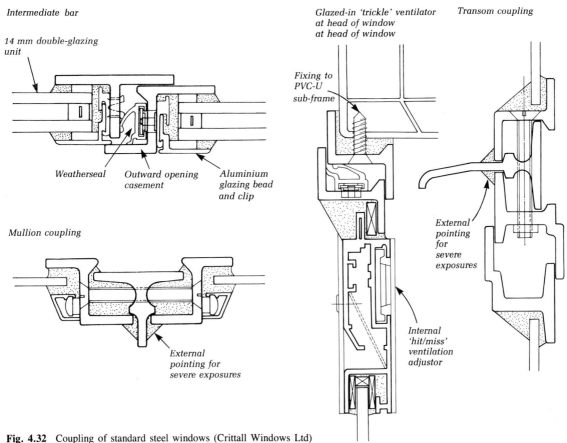

Intermediate bar

14 mm double-glazing unit

Weatherseal   *Outward opening casement*   *Aluminium glazing bead and clip*

Mullion coupling

External pointing for severe exposures

Glazed-in 'trickle' ventilator at head of window at head of window   *Transom coupling*

Fixing to PVC-U sub-frame

External pointing for severe exposures

Internal 'hit/miss' ventilation adjustor

Fig. 4.32 Coupling of standard steel windows (Crittall Windows Ltd)

outside window size and the basic dimensions of the openings to receive the window units.

Side-hung casements These are hung on projecting friction hinges without a stay. They are made of steel and are welded or riveted to the frames. The hinge pins are either stainless steel, brass or aluminium alloy. The friction hinge is adjusted by the manufacturer at his works to require a given pressure on the handle to move the casement; the hinges can be adjusted *in situ* or are designed to retain a constant level of friction. When the casement is open to 90 degrees the projecting arm gives a clear distance between the frame and the casement of not less than 95 mm as shown in fig. 4.12. This will allow both sides of the glass to be cleaned from inside the window which is an important factor in reducing maintenance costs when the window is used in multi-storey buildings. If required, side-hung casements are also provided with a lever handle that is forked so it can engage on a striking plate in two positions, either shut or slightly open to provide limited night-time ventilation. Side-hung and top-hung casements are not always weather-

stripped as standard, but if specified a PVC weatherstrip is fixed by studs to the inside of the casement section. A variety of security and safety fittings have been designed specifically for steel windows. Side-hung casements can have a safety catch at the head of the window, out of reach of children, that limits the initial opening to around 100 mm.

Horizontally pivoted casements These are fully reversible and are weatherstripped by a synthetic rubber extrusion bonded or clamped into a groove within the frame. This type of window also permits cleaning of the building and all the adjoining glass areas within arm's reach. So with careful design the whole of the glazing to a multi-storey building can be cleaned with safety from the inside. The glass in a reversible window can also be replaced from within the building. The hinges for reversible casements are of the friction type, adjusted to hold the casement in any position. A safety catch (releasable by hand) limits the initial opening of the casement to approximately 15 degrees which, depending on the height of the window, will mean that the window projects into the room from 100 to 150 mm. When it is released, the window can be reversed and pivoted through 180 degrees. The catch then re-engages to hold the window firmly in the reversed position for maintenance or window cleaning to take place. This is shown in fig. 4.13.

For the standard range of windows, handles are made in the following alternative finishes: forged brass; nickel chromium plated on brass, or on zinc-based alloy; and various aluminium alloys. Handles are detachable and can be replaced without disturbing the glass.

The whole range of windows to this British Standard is manufactured from only 11 basic steel sections; by standardization, and the application of industrial techniques of large-scale manufacture, an acceptable and comparatively inexpensive range of windows offers a choice from a range of types and sizes.

4.4.3 Fixing standard steel windows

Steel windows are usually installed as the building work proceeds and must be protected from damage (they are always site glazed after fixing). However windows with factory-applied finishes are best installed into previously prepared openings. If subframes are required, they may be factory-coupled to the windows or are built-in first, the steel windows being fixed later.

For maximum weather performance, windows should be set back at least 75 mm from the face of the wall. When building-in, care must be taken to keep the window both plumb and square. In the case of composite windows, attention must also be paid to alignment across the couplings; and fixings should be at the holes adjacent to the couplings, care must be taken to adequately seal the joint between the window and coupling.

The number of fixings required will vary according to the size of the frame, a minimum of 2 per jamb and at 450 mm centres. Countersunk screws are fitted through factory-punched holes in the web of the sections, windows over 900 mm wide will need to be centre-screwed at top and bottom. The types of fixings are as follows:

- Rust-proofed wood screws, not less than 10 gauge (3.25 mm) for fixing into proprietary plastics or fibre plugs in pre-drilled holes in precast concrete surrounds or *in situ* openings;
- Short countersunk screw and nut for securing the frame before building-in to steel lugs set in the joint of brick or masonry openings. The lug has elongated slots to allow adjustment to accommodate variation in joint position (fig. 4.33).
- Self-tapping screws for fixing to plastic subframes and steel combined lintels.

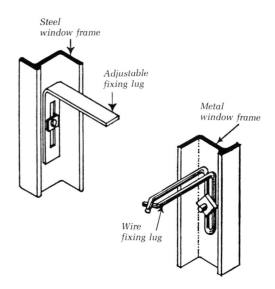

Fig. 4.33 Alternative types of adjustable fixing lug

Typical fixing details are shown in fig. 4.34. Where fixed direct into the opening, the metal windows are set in a waterproof cement fillet, which is butted to the jambs of the opening before the window is offered into position and connects with the wall DPC. The space between the frame and the opening is then pointed in a suitable mastic, and the inside reveal usually plastered. As the window height dimension may not always coincide with brick courses, the window should be designed to line through with a brick course at the head; the brickwork under the sill (if pressed metal) can be cut or adjusted to suit. In cavity brickwork the relationship between the window frame and the DPC is important. The DPC should extend 15 mm into the opening, the window being positioned so the front leg is against the DPC to limit its exposure to wet brickwork (the depth of the reveal is as a consequence largely predetermined when using steel windows in cavity construction). The frame should then be sealed to the sill with bedding compound and external pointing is recommended.

Examples of the installation of metal windows are shown in fig. 4.35. Figure 4.35(a) shows the *conventional use of standard windows* within a cavity wall where the windows are fixed after the opening has been constructed. The brickwork and inner leaf of block are supported by a steel combined lintel with internal insulation. As described above, for the DPC to coincide with the front of the frame, the inside face of the front leg of the section is fixed to align with the back face of the brickwork. The inside wall and the reveals are finished with plaster. The jambs and pressed metal sill are fixed with galvanized lugs that are

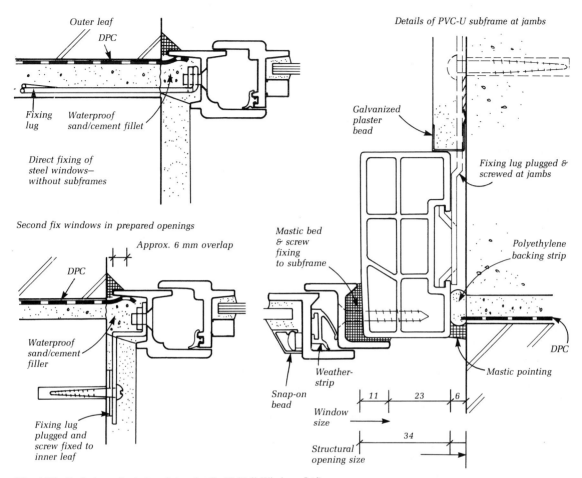

Fig. 4.34 Typical metal window fixing details (Crittall Windows Ltd)

plugged and screwed back to the internal blockwork. For the window to fill the opening, bricks at the sill are cut to form a brick-on-edge course; the cut bricks are obscured by the metal sill which is bedded in mortar.

The *W20 double-glazed, top-hung casements* in fig. 4.35(b) are fixed at the outside of an opening formed within a stud wall faced in horizontal profiled aluminium cladding. At the head, the window frame is screwed to the 100 × 50 mm preservative treated rail that forms the opening and jointed with bonding compound. The projecting bottom edge of the cladding weathers the head of the window over which is lapped the breather paper to the insulated stud wall. The jambs are screwed to vertical wrot timber posts to which a polyester powder-coated T-section is face fixed, forming mullions within the run of windows (see photo 4.8).

The windows in fig. 4.35(c) are fixed at the front of the opening, this time the walls are *externally insulated blockwork*. The system of factory-formed insulation panels, which are fully bonded to the concrete frame, incorporate a 2.0 mm thick self-coloured acrylic resin external render. The large W20 outward opening top-hung lights are fixed at the head by lugs countersunk bolt anchored into the concrete soffit. Weathering to the head and sill is provided by a W20 transom section screwed to the frame. At the jambs the frame is screwed to a 45 × 5 mm mild steel flat. The window is sealed all round to the insulation panels by mastic on a compressible backing strip.

Subframes Steel frames used commonly to have timber subframes. BS 1285: 1980 *Specification for wood surrounds for steel windows and doors* describes their construction and details preferred cross-sections. Timber surrounds are rarely employed now (other than when replacing existing windows). Cellular PVC-U sections are generally used instead; for refurbishment work they are available in sizes corresponding to those of BS timber subframes. Windows can be more easily replaced when they have been fixed within a subframe.

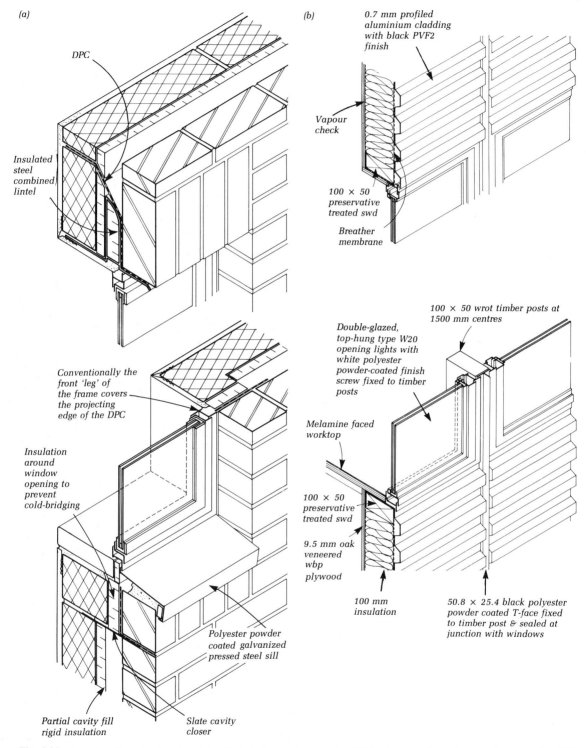

(a)

DPC

Insulated steel combined lintel

Conventionally the front 'leg' of the frame covers the projecting edge of the DPC

Insulation around window opening to prevent cold-bridging

Partial cavity fill rigid insulation

Slate cavity closer

Polyester powder coated galvanized pressed steel sill

(b)

0.7 mm profiled aluminium cladding with black PVF2 finish

Vapour check

100 × 50 preservative treated swd

Breather membrane

Double-glazed, top-hung type W20 opening lights with white polyester powder-coated finish screw fixed to timber posts

100 × 50 wrot timber posts at 1500 mm centres

Melamine faced worktop

100 × 50 preservative treated swd

9.5 mm oak veneered wbp plywood

100 mm insulation

50.8 × 25.4 black polyester powder coated T-face fixed to timber post & sealed at junction with windows

Fig. 4.35 Examples of constructions to receive steel windows: (a) conventional fixing into a cavity wall — see the note to fig. 4.22 regarding the use of sub-sills which do not project beyond the face of the wall below; (b) details of a house at Happisburgh, Norfolk (see photo 4.8(a)); (c) external rendered wall of Hospital of St. John and St. Elizabeth (see photo 4.8(b))

(c)

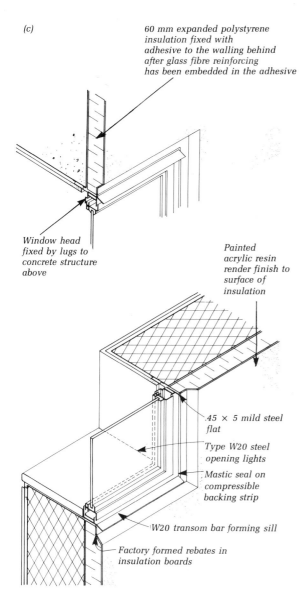

60 mm expanded polystyrene insulation fixed with adhesive to the walling behind after glass fibre reinforcing has been embedded in the adhesive

Window head fixed by lugs to concrete structure above

Painted acrylic resin render finish to surface of insulation

45 × 5 mild steel flat

Type W20 steel opening lights

Mastic seal on compressible backing strip

W20 transom bar forming sill

Factory formed rebates in insulation boards

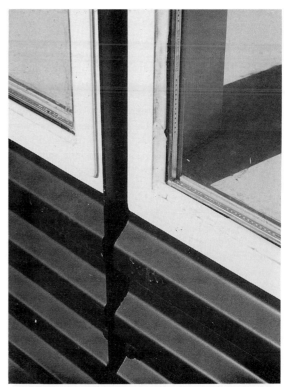

Photo 4.8 (a)   House at Happisburgh, Norfolk. Architects: John Winter and Associates, 1989. Construction details − fig. 4.35 (b).
Photo 4.8 (b)   Hospital of St. John and St. Elizabeth. Architect: David Morley, 1991. Construction details − fig. 4.35 (c).

Subframes can help make up the dimensions of standard windows to the size of brickwork openings in cavity walls. They also give greater 'weight' to the elevation of the windows though in contradiction to a principal advantage of steel frames — the relatively small obstruction they present to the admission of daylight and view. Their use is recommended in the BRE report *Thermal insulation: avoiding risks* to deflect water that is likely to condense on steel frames away from the plaster finish or paintwork of the surrounding reveals. Unlike timber, PVC-U subframes require no site painting and little subsequent maintenance but they have a similarly low thermal conductivity which helps prevent cold-bridging and condensation at the perimeter of the frames.

PVC-U subframes are constructed from PVC-U multi-cell extruded sections, corners are mitred and welded. They can be assembled with the windows at the factory and help protect the window during transport to site; or the subframes can be built into the opening as work proceeds (in which case they must be braced to keep them plumb and square) forming a permanent template for a factory-finished window to be installed at a later time. The joint between the subframe and its masonry opening is sealed with mastic gunned on to a compressible backing strip or the joint is filled solid with polyurethane construction foam and then pointed in mastic. It should be noted that the neatness of these perimeter joints is very dependent on the regularity of the bricks specified. Details of the arrangement of a steel window in a standard PVC-U surround are shown in fig. 4.36. Galvanized steel straps are used for building subframes into new work or they can be installed by screwing through the subframe directly into the wall; the steel frames are set in bedding compound on to the subframes and then self-tap screwed through the cellular divisions.

Mastic jointing A mastic seal to a joint is used where some degree of movement is likely to occur, usually between dissimilar materials, such as between metal and plastic or brickwork. The function of mastic is to accommodate movement, provide weather-proofing and seal against draughts, dust or fumes. Steel window technology is very reliant on the use of mastic particularly in the construction of composite assemblies. A mastic will only fulfil these functions if it satisfies an exacting set of requirements. The various types of mastic and their uses are discussed in *MBS: Materials* chapter 16 and for the use of mastics in curtain walling, etc., see *MBS: Structure and Fabric Part 1*. Having made the correct choice of mastic it is essential that the joint is designed so that unreasonable demands are not made on the jointing material in its effort to accommodate movement. The application of mastic seals for bedding and fixing window frames is shown in fig. 4.34.

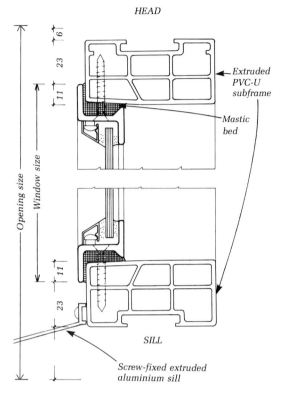

VERTICAL SECTION:

HEAD

Extruded PVC-U subframe

Mastic bed

SILL

Screw-fixed extruded aluminium sill

Fig. 4.36 Standard metal window in a PVC-U subframe (Crittall Windows Ltd)

The method of applying the mastic is dependent upon the type of joint and the materials to be connected at the perimeter of the window and the exposure rating. Where metal frames are to be set into a PVC-U subframe a continuous ribbon of mastic applied to the external and internal edges of the surround will ensure that the frame is fully bedded once placed in position. The external vertical joint between a PVC-U subframe or surround and the brick or masonry jambs of the window opening are particularly subject to differential rates of expansion and contraction. It is essential that the joint is pointed with a mastic which will be able to accommodate this movement. Figure 4.37 shows the use of mastic for the joints between frames and loose mullions (or transom rails) in metal windows.

Composite steel windows are inevitably subject to a certain amount of movement at the junction of the fixed frames and the mullions and transoms. The joints should be solid bedded with oil-based hand grade bedding compound. A ribbon of mastic for the internal and external joint should be applied either to the fixed frame or to the mullion or transom as convenient during assembly. Windows finished in polyester powder paint require vertical

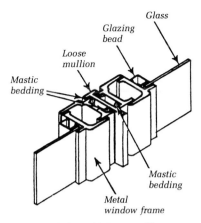

Fig. 4.37 Metal window with loose mullion

Table 4.3 Extreme sizes of ventilators made of steel W20 Universal Section (mm)

Method of opening	Size of section	Height plus width	Height	Width
Side-hung	Normal	2600	1900	700
	Heavy	3300	2600	900
Folding	Normal	3200	1900	1300
	Heavy	3900	2400	1800
Vertically	Normal	2900	1900	1100
Pivoted	Heavy	3900	2600	1400
Folding vertically	Normal	3600	1800	1800
Pivoted	Heavy	4700	2400	2300
Top hung	Normal	2600	1500	1500
	Heavy	3200	1800	1800
Horizontally	Normal	2600	1500	1500
Centre hung	Heavy	3200	1800	1800
Bottom hung	Normal	2600	1500	1500
	Heavy	3200	1800	1800

and horizontal joints subject to severe exposure conditions, above third floor level, to be pointed on the outside with a self-curing sealant, such as one-part polysulphide rubber.

4.4.4 Purpose-made

A further range of universal steel sections is produced known as W20 from which purpose-made windows are produced. The hot-rolled sections are heavier than those used for type F windows (see fig. 4.38), being intended for commercial and public rather than domestic buildings. There is not a standard range but for the dimensional co-ordination of W20 windows the modular spaces proposed in BS 6750: 1986 *Specification for modular coordination in building* are shown in fig. 4.39 and the maximum manufactured sizes of window to fill them are as indicated in table 4.3

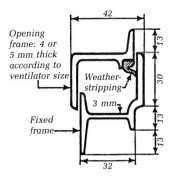

Fig. 4.38 W20 steel section for purpose-made windows

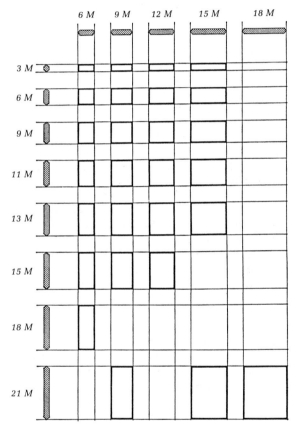

M = 1 module of 100 mm and refers to the size of opening into which the window fits

Fig. 4.39 Basic spaces for purpose-made steel windows

Purpose-made windows are coupled together by mullions and transoms, in the same way as standard steel windows, to form composite units that may span from floor to floor. The mullions are usually the primary structural members (as in curtain walling); they are mechanically joined with the windows to combine the structural rigidity of steel with the integral strength of the frame bars.

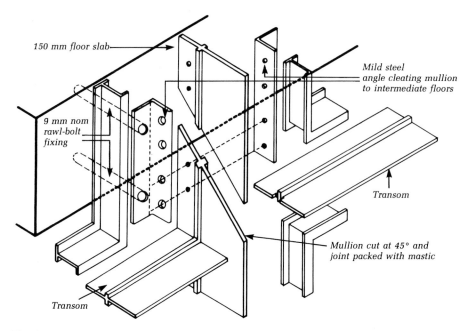

Fig. 4.40 Coupling of large metal windows

Figure 4.40 shows such a window detailed to have substantial mullions 175 mm deep with a total height of 7.8 m. The problems of jointing mullions of this kind can be solved in several ways. In the example shown, the stops are concealed so that the mullion is built up from two 9 mm thick steel flats lapping each other and screwed together. This method reduces the parts to a convenient dimension for galvanizing whilst enabling assembly to the length required. The feet and tops of the mullions have fixing plates welded across them that are packed off the concrete with galvanized shims. The base plates are bolted to the concrete structure at the sill and head. The fixings are hidden inside pressed steel casings. It is, of course, possible to joint mullions at intermediate supports such as floor and landing levels as in curtain walling technology. The fixing plates have to be designed to resist the particular wind and other structural loads that are anticipated.

One design is shown on the detail in fig. 4.40 — the splayed cut gives a neat watertight joint which must be well sealed with mastic. When constructing large windows of this kind, sashes may be coupled together with transoms if they are not structural, and built from the sill up, fixing to one side. The first mullion is placed and then the next vertical range of sashes and so on across the total width. Expansion in large metal windows is a problem which must be anticipated. The larger structural members are the main difficulty as the expansion of the smaller members, such as the sashes, can be taken up in the many mastic-bedded joints. Box stanchions can be used (as for curtain walling)

but made from pressed metal or rectangular hollow section steelwork. These can be designed to allow for expansion in the length of the window by providing a sliding fixing at the tops of mullions. For further information on the construction of window walls see *MBS: Structure and Fabric Part 1* chapter 5 *Curtain walling*. BS 5516: 1977 (1991) *Patent glazing* Appendix G: 'Safety of vertical patent glazing' gives recommendations about the risk areas associated with vertical glazing, including door side panels that may be mistaken for doors or which contain more than one pane of glass; glass wholly or partially within 800 mm from floor level; and glass in bathing areas (see chapter 7). These requirements have now been incorporated into Part N of the *Building Regulations 1991*.

4.4.5 Pressed steel

Metal pressings were first used by the building industry in the manufacture of standard door frames, which proved strong, quick to erect, and clean in outline. The material has since proved suitable for skirtings, stair treads, risers, window frames, shelving and lift doors. Mild steel sheet is used in a range of gauges from 26 gauge for angle bead, 20 gauge for door frames, 12, 14 and 17 gauge for window frames to sometimes 10 gauge, approximately 3 mm thick, where greater strength and rigidity are required. Sheets up to 4 × 1.5 m can be fabricated, depending on the size of the break press the manufacturer has. Contemporary machines are controlled by computer, making a greater

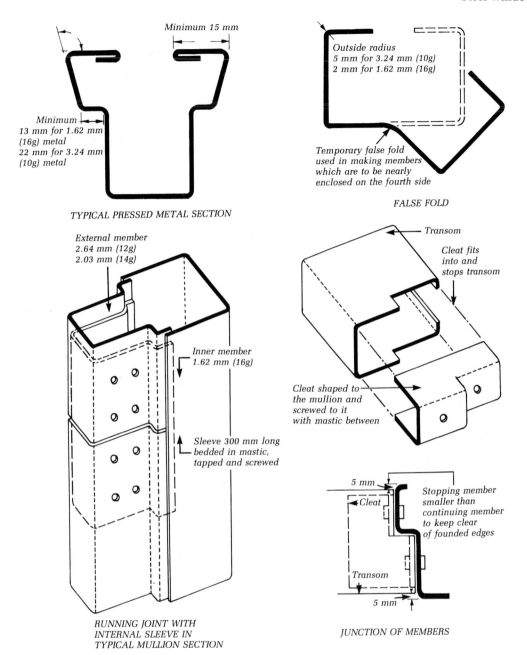

Minimum 15 mm

Minimum
13 mm for 1.62 mm
(16g) metal
22 mm for 3.24 mm
(10g) metal

TYPICAL PRESSED METAL SECTION

Outside radius
5 mm for 3.24 mm (10g)
2 mm for 1.62 mm (16g)

Temporary false fold
used in making members
which are to be nearly
enclosed on the fourth side

FALSE FOLD

External member
2.64 mm (12g)
2.03 mm (14g)

Inner member
1.62 mm (16g)

Sleeve 300 mm long
bedded in mastic,
tapped and screwed

RUNNING JOINT WITH
INTERNAL SLEEVE IN
TYPICAL MULLION SECTION

Transom

Cleat fits
into and
stops transom

Cleat shaped to
the mullion and
screwed to it
with mastic between

5 mm

Cleat

Stopping member
smaller than
continuing member
to keep clear
of founded edges

Transom

5 mm

JUNCTION OF MEMBERS

Fig. 4.41   Pressed metal box mullions

complexity of edge profiles possible.

Typical details of pressed steel sections used in combination with steel windows are shown in fig. 4.41. Folds can be made through any angle up to 105 degrees (that is by forming an angle of not less than 75 degrees) but a stiffened edge can be made by pressing a flat fold down; this is called a bar fold. Surfaces must be flat planes

and it is not possible to form sharp angles, the minimum outside radius being usually 2 mm for 16 gauge sheet to 20 mm for 10 gauge. Sections are usually open for most of one side, as required for connections, fixing cross-members to stiffen mullions etc. If a member is to be formed with the fourth side mostly closed, it is necessary to form it with a false fold as shown.

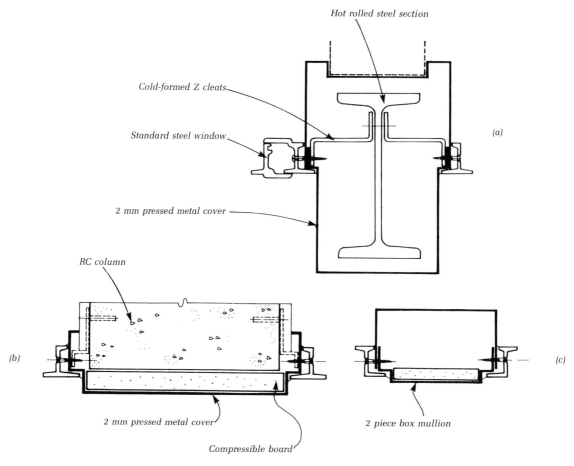

Hot rolled steel section

Cold-formed Z cleats

Standard steel window

(a)

2 mm pressed metal cover

RC column

(b)

2 mm pressed metal cover

Compressible board

2 piece box mullion

(c)

Fig. 4.42 Pressed metal ancillary items: (a) steel column casing; (b) concrete column casing; (c) box mullion (Crittall Windows Ltd)

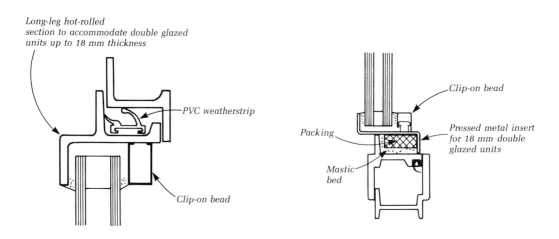

Long-leg hot-rolled
section to accommodate double glazed
units up to 18 mm thickness

PVC weatherstrip

Clip-on bead

Clip-on bead

Pressed metal insert
for 18 mm double
glazed units

Packing

Mastic
bed

Fig. 4.43 Manufacturers' methods of double glazing steel windows

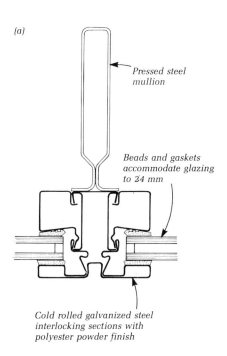

(a)

Pressed steel
mullion

Beads and gaskets
accommodate glazing
to 24 mm

Cold rolled galvanized steel
interlocking sections with
polyester powder finish

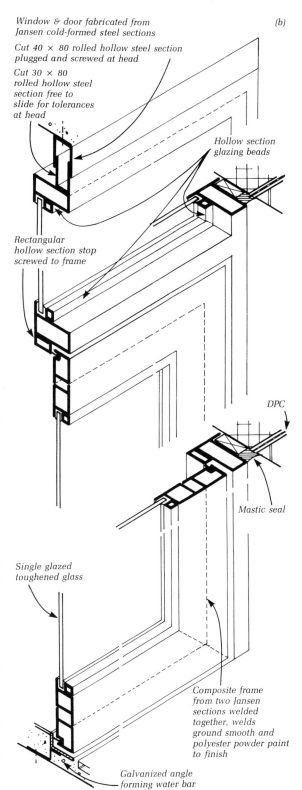

Window & door fabricated from
Jansen cold-formed steel sections

(b)

Cut 40 × 80 rolled hollow steel section
plugged and screwed at head

Cut 30 × 80
rolled hollow steel
section free to
slide for tolerances
at head

Hollow section
glazing beads

Rectangular
hollow section stop
screwed to frame

DPC

Mastic seal

Single glazed
toughened glass

Composite frame
from two Jansen
sections welded
together, welds
ground smooth and
polyester powder paint
to finish

Galvanized angle
forming water bar

Fig. 4.44 Cold formed sections for steel windows (Secco Windows)

Cutting is done by mechanical saws or by grinding wheels, or by burning. Sheet metal is also cut by knife in a guillotine. Burning is precise enough to be used for cutting square mortices for a square member to pass through. Holes, including small mortices, are formed by drilling or punching. The latter may cause some deformation of the metal which, if there are many holes, may accumulate to measurable increases in length and width. Running joints are made with sleeves, bedded in mastic and tapped and screwed with countersunk screws, as shown in the diagram. Right-angled junctions between members (as for example between mullion and sill) are formed by scribing the end of the stopping member to the profile of the continuing member, filling it with a shaped cleat and screwing it to the continuing member, with a generous amount of mastic packed in the joint. A plain rectangular profile simplifies the junction. Complicated profiles call for a complicated cleat and may present difficulties in screwing up. Similarly, the scribing can only be done against flat planes but most importantly, the stopping element should be smaller than the continuing one it is adjoining so that it is quite clear of the slightly rounded angles of the continuing element. All this is illustrated in fig. 4.42 which shows typical column cover and mullion sections.

Where long members are used — tall mullion or a long sill — it is important to ensure that the arris is dead straight and the planes are true. What may look satisfactory in elevation may look very poor when seen from directly below or from one side, where it is easy to get 'an eye along the edge'. To get a true line it is essential that the metal is thick enough to keep its folded shape and that it is carefully fixed. The designer can get the latter put right, but if the metal is too thin any remedy is rather expensive.

Pressed steel sections are supplied galvanized and primed ready for site painting, or galvanized and factory finished with polyester powder coating to BS 6497 (see p. 88).

4.4.6 Double-glazed steel windows

Steel windows to BS 6510: 1984 can be double glazed. Type F windows can accommodate single glazing up to 6 mm thick and double-glazed units of 14 or 18 mm thickness (although not all of the Steel Window Association's standard range are available double glazed). W20 windows allow single glass panes up to 10 mm and double glazing of 24 mm thickness. The methods by which the thicker double-glazed units are fixed varies between manufacturers, either by special aluminium inserts that are fitted into the steel frames (but reduce the extent of sight lines as a result), by an additional component that projects beyond the face of the frame, or by the use of extra frame sections not within the BS 6510 range (see fig 4.43).

Similarly other components can be 'glazed-in' to the frames. Trickle ventilators as required by Part F of the *Building Regulations 1991* can be made from an aluminium extrusion polyester powder coated to match the windows, they are glazed-in at the top and sides and have a rebated bottom edge to receive the glass (see fig 4.32).

4.4.7 Cold-formed sections

In addition to the long-standing tradition of steel windows being made from hot-rolled sections, they can also be fabricated from cold-rolled steel. One continental system (fig. 4.44(a)) available in the UK is made with 'open' joints (see timber windows, p. 60) on the outside which results in a flush appearance on both the outside and inside. Double glazing up to 24 mm thick can be accommodated but as for other steel windows, the frames are not thermally broken. The addition of an internal pressed steel mullion enables construction of composite assemblies up to 4 m high.

Shown in fig. 4.44(b) is a door/window constructed from Jansen hollow steel sections. These are manufactured in a wide variety of profiles and sizes, basically box sections but with projecting flanges and also rebates. When welded up to form frames, they can be fitted with steel beads and stops to form strong and relatively slender doors and windows. The rebates can accommodate weatherstrips and thermal breaks.

4.5 Aluminium windows

In contemporary construction very considerable use is made of aluminium windows. Aluminium alloy is an adaptable material which produces windows to a very high degree of accuracy and with a high standard of finish. Because it is light in weight, aluminium is particularly suitable for the manufacture of both horizontal and vertical sliding windows, 'tilt and turn' hinged windows and reversible pivot windows. Aluminium sections can be very easily weatherstripped. They can as a result achieve efficient draught and weather-proofing as well as a good degree of sound control. Double aluminium windows are used to meet particularly onerous requirements for sound reduction.

Aluminium windows should conform with BS 4873: 1986 *Specification for aluminium alloy windows*, and some manufacturers also participate in BSI Quality Assurance Schemes. The windows are constructed from extruded aluminium to BS 1474: 1987 *Wrought aluminium and aluminium alloys for general engineering purposes*, and as mentioned for timber windows, should be tested for compliance with BS 6375. Aluminium window sections used in the construction of frames excluding glazing beads, nibs, interlocks and similar features, must be not less than 1.2 mm thick. Because the slenderer the walls of the section, the greater the likelihood of distortion during extrusion, the depth to wall ratio should preferably be in the region of 3:1 and the wall thicknesses should be as uniform as possible. Because aluminium is less stiff than steel, the profiles of aluminium frames are considerably larger than those for equivalent steel windows.

Sections are extruded by forcing, under extreme pressure, a heated billet of aluminium through a steel die of the desired profile that is inserted into a hydraulic press. It is a simple matter to incorporate grooves in the section during extrusion to accommodate weatherstripping. The latter can be solid or cellular chloroprene rubber, cured ethylene propylene diene monomer (EPDM), polypropylene pile, or plasticized PVC. The corners of the frames are made by cutting matching mitred ends, inserting a close-fitting L-shaped aluminium cleat and then mechanically joining the lengths together; a sealant is applied at the joints. Alternatively the mitres are self-tap screwed together or mortice and tenoned. The latter two are more suited to dissimilar cross-sections. Mechanical mortice and tenon joints are used to attach glazing bars and intermediate members and riveting or screwing are employed to fix the fittings. The maximum sizes for primary ventilation will depend upon the cross-sectional strength of the extruded sections. The mechanical joints used to fabricate aluminium windows are less strong than welded corners and this limits the feasible maximum size of pane compared with steel frames.

Aluminium is a good conductor of heat and this must be taken into account in heat loss calculations. Most UK manufacturers still produce frames without thermal breaks as an economical (in first cost) alternative to their ranges of thermally broken windows. BS 4873 requires that where aluminium window frames are thermally improved by the inclusion of a barrier between independent internal/external sections (to reduce 'cold-bridging'), the insulating material should be stable under the conditions of service, e.g. under wind and dead loads and within the likely surface temperature range of the frames. The thermal barrier may be of polythene resin, neoprene extrusion, or PVC-U, nylon, polypropylene or polyamide extrusions, or rigid foam plastic. The thermal break is fitted either by rolling it into matching rebates within the internal and external parts of the frame, or the extrusions are made with detachable legs that enable them to be placed together, forming a box, into which liquid foam plastic is poured; the legs are broken out after the foam has solidified.

Finishes normally available are: *mill finish* (untreated aluminium as it comes from the die), *anodized* (a protective oxide coating produced by electrolytic oxidation), *organic* (stoved acrylic and polyester powder coatings which are produced in many more colours than anodizing), and *claddings* (usually of stainless steel). Mill finish will weather naturally. In the past, if the atmosphere was not industrially corrosive or marine, this was considered satisfactory. Naturally finished aluminium can be painted provided that a zinc chromate primer is used, but electro-statically applied polyester powder coatings have been most usual in recent years.

Mill finish, although it only corrodes superficially, is however not very durable. The closing edges of windows become pitted and corrosion induces stresses in the profile of thin extrusions; eventually sliding windows no longer slide and side-hung and top-hung casements twist. Care is required if using aluminium windows in conjunction with stone or concrete cladding. Both mill-finish and anodized aluminium are adversely affected by water running off limestone and cementaceous materials. Large overhangs above windows help obviate this problem. Naturally finished aluminium can be painted provided that a zinc chromate primer is used but electrostatically-applied polyester powder paint has become the predominant type of applied finish in recent years.

Anodizing should be to BS 3987: 1974 *Specification for anodic oxide coatings on wrought aluminium for external architectural applications*, and although expensive, produces an attractive finish on polished aluminium. Certain colours will not remain light-fast and manufacturers' recommendations should be sought. Organic and stove acrylic finishes should comply with BS 4842: 1984 *Specification for liquid organic coatings for application to aluminium alloy extrusions, sheet and preformed sections for external architectural purposes, and for the finish on aluminium alloy extrusions, sheet and preformed sections coated with liquid organic coatings*, and polyester powders to BS 6496: 1984 *Specification for powder organic coatings for application and stoving to aluminium alloy extrusions, sheet and preformed sections for external architectural purposes, and for the finish on aluminium alloy extrusions, sheet and preformed sections coated with powder organic coatings*. These finishes are now widely used, are proving to be very durable and are available in a wide range of colours. A few manufacturers produce aluminium window frames (usually for shopfronts), which are clad in stainless steel to produce relatively economic frames which take advantage of the properties of both materials.

The aluminium alloys used in windows are very durable, and good quality anodizing or organic coatings will maintain appearance for 15–20 years. Aluminium can be attacked by atmospheric pollutants and cleaning should take place with mild detergent or soap and water every 6 months on rural sites, and every month on urban and marine sites. Mill finished aluminium windows should not be installed on sites near the sea.

Wheels and rollers fitted at the head or sill of horizontal sliding aluminium windows to support the weight of the sashes and the ironmongery used on aluminium windows need special consideration (to minimize wear and eliminate the likelihood of bimetallic corrosion). BS 4873 lists a range of suitable materials, including aluminium, in compliance with BS 3987, or BS 5466: *Methods for corrosion testing of metallic coatings* Part 1 1977 *Neutral salt spray test* (*NSS test*); and zinc die casting alloy complying with BS 1004: 1972 *Specification for zinc alloys for die casting and zinc alloy die castings*, or steel finished in accordance with BS 6338: 1982 *Specification for chromate conversion coatings on electroplated zinc and cadmium coatings*.

Mill finish windows will normally be dispatched from the factory unprotected and must be cleaned down by the contractor just before the scaffolding is dismantled. When the windows are anodized, organic, stoved or powder coated, they should arrive at the site covered by a film of wax and then protected by strong self-adhesive tape on all the visible and exposed surfaces. The contractor can then easily remove the tape and polish the wax away to give a finished surface. This should not be done, of course, until there is no risk of damage by following trades.

4.5.1 Standard

BS 4873 does not contain preferred modular sizes for aluminium windows. However, many manufacturers offer standard size ranges: some are dimensionally co-ordinated

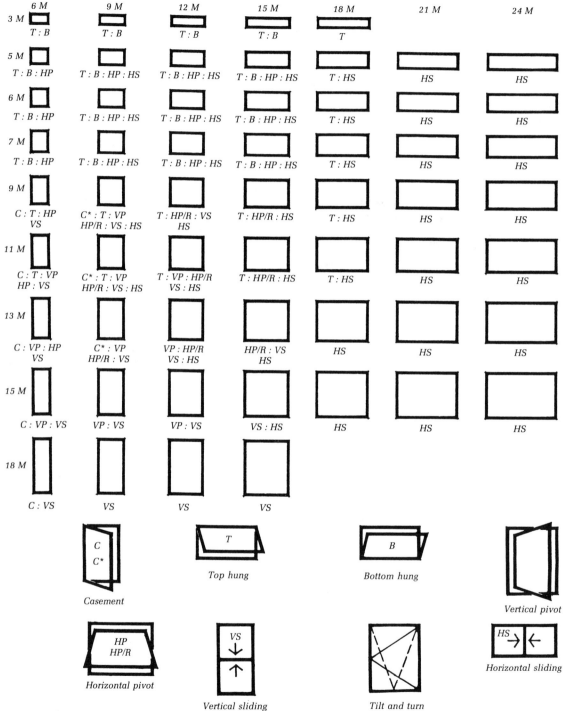

Fig. 4.45 Basic spaces and ranges of aluminium windows

but only as an indication of what is available, since many windows are made to order.

The Aluminium Window Association has produced a chart of the basic spaces and ranges in which dimensionally co-ordinated windows should be made. Figure 4.45 shows these spaces which indicate aperture sizes and not the working size of the window. The dimensions of an assembled frame should be within a permissible deviation of ±1.5 mm from the work size and the maximum difference between diagonal measurements should be 4 mm. In addition to the sizes shown there is a range 2100 mm high suitable for vertical sliding sashes, single or double doors and sliding doors. Modular windows can be installed in non-modular brickwork openings by using filler pieces or subframes (see *MBS: Internal Components* section 1.8 *Modular coordination*).

Fixing aluminium windows Aluminium windows are normally fixed into prepared openings as second fixings. Most manufacturers also provide a fixing service so that the window and its installation are the responsibility of one company. All straps, clips, brackets, lugs, screws, bolts, metal washers and shims and other fixings must be of suitable strength and of materials which minimize corrosion arising from electrolytic reaction (as previously described for bearings and hardware, e.g. zinc plated and hot-dip galvanized steel or aluminium, etc.). It should be established that any timber used in the construction adjacent to aluminium windows and any chemicals used when preservative treating the timber will have no harmful effects on aluminium, in accordance with advice given in CP 153: *Windows and rooflights* Part 2 1970 *Durability and maintenance*. There should be no direct contact between aluminium and oak, sweet chestnut or western red cedar.

Great care must be taken during handling and installation on site. The frames are not as strong as those made out of steel and will not support scaffold poles or boards, etc. Aluminium windows should preferably not be fixed until all the structural work and wet finishes are completed and should be wrapped for protection before dispatch and carefully stored on the site. They must be kept very clean during the progress of the work as cement or plaster will adhere to their surfaces and leave marks on the bright aluminium. Allowance has to be made for the large thermal expansion of aluminium compared with other metals.

4.5.2 Horizontal sliding

Figure 4.46 shows a purpose-made horizontal sliding window in aluminium which is made in a range of sizes to comply with the basic space recommendations for aluminium windows shown in fig. 4.45. The horizontal sliding window is a type much used in commercial buildings, schools, and hospitals, being particularly suitable for taller buildings. The window is detailed so that both sashes can slide past one another and be safely cleaned from the inside of the building. These sashes run on ball-bearing rollers (although some manufacturers use nylon rollers). The windows are fitted into the prepared opening either by screw fixing to a subframe (which simplifies installation into uneven openings) or by the use of purpose-made brackets which twist lock into position in the frame and are then screwed or shot fired into the concrete or brickwork. The tracks are fixed into the opening first and then the pre-assembled window at a later date. When fixing aluminium windows, ordinary steel or brass screws which cause bimetallic corrosion should not be employed. The window can optionally be fitted with a catch that limits the window to being only slightly open, an important security consideration when intending to use horizontal sliding windows.

The windows are to be factory glazed, the glass being sealed into the frames by one-piece PVC gaskets, the stiffness and type of glass used being an important factor in the overall rigidity of the component. As shown, the same frame sections can accommodate either single or double glazing by changing the type and thickness of the gaskets. When channel section gaskets are used, the window is built around the glass and gasket assembly; this is a usual method in the construction of aluminium-framed windows. To repair a broken pane, the window frame has to be taken to pieces and reassembled.

Weather resistance is a difficult problem to overcome in the design of horizontal sliding windows. This example conforms with the 'moderate' exposure rating of BS 6375: Part 1: 1989 *Classification for weathertightness*. Polypropylene pile weatherstripping is incorporated at the head, sill and meeting style of the window panes. Although able to be double glazed the economically priced window shown here does not have a thermally broken frame and may attract condensation if used in rooms with hot, humid conditions. There will also of course be significant heat loss through the aluminium frames.

4.5.3 Vertical sliding sash

The alloy used for the frame of this window (fig. 4.47) is the same as for the extrusions for the bottom hung sash illustrated previously. However, this is a higher performance window which is double glazed as standard and each section incorporates a thermal break made from rigid polyurethane. The corners of the frames are mitred, sealed with mastic and screwed together and butt jointed and screwed to the sill. The thickness and stiffness of the glass contributes to the rigidity of the frame.

The sashes are hung on spring-loaded sash balances and the meeting sections of the frame are made from PVC-U

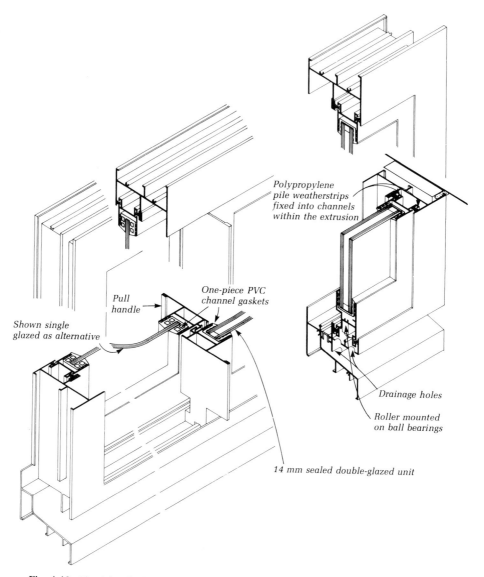

Polypropylene
pile weatherstrips
fixed into channels
within the extrusion

Pull
handle

One-piece PVC
channel gaskets

Shown single
glazed as alternative

Drainage holes

Roller mounted
on ball bearings

14 mm sealed double-glazed unit

Fig. 4.46 Aluminium horizontal sliding window

which reduces friction and is better resistant to wear than the finished surface of the aluminium extrusions. The polypropylene weatherstripping is clipped into the fixed frame in such a way that it can be replaced after damage or wear during use. The meeting rails have a positive interlocking action which ensures good security by preventing the release of the catch from the outside.

The extruded sash sections are designed to accommodate proprietary double glazing units of 20 mm overall thickness with a 12 mm air gap. As is now quite usual in the design of vertical sliding windows, the sashes can be released to swivel open inwards so the glass can be cleaned or replaced

from within the building. Vertical sliding windows are otherwise very dangerous to maintain (unless on high buildings provision is made for external cleaning cradles, etc). This dual action operation requires quite sophisticated ironmongery. These windows are fitted with a rotary action catch which is built into the frame and is more secure than the traditional sash fastener. An optional deadlock can be fitted to the catch as well as stops that engage the sashes in a fixed partially open position. Incorporated into the horizontal rails are projecting bars for opening and closing the sashes.

A window of this type could be made up to a maximum

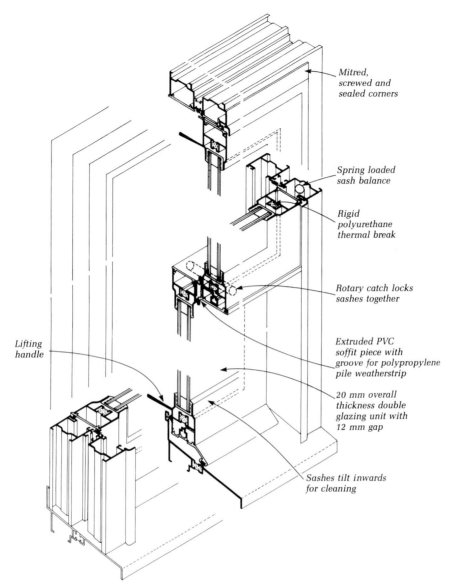

Fig. 4.47 Aluminium vertical sliding/tilting sash window

height of 2440 mm and a maximum width of 1525 mm with
the limitation that the perimeter measurement must not
exceed 7625 mm, which means that the maximum width
and height cannot be used in the same window. The basic
spaces relative to this type (VS) of window are shown in
fig. 4.45.

4.5.4 Bottom hung

A bottom-hung opening inwards aluminium window is
shown in fig. 4.48. This type of window is useful where
draught-free ventilation at low level is desirable. The

sections are extruded from aluminium hollow sections of
6063T6 alloy in accordance with BS 1474: 1987
*Specification for wrought aluminium and aluminium alloys
for general engineering purposes*. Casements that open
inwards are harder to seal against the weather and so this
design is double weatherstripped, with an external
rainscreen gasket and a central rainshield 'flipper' gasket.
Both of them fit into grooves extruded in the flanges of
the extrusions to provide air and water seals between the
fixed and moving frames. The lengths of weatherstripping
clip in and can be removed for replacement if they become
damaged or worn over the life of the window. The opening

Labels (fig. 4.47):
- Mitred, screwed and sealed corners
- Spring loaded sash balance
- Rigid polyurethane thermal break
- Rotary catch locks sashes together
- Extruded PVC soffit piece with groove for polypropylene pile weatherstrip
- 20 mm overall thickness double glazing unit with 12 mm gap
- Sashes tilt inwards for cleaning
- Lifting handle

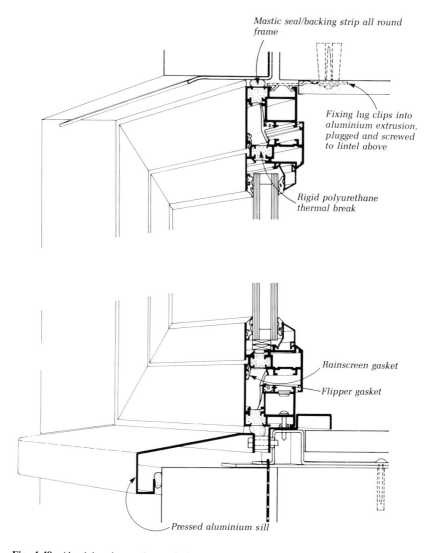

Mastic seal/backing strip all round frame

Fixing lug clips into aluminium extrusion, plugged and screwed to lintel above

Rigid polyurethane thermal break

Rainscreen gasket

Flipper gasket

Pressed aluminium sill

Fig. 4.48   Aluminium bottom hung window (Crittall Windows Ltd)

casement is hung on butt hinges and friction arms that hold the window at whatever position is required, and the head of the opening light is fitted with a spring catch.

The window can be site or factory glazed with insulated glass units up to 24 mm thick and the sections have a thermal break of rigid polyurethane poured in place. These windows have been designed not only to be fixed within wall openings but also to be dimensionally co-ordinated with the manufacturer's curtain wall system. They can consequently form fixed lights within either stick construction or unitized curtain walling. The window is mastic sealed to the pressed aluminium sill; the joint is continuous with the vertical joint between the jambs of the window and the enclosing wall. Both frame and sill are fixed by lugs that are plugged and screwed to the inner leaf of masonry.

It is interesting to compare the profile of the extruded aluminium sections used in this window with the mild steel rolled sections used in the standard steel window illustrated in fig. 4.32.

4.5.5 Tilt and turn

This type of window (fig. 4.49) originated in Germany and consists of a window which is inward opening with a double action sash — bottom hung inward opening (tilt) for

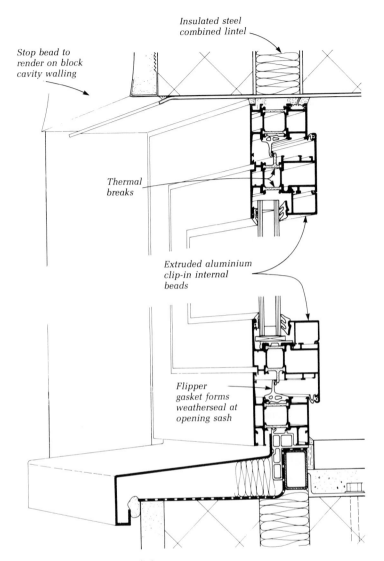

Insulated steel combined lintel

Stop bead to render on block cavity walling

Thermal breaks

Extruded aluminium clip-in internal beads

Flipper gasket forms weatherseal at opening sash

Fig. 4.49   Aluminium tilt/turn window

draught-free ventilation and side-hung opening-in (turn) for cleaning. They are very sophisticated windows which are now widely used in the UK. Advances in tilt and turn fittings have improved their safety: these include the provision of security stays, crack (narrow gap) ventilation in either the tilt or turn mode, and the introduction of switch barriers to ensure the windows cannot be put into both modes at the same time. The espagnolette bolt mechanism that is continuous around the edges of the sash help make them rigid. Roller shutters can be fitted in combination with this design to give added security and thermal insulation.

Several other dual action windows are also available, the most familiar of these, which many manufacturers produce

in aluminium, is the *projected top-hung reversible* type (also made in timber see p. 87). Similarly, *horizontal sliding and tilting* windows enable the casement to open inwards, hinged from the bottom for draught-free ventilation and for the sashes to slide past one another, on stand-off hinges, so the glass can be cleaned.

4.5.6 Built-up

Composite window walls and continuous runs of window can be constructed in aluminium in a similar way to that employed to form composite steel windows. Mullions are coupled together by H- or T-shaped bars or box sections

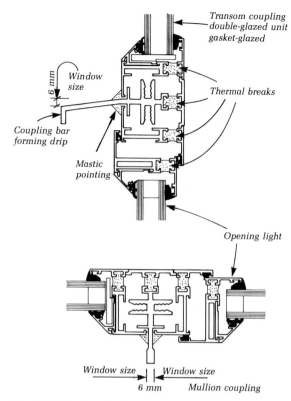

Transom coupling
double-glazed unit
gasket-glazed

Thermal breaks

Window
size

6 mm

Coupling bar
forming drip

Mastic
pointing

Opening light

Window size   Window size

6 mm   Mullion coupling

Fig. 4.50 Coupling aluminium windows (Crittall Windows Ltd)

(see fig. 4.50). Transom couplings can form a projecting drip when a deep section is required for long spans and depending on the anticipated loadings. Long runs of coupled aluminium windows are subject to greater thermal movements than would be the case if steel windows were used.

4.5.7 Louvred

This type of window (see fig. 4.51) originated in the tropics and became very widely used in the UK. Standard louvre windows are, however, not very secure and their use is as a consequence less than previously. Available measures to upgrade their security include steel burglar bars, built into the channel frame on the inside, and metal inserts which are designed to be glued into the bladeholders so the glass cannot be forced out.

The louvre consists of a number of horizontal panes of glass gripped in a U-shaped aluminium or plastics extruded section at each end, and pivoted on an aluminium vertical channel which is secured within the window opening. The blades of glass are connected at the top and bottom to a lever bar for opening. Ventilation can be varied from 1 to 95 per cent of the net louvred area.

The weathering of a louvre window depends on the overlap of the glass blades and the precise interlocking of the mechanism which holds the blades in position. Under wind pressure the blades may flex perhaps leading to the penetration of driving rain. Clear widths should not exceed 1220 mm to minimize this risk in sheltered situations reducing to 865 mm in more exposed places; manufacturers publish tables that determine maximum lengths for particular exposure categories. Louvre windows can enable cross-ventilation between rooms if placed over a door, or in a partitioning system.

4.5.8 Curtain walling

Aluminium frames of various purpose-made profiles can be used for patent glazing and curtain walling. Load-bearing aluminium alloy bars to BS 1474: 1972 *Specification for wrought aluminium and aluminium alloys for general engineering — bars, extruded round tube and sections* should be employed, extruded to a profile that incorporates water channels and other features. Bars are available in which aluminium or plastics caps or wings provided separately are fitted after the glass is in position, or extruded lead wings are drawn into the profile and become integral with the bar. Also a grid of box section aluminium mullions and transoms can be designed as a curtain walling system, the glass being held in place by pressure plates and/or gaskets. For further information on the construction of glazed walling see section 5.13, and curtain walling in *MBS: Structure and Fabric Part 2* chapter 4 *Walls and piers*.

4.6 PVC-U windows

The first PVC-U windows imported into this country from the continent were tilt/turn windows. As the material's market share has increased so has the range of types available until now the largest range of windows is available in PVC-U (with the exception of close-coupled windows which are not included in standard ranges). All casement types can be supplied pre-glazed or for glazing on site: normally the windows are built into prepared openings and require careful protection from damage by other trades. PVC-U has found particular application for the replacement of rotted or corroded timber and steel windows though their often radically different appearance has led to widespread disquiet. The Department of the Environment circular 8/87 warns against the use of unsuitable PVC-U frames in 'elevations of value'. Plastic window use has increased for new housing, with a resulting reduction particularly in the number of aluminium windows being sold.

PVC-U windows were not covered by a British Standard until the introduction of BS 7412: 1991 *Specification for plastic windows made from PVC-U extruded hollow*

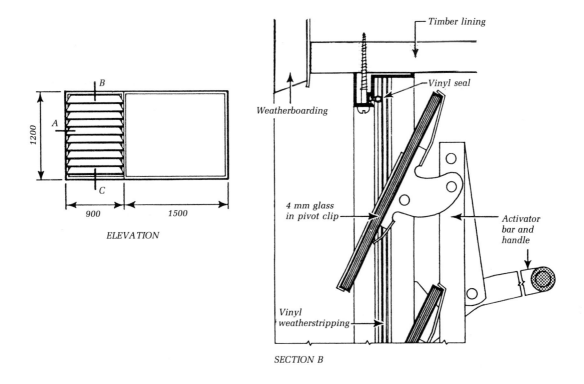

ELEVATION

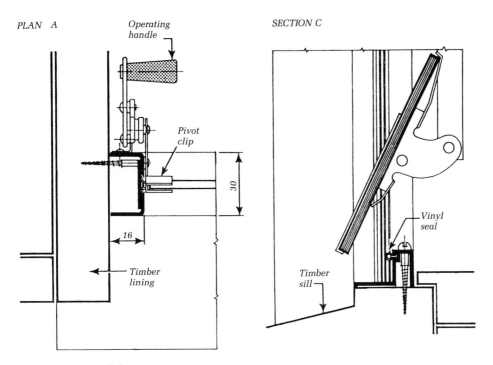

Fig. 4.51 Louvre window

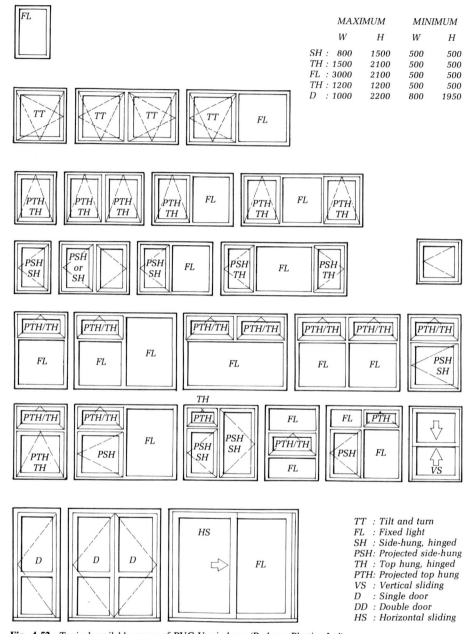

	MAXIMUM		MINIMUM	
	W	H	W	H
SH :	800	1500	500	500
TH :	1500	2100	500	500
FL :	3000	2100	500	500
TH :	1200	1200	500	500
D :	1000	2200	800	1950

TT : Tilt and turn
FL : Fixed light
SH : Side-hung, hinged
PSH: Projected side-hung
TH : Top hung, hinged
PTH: Projected top hung
VS : Vertical sliding
D   : Single door
DD : Double door
HS : Horizontal sliding

Fig. 4.52 Typical available range of PVC-U windows (Radway Plastics Ltd)

profiles. This specifies requirements for materials, construction, safety, weathertightness, operation and strength of plastics windows with frames up to 3 m in length. The delay in introducing the British Standard was due to an extended debate within the industry as to the best chemical formulation for the material. Some manufacturers use a PVC resin which is polymer modified to give improved impact resistance, others regard this as unnecessary. In the event this has been resolved by referencing the new standard to BS 7413: 1991 *White PVC-U extruded hollow profiles with heat welded corner joints for plastic windows: material type A (impact modified PVC-U)* and BS 7414: 1991 which deals with *material type B (unmodified PVC-U)*. The British Plastics Federation Windows Group consider the type A material superior for use in this country.

Despite the long-standing lack of a British Standard, the form of construction is quite similar amongst the major manufacturers no matter whether the extrusions are imported or made in the UK. Prior to introduction of the BS, the fabrication and performance of plastics windows were defined by a trade standard established by the British Plastics Federation (BPF) and the GGF. This is now being withdrawn but new trade standards are to be introduced covering aspects not governed by the British Standard (e.g. metal reinforcement of frames). Most manufacturers have their windows tested to BS 4315 for grading to BS 6375; Part 1. Many manufacturers have also obtained British Board of Agrément test certificates and participate in the BSI Quality Assurance Scheme (see *MBS: Internal Components*).

There is no standard range of types or sizes, a typical manufacturer's list and the maximum dimensions for each type being shown in fig. 4.52. Combinations of window units can be used by incorporating mullions and transoms which are connected to the outer frame and where relevant, to each other, by means of welded joints. Replacement windows are made to match the dimensions of old BS 644 timber windows so they fit within existing openings. Although some joints are made mechanically, the principal method employed for fixing the mitred corners of frames is by a machine which clamps and simultaneously hot fusion welds the joints.

The basic colour of PVC-U is white although grey and brown (also woodgrain!) frames are produced. Many manufacturers coat their profiles with grained acrylic foil which is available in a number of colours and is claimed to be resistant to weathering and UV light (these laminated coatings are available with a 7-year guarantee). Plastics may be painted, but it is advisable to avoid mixing light and dark colours in the same window in order to avoid differential expansion.

Unlike aluminium extrusions, PVC-U is formed into multi-chambered profiles, the outer wall thickness being around 3 mm. The cellular structure increases the thermal insulation capability of the frames, indeed the *U*-value increases with the number of cells — there are usually two or three across the width of a frame. The thermal insulation characteristics of plastics windows are roughly equivalent to timber windows, but their weatherstripping is superior. The tight-fitting weatherseals give relatively good sound reduction performance. The cross-wall structure also imparts strength to the section. Compared with metals, PVC-U is lacking in stiffness so depending on the anticipated wind loading, large frames have to be reinforced. Aluminium extrusions, or more usually, cold-formed galvanized steel stiffeners are used. To avoid their eventual corrosion, the stiffeners are contained within a sealed chamber of the extrusion. The chambers at the

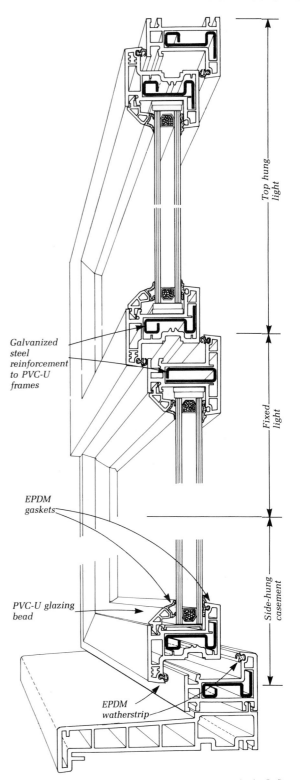

Fig. 4.53 Typical PVC-U window sections (Radway Plastics Ltd)

outside of the frame are interconnected to provide a drainage route for any water that might bypass the gaskets and to ensure that the edge seals of the double glazing unit remain dry. The water escapes through slots cut at the bottom of the sash and pressure equalization holes are provided to ensure free flow. Larger-sized windows are also reinforced at hinges and fastening points.

PVC-U windows always employ gaskets to retain the glazing and are able to accommodate glazing between 4 and 36 mm in thickness. The windows can be either factory glazed or glazed on site but are fixed after the openings are built. Window openings have to be carefully constructed with a template to ensure that they are properly dimensioned and square. PVC-U has quite a high coefficient of expansion so a clear gap of 5 mm should be allowed at the jambs between the frame and the structural opening. This is sealed with a silicone sealant and the space between the frame and the wall is filled with construction foam. A number of systems include reveal liners and window boards made from cellular PVC-U or rigid PVC-U foam. These are used to provide a self-finished lining to the reveals, insulating the underside of steel combined lintels and eliminating plastering or cutting and fitting plasterboard. Windows are fixed at maximum 600 mm centres around all four sides of the frame, lugs are clamped on to the edge extrusions of the frame and plugged and screwed to the masonry surrounds.

Plastic frames do not offer as strong a fixing for ironmongery as either timber or metal. Where the PVC-U sections are steel reinforced, the ironmongery is screw fixed through to the reinforcement. The chambered structure does, however, readily accommodate multi-point locking using espagnolette bolts that has become usual for larger windows. Some systems instead of employing metal ironmongery, use hinges made from polyamide or another plastic to be compatible with the frames. As a result of accelerated testing procedures, the anticipated life of plastics windows is stated to be more than 50 years. They should be cleaned regularly to maintain a bright appearance, although they may eventually require painting. Fittings may require replacement after 20 years and gaskets after about 15. Figure 4.53 indicates typical sections and details.

4.7 Composite windows

Of the several materials commonly used for the construction of windows, none is without its disadvantages:

Timber

- Can be machined to fairly complex profiles but is not capable of the precision detailing possible in aluminium or PVC-U. Intricate timber sections entail increased fabrication costs whereas extrusions can be made at one operation.
- Requires regular maintenance particularly when used externally.
- Moves with changes in the level of moisture in the atmosphere.
- Requires less energy than the others to convert the raw material into the finished component but softwoods need preservative treatment with toxic chemicals, some hardwoods are obtained from non-renewable sources.

Steel

- Hot-rolled steel frames are strong and consequently thinner than any of the alternatives; they offer as a result less obstruction to the admission of daylight.
- It is only made into a restricted range of sections and the types of window available are relatively few.
- Steel is highly conductive of heat and steel frames cannot be thermally broken; they are as a result subject to condensation internally and loss of heat from within rooms.
- Steel windows have had a bad reputation because of the widespread corrosion of pre-war windows. Since 1950 all steel windows have been galvanized and the introduction of polyester powder coatings in recent years has greatly increased their durability.

Aluminium

- Aluminium can be formed into complex sections allowing sophisticated glazing and weatherstripping techniques.
- The natural corrosion resistance of the material is enhanced by the variety of finishes which are available.
- The metal is lightweight and relatively soft so the sections need to be considerably larger than for steel frames with internal stiffeners adding to the complexity of the extrusion.
- Aluminium is very highly conductive of heat and even when thermally broken has a higher U-value than PVC-U or wood.
- It is one of the more difficult metals to obtain from ore and this is achieved only with a very considerable input of energy.

PVC-U

- PVC-U is durable, fairly insulative and can be formed into complex profiles but in a limited range of integral colours.
- The material has a low modulus of elasticity; this lack of stiffness is overcome in the design of windows either by multi-chambered construction or by stiffening the frames with internal cold-formed galvanized steel sections which lessen the insulative advantages of the PVC-U.
- PVC-U offers a less rigid fixing for ironmongery and

affords as a result less security than steel frames for example.

- The material can be recycled but its manufacture requires a large input of energy and environmentally damaging substances such as cadmium are involved in its production.

All windows involve the use of several materials in their construction. Timber windows nowadays incorporate metal components not only for ironmongery but also trickle ventilators and the bottom horizontal glazing bead is likely to be in aluminium. Weatherstrips, gaskets and glazing materials generally are made from synthetics. Steel windows have aluminium glazing beads and carriers for double glazing, aluminium frames have plastic thermal breaks and PVC-U windows are often steel reinforced.

Composite windows are an emerging type that cannot readily be identified by the predominant use of any one material. They fall into various types:

1. *Plastic-faced aluminium windows* (fig. 4.54). Aluminium windows that instead of being thermally broken have an internal cladding of 6 mm extruded

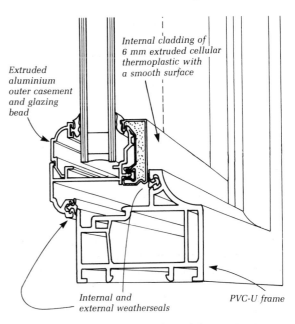

Fig. 4.54   GKN Window Systems 'Thermaclad' composite window

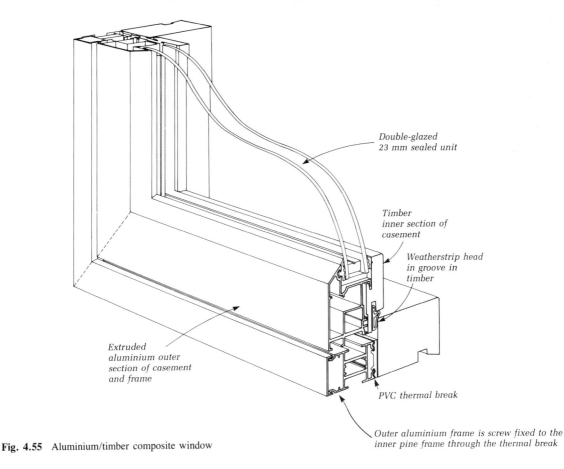

Fig. 4.55   Aluminium/timber composite window

cellular thermoplastic. The aluminium and plastic sections are extruded with interlocking profiles, and the foam core of the cladding has an integral, smooth and impermeable outer skin.

2. *Aluminium/timber composite windows* (fig. 4.55). Similar to (1) but instead of plastic, the internal visible surface is a timber section which is machined to fix into the outer aluminium extrusion.

3. *Composite PVC-U/aluminium sashes* (fig. 4.56). Because both aluminium and PVC-U frames are made by extrusion, interlocking sections can be produced where the outer glazed part of the sash is made from PVC-U and the inner from aluminium. The PVC-U provides an insulative and durable face to the exterior which is reinforced by the aluminium box section inside. The combined capabilities of the two materials make complex thermal breaks unnecessary.

4. *Plastic-faced timber windows*. A number of window systems have been developed using a timber core to provide reinforcement to PVC-U extrusions. The timber, which has to be of suitable moisture content, can as a result be machined to fairly simple sections. The PVC-U casing affords a durable surface requiring little maintenance and the extrusions can be sufficiently intricate to incorporate weatherstripping and gaskets for glazing (see fig. 4.57).

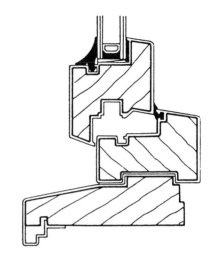

Fig. 4.57 PVC-clad timber window (Blacknell Buildings Ltd)

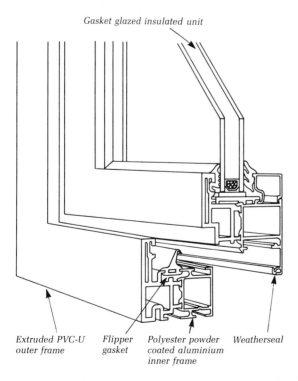

Gasket glazed insulated unit

Extruded PVC-U outer frame   *Flipper gasket*   *Polyester powder coated aluminium inner frame*   *Weatherseal*

Fig. 4.56 Schuco plastic/aluminium composite window

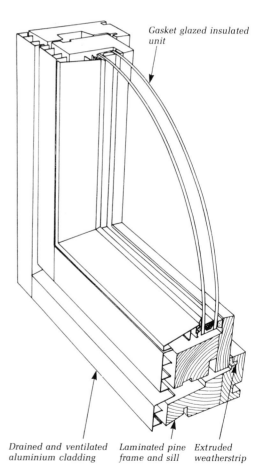

Gasket glazed insulated unit

Drained and ventilated aluminium cladding   *Laminated pine frame and sill*   *Extruded weatherstrip*

Fig. 4.58 Aluminium-clad timber window (Scandanavian Window Systems Ltd)

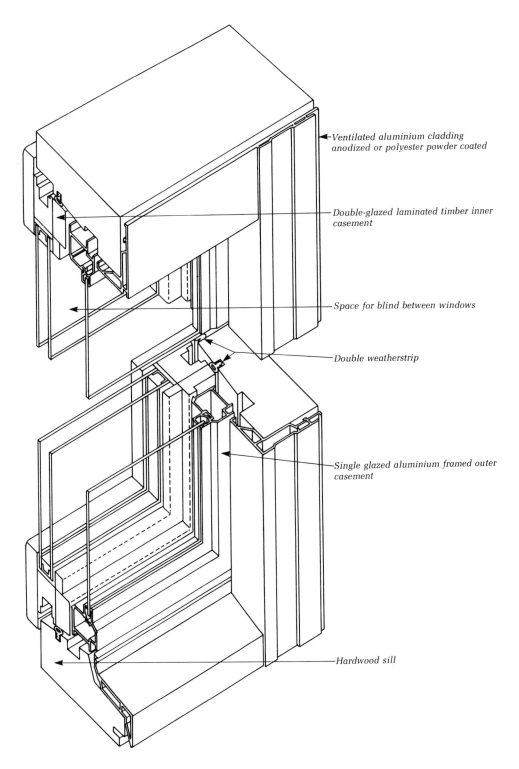

Ventilated aluminium cladding anodized or polyester powder coated

Double-glazed laminated timber inner casement

Space for blind between windows

Double weatherstrip

Single glazed aluminium framed outer casement

Hardwood sill

Fig. 4.59 Aluminium/timber coupled window (Reventa Products UK Ltd)

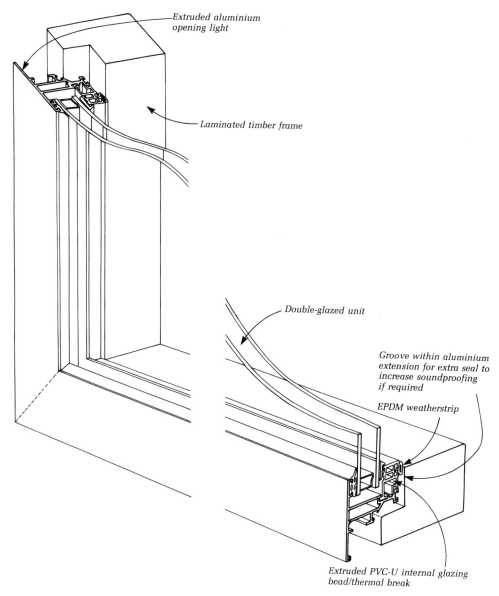

Fig. 4.60   Aluminium sash/timber frame window (Velfac Windows Ltd)

5. *Aluminium-faced timber windows*. In the manufacture of timber windows increasing use is being made of external anodized or polyester powder coated aluminium glazing beads, flashings, drips and sill cladding to provide a protective face to those parts of a windows most heavily exposed to the weather. High performance Scandinavian and some UK manufactured windows are now available totally faced in aluminium, stainless steel or copper on the outside whilst retaining the 'warmth to the touch' of timber internally. The strength and insulative properties of timber obviate the need for complex thermal breaks. In the example shown (see fig. 4.58) the metal facing is spaced away from the timber to form a drained cavity behind the double glazing edge spacers, so the timber can breathe and any condensation forming on the back of the aluminium can be dispersed. The external gaskets are clamped in place by the aluminium extrusions.

6. *Coupled sash 'double windows'* (fig. 4.59) with the outer sash in aluminium and inner sash in timber. These most

high performance of windows have a single-glazed outer window in an aluminium frame without a thermal break. This is coupled to the internal double-glazed timber window with a ventilated void between them as described under coupled windows (see p. 87). The outer window provides a durable weather-proof envelope, and the internal timber window is protected from the weather to form an insulated inner skin. The example illustrated can achieve a U-value of 1.8 W/m²K and a sound reduction of 36 dB.

7. *Aluminium opening casements within timber frames*. This type has a double-glazed pane gasket glazed into a polyester powder coated aluminium opening sash. The main timber frame is shielded from the weather and machined to a simple profile. The internal glazing bead, made of PVC-U, doubles as a thermal break and is rebated to fit the EPDM weatherstrip, the section can receive an additional seal for enhanced soundproofing. The timber provides stiffness to the window and its internal finish; the sections are relatively thin and the frame width is consistent in elevation no matter what type of opening light is employed (see fig. 4.60).

Such has been the progress in the development of glazing and particularly low-emissivity coatings that the U-values achievable by the glass and frame are now comparable. Timber frames have only the same U-value as double glazing, and thermally broken aluminium frames are in effect only designed to prevent condensation not to limit heat loss. The use of composite construction is an approach to the creation of higher performance windows for the future.

5 Rooflights and patent glazing

5.1 Introduction

Although the top-lighting of buildings has been common since antiquity, most famously at the Pantheon in Rome, its prevalence in contemporary architecture is a product of the development of glass manufacture and glazing techniques since the eighteenth century.

Increasingly, there is a trend towards buildings having deeper plans to maximize site usage. Perimeter accommodation is used for primary activities so rooflights are required to provide natural light and ventilation at the centre of buildings. Rooflights are particularly useful above circulation spaces and internal rooms used for service functions or where privacy or even protection is required. Unobstructed light from the sky is the most efficient and intense source of daylight, so the design of buildings that are to be rooflit (as are many factories, art galleries, etc.) may have to incorporate environmental controls to guard against solar gain, glare and daylighting of too great an intensity. Also, apart from purely practical considerations, rooflights can considerably enhance the perception and modelling of spaces in buildings (see photo 5.1).

5.2 Performance requirements

Rooflights are usually constructed by dry glazing techniques and take the form of domes, vaults, monitors, skylights and lantern lights; their design has become a specialist matter. Apart from some general guidance on performance given in CP 153: *Windows and rooflights* Part 1 *Cleaning and safety*, Part 2 *Durability and maintenance*, and Part 3 *Sound insulation*, at present there is no British Standard covering the detailed requirements of rooflights. On-site procedures for roof glazing are described in BS 8000: 1990 *Workmanship in building*, Part 7, 1990: *Code of practice for glazing*. Also, the comments given under section 3.2 are applicable to rooflights. Guidance is available from the Association of Rooflight Manufacturers and also from

Photo 5.1 Foyer, Sydney Opera House. Architect: Jorn Utzon, 1957.

This famous example well illustrates the architectural potential of roof lighting in enhancing 'the perception and modelling of spaces in buildings'.

publications of the GGF (including one on overhead glazing).

Since rooflights are situated at the most vulnerable and exposed part of a building, usually have joints within their construction and may require integration with different roof finishes, formidable problems of weather resistance have to be overcome. In addition to being completely weatherproof, a system of rooflights must be strong enough to withstand wind and dead loads; act as a defence against the external penetration of fire; be able to resist thermal losses and solar heat gains; perhaps provide ventilation and escape for smoke (PVC rooflights and some polycarbonates are designed to form a vent under fire conditions) and humans in the event of a fire whilst not causing condensation and draughts; and be as far as possible burglar-proof as well as easily maintained or replaced.

The contribution made by rooflights towards the overall heat loss from a building and the associated requirements of the *Building Regulations 1991* are discussed in section 4.2.5. Solar control glass, insulating glass and plastic double-skinned rooflights are available, complete with upstand kerbs in insulated metal or plastic, to avoid cold-bridging. Also, electronically operated blinds, either external or internal, are sometimes incorporated and can be automatically activated by external sensors to reduce solar heat gains and glare from sunlight. Manually or electrically operated louvres for ventilation are also available.

The principles involved in a roof construction providing resistance to fire are covered in section 8.2.5. According to the size and use of a building and the proximity of its roof to the site boundary, the *Building Regulations 1991*: Document B classifies the performance of various roof coverings (including glazing) with regard to their ability to withstand external fire penetration and their resistance to surface spread of flame. The covering is designated by the letters A, B, C, and D in each respect according to their performance when tested to BS 476: Part 3: 1975 *External fire exposure test*. The classification AA is the best result for resisting external fire penetration and in resisting surface spread of flame. Roof coverings designated AA, AB or AC are allowed to be placed close to a boundary, which is defined as the line of the land belonging to the building, including up the centre of an abutting street, railway, canal or river. Glass achieves class AA so both rooflights with mild steel, aluminium or plastic frames and frameless types employ wired annealed glass, or fire-resisting glass (a nominal thickness not less than 4 mm is generally acceptable).

The basic types of plastic sheet glazing materials used for rooflights have varying surface spread of flame characteristics when tested to BS 476: Part 7: 1987 *Method for classification of the surface spread of flame of products*; and although some are self-extinguishing, they are combustible at different temperatures. Thermoplastic materials can be tested under BS 476 or alternatively if they are to be used for rooflights, as described in the *Building Regulations 1991*, they can be divided into three categories of which the principal representatives are:

1. *TP(a) rigid* incorporating rigid PVC, polycarbonate at least 3 mm thick and double or multiple skin polycarbonate or PVC-U sheet.
2. *TP(a) flexible*, meaning flexible sheet less than 1 mm thick.
3. *TP(b)* incorporating rigid polycarbonate less than 3 mm thick or multi-skin polycarbonate sheets which do not qualify as TP(a) by test.

Acrylic rooflights cannot be tested to BS 476 Part 3 and since they also propagate fire, the area of acrylic rooflight, in relation to the floor area of the building, is limited by the *Regulations*; 1.5 mm polycarbonate corrugated rooflight panels have now been tested to comply with BS 476 Part 7, class 1.

Rooflights made from TP(a) rigid thermoplastic may be used in any space other than over a protected stairway; they should be at least 6 m from a boundary unless the rooflights are part of a balcony, verandah, carport, covered way or loading bay with one side permanently open or within the roof of a detached swimming pool, garage, conservatory or outbuilding with a floor area not exceeding 40 m^2.

Alternatively, rooflights made from any plastic with a class 3 rating to BS 476, Part 7 or type TP(b) thermoplastic can be used in rooms and circulation spaces other than protected stairways, provided that the maximum area of each rooflight is no greater than 5 m^2, that the maximum total area as a percentage of the floor area is limited and that there is a maximum separation between rooflights of 3 m. These rooflights should generally be at least 6 m from a boundary but if constructed from plastic with an external surface spread of flame classification DA, DB, DC or DD to BS 476: Part 3: 1958 *External fire exposure roof test* (acrylic rooflights for example), should be at least 20m from a boundary. In order for designers to ensure compliance with these conditions, it is essential that fire test results are obtained from manufacturers.

Table 5.1 indicates the fire performance characteristics of thermoplastic rooflight glazing. BS 2782 *Methods of testing plastics* has 11 parts with 143 main subdivisions which provide information for testing the thermal, electrical, mechanical, chemical, optical and colour, dimensional, rheological, and other properties of plastics. *MBS: Internal Components* chapter 3 *Suspended ceilings* describes the minimum internal surface spread of flame classifications for rooflights when they form part of a ceiling.

5.3 Acrylic roof tiles

Small fixed lights in pitched roofs can be formed by acrylic roof tiles. These are made to match the profiles, shapes and sizes of several manufacturers' ranges of single lap interlocking roof tiles. Acrylic tiles are fixed by nailing them, through factory-formed holes, to the roof battens in the conventional way. They are very efficient in terms of light transmission and are relatively shatterproof but their poor performance in fire limits the area of roof that can be glazed with them and the minimum permissible distance of the rooflights from site boundaries (as described above). Given the current requirements for roofs to be heavily insulated, care has to be taken to counter the problem of condensation which may arise due to the very different thermal characteristics of acrylic tiles compared with the construction of the rest of the roof.

Table 5.1 Thermoplastic glazing materials for rooflights tested in accordance with relevant British Standards (Williaam Cox Ltd)

Test	Polycarbonate	Standard acrylic	PVC-U	Reinforced PVC-U	10 mm Twin-wall Polycarbonate
BS 2782: 1970 *Method 508A*	Flame retardant	Not flame retardant	Flame retardant	Flame retardant	Flame retardant
BS 2782: 1970 *Method 508D*	Very low flammability	Flammable	Very low flammability	Very low flammability	Very low flammability
BS 476: Part 4: 1970 *Non-combustibility*	† Combustible	† Combustible	† Combustible	† Combustible	† Combustible
BS 476: Part 6: 1981 *Fire propagation test for products*	Index of performance I = less than 12 i_1 = less than 6	Not applicable	Not applicable	Not applicable	$I = 9$ $i_1 = 4.5$ $i_2 = 2.7$ $i_3 = 1.74$
BS 476: Part 7: 1971 *Surface spread of flame*	Class 1	Class 3	Class 1	* Class 1	Class 1
BS 476: Part 8: 1972 *Fire resistance of elements of building construction*	† Less than $\frac{1}{2}$ hour	† Less than $\frac{1}{2}$ hour	† Less than $\frac{1}{2}$ hour	† Less than $\frac{1}{2}$ hour	† Less than $\frac{1}{2}$ hour
BS 476: Part 3: 1958 *External fire exposure roof test*	Not applicable	DDX	Not applicable	F.AA	Not applicable

* When tested in 3 mm thickness and over.
† Assumed results.

5.4 Profiled plastics sheet rooflights

Corrugated translucent plastic sheets are made to match and interlock with the standard profiles of aluminium, steel or fibre cement sheet cladding. They are manufactured from translucent GRP or PVC and polycarbonate and are used in single, double or triple thicknesses. Single-skin rooflights are only suitable for use in unheated buildings since they only achieve an approximate *U*-value of 5.7 W/m²K, double rooflights attain 2.8 W/m²K and triple 2.0 W/m²K. The specification of rooflight panels and their fixings has to be in relation to the slope and form of the roof and the anticipated exposure conditions; reference should be made to BS 5427: 1976 *Code of practice for performance and loading criteria for profiled sheet in building* (see also pitched roofing sheets, p. 237). To avoid differential movement between the plastic profiled sheeting and the adjacent roof cladding, it is preferable if the roof is of a light colour otherwise thermal stress at the junction can cause the plastic profile to distort and the roof to leak. This is particularly a difficulty if PVC rooflights are to be used. GRP sheets can achieve an AA fire rating and class 0 surface spread of flame when tested to BS 476 Part 3, they can consequently be employed without restriction of the *Building Regulations* with regard to area, proximity of rooflights one to another, or distance from boundaries. PVC is a rigid thermoplastic and should generally not be used closer than 6 m to the site boundary.

5.5 Metal-framed wired glass rooflights

These simple rooflights are generally made from mitred, welded aluminium sections to 300 mm modular sizes that are screwed either to a timber builders' work upstand or a timber plate to level a concrete upstand. Alternatively, separate metal upstand sections are made for use with them (fig. 5.1). The height of upstand required will be determined by the site location and the anticipated exposure conditions.

The wired glass oversails the frame and is laid flat on it, and glass to metal contact is avoided by a separating strip of butyl mastic. The glass is held in place either by spring metal clips or a continuous section round all four sides. Over about 1200 × 1200 mm, glazing bars are used to divide the panes.

Although principally used as fixed rooflights, they can be provided with permanent ventilation by raising the rooflight to create a slot above the kerb, or adjustable ventilators can be accommodated within the upstands.

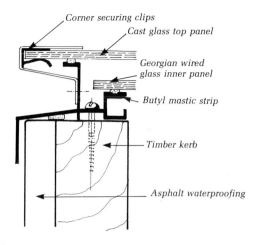

Corner securing clips
Cast glass top panel
Georgian wired glass inner panel
Butyl mastic strip
Timber kerb
Asphalt waterproofing

Fig. 5.1 Upstand detail for metal framed skylights

Although useful in summer, ventilators even when shut can present a condensation risk during cold weather. This can be overcome by local heaters or the installation of lights within the rooflight kerbs (having the further advantage of co-ordinating natural and artificial lighting). Some are available with hinged opening lights operated by long arm rod and screw winding gear. It is always difficult to guarantee that rain will not be blown into a building when an opening rooflight is used and for this reason the hinged side of the rooflight should always be placed facing towards the direction of the prevailing wind.

Thermal breaks, now usual in aluminium windows are not generally found in these simple aluminium-framed rooflights but they are available double glazed with two layers of wired glass.

5.6 Plastic domes

Although domes can be made of glass, circular or rectangular on plan (see p. 141), most are now made less expensively and with less complex joints when moulded one-piece in plastic. They are available clear, opal and tinted in polycarbonate (PC), polymethyl methacrylate acrylic (PMMA), unplasticized polyvinyl chloride (PVC-U), GRP and wire-reinforced PVC. When making a choice, it is very important to check the performance characteristics of each material to ensure compliance with design, technical and legislative requirements. The varying capabilities of thermoplastics in fire conditions make it important to obtain a fire test certificate.

Figure 5.2 gives a selection from a range of 300 mm modularly co-ordinated dome lights. Preformed plastic gutter sections enable these rooflights to be joined together in continuous lengths. Barrel rooflights are made with

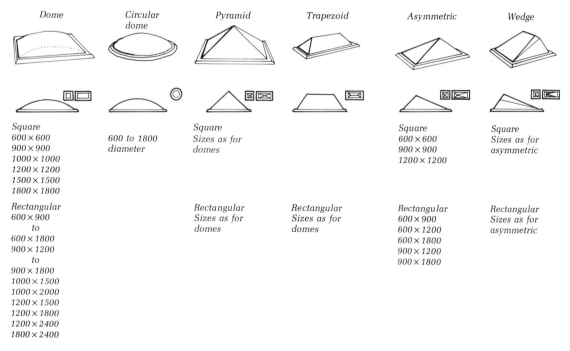

Dome	Circular dome	Pyramid	Trapezoid	Asymmetric	Wedge

Square
600 × 600
900 × 900
1000 × 1000
1200 × 1200
1500 × 1500
1800 × 1800

Rectangular
600 × 900
to
600 × 1800
900 × 1200
to
900 × 1800
1000 × 1500
1000 × 2000
1200 × 1500
1200 × 1800
1200 × 2400
1800 × 2400

600 to 1800
diameter

Square
Sizes as for
domes

Rectangular
Sizes as for
domes

Rectangular
Sizes as for
domes

Square
600 × 600
900 × 900
1200 × 1200

Rectangular
600 × 900
600 × 1200
600 × 1800
900 × 1200
900 × 1800

Square
Sizes as for
asymmetric

Rectangular
Sizes as for
asymmetric

Fig. 5.2 Manufacturer's range of plastic rooflights (Transplastix Ltd)

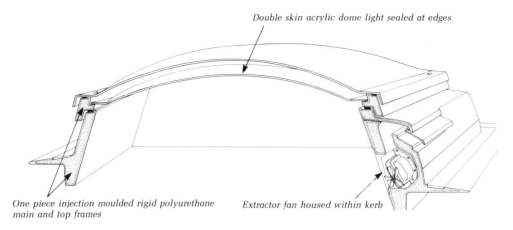

Double skin acrylic dome light sealed at edges

One piece injection moulded rigid polyurethane main and top frames

Extractor fan housed within kerb

Fig. 5.3 Double skin acrylic dome rooflight (Alwitra)

interlocking flanges so they can be joined to make a continuous and uninterrupted length of rooflight.

Most manufacturers allow rooflights either to be built directly off builders' work kerbs or supply kerbs made from steel or welded aluminium, or increasingly from one-piece moulded plastic with an insulated core. Figure 5.3 shows an example of a kerb-mounted double-skin acrylic opening rooflight. These can also be obtained with an inner panel of glass laminated to a thin translucent insulating sheet of fibreglass. The whole rooflight unit can then have a heat conductivity sufficiently low for installation in fully air-conditioned buildings.

The dome lights are sealed together and mechanically fixed to the frame, which like the kerb, is made from one-piece injection-moulded polyurethane with a foam insulation core. These rooflights allow the optional installation of electrically operated blackout blinds and a specially formed kerb section is available to house an electrically operated extract fan. For sealing to single-skin roof membranes the kerbs come with a jointing strip to which the roofing can be either welded or glued with solvent-based adhesive (saving the complex and often unsightly cutting of sheets round the upstands). The kerb construction is, however, sufficiently temperature-resistant for conventional roof coverings requiring the use of hot bitumen.

Hinges and locks are composite in construction with the frame and upstand sections and ensure that the rooflights can only be opened from the inside. Both insurers and manufacturers are concerned at the security risk posed by rooflights on flat roofs and considerable design ingenuity goes into the elimination of that risk. Rooflights with metal kerbs allow a grid of 'burglar bars' to be fixed within the roof opening and so provide a second line of defence.

Figure 5.4 indicates a double-skin plastics rooflight fixed to a galvanized steel upstand. This is made in two dimensionally co-ordinated sizes, 900 and 1800 mm square.

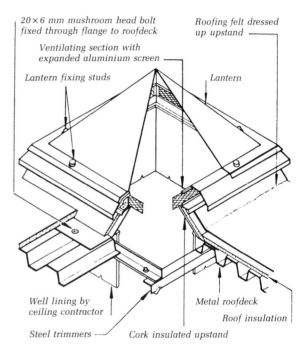

20 × 6 mm mushroom head bolt fixed through flange to roofdeck

Roofing felt dressed up upstand

Ventilating section with expanded aluminium screen

Lantern fixing studs

Lantern

Well lining by ceiling contractor

Metal roofdeck

Roof insulation

Steel trimmers

Cork insulated upstand

Fig. 5.4 Single skin plastic pyramid

The rooflights are available in clear, opal and solar control PVC, or in GRP with an expanded aluminium ventilation screen and a projecting aluminium weather baffle fixed to the upstand. Additional ventilation can be provided by simply adding more screens and baffles, or they can be omitted altogether if no ventilation is required. 'Hit and miss' ventilating screens and baffles can be used for controlled ventilation. Note that the upstand is bolted to the metal deck. As openings for rooflights have an effect on the structural performance of roof decking, there are

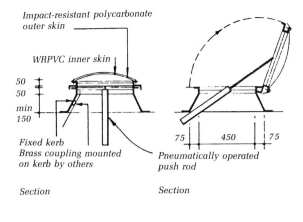

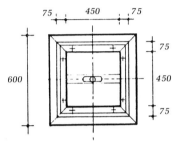

Section

Plan
600 × 600 mm Rooflight in closed position

Fig. 5.5 Smoke venting rooflight: high impact polycarbonate outer skin plus wire-reinforced PVC inner skin. Pneumatic cylinder opening mechanism, mounted between rooflight and galvanized steel kerb, operated by smoke/heat sensor or other alarm system which activates airline (Transplastix Ltd)

some limitations on the number of rooflights occurring within a given area of roof.

Figure 5.5 shows a high impact resistant rooflight with a polycarbonate outer skin and wire-reinforced PVC inner skin. This incorporates an automatic opening mechanism which, when activated, allows smoke from a fire to be vented and so reduces the likelihood of the occupants being asphyxiated whilst trying to escape from the building. These automatic fire vent rooflights are opened by gas rams switched by connection into the fire detection system or activated by a fusible link. They are usefully employed for the automatic fire venting of warehouses, etc. Gas vents operate more smoothly than springs and are strong enough to function even when loaded with a full 0.75 kN/m^2 of snow. Opening rooflights for access to the roof or for ventilation can be similarly constructed, either unlocked manually or opened by a telescopic screw jack remotely operated by a mechanical arm, cord, crank drive or

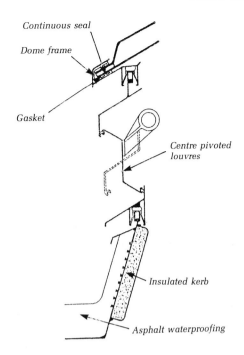

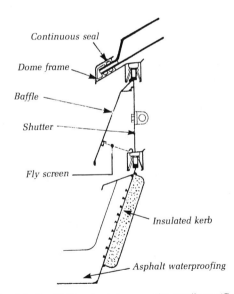

Fig. 5.6 Rooflight upstands incorporating ventilators (Coxdome Ltd)

switched to an electric motor. Ventilators of various types: centre pivot louvres, hit and miss ventilators and hinged flaps shielded beneath a weatherhood can all be incorporated within the upstands and operated in the same way as opening rooflights (fig. 5.6).

The risk of condensation is greatly reduced by double

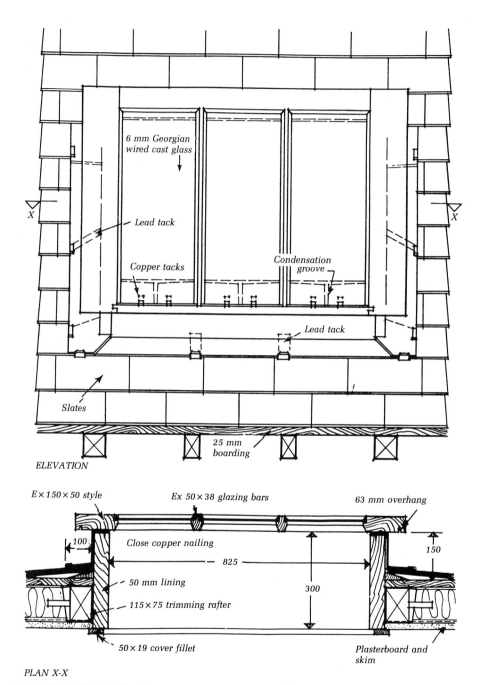

Fig. 5.7 Traditional timber framed skylight

glazing and insulated kerb construction that eliminates cold bridges. The internal environmental conditions, particularly a combination of high internal temperature and humidity in conjunction with low outside air temperature, will make it more likely that condensation will occur. Local ventilation, by one of the methods previously described, will help dissipate warm, humid air as it collects at the rooflight. Some manufacturers' kerb sections include condensation gutters at the bottom of the slope of the glazing designed to collect condensation and drain it to the outside.

5.7 Traditional timber skylights

Although rare nowadays, traditional timber skylights are sometimes used on new buildings as well as existing in large numbers on older buildings that may undergo refurbishment. Figure 5.7 shows a fixed skylight suitable for a pitched roof. The roof members have to be trimmed to form the structural opening, the 50 mm wide timber frame is housed together at the corners and fixed to the trimming rafters. The frame has to be wide enough to stand well up above the line of the finished roof and form a deep enough gutter at the top edge. If not, water or particularly snow may enter the building at the joint between the frame and the skylight; opening lights of this sort are best avoided. In the detail shown, the structure of the roof is not very deep and so it has been possible using a frame of ex. 300 by 50 mm timber to make the bottom edge coincide with the line of the ceiling; the frame can then also form the lining to the opening. The light itself is 50 mm softwood with 150 mm nom. styles and top rail and 175 × 38 mm nom. bottom rail.

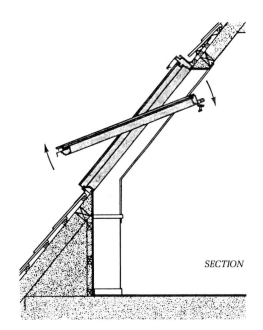

SECTION

Fig. 5.8 Modern timber framed skylight

5.8 Sloping roof windows

An alternative to the dormer window (described on p. 82) is the sort of skylight shown in figs 5.8–5.11. Roof windows are more efficient than dormers, for example a roof window in a 60 degree slope will admit twice as much light as a dormer and in a 15 degree slope will admit three times as much.

Roof windows are in increasingly common use for attic conversions and new construction built into the slope of a roof. They can be obtained either made from timber (fig. 5.10) with an outer pre-finished aluminium cladding (to reduce maintenance) or in moulded polyurethane with a laminated timber core (suitable for areas of high humidity such as kitchens and bathrooms).

Modern roof windows are combined with different flashings for use with different roofing materials (roof slopes down to 10 degrees are possible). Their design amalgamates features found in both rooflights and in contemporary windows: friction hinges enable opening lights to rotate for easy cleaning from inside the building and barrel bolts can be locked in different positions for safety. High-level windows can be operated by rod, cord or electric control which can also be activated by sensors or linked to a fire alarm system for the skylights to be used as smoke vents.

Within one system, a sloping skylight can be coupled to a vertical window or door at the eaves to increase the area of glazing or to form a balcony door (fig. 5.11 and photo 5.2). Under the *Building Regulations 1991*, 'Mandatory rules for means of escape in case of fire', all buildings of

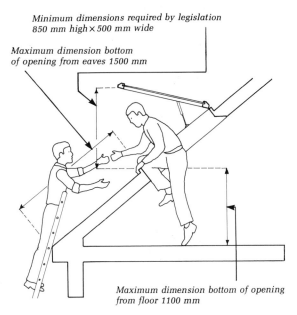

Minimum dimensions required by legislation 850 mm high × 500 mm wide

Maximum dimension bottom of opening from eaves 1500 mm

Maximum dimension bottom of opening from floor 1100 mm

Fig. 5.9 Dimensional requirements for escape from rooflights (*Building Regulations 1991*)

three or more storeys (including habitable spaces within the roof) must have provision for means of escape from each separate room. As a consequence, loft conversions require a window or skylight with the bottom line of opening to be a maximum of 1100 mm high from the floor.

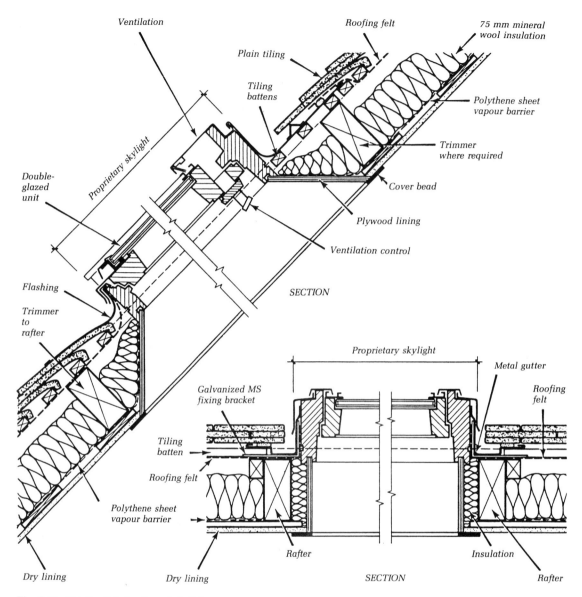

Fig. 5.10 Details of timber-framed skylight

The window or skylight must be at least 500 mm wide and 850 mm high to allow people to escape and must be positioned for ease of rescue by ladder from the ground (fig. 5.9).

The maximum distance from the bottom of the window to the eaves must not exceed 1500 mm and the height of the bottom of the window from the floor must not exceed 1100 mm. For rapid opening some manufacturers fit these fire-escape windows with gas pressure operated pistons.

5.9 Large-area glazed roofs using plastics rooflights

Plastics domes or pyramids can be linked together by a grid of gutters to form large areas of glazed roof; subsidiary gutters run in one direction, deeper main gutters in the other. The latter are also designed to be the structural lines of support corresponding dimensionally with the roof structure, for example a space frame or parallel lattice

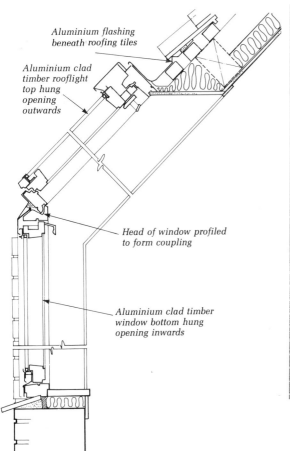

Aluminium flashing
beneath roofing tiles

Aluminium clad
timber rooflight
top hung
opening
outwards

Head of window profiled
to form coupling

Aluminium clad timber
window bottom hung
opening inwards

Fig. 5.11 Details of coupled timber window and skylight (The Velux Co Ltd)

Photo 5.2 Velux coupled windows (see section 5.8). Photo courtesy of the Velux Company Ltd.

trusses. The overall span and gutter lengths must be calculated for each project in accordance with local wind loadings and rainfall.

Common in recent construction are continuous and large-area glazed roofs employing linked barrel vaults or dome lights. The curved glazed panels are made from plastics; both single skin and double are available in clear, diffused or tinted polycarbonate, PVC-U and wire-laminate PVC-U. One proprietary system accommodates sheets up to 1200 mm in width are clamped at their edges by aluminium inner and outer glazing bars with integral seals. At the end of the bars, holding them in position, are brackets threaded to receive stainless steel set screws that are tightened to the correct torque to exert a uniform inward pressure across the vault, eliminating the necessity for intermediate fixings (fig. 5.12). Double-glazed types have an inner bar to space the sheets apart and are similarly clamped into position at the edges. Barrel vaults up to 4 m across can be constructed in this way (though spans in excess of 2.1 m may require

a central ridge bar). The rooflights are suited to fixing either to builders' work kerbs or insulated aluminium upstands fixed to the roof deck. The vertical ends of the vault can be glazed in the same material. Panels within the vault can be openable by electric actuators if at high level, or if to act as a smoke vent, the actuators are triggered by smoke sensors or by the fire alarm system.

Curved plastics cut into segments can be used to make dome lights between 1100 and 6500 mm diameter without any secondary structure.

Plastics glazing systems can be used to form extensive fully glazed roofs. Semicircular vaults up to 10 m across can be achieved without a secondary structure by using deeper glazing bars. A more substantial base bracket is required with the kerb structure designed to resist the thrust imposed by the vault. Across these wider spans capping screws may be needed to clamp the glazing bars together (fig. 5.13). Although yet larger single spans can be made by incorporating steel reinforcement within the aluminium sections, a series of smaller vaults linked by structural gutters is often a more economical solution. In this way using whole or half vaults entirely glazed roofs can be constructed.

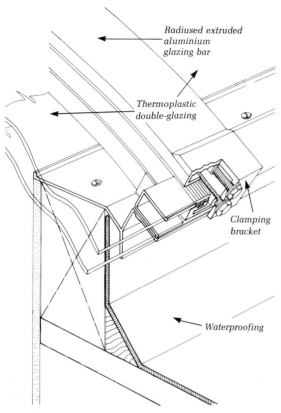

Fig. 5.12 Kerb detail for proprietary double-skin plastic aluminium-framed rooflight (Coxdome Ltd)

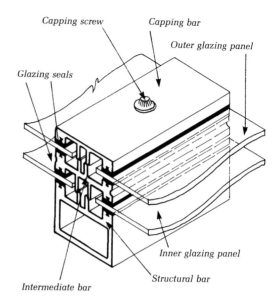

Fig. 5.13 Glazing bar detail for proprietary double-skin plastic aluminium-framed rooflight (Coxdome Ltd)

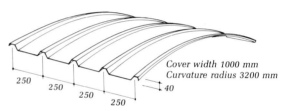

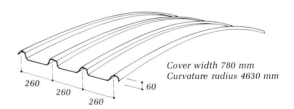

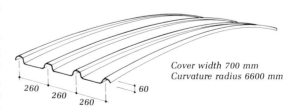

Fig. 5.14 Manufacturer's range of profiled GRP translucent vaults (Skyvault: Daylight Insulation Ltd)

Consideration must be given when specifying the use of plastics as to their durability. The effects of UV radiation in conjunction with local pollution or a marine environment can greatly shorten the life expectancy of plastics, particularly because rooflights are inevitably fully exposed to the elements. Ageing can take the form of crazing, discolouration, increased brittleness and opacity.

5.10 GRP rooflight vaults

An economical alternative, not requiring the use of metal glazing bars, is a shallow vault made from overlapping corrugated translucent GRP sheet. This is available in several radii of curvature and depths of corrugation (fig. 5.14). Apart from the basic opal translucent sheeting, the material can also be obtained tinted and in fire-resisting grades. The ends of the curved sheets are end-fixed at the kerbs using self-tapping or self-drilling screws with a neoprene washer to seal against the weather. Opening panels (overlapping adjacent sheets to weather the joint) can be activated by manual control or gas rams, and the panels are hinged at one end and lifted at the other.

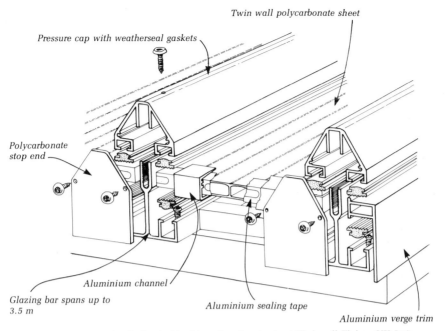

Pressure cap with weatherseal gaskets

Twin wall polycarbonate sheet

Polycarbonate stop end

Aluminium channel

Glazing bar spans up to 3.5 m

Aluminium sealing tape

Aluminium verge trim

Fig. 5.15 Glazing bar details for double-skin polycarbonate sheet (Twinwall Fixings UK Ltd)

Double-skin and insulated vaults can be site assembled using two separate GRP skins held apart by galvanized steel or aluminium spacers that are fixed to the curb through the lower sheet and screw fixed to the weathering sheet. Alternatively, the composite double-skin panels can be made up in the factory (but are not available hermetically sealed). To make a more highly insulated roof whilst maintaining its daylighting qualities the void between the two skins can be filled with translucent extruded acrylic insulation to achieve a *U*-value below 1.0 W/m²K.

As in all rooflighting systems where the double-glazed construction is not factory bonded together, condensation may form between the sheets where warm moist internal air comes into contact with the outer cold surface. In continuously heated buildings with normal humidity, condensation is not usually a problem. If, however, the heating is intermittent or if the building has high levels of humidity it may be necessary to install additional ventilation.

5.11 Multi-walled rooflight panels

In addition to plain plastics sheet used single or double glazed, PVC-U, acrylic and polycarbonate are available extruded into twin, triple, or quadruple walled sheets with the external and internal skins connected by diaphragms (fig. 5.15). Because the diaphragms are in the same material which has good insulating properties, a double skin panel typically has a *U*-value around 2.6 W/m²K. The

Photo 5.3 Liverpool Garden Festival Hall. Architects, engineers and quantity surveyors: Arup Associates, 1984. Photo courtesy of Arup Associates.

Photo 5.4   Liverpool Garden Festival Hall. Architects, engineers and quantity surveyors: Arup Associates, 1984. Photo: Carter Matanle Southern Ltd
   The roof glazing of the Festival Hall is 16 mm, double-skin polycarbonate sheet. The panels are one metre wide and clamped at their edges on to an aluminium glazing bar with neoprene weatherseals. The apsidal ends of the building are roofed with tapering aluminium profiled sheeting.

composite section also has considerable strength and spanning capabilities, and some types are made with diaphragms both diagonal and vertical to the face of the external sheets. This gives the panel the profile of a truss in section and enables a 16 mm panel to have a width between supports up to 1200 mm and length up to 3 m, with an overall sheet length up to approximately 8 m. Sheets are joined using aluminium glazing bars derived from traditional patent glazing practice (see below), for wide panels; spot fixings may also be required to resist wind uplift (see figs 5.16 and 5.17).

Deeper sections are made from both PVC-U and polycarbonate as multi-walled planks that join together at their edges using plastic or aluminium locking bars (fig. 5.18); they are available both flat and curved. The U-value of a 60 mm thick panel is 1.85 W/m^2 K, and panels can be obtained tinted or opal as well as clear. The system incorporates a variety of aluminium fixings and edge profiles for jointing panels one to another and to the adjoining structure.

5.12 Inflated ETFE roof 'cushions'

A new form of double-skin plastics roofing combining light weight with high transparency and light transmission has become available as a result of developments in polymer chemistry. ETFE is a modified fluorinated copolymer that can be processed like a thermoplastic to make a film that has good load-bearing properties, is tear-resistant and self-extinguishing with no noxious fumes; at 200°C it simply disintegrates. The film also has good light transmittance (94 per cent), is highly transparent to UV light, has a high resistance to weathering, is non-stick and has no anticipated tendency to discolour over time.

Several layers of copolymer membrane are electro-thermally welded to a perimeter gasket that is clamped into an aluminium or steel frame. Dried air (to reduce condensation) is supplied between the layers by small centrifugal fans that inflate the 'cushions' to a constant low pressure of 400 or 500 Pa; the fans are activated as required to periodically top up the pressure (fig. 5.19). This

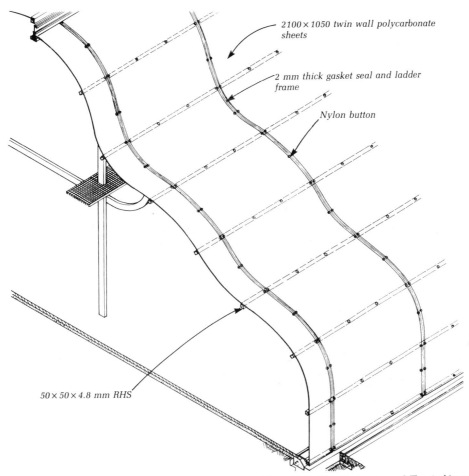

Fig. 5.16 Part section through Clifton Nurseries greenhouse. (Architect: Terry Farrell. Courtesy of *The Architects' Journal*)

Labels in figure:
2100 × 1050 twin wall polycarbonate sheets
2 mm thick gasket seal and ladder frame
Nylon button
50 × 50 × 4.8 mm RHS

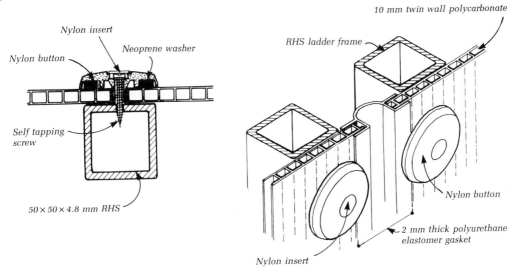

Labels in figure:
Nylon insert
Nylon button
Neoprene washer
Self tapping screw
50 × 50 × 4.8 mm RHS
10 mm twin wall polycarbonate
RHS ladder frame
Nylon button
2 mm thick polyurethane elastomer gasket
Nylon insert

Fig. 5.17 Details of Clifton Nurseries greenhouse. (Architect: Terry Farrell. Courtesy of *The Architects' Journal*)

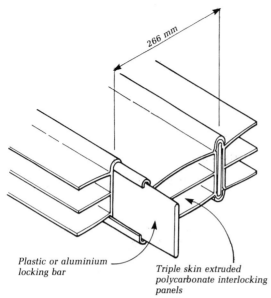

266 mm

Plastic or aluminium
locking bar

Triple skin extruded
polycarbonate interlocking
panels

Fig. 5.18 Triple-skin polycarbonate interlocking roof panels (Everlite: Plastmo)

Photo 5.5   Roof glazing, Lion Yard, Cambridge. Architects: City of Cambridge Architects Dept., 1969.
A fully glazed roof constructed from sloping and vertical patent glazing.

pneumatic system forms an insulating blanket with a U-value, if three layers thick, of 2.1 W/m²K. The thermal insulation value depends on the number of layers and air thicknesses between. To reduce solar gain the sheets can be printed with a pattern of reflective dots. Small holes in the corners of the intermediate sheets ensure that the whole assembly is inflated to the same pressure but also prevents excessive convection, the individual layers act as cushions of still air. Intermediate sheets can also form a temporary seal should one of the outer membranes fail. The inflated panels are supported on the underside by tensioned stainless steel wires that provide additional support for snow loads should the roof deflate for some reason (fig. 5.19).

Despite the undoubted efficiency of rooflights in providing a high level of illuminance, many building owners are concerned at the risk of burglary, the potential problems of solar control and the complications in roof construction they present. For factory applications, the monitor form of roof was invented to impart a consistent and shadowless pattern of lighting at the production line. The configuration of roof lights is capable of infinite variation to accommodate different requirements, for example the 'haystack' type used to vent fly towers in theatres. Recently, increased use has been made of concrete and glass pavement lights to form a flat and relatively secure method of rooflighting.

5.13 Patent glazing

Contemporary patent glazing is a direct descendant of the technology that was responsible in the last century for the Palm House at Kew, the Crystal Palace and the glazed roofs of London's railway stations. Patent glazing is a dry, site-assembled system of glazing that consists of supporting members parallel to the slope (glazing bars) with an infill of glass or other sheet material. The bars may now be made from aluminium and are available double or triple glazed with solar control or laminated glass or plastics but the general principles of its assembly are unchanged.

It is an economic and flexible system, which if designed within the limitations of the method and properly installed, requires very little maintenance. BS 5516: 1991 *Code of practice for design and installation of sloping and vertical patent glazing* deals with single and double patent glazing in sloping and vertical positions, and provides information on types of bars, glass, thermal insulation and fire resistance; it should be consulted in conjunction with BS 6262: 1982 *Glazing for buildings*. The earliest type of patent glazing, used extensively in the nineteenth century but still available, supports glass only on two edges (see photo 5.5). Systems with intersecting bars supporting the glass on all four edges were introduced to allow wider spacing of bars for vertical glazing applications. The British Standard gives design recommendations for two-edge and

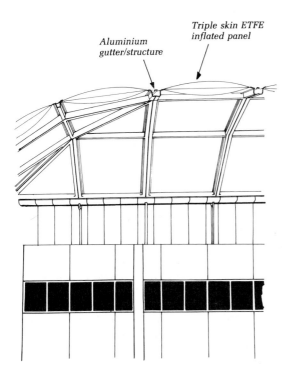

Aluminium gutter/structure

Triple skin ETFE inflated panel

four-edge supporting systems, and standards for safety both in terms of accidental human impact and other structural considerations.

Patent glazing has to be designed to resist the combined loads imposed by the wind, snow (on roofs) as well as the self-weight of the components. When the glass is sloping, the likelihood of frame or support bar deflection is also greater and, because some of these factors can impose a sustained stress on the glass, it cannot be dealt with in the same manner as for '3 second wind loads'. BS 5516 discusses these considerations and includes a method by which working pressures can be determined for glass and glazing bars at various pitches as well as bar sizes and spacings and the maximum sizes of glass. The values for imposed loads to be used in calculations are given in CP 3: Chapter V: Part 2 (see section 3.2.2) and BS 6399: Part 1. In very severely exposed situations it may be advisable to use a four-edge supported system with a screw-on pressure cap. The minimum pitch for patent glazing is usually 15 degrees, sometimes lower pitches are used but wings, cappings and other weatherings to the glazing must be sealed with mastic. The relatively high thermal expansion of aluminium is a severe challenge to the flexibility of mastic and is a common cause of leaks. Ambitious layouts which require complex geometry at junctions are difficult to make

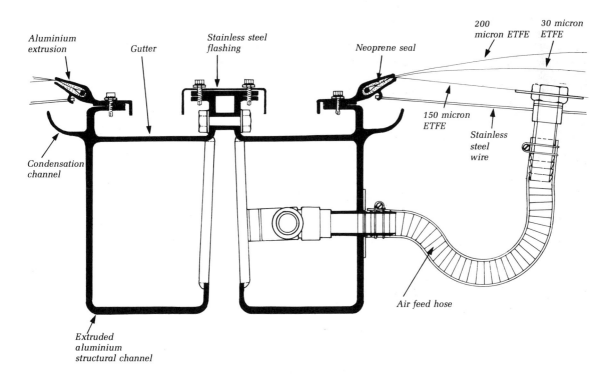

Aluminium extrusion

Gutter

Stainless steel flashing

Neoprene seal

200 micron ETFE

30 micron ETFE

150 micron ETFE

Stainless steel wire

Condensation channel

Air feed hose

Extruded aluminium structural channel

Fig. 5.19 Detail of inflated ETFE rooflight panels (Westminster and Chelsea Hospital. Architects: Sheppard Robson)

watertight when using the simplest forms of patent glazing.

Additional information is contained in MBS: *Structure and Fabric*, Part 2 'Curtain walling,' and 'Walls and piers'.

Most of the glasses described in BS 952: Part 1 1978 may be used for patent glazing either single or double glazed and subject to thermal safety checks and safety recommendations being observed. The most commonly used types are float glass, solar control glass, wired glass, patterned and rough cast glass, laminated glass, toughened glass (normally 'heat-soaked' — see section 6.3.2), and insulating glass units including double glazing. Plastics in various types and forms are available both single skin and extruded to form separate skins with internal diaphragms. They have higher impact strengths and are less rigid than glass, but require more positive edge restraint to prevent 'spring-out' or displacement under load due to excessive deflection. Opaque infill panels, including metal-faced composite panels and opaque glass or plastics with or without insulation, can be incorporated within clear patent glazing. Single-glazed sections can usually accommodate panels up to 7 mm thick whilst double glazing sections allow thicknesses up to 24 mm. If the edges of panels are rebated they can be made to any thickness up to the load limit that can be supported by the bars.

Bars for supporting the glazing can be either extruded aluminium alloy sections to BS 1474, or hot-rolled mild steel to BS 18, either hot-dip galvanized finish, covered or clad with lead or coated in PVC. In essence a patent glazing bar is a T- or I-shape of sufficient depth to span between supports, incorporating seatings at the edges of the glazing panels, wings or a capping section that holds in the panel whilst allowing thermal movement and resisting the ingress of rainwater. Cappings and wings can be of aluminium, copper, zinc or plastics to be compatible with the bars themselves. Various types of fixing are used (snap, bolted, etc). Alternatively, the supporting bars can have wings which are dressed down to form a barrier against water penetration. These should be integral with the bars and made from either lead or plastics.

Traditionally the depth of the bar projected to the outside of the glass which was retained either by wings or caps. Alternatively the bars can be inverted to project into the building, to avoid cold-bridging thus reducing heat loss and the likelihood of condensation forming on the inner surface of the bar.

The design of a patent glazing installation must include provision for cleaning and maintenance access both inside and out. Regular cleaning of glass is essential to maintain levels of light transmission and to avoid or reduce the risk of glass breaking due to thermal stress; metal components with factory-applied finishes must be regularly maintained to comply with the terms of their guarantees. The necessity for cleaning gantries can be a significant additional cost

to a glazed roof installation. The layout needs to be fully accessible for maintenance otherwise, in the future, scaffolding may be required.

The accuracy of the assembly on site is obviously dependent on the accuracy that can be achieved within the supporting structure. BS 5516, Section 4 contains guidance on structural tolerances which are explained diagramatically in *Setting the standard* published by the Patent Glazing Contractors' Association.

Vertical patent glazing forming external walls (which includes roof glazing at 70 degrees or more) is subject to the provisions against the spread of fire contained in the *Building Regulations 1991* as described in section 4.2.6 as they are classified as totally unprotected areas and distance from boundaries is important. Patent glazing used to separate accommodation from an internal atrium void may form part of a separating wall and require a high fire resistance rating. For roofs, the principal fire protection requirement is that the roof deck and its covering shall prevent fire from entering the building. Roof patent glazing is considered to be an adequate barrier provided that the glass, whether wired or unwired, has a nominal thickness of at least 4 mm (see also under 'Rooflights'). The factors outlined in sections 3.2.8 and 3.2.9 are also important when considering the use of patent glazing and the advice of specialist manufacturers should always be adhered to.

Other design aspects to be considered include:

- Appearance, shape and form.
- Strength and stiffness of glazing bars.
- Span between points of attachment to structure.
- Span between glazing bars for glass or plastics sheets.
- Points of connection with ground, wall or roof.
- Weathertightness.
- Thermal insulation requirements.
- Condensation and the need for ventilation.
- Doors, windows and rooflights.
- Points of penetration (service pipes, etc).
- Method of drainage (if any).

Patent glazing is still available in traditional patterns with glazing bars in either mild steel or aluminium alloy. They are now also made to be double glazed, with an inner layer of Georgian wired or laminated/toughened glass, in hermetically sealed or dry 'built-up' units. Lead flashings are used at edge junctions with walls or the enclosing structure. The industry standard is for the glazing bars to be spaced at 622 mm centres to accommodate 600 mm wide (typically 6 mm thick) glass panels. Bars are made in different depths for varying distances between purlins or other structural supports; the usual maximum span distance is 3 m. The profiles have grooves for the water which may penetrate the wings, and also to collect condensation from the underside of the glass (fig. 5.20).

Patent glazing alternatives:

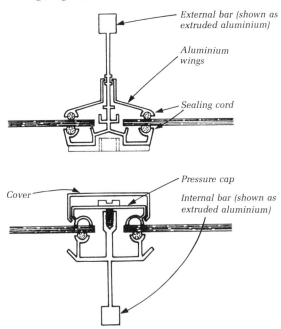

External bar (shown as extruded aluminium)

Aluminium wings

Sealing cord

Pressure cap

Cover

Internal bar (shown as extruded aluminium)

Fig. 5.20 Traditional patent glazing with internal and external bars. (Courtesy of *The Architects' Journal*)

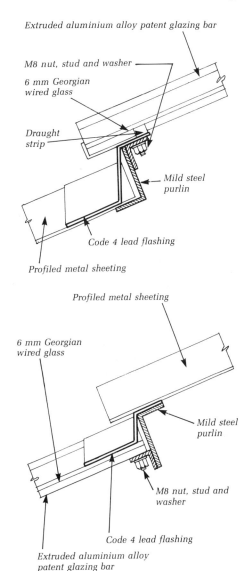

Extruded aluminium alloy patent glazing bar

M8 nut, stud and washer

6 mm Georgian wired glass

Draught strip

Mild steel purlin

Code 4 lead flashing

Profiled metal sheeting

Profiled metal sheeting

6 mm Georgian wired glass

Mild steel purlin

M8 nut, stud and washer

Code 4 lead flashing

Extruded aluminium alloy patent glazing bar

Fig. 5.21 Junction of patent glazing with profiled metal roof sheeting (Universal Glazing Ltd)

Because conventional patent glazing is a two-edge supported system, at junctions within the length there is normally a change of level at the junction between glass panels or with other roofing materials. This is achieved using a steel angle or Z-section glazing purlin (fig. 5.21). A shoe fixed to the structure at the foot of each bar prevents the glass from sliding downwards. Opening lights can be incorporated and if at high level can be controlled by simple rack and cord or a remote control mechanism. In vertical glazing applications top-hung, pivot-hung lights, louvres and entrance doors can be accommodated. Various anodized (aluminium) or polyester (steel or aluminium) finishes are used on bars, flashings or ventilators.

Patent glazing is a factory made system which depends on the correct use of proprietary components (see fig. 5.22). Designs which rely on a single span are much easier to install and maintain than stepped sections of fully glazed roof. They also have a reduced risk of leaks at junctions and fewer problems of cold bridging at structural connections. Some systems include a partial thermal break (fig. 5.23), a gasket separating the capping from the glazing bar. Increasingly double-skin polycarbonate is used, the panels being supported off aluminium box sections rather than glazing bars. The construction can then be fully thermally broken by a plastic connector used between the inner and outer aluminium sections. Throughout its history, patent glazing has proved a flexible technology that has lent itself to a variety of roof forms, particularly those derived for the daylighting of industrial buildings such as monitor roofs, north lights and ridge lights.

Continental systems draw upon a rather different tradition of roof glazing. One type available in this country (see figs 5.24 and 5.25) utilizes steel or aluminium box-section bearing members with projecting threaded studs. Inner and outer glass panes are separated by a resilient buffer and held in place by gaskets compressed between the face of the box section and the outer copper or aluminium clamping plate. The plates are drilled to suit the projecting bolts and secured by an exposed plastic nut and neoprene washer.

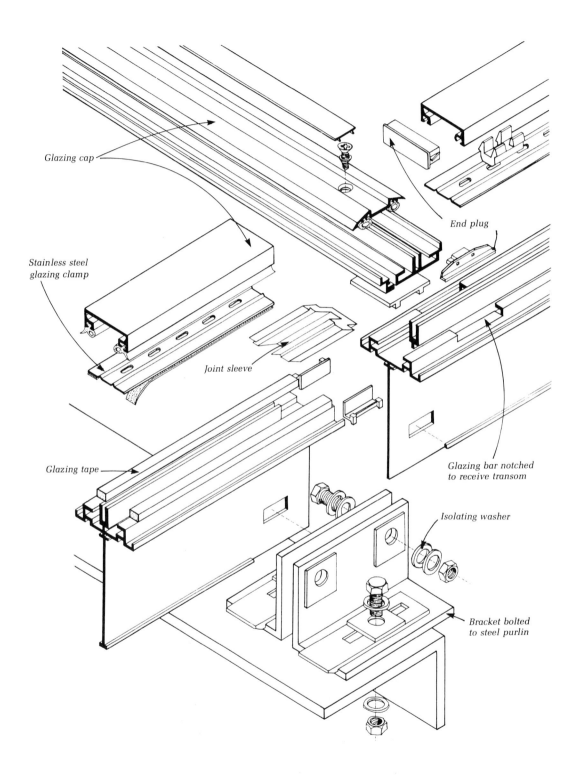

Fig. 5.22 Component assembly for patent glazing with internal bars

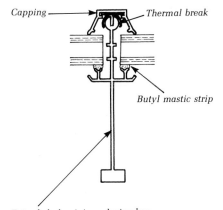

Fig. 5.23 Aluminium patent glazing bar with thermal break

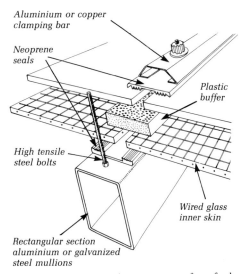

Fig. 5.24 Continental proprietary system of roof glazing (Eberspacher)

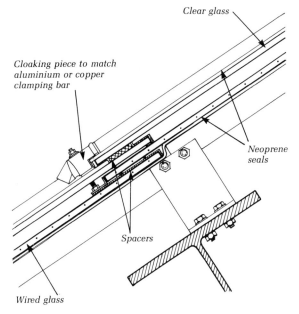

Fig. 5.25 Continental proprietary system of roof glazing (Eberspacher) junction detail between glazing panels

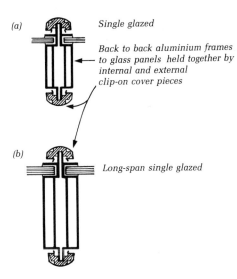

Examples

(a) Single glazed

Back to back aluminium frames to glass panels held together by internal and external clip-on cover pieces

(b) Long-span single glazed

(c) Long-span double-glazed

Fig. 5.26 Aluminium framed glazed roof panels bar types (Vitral UK Ltd)

At junctions within the slope rather than the 'step' familiar in traditional UK patent glazing, the glass panes are interleaved (fig. 5.25).

Another system from the continent essentially consists of linked roof windows for slopes of 30 degrees and over and for use as vertical glazing. Each glazed panel comprises one or two layers of clear, tinted or solar control glass mounted into an aluminium framework; glass and frame form a self-supporting unit which can span between supports on two opposite sides. The joints between panels are weathered by cover pieces that clip on to the aluminium frame sections internally and externally. Framed panels can be hinged and weathered by neoprene gaskets to form opening lights or smoke exhaust ventilators activated by 'gas-guns'. The finished sight lines are very slim and the same for both fixed and opening lights (fig. 5.26). Within the standard sections insulated opaque panels can be incorporated rather than glazing.

5.14 Four-edge supported and structurally glazed rooflights

In response to increasing performance expectations, more sophisticated methods of roof glazing have been introduced that owe more to the technology of curtain walling than to traditional patent glazing (fig. 5.27). These systems may be pressure-equalized, drained, four-edge supported and thermally broken.

Rainscreen systems, fully air-sealed at the internal junctions with rainscreen gaskets externally, are the most sophisticated. They mostly originate from continental or American practice and can achieve the high expectations of contemporary building envelope design — an internally controlled and consistent environment whatever the vagaries of the external climate. Factors to be controlled include wind-driven rain, hot humid air, pollutants, sound, cold dry air and interstitial condensation. Of the mechanisms of rain penetration — capillary action, kinetic energy, gravity and differential air pressure between inside and outside — it is the last which is the most hazardous. Rainscreen systems rely on internal air-sealing of all joints to obviate this potential source of leaks. The seals also prevent moist air entering the construction and forming condensation. Airtightness also reduces energy consumption, because sensible and latent losses are a direct result of greater air infiltration.

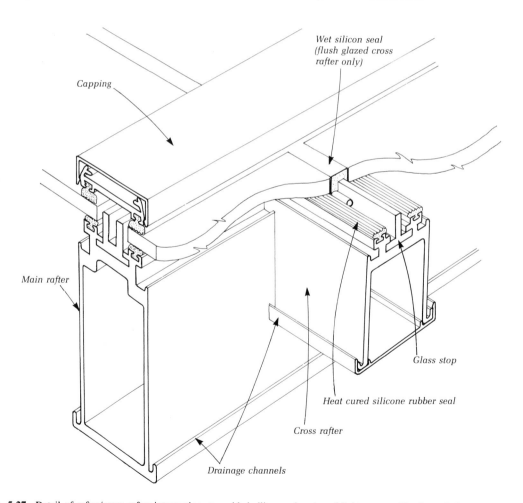

Fig. 5.27   Detail of rafter/cross rafter intersection, two-sided silicone glazed roof light system (Coxdome Ltd)

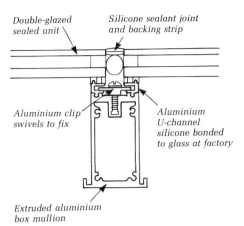

Double-glazed sealed unit

Silicone sealant joint and backing strip

Aluminium clip swivels to fix

Aluminium U-channel silicone bonded to glass at factory

Extruded aluminium box mullion

Fig. 5.28 Rafter detail for silicone-glazed aluminium framed rooflight system (Super Sky Products Inc)

Photo 5.6 Bentall's Department Store, Kingston-upon-Thames, Surrey. Architects: Building Design Partnership, 1992. Photo courtesy of Space Decks Ltd.

The two straight vaults, one of 15 m and the other of 10 m span, were assembled in 7 m wide half-arch sections. Overall the barrel vault is 120 m long. The completed arched space frame structure is here being fitted with the aluminium channels to which will be clamped structurally glazed roof panels (see detail fig. 5. 31).

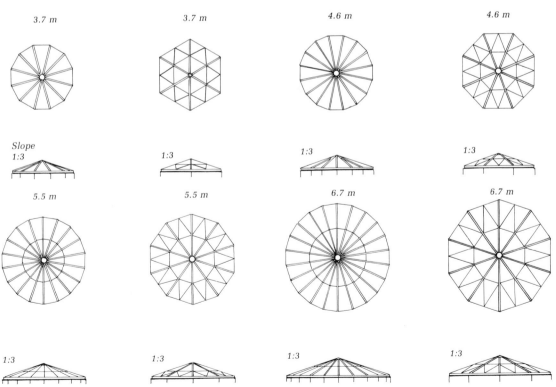

Fig. 5.29 Manufacturer's range of aluminium-framed rooflights and range of sizes (Super Sky Products Inc)

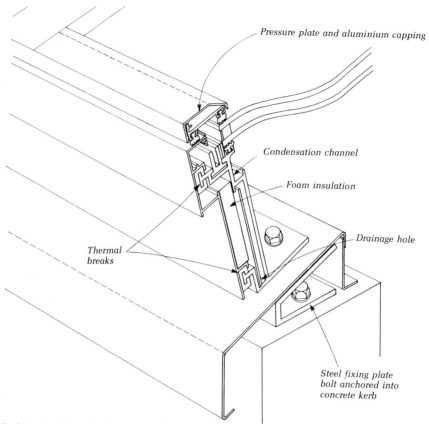

Fig. 5.30   Detail of insulated/thermally broken kerb (Coxdome Ltd)

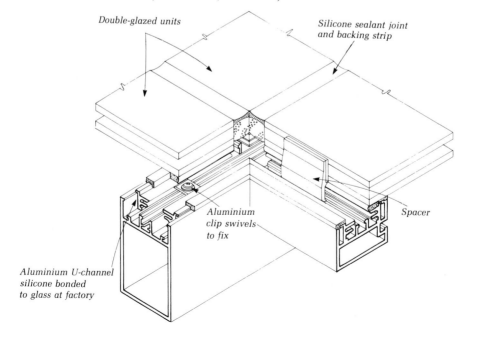

Fig. 5.31   Four-sided structurally glazed rooflight system (Space Decks Ltd)

Photo 5.7 Eaton Centre, Toronto. Architects: Zeidler Partnership, 1973–78.

The glazing panels, being supported at each edge, can be larger than is usual in patent glazing, up to 900 or 1200 mm. The transoms are designed to allow thermal and structural movement and for their integral drainage channels to interconnect with and discharge into the sloping glazing bars. The system may incorporate brackets for window cleaning machines or ladders, snow barriers and integral insulated gutters.

Figure 5.28 shows a rooflight system also derived from curtain walling technology with aluminium box-section bars and transoms supporting the glass on all four edges. The relatively new technology of structural glazing is used to make a flush detail at the transoms which consequently do not trap water as do horizontal projecting cappings. The sealant that is site applied to form these joints is low-modulus silicone (see photo 5.6). Square, circular and polygonal rooflights can be manufactured up to 7.5 m across without secondary structural framing. Polygons are constructed with a 30 degree pitch, pyramids with a 20 degree pitch. All sections including the kerbs can be thermally broken and the kerbs are designed to contain all horizontal thrusts so the structure only has to support the vertical component of the self-weight and imposed loads on the rooflight (see fig. 5.29).

Photo 5.8 Yorkdale metro station, Ontario, Canada. Architects: Arthur Erickson, 1975.

The barrel vaulted form of rooflight has recently found application for a wide variety of building types. Small span rooflights with aluminium glazing bars can be self-supporting, though for greater distances a subsidiary structure is required. At Yorkdale, rolled hollow section arches span between steel edge beams. The glazed roof of the Eaton Centre, a covered street three blocks in length, is supported by arched lattice trusses.

Generally imported into this country from the USA are totally flush rooflights where no retaining cap sections are used at all, the glass retained only by structural sealant (see chapter 6). Adhesive technology for glazing is a relatively new development. The long-term performance of the installation is dependent on the correct design of the joints and the compatibility of the sealant with the cleaners and primers to be used. The structural glazing of each glass panel to its aluminium supporting frame is carried out in the factory. On site the panels are secured to box-section aluminium glazing bars with swivel clips and the weathering joint between the panels is formed by a plastics backing strip and gun-applied silicone sealant (fig. 5.31).

6 Structural glazing

This technology relies on the characteristics of *elastomeric sealants*. Silicone, like carbon, is able to join with other elements to form large polymer molecules. In the case of elastomeric materials the molecular chains are zigzag in pattern, making the sealant highly extensible, the chains straighten when the joint is put in tension. These synthetic polymers have other unique capabilities. Not only do they adhere to both organic and inorganic materials but because of the chemical stability of the polymer, the sealants retain their properties under exposure to varying environmental conditions. Silicone sealants are resistant to sunlight, rain, UV radiation, pollution, ozone and extremes of temperature. Their elongation, tensile strength, hardness and adhesion do not change significantly with age, up to in excess of 20 years. Clear silicone rubber sealant for glass-to-glass joints should comply with BS 5889: 1980 *Specification for silicone-based building sealants*, type B.

In the United States the term *structural glazing* is used interchangeably with *silicone glazing* to mean any glass system that employs elastomeric sealants, a common classification of framed structural glass systems is as given under sections 6.1 and 6.2.

6.1 Direct and two-sided silicone glazing

Direct and *two-sided* glazing is shown in fig. 6.1. Silicone sealant can be used as a structural adhesive at two opposite edges of a glass panel, the other sides being mechanically fixed in position (see photo 6.1), as for example, in the case of some rooflight glazing systems where glass-to-glass joints across the slope are silicone sealed, whilst down the slope the glass is held in position by pressure plates and gaskets (see fig. 5.27).

Similarly at the top and bottom edges of vertical glazing, the glass can be directly glazed into rebates or grooves in stone, concrete and similar surrounds (as leaded lights were traditionally fixed in place with mortar fillets). Grooves should be at least 20 mm deep (larger at the top so the glass can be lifted into position and for its future replacement if necessary). This allows for construction tolerances and for the glass to project at least 12 mm into the groove with a 3 mm minimum separation between the edge of the glass and its surround (using setting blocks of solid neoprene, PVC or hardwood).

For direct glazing into masonry the manufacturers of glazing compounds should be consulted as to the need for alkali-resistant priming, the compatibility of their material and the width of rebate required. Non-setting compounds need not be painted, but distance pieces should be used.

Photo 6.1 Centre for Contemporary Art Studies and Miró Foundation, Barcelona. Architect: Josep Lluis Sert, 1972.
Storey height glass panes are glazed into the concrete structure at top and bottom and sealant glazed at the vertical joints between panels. An example of two-sided structural glazing.

Providing that grooves can be accurately formed, using recessed aluminium channel for example, dry glazing with gaskets may be suitable. In this case a width of at least 3 mm should be allowed at each side of the glass, which is secured at the centre of the channel by continuous guides of solid neoprene or PVC.

6.2 Four-sided silicone glazing

Four-sided or *flush glazing* uses structural adhesive to join all four edges of the glass back to a frame. The sealant is then the only structural link between the glass panel and its support and becomes an integral structural element of the building.

The rooflight shown in fig. 5.31 is a system of glass panels that are factory-fixed back to internal aluminium channels using silicone sealant. The silicone waterproof joints between the edges of the glass are gunned into place

on site. The pyramids at the Louvre (architect: I.M. Pei) are the most notable of recent examples of this technology.

Four-sided curtain walling systems are now widely available, the glass panes being factory bonded to an aluminium box section frame. In some types opening lights are incorporated using preformed silicone hinges.

The narrower definition of structural glass assemblies given in the *National Building Specification* is: 'Walling of glass or toughened glass sheets, in which the self-weight of the glass and wind or other loads are carried by the glass itself rather than by glazing bars, mullions or transoms. The glass sheets and any glass fins are joined directly together by adhesive sealants and/or mechanical connectors' (see photo 6.2). The earliest versions of this technology employed 'patch' connectors, these are still widely used for shop and display windows but for multi-storey applications 'Planar' (Pilkington's patented trade name) and other bolted systems are now more common.

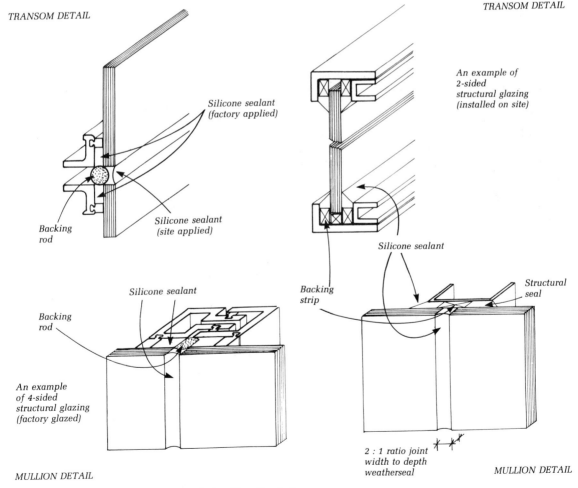

TRANSOM DETAIL

Silicone sealant (factory applied)

Backing rod

Silicone sealant (site applied)

Silicone sealant

Backing rod

An example of 4-sided structural glazing (factory glazed)

MULLION DETAIL

TRANSOM DETAIL

An example of 2-sided structural glazing (installed on site)

Silicone sealant

Backing strip

Structural seal

2 : 1 ratio joint width to depth weatherseal

MULLION DETAIL

Fig. 6.1 Two- and four-sided structural glazing (Dow Corning)

Photo 6.2 Conservatory at Science Museum Parc de La Villette, Paris. Architects: Rice/Francis/Ritchie, 1986.

The conservatories are single glazed, sheer glass enclosures 32 m high and 8 m deep divided into 8 m square bays. The glass panels are bolted at their corners and jointed with silicone sealant. The slender internal structure has to accommodate large deflections due to wind loadings but without rigid connections which would cause the glass to crack. Instead, the glass walls are designed as a membrane relying on the flexibility of toughened glass, and stabilized by a system of steel spars and cables that withstand both positive and negative wind loads. The countersunk fixing bolts through the glass are ball-jointed for freedom of movement. The connection at the top of the wall has springs like motorcycle forks. In a 'once in every 50 years' wind gust, the glazing is designed to flex by up to 60 mm in both directions.

6.3 Frameless glazing using patches

Single height sheets of toughened glass (for shop windows, etc) can be butted at their vertical edges and if necessary, stiffened by full height 12 mm minimum thickness glass fins. The three-way junction formed on plan is bonded with silicone sealant. Toughened glass is up to five times stronger than ordinary glass and impact resistant, thus making its

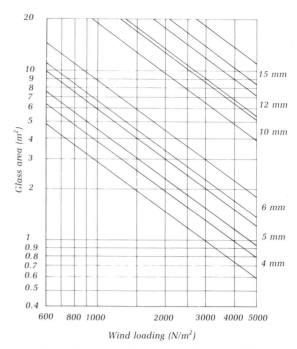

Fig. 6.2 Wind loading graph for toughened float glass (Pilkington Glass Ltd)

use appropriate for both shop windows and multi-storey structural glass walls.

Figure 6.2 shows recommended thickness and areas for the main panes of toughened float glass. Fixing methods should be in accordance with BS 6262: 1982 *Code of practice: glazing for buildings*.

6.3.1 Suspended multi-panel assemblies

Structural glazing with patch fittings (fig. 6.3) is used for vertical glazing assemblies where the height necessitates the use of panes mechanically fixed together. The patch fittings join the corners of adjoining glass panels by being bolted together through holes drilled in the glass. Hung from the building structure, sheer glass walls up to 23 m in height and of unlimited length can be made. The adjustable suspension system, as well as the perimeter channel sections, allows the glazed wall to move independently from the building frame and accommodates tolerances between the highly accurate glazing assembly and the less accurate main structure.

A variety of patch fittings incorporating door rails and locks are available including pivot fittings for hinged doors (fig. 6.4). Inside these patches are metal castings made in two halves to clamp either side of the glass and incorporate pivot or other components. At front and back a flush aluminium or stainless steel cover is fixed in place with

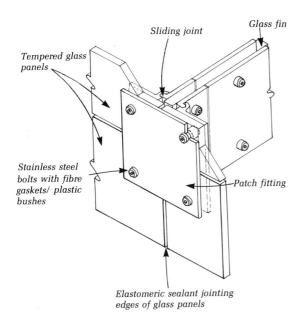

Glass fin

Sliding joint

Tempered glass panels

Stainless steel bolts with fibre gaskets/ plastic bushes

Patch fitting

Elastomeric sealant jointing edges of glass panels

Fig. 6.3 Structural glazing 'patch' fitting showing four-panel junction (Pilkington Glass Ltd)

grub screws to give the assembly its finished appearance. Transparent and translucent toughened glass doors for exterior and interior use are normally fitted with rails and/or patch fittings at the top and/or bottom edges, from which the doors are pivoted or hinged. Plastics can also be used for completely frameless doors. With both materials, edges should be rounded and hinges and locks bolted through using preformed holes. Holes in glass should not have a diameter less than the thickness of the glass and must not be located too close to the edges and corners of the pane. It is important that the fittings are isolated from the glass or plastics by a gasket and that the bolts are not over-tightened. Recent examples have refined this technology to achieve an extreme minimalism of detail.

The glass used is 12 mm tempered clear float or tempered solar control glass to meet the safety requirements of BS 6262: 1982 and BS 6206: 1981 (class A). Glass mullions are usually of 19 mm toughened clear float. The corners of the glass panels are drilled and clamped into position by the patches and stainless steel bolts, with fibre gaskets and plastics bushes to eliminate glass-to-metal contact and to allow thermal movement. The glass edges are made waterproof by site-applied flexible silicone sealant to BS 5889, at the perimeter the glass wall can be directly glazed into channels recessed into the surrounding structure

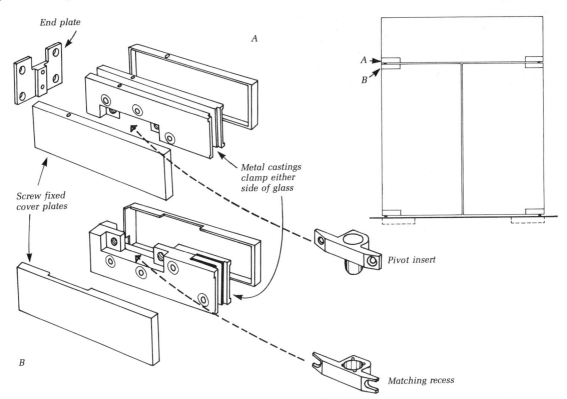

End plate

A

Metal castings clamp either side of glass

Screw fixed cover plates

Pivot insert

B

Matching recess

A

B

Fig. 6.4 Patch fitting used to form the hinge of glass doors (Pilkington Glass Ltd)

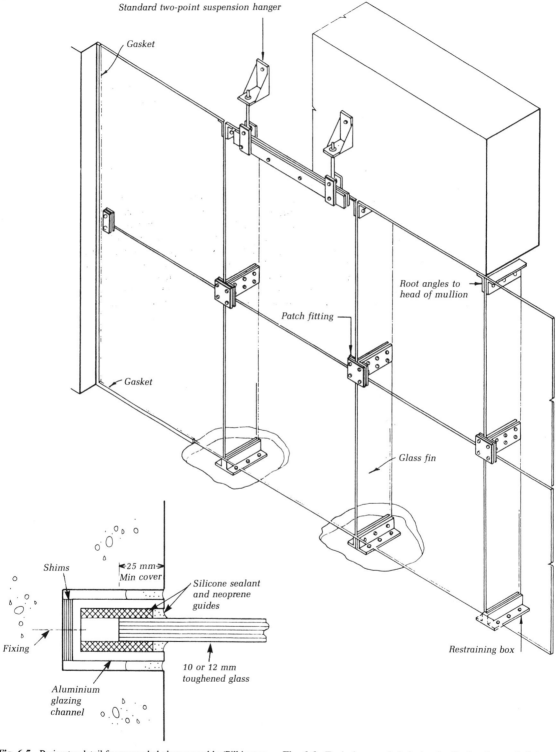

Standard two-point suspension hanger

Gasket

Root angles to
head of mullion

Patch fitting

Gasket

Glass fin

Shims

←25 mm→
Min cover

Silicone sealant
and neoprene
guides

Fixing

10 or 12 mm
toughened glass

Aluminium
glazing
channel

Restraining box

Fig. 6.5 Perimeter detail for suspended glass assembly (Pilkington Glass Ltd)

Fig. 6.6 Typical suspended glazing details showing patch fittings (Pilkington Glass Ltd)

(see fig. 6.5). Patch fittings can also be used where the glass wall is to have a continuous background support such as a vertical space frame, structural steelwork or masonry allowing attachment at the corners of panels.

Where the live load movements are likely to be extensive a fully suspended system may be employed. The glass is clamped at the top by suspension hangers (fig. 6.6) bolted back to a horizontal structure capable of supporting the overall weight of the glass wall. Wind loads will need to be resisted either by glass fins, steel sections or a space frame coinciding with the joints between the glass panes.

6.3.2 Ground-based multi-panel assemblies

Alternatively, the glass assembly can be supported from the ground, up to a usual maximum height of 9 m. Once again the necessity for lateral restraint will require glass fins (or another structural system) set out at the same centres as the glass itself. Combinations of these methods can be used where the glazing is rigidly fixed and cantilevers both at top and bottom, or from the middle if an intervening floor structure can be used for attachment. Alternatively the assembly can be rigidly fixed at the bottom and have a sliding joint at the top (if the roof has a flexible lightweight structure allowing pin-jointed connections only).

Patch fittings have a sliding connection to the bracketry tying the glass wall back either to glass fins or other restraining structures, to allow for differential movement between the two. The example shown in fig. 6.7 uses this technology to good effect; lateral restraint is given to the glass by fins fixed to the projecting edges of concrete cantilevers (that are subject to long-term creep).

The glass used in this sort of construction has to be precisely toughened to attain a reliable strength level and to comply with BS 6206 for grade A impact strength. The maximum size of pane is governed by limitations of allowable deflection and stress in the glass as well as the available size of toughening oven, which for clear horizontally toughened glass is 4200 × 2000 mm in the UK (though larger sizes are available from continental manufacturers). These dimensions may be further modified if coated or tinted glass is required. Toughened glass is manufactured by heating and then rapidly air-cooling the glass sheets; the resulting balanced stresses built into the material make it resistant to a temperature difference of approximately 200°C between the front and back of the glass and therefore resistant to *thermal shock*.

A problem in glass manufacture is the inclusion of particles of nickel sulphide which are impossible to detect visually or to eliminate totally. Fluctuations in ambient temperature result, over a period of time, in the expansion of the *nickel sulphide inclusions*, the rate depending on the extent of heating. The growth of the inclusions upsets the balance of stresses in the glass, causing it to shatter. Much effort has been expended by glass manufacturers to reduce the presence of nickel sulphide inclusions to a very low level, making spontaneous breakage a rare occurrence. It is essential, however, that glass for structural glazing should be both *stress-tested* and *heat-soaked* (the destructive test devised to detect the presence of nickel sulphide inclusions which is carried out after the glass has been toughened).

6.4 Planar glazing

Planar glazing is a further development of frameless glazing. To reduce the visual impact of the fixings, toughened glass sheets (as described above) are bolted back through countersunk holes in the glass (figs 6.8 and 6.9 and photo 6.3). The cutting and drilling of the glass is carried out before it is tempered. The *spring plates* to which the countersunk bolts are attached and the silicone-sealed joints between panels produce a flush, all-glass façade. The spring plates are usually made from type 316 stainless steel to BS 970 Part 1 or aluminium to BS 1474 and can range from a simple 80 × 80 mm angle to a sophisticated cast fitting designed to minimize the visual impact of the fixings. Spring plates are engineered to match the stiffness of the glass, allowing the whole assembly to flex under wind load. They also cushion the glass from the greater rigidity of the internal structure and allow rotation, thereby reducing the stress in the glass around the holes and the extent of the restraining structure. The spring plate is separated from the glass by a pressure nut separated from the glass by a fibre washer; the backnut is only 'finger-tightened' so the glass is free to move independently of the supporting structure. In fact the fastening has to be made within precise limits and only calibrated torque wrenches should be used for the assembly. As with patch fittings, bushes and gaskets of non-compressible vulcanized fibre to BS 6091 and nylatron polyamide are employed throughout to eliminate glass-to-metal contact, to separate dissimilar metals and allow for thermal movement. Generally, the fastenings used are stainless steel grade 316 and the whole installation should comply with BS 6262: 1982 *Glazing for buildings* although the stresses in the glass can only be accurately predicted by reference to the system manufacturer's proprietary design data.

Planar curved glass has now been developed, the maximum panel size being girth 2.0 m, length 3.6 m and minimum radius 1.0 m. The glass is toughened to achieve similar properties to those defined by BS 6206 class A, although there is not a British Standard for toughened curved glass. Special fittings have been developed for this material and if of sufficient radius the glass can be formed into insulated spandrel panels.

Within the overall logic of the system, the method of

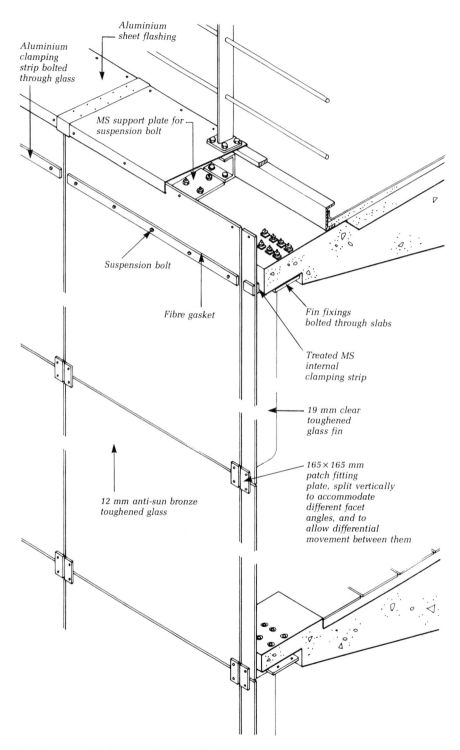

Fig. 6.7 Isometric of the structurally glazed cladding of Willis Faber Dumas, Ipswich. (Architects: Foster Associates. Courtesy of *The Architects' Journal*)

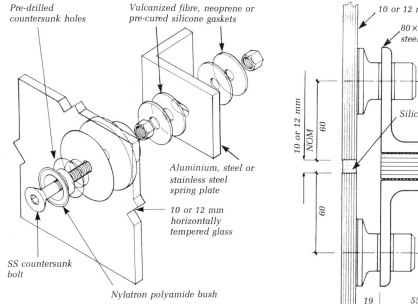

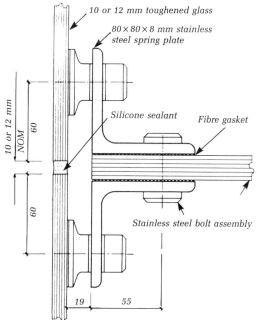

Fig. 6.8 Planar spring plate attachment for single glazing (Pilkington Glass Ltd)

Fig. 6.9 Planar type 902 bolt with spring plate attachment to glass fin (Pilkington Glass Ltd)

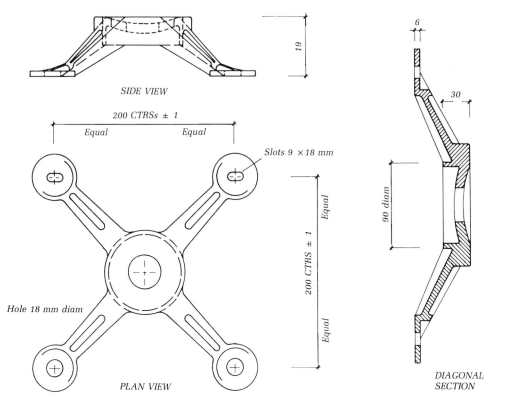

Fig. 6.10 An example of a cast spider bracket (Pilkington Glass Ltd)

Photo 6.3 Renault Distribution Centre and Warehouse. Architects: Foster Associates. Photo courtesy of Pilkington Glass Ltd. An early example of Planar glazing, the panels are supported by internal hollow section steel transoms.

support can be designed to specific requirements and the spring plates are often custom made by amalgamating all four corner fixings into one, often quite complex, stainless steel or aluminium casting (fig. 6.10). The geometry of castings has to be within the limitations of the technology, the molten metal inlets, for instance, have to be located for easy removal and to enable grinding smooth at this point and between the two halves of the mould. Three-dimensional assembly tolerances have to be accommodated,

bearing in mind that this is a fully factory-made system that cannot be modified on site in the traditional way. A variety of finishes are possible, usually anodizing or electrolytic polyester powder paint on aluminium and satin or polished finish to stainless steel.

Where large live loads are anticipated, toughened glass panels can be suspended from a horizontal beam or other structure by means of suspension hangers. Suspended Planar walls can be built to a height of 21 m, but for tall walls the glass module may have to be reduced (according to the weight of glass and the restriction imposed by the shear strength of the supporting bolts). Even larger assemblies can be constructed, but require intervening horizontal structures to divide the wall into practicable dimensions. Glass fins, a space frame or structural columns or beams with sliding joint connections to the glazing can be used to give lateral restraint.

Two structural systems are possible where the glass wall is to be self-supporting and independent of intermediate floors or any other part of the main structure that might afford lateral restraint:

1. A *cantilever* where one end of the fin is bolted to a rigid structure (made from reinforced concrete, for example) that is capable of resisting the turning moment transferred from the fin as a result of the wind load acting upon the glass wall. The other end of the fin is not attached to the structure.
2. A *pin-pointed fin* where each end of the fin is restrained in the horizontal plane but it is free to rotate. Usually the top is suspended from a pivot and the bottom end slides vertically in a box. The fin acts in bending as a simply supported beam, transferring wind loads equally into the structure.

The maximum size of single-glazed panels is, as noted above, 4.2 × 2 m and double-glazed 2 × 4 m maximum. For glass panel sizes up to 2 × 2 m, fixings are usually only required at the corners. As the glass size increases so does the number of bolt fixings that are required (along the long edges only), according to the following design conditions:

- Design wind load.
- Glass thickness (minimum 10 mm).
- Panel size.
- Aspect ratio.
- Displacement (deflection) limitation.
- Stress limitation.

The corner bolts are designed to hold the weight of the glass and the intermediate fixings resist the wind loads (the precise format depending on whether, and how, the assembly is to be suspended). Where glass fins are used, they are usually 19 mm in thickness and spliced together along their length, the splices coinciding with glass panel joints.

As a general guide, in a sheltered urban area: 2.0 m square panels will require 4 bolts (at corners), 2.0−3.5 m panels will require 6 bolts and >3.5 m panels will require 8 bolts. Although tall façades may require more bolts and fins, and panels smaller than 2 m square.

Panel sizes and fixing positions are factory made to precise tolerances:

- *Panel size tolerance:* ±1 mm of figured dimensions.
- *Squareness*, the difference in the measurement of the diagonal dimensions, within 3 or 4 mm where the diagonal measures over 4 m.
- *Glass thickness* with the range 9.7−10.3 mm for nominal 10 mm glass and between 11.7 and 12.3 mm for 12 mm.
- The extent of *bowing* to be not more than 0.2 per cent.
- *Location tolerance* of holes within ±0.5 mm of a datum position for measurement.

Butt joints at the edge of glass panels should be a minimum of 8 mm in width but a nominal thickness of 12 mm is preferable (the allowable tolerance for movement is then ±4 mm, ±2 mm manufacturing tolerance and ±2 mm erection tolerance). On a wide façade, where the primary structure is to be broken by *expansion joints*, vertical weathertight expansion joints can be detailed into the glass wall without interrupting its surface. The preferred detail to form entrances within a Planar-glazed wall is to construct a steel 'goal-post' portal frame into which the entrance doors are fixed and on to which the glass fins and wall are attached. A variety of door systems are then possible, i.e. revolving and side pass doors, automatic sliding, etc. The portal helps define the entrance though, if desired, it is possible to reduce the visual impact of the steelwork, for example by constructing a projecting glass lobby and insetting the doors within it.

Glazed wall technology is advancing rapidly with the incorporation of components and design techniques from the yacht-rigging industry. The use of tensioned trusses of small-diameter high-strength stainless steel rod and braided steel cable results in an unobtrusive secondary structure which complements the transparency of a frameless glass wall (see photo 6.2).

Alternatively, Planar glazing can be fixed to a fully supporting structure, whether of masonry or concrete, metalwork mullions and transoms or columns or beams with outriggers to the bracketry at the corner connections of the glass panels (see photo 6.4). One of the great advantages of Planar, belied by its simplicity, is the variety of designs for spring plates and supporting structure that is possible. However, it is an engineered cladding system which requires that anticipated stress levels are matched to the characteristics of the glass; the manufacturer's input is consequently required at the design stage.

The use of laminated glass in Planar construction (for greater security and safety, as required for 'overhead' glass in the USA and Germany), is another recent development. Possible glass combinations are 10 mm plus 6 or 4 mm, and 12 mm plus 6 or 4 mm with a 1 or 2 mm interlayer; a variety of tinted and coated glasses can be incorporated, single or double glazed.

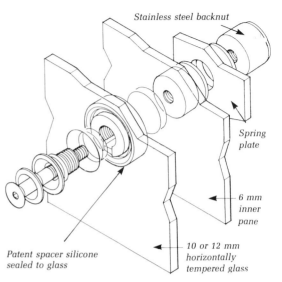

Stainless steel backnut

Spring plate

6 mm inner pane

10 or 12 mm horizontally tempered glass

Patent spacer silicone sealed to glass

Photo 6.4   Financial Times Printing Works, Isle of Dogs, London. Architects: Nicholas Grimshaw and Partners, 1988. Photo courtesy of Pilkington Glass Ltd.

Fig. 6.11 Planar spring plate attachment for double glazing (Pilkington Glass Ltd)

The external steel columns were fabricated from steel plates with welded ends which are half-tubes in section. The outrigger arms resist wind loads against the glass wall. The circular plates attached at the ends of the outriggers accommodate Planar bolt fixings through to each of the four intersecting corners of the 2 × 2 m glass panels.

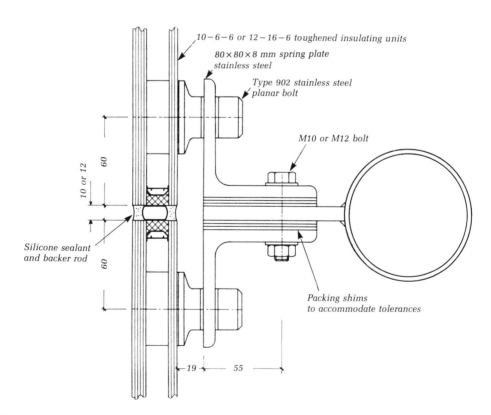

10 − 6 − 6 or 12 − 16 − 6 toughened insulating units

80 × 80 × 8 mm spring plate stainless steel

Type 902 stainless steel planar bolt

M10 or M12 bolt

10 or 12

60

60

Silicone sealant and backer rod

Packing shims to accommodate tolerances

19   55

Fig. 6.12 Planar type 902 bolt assembly for attaching double-glazed panels to internal structure (Pilkington Glass Ltd)

6.5 Planar glazing — double-glazed

The system can also be double-glazed (figs 6.11 and 6.12). Normally the double glazing units comprise 10 or 12 mm outer glass tinted if required, 16 mm air space and 6 mm clear float glass inner pane. Alternatively the inner leaf can be a 6 + 6 mm laminate. The outer glass is pre-drilled with between four and eight countersunk holes (depending on the design conditions), the inner sheet having corresponding plain drilled holes. Continental companies install similar systems but with face rather than countersunk fixings. The edges are factory-sealed with a butyl backing strip and silicone caulking. The cylindrical bushes separating and hermetically sealing the glass panels at the bolt positions are themselves silicone-glued to the glass (the method of flush bolt fixing through the double glazed panels is patented by Pilkingtons).

Fully suspended double-glazed walls are limited only by the capabilities of the suspended secondary structure. The joint with the rest of the building enclosure at top and sides can be made by either fully bedding the edges of the glass in silicone compound within a rebated channel or using neoprene gaskets, depending on the amount of differential movement that is anticipated. Care has to be taken when detailing the abutment of the relatively precise Planar wall and other materials, such as brickwork, that are constructed to more generous tolerances. A channel recessed within the brickwork is more likely to accommodate these

tolerances than a face-fixing detail. If the glass is not suspended but attached to structure (for example mullions or floor slabs) then there is no height limitation and the system can be used as a complete method of cladding/curtain walling.

Double-glazed Planar panels can achieve a U-value of 2.7 W/m^{2}K and with the addition of a low-emissivity coated outer surface of the inner pane the U-value may be reduced to 1.8 W/m^{2}K. The thickness of glass makes the system relatively good at excluding external noise. The edges of the double-glazed units must not be allowed prolonged exposure to moisture (which causes their deterioration). If a leak should occur, it will generally be due to poor adhesion; defective joints can simply be cut out and regunned. Cruciform gaps are left at the inside junction of four panels across the face of the Planar wall when the elastomeric sealant is applied. When the wall is water-tested, any leakage can be detected at the cruciform gaps. Although it is relatively easy to detect and correct leaks in Planar glazing, the jointing is highly workmanship dependent and only specialist sub-contractors and experienced operatives should carry out the work.

A recent introduction to the Planar system is a hook bracket that enables the glass panels to be hung off conical steel pegs attached to box-section mullions (fig. 6.13 and photo 6.5). The peg to bracket connection accommodates horizontal tolerances and, in conjuction with the silicone joints, thermal movement. The peg to mullion junction is

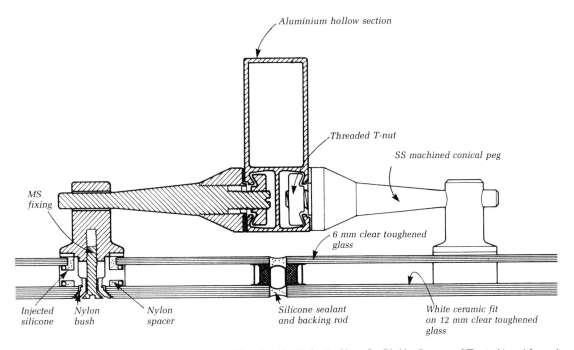

Fig. 6.13   Planar type 902 fittings as used at the B8 Building, Stockley Park. (Architect: Ian Ritchie. Courtesy of *The Architects' Journal*)

Photo 6.5 Pilkington's type 902 brackets with steel lug attachments welded to the rectangular hollow section frame. Photo courtesy of Pilkington Glass Ltd.

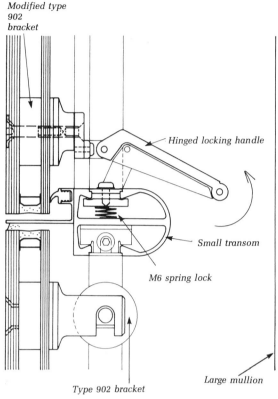

Modified type 902 bracket

Hinged locking handle

Small transom

M6 spring lock

Large mullion

Type 902 bracket

Fig. 6.14 Section through opening light using Planar 902 brackets (Pilkington Glass Ltd)

made with a nut wedged within a channel inside the mullion section but able to slide, to allow for vertical tolerances, before final fixing. The mullions are bolted into a curtain walling shoe at the ground floor, lateral support being provided at the upper floors by bolting through vertical slotted holes to allow differential movement. Alternatively these hook brackets can be fixed to plate flanges welded to a steel truss (see photo) or glass fins (though in this case the channel fitting would not be used, instead tolerance being provided by vertical slotted holes).

Opening lights with the same frameless external appearance can be incorporated, modified brackets being used with hinges that form the attachment to the glass casements (fig. 6.14). Spandrel panels can be formed by bonding insulation to the back of obscured glass panels and where solar control is required the outer pane can be tinted or partially obscured with ceramic frit (see photos 6.6 and 6.7). Consequently a complete system can be assembled that is analogous to curtain walling but is constructed from panels.

6.6 Planar glass roof details

The Planar system can also be used to construct completely glazed flat glass roofs (see photo 6.8). The roof structure has to provide support to each fixing location at the intersecting corners of glass panels. Planar glazed roofs should be designed to a module of approximately 1.5 × 1.5 m with corner fixings only. The standard dimension for the spacing of fixings from the edges of the panels is 60 mm, but this may be increased to 75 mm (using oversized spring plates) to accommodate special conditions and special panel sizes that may be required at roof edges and junctions.

Larger panels are liable to sag resulting in puddling; in any case the roof should be laid to a minimum pitch of 3−4 degrees to overcome surface tension between glass and water sufficient to shed rain. A 1.5 × 1.5 m grid can often be suitable for space frame roof construction and so the two systems are highly compatible. The Planar spring plate is modified so the assembly can be attached to the node

Photo 6.6   B8 Building, Stockley Park, Heathrow. Architect: Ian Ritchie, 1989. Photo courtesy of Pilkington Glass Ltd.

Photo 6.7   B8 Building, Stockley Park, Heathrow. Architect: Ian Ritchie, 1989. Photo courtesy of Pilkington Glass Ltd. Detail showing the countersunk planar bolts at the corners of the glass panels and the internal rectangular hollow section steel mullion within (see fig. 6.13). The stripes are ceramic frit, a grid of solar reflective dots silk-screened on to the surface of the glass, standard patterns can achieve between 20 and 80 per cent light exclusion depending on the density of application of the dots.

Photo 6.8   Waverley Market, Edinburgh. Architects: Building Design Partnership. Photo courtesy of Pilkington Glass Ltd. A planar glazed roof.

connections of the space frame. Alternatively a one-way structure, steel beams or lattice trusses to which the spring plates are attached, should be set out at the same centres as the glass panels. Smoke vents, fans, etc. may be substituted for glass panels as required.

Whilst it is not recommended that Planar roofs be used as a walking surface (because of the danger of heavy pointed tools or equipment being dropped, resulting in breakage of the glass), crawler boards must be employed for maintenance. Soft shoes should be used and care taken not to walk grit on to the glass. Because of their flush surface, most Planar roofs can, however, be cleaned using telescopic squeegees (also avoiding the installation of costly cleaning gantries).

Recent developments enabling laminated glass to be used in the Planar system provide an additional safety measure since even after breakage of both panes, the laminated Planar panels remain in place overhead.

Structural glazing has the advantages of simplicity, relative ease of installation, durability of the glass and, if kept clean, the stainless steel components (though the silicone joints may have to be replaced after 25–30 years). The torque of fixing bolts and the adhesive used to secure them should be checked approximately every 5 years. Being relatively low maintenance, the 'whole-life' costs of the construction are relatively favourable. The system is developing into a complete method of wall and roof cladding including solar control glasses, double glazing and bonded insulated spandrel panels.

Photo 6.9   Reina Sofia Museum Towers, Madrid. Architect: Ian Ritchie, 1991. Photo courtesy of Pilkington Glass Ltd.

The structural glazing of these lift towers is a development of Rice/Francis/Ritchie's system developed for the conservatories at la Villette. The planar bolt connections are attached to external rocker arms that are hung at their pivot points by rods and restrained by 'tie-down' rods that are fixed at the ground and to the ends of the projecting arms. The same sort of spring (similar to motorcycle forks) that was used at La Villette was also employed here.

7 External doors

7.1 Introduction

Doors are one of the most heavily used and abused parts of a building. Because they may be subject to frequent movement and heavy wear, their construction and method of assembly must be suitably designed (see photo 7.1). In the case of an external timber door, conditions of temperature and humidity will often be different on each side of the door which will cause it to either warp or twist (unless anticipated by the component's detailing). The joinery sections of external doors need to be robust and sized to ensure that adequate joints can be formed. There are many locations where, rather than timber, metal or glass doors will be appropriate.

The usual types of external timber doors are *unframed*, *framed* and *flush* (as shown in fig. 7.1). An unframed door is made of tongued and grooved boarding fixed to timber bracing at the inside face of the door. Framed construction has an outer frame of timber with subsidiary framing outlining panels that may be either solid or glazed. Flush doors have an outer facing of sheet material such as veneered plywood bonded to a solid or honeycomb core.

7.2 Performance requirements

The most significant requirements for external domestic doors are in terms of their *appearance, durability, strength, weathertightness, security* and *safety*. In public buildings, external doors may form the termination of a *means of escape from fire*.

There is a lack of British Standards applicable to doors, the most useful being BS DD171 *Guide to specifying performance requirements for hinged and pivoted doors*. This advises on specification and selection in terms of strength, operating requirements under particular circumstances and methods of testing. It also identifies four categories requiring different levels of performance (as

Photo 7.1 Doors to the Pantheon in Rome, 123 AD.
The original cast bronze doors and hinges are still functioning.

shown in table 7.1) against which particular components can be evaluated.

7.2.1 Appearance

The appearance of an external door has both to signify a building's entrance and also contribute to the form of the external elevation. The type of materials to be used as well

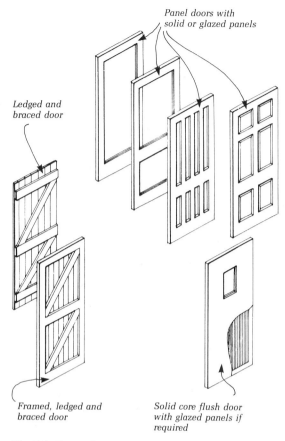

Panel doors with solid or glazed panels

Ledged and braced door

Framed, ledged and braced door

Solid core flush door with glazed panels if required

Fig. 7.1 Types of external timber doors

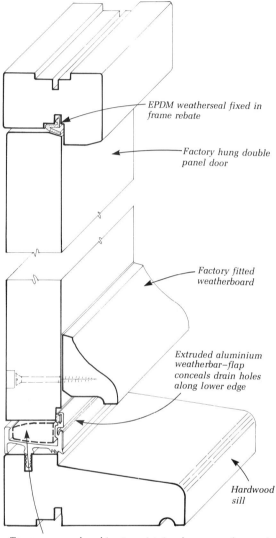

EPDM weatherseal fixed in frame rebate

Factory hung double panel door

Factory fitted weatherboard

Extruded aluminium weatherbar–flap conceals drain holes along lower edge

Hardwood sill

Temporary wood packing to maintain edge gaps and prevent distortion whilst door is being built in

Fig. 7.2 Inward opening panel door

as the relationship of the door to the surface of the outer wall and the general pattern of fenestration will have to be considered; also the choice of ironmongery and proportion, scale and detailing both of the door and its frame.

7.2.2 Durability

As in the case of building components generally, to ensure durability, high initial standards of design and specification have to be accompanied by satisfactory workmanship and followed by adequate maintenance. External timber doors may need special consideration, regular painting or clear treatment is necessary at a frequency depending on the extent of exposure to the weather.

Domestic doors are available made from both hardwood and softwood; nowadays it is usual practice for non-durable timbers to be preservative treated. Softwood doors may be delivered to site 'in the white' for site painting or primed or base stained ready to receive finish coats only. Hardwood doors are usually intended to be finished either clear or

translucent (see chapter 1 concerning the use of hardwoods in building construction).

7.2.3 Weather protection

As in the case of windows, the junction between an external door and its frame needs to exclude both air and water. There are no British Standard tests specifically for the weather-proofing of doors. Manufacturers often apply the same weathertightness categories used for windows (to BS 6375 Part 1: 1989) by testing doors to BS 5368 *Methods of testing windows*. Timber doors may be factory weather-

Table 7.1 Categories of duty, description and examples

Category of duty	Description	Examples
Light duty (LD)	Low frequency of use by those with a high incentive to exercise care, e.g. by private house owners, i.e. where small chance of accident occurring or of misuse	Internal doors in dwellings. External doors in dwellings providing secondary access to private areas
Medium duty (MD)	Medium frequency of use primarily by those with some incentive to exercise care, i.e. where some chance of accident occurring or misuse	External doors of dwellings providing primary access. Office doors providing access to designated public areas but not used by public or by people carrying or propelling bulky objects
Heavy duty (HD)	High frequency of use by public and others with little incentive to exercise care, i.e. where a chance of accident occurring and of misuse	Doors of shops, hospitals and of other buildings which provide access to designated public areas and which are used by the public and others frequently carrying or propelling bulky objects
Severe duty (SD)	Subject to frequent violent usage	Doors of stockrooms, etc. commonly opened by driving trolleys against them; doors in educational establishments subject to frequent impact by people

Source: BS DD 171: 1987 *Draft for development: Guide for specifying performance requirements for hinged or pivoted doors.*

stripped or delivered with rebates cut to receive weather-stripping fixed on site.

The bottom of a door is particularly vulnerable so weatherboards, weather bars and special seals are used at this junction. Figures 7.2 and 7.3 show typical manufacturer's details of a standard external, pre-hung door

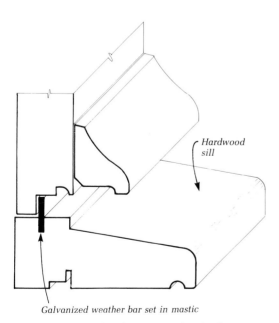

Hardwood sill

Galvanized weather bar set in mastic

Fig. 7.3 Inward opening door — water bar detail

set. A potential source of difficulty is the tenon joint where the jamb of the door is let into the sill. Continental manufacturers tend to design the sill with a step at this vulnerable location or to make the sill out of artificial stone with haunched ends and a cast-in metal dowel to locate the bottom of the timber frame. In exposed locations, inward-opening doors can be problematic and are best protected by a porch or overhang.

Outward-opening doors should, wherever possible, be set back into the opening and detailed with a projecting drip moulding to weather the head of the frame. If practicable, the edges of the meeting styles of doors hung in pairs should be rebated; weatherstripping at the rebate is an additional wise precaution. Weatherstripping helps both exclude rainwater and reduce air infiltration which causes draughts (as buildings have become better heated, occupants tend to feel draughts where previously there was no discomfort). Methods of weatherstripping timber doors are similar to those used for windows (see fig 4.7).

A double swing external door presents a particularly difficult situation since the weatherstripping must not impede the smooth action of the door. Figure 7.4 shows a method of weatherstripping at the jambs of double-swing timber doors. The EPDM insert is fixed in a PVC-U channel glued into a rebate within the frame. The projecting tongue is adjustable to seal gaps from 1 to 6 mm wide. It closes and compresses against a radiused aluminium buffer plate screw fixed to the edge of the door. The weatherstripping at the centre junction between the styles

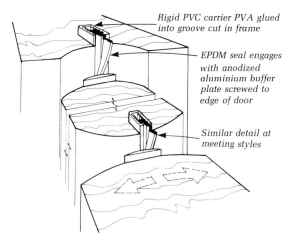

Rigid PVC carrier PVA glued into groove cut in frame

EPDM seal engages with anodized aluminium buffer plate screwed to edge of door

Similar detail at meeting styles

Fig. 7.4 Weatherseal for double-swing doors (Wintun Ltd)

of the two leaves is fixed in the same way. Weatherstripping the top rail of the door can be similarly achieved. Figure 7.5 shows a proprietary threshold weatherstrip that will withstand heavy foot traffic and has a minimal height of upstand that incorporates the internal floor finish. An alternative is shown in fig. 7.6 where the finished level is the same on both sides of the door. These thresholds are suitable for heavy wheeled traffic and provide, within the extruded aluminium section, a small cable or pipe duct across the opening.

Toughened glass doors can be weatherstripped by fixing an extruded channel with a neoprene insert around the closing edges of the doors. The channel is secured by adhesive and is cut away to accommodate locks or a kickplate to the base of the door. The neoprene tongue would seal against the glass, but the plain channel provides a symmetrical detail, as shown in fig. 7.7.

7.2.4 Thermal control

The loss of heat through a door can be high although the provision of weatherseals may improve the situation. Doors form a small proportion of the external envelope of domestic buildings and little consideration has been given in the UK to their thermal performance; insulated timber doors are, however, imported from Scandinavia (see fig 7.8). These are typically made with inner and outer skins of plywood or tongued and grooved boarding with a core of rigid foam insulation. If required, these high-performance doors are available with factory-fitted double or triple glazing. Also for domestic applications, steel-faced and moulded plastics doors are made with an insulating foam core forming a rigid composite panel with good thermal characteristics.

Industrial doors may, however, be significant in area and in the extent of heat loss. Increasing *Building Regulations* requirements for the insulation of industrial buildings has led to more widespread use of steel-faced composite panels with a core of foamed plastic insulation. These are made as horizontal hinged planks for mounting on 'up and over' door track.

7.2.5 Fire precautions

Means of escape requirements in public buildings specify the adequate width of door, correct direction of opening and method of hanging in order to enable occupants to leave the building in safety. There are various legislative requirements appropriate to different types of buildings; for particular situations the advice of local fire officers should be sought.

7.2.6 Strength and security

A door is called upon to resist a number of stresses that

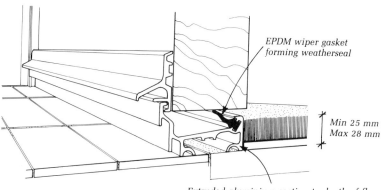

EPDM wiper gasket forming weatherseal

Min 25 mm
Max 28 mm

Extruded aluminium section to depth of floor finish screw fixed to sub-floor

Fig. 7.5 Aluminium threshold for inward-opening doors

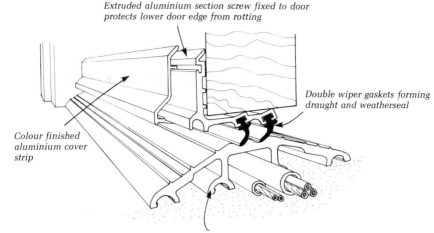

Extruded aluminium section screw fixed to door protects lower door edge from rotting

Double wiper gaskets forming draught and weatherseal

Colour finished aluminium cover strip

Extruded aluminium threshold forming wireway

Fig. 7.6   Aluminium threshold incorporating wireway

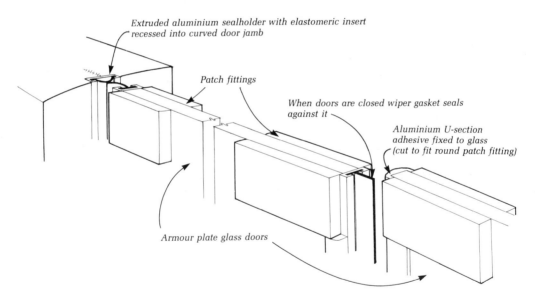

Extruded aluminium sealholder with elastomeric insert recessed into curved door jamb

Patch fittings

When doors are closed wiper gasket seals against it

Aluminium U-section adhesive fixed to glass (cut to fit round patch fitting)

Armour plate glass doors

Fig. 7.7   Weatherseals for glass doors (Pilkington Glass Ltd)

will vary according to its use and position: closing and opening, banging, slamming and impact from the passage of articles and people. Also a door must withstand stresses due to the variations in humidity that occur through changes in weather and environmental conditions within the building.

The strength of a door is dependent on its method of construction, and in the case of framed timber doors its strength is dependent on the joints used. Large sections of timber may impart general stability, but the jambs will have to withstand greater internal stresses due to moisture movement. Flush doors, on the other hand, depend upon

their total construction, since the facing forms a *stressed skin*.

Requirements for security vary according to a building's use and location. External timber door frames designed for security must be sufficiently robust to resist the door being forced open. Frame rebates should be at least 18 mm deep, preferably formed from solid timber, but if planted stops are used they should be glued and screwed into position and the fixing holes pelleted. External doors should be at least 44 mm thick and flush doors should be constructed with a solid core. Secure external doors are best designed as outward opening to prevent forcing of locks and rebates.

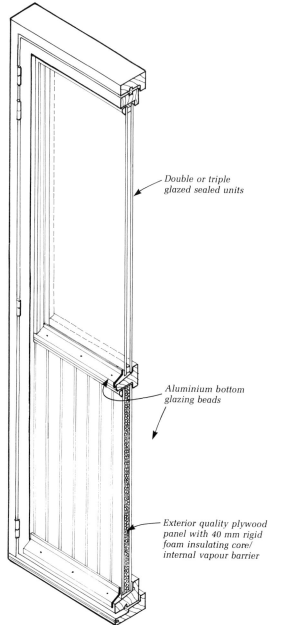

Fig. 7.8 Scandinavian insulated door that can be double or triple glazed

- Double or triple glazed sealed units
- Aluminium bottom glazing beads
- Exterior quality plywood panel with 40 mm rigid foam insulating core/internal vapour barrier

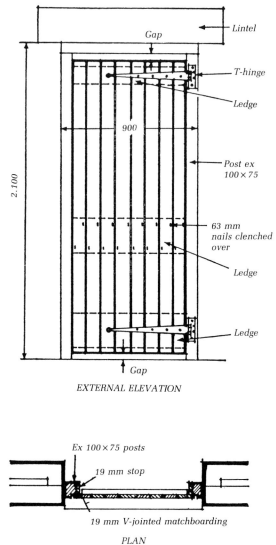

EXTERNAL ELEVATION

- Lintel
- Gap
- T-hinge
- Ledge
- Post ex 100 × 75
- 900
- 2.100
- 63 mm nails clenched over
- Ledge
- Ledge
- Gap

Ex 100 × 75 posts
19 mm stop
19 mm V-jointed matchboarding

PLAN

Fig. 7.9 Ledged and battened door: showing typical installation of unframed doors

It may be sensible, depending on the level of protection required, to use steel channel for the construction of frames with long anchor fixings back into the wall and for the door to be steel-faced on the outside.

Unlike windows where the ironmongery is usually an integral part of the assembly, door ironmongery is scheduled separately for fixing on site. The specification and fixing of the ironmongery have an important effect on overall security, for example, morticed door locks should not be located so as to weaken the framing of the door. Wide, heavy doors are best hung on spring-operated pivots rather than hinges, see *MBS: Internal Components* section 7.5. BS 8220 *Guide for security of buildings against crime* gives advice for different building types and reference should be made to *MBS: Internal Components*. Security of dwellings is also subject to guidance by the National House Building Council for dwellings covered by its warranty scheme.

7.3 Unframed doors

These are made from a number of vertical tongued and grooved V-jointed boards (known as *matchboarding*), which is held firm by horizontal members termed *ledges* that may be strengthened by diagonal members called *braces*. They are a traditional and much used method of construction for inexpensive exterior doors and for temporary doors. Additional strength can be given by extra framing, in the form of *styles* and a top *rail*. This type of construction has been proven over the years, having been employed particularly in industrial and agricultural buildings. Matchboard doors are only suitable for external situations where there is minimal temperature difference on either side of the door (i.e. to unheated buildings) since the doors being boarded on one side only will otherwise warp and twist.

7.3.1 Ledged doors

This is the simplest form of door, constructed from vertical tongued and grooved boards strengthened by horizontal ledges. *Ledged doors* are mostly used for temporary work since after a time, if subjected to heavy use, they tend to distort for lack of diagonal bracing. This type (shown in fig. 7.9) is not included within the provisions of the current British Standard.

BS 459: 1988 *Specification for matchboarded wooden door leaves for external use* specifies the quality, construction and sizes of *ledged and braced* doors, *framed and ledged* doors and *framed, ledged and braced* doors.

7.3.2 Ledged and braced doors

This is a more satisfactory form of construction since the diagonal braces prevent the distortion of the door. Figure 7.10 shows this type, which is often used for agricultural outbuildings. The matchboard which must be tongued and grooved and jointed on both sides should be not less than 15 mm thick and in the case of ledged and braced doors (see fig. 7.10(b)) there must be three horizontal ledges and two (diagonal) braces nailed or stapled together. Each board has to be fixed at least twice to each ledge and once to each brace, traditionally by clench nailing through the timbers. Alternatively the boards may be stapled to the ledges and braces with 18-gauge clenching staples a minimum of 30 mm long driven by a mechanically operated gun. The width of the matchboard (excluding the tongues) must be not less than 70 mm and no more than 115 mm. A ledged and braced door is made with the braces parallel, sloping diagonally upwards away from the hinged side of the frame (the brace works in compression but not in tension, if the door is hinged on the wrong side the joints open and the

door distorts). Instead the door can be constructed with the braces forming an 'arrowhead' allowing the hinges to be fixed on either side (fig. 7.10(d)). Since there is not enough thickness of timber on the edge of the door to accommodate screws for butt hinges, this type of door is hung on T-hinges fixed over the face of the boarding.

7.4 Framed doors

7.4.1 Framed and ledged doors

If the door is framed and ledged (unbraced), the rails must be through tenoned into the styles, with haunched and wedged tenons to top and bottom rails and further secured by adhesive. The framing can be made up by dowel jointing rather than mortice and tenon construction, provided that the dowels are at least 15 × 110 mm glued in place. The matchboarding is framed into the top rail and styles either by tongued joints or by rebating the boards a minimum of 12 mm. Figure 7.10(c) describes a framed and ledged door to BS 459.

7.4.2 Framed, ledged and braced doors

Framed, ledged and braced doors have styles, top, middle and bottom rails and two braces (arranged in either of the formats described above). The braces are cut to fit tightly into the corner formed by the junction of the styles and rails which are mortice and tenon jointed or dowelled together as described previously; an example is shown in fig. 7.10(a). The styles and top rail are the same thickness, which is equal to the bottom and middle rails plus the thickness of the boarding. The matchboarding is framed into the top rail and fixed to the face of the middle and bottom rails so as to shed rainwater. The door can be hung on strap or T-hinges (which are particularly suitable for larger doors, over 900 mm wide), but since there is an outer frame, narrow doors can also be hung on butt hinges. Reference should be made to *MBS: Internal Components* section 7.4. Large garage or warehouse doors are traditionally made using the same method of construction.

7.4.3 Framed/panelled doors

A *framed* or *panelled* door is a traditional form of joinery construction commonly used for the front doors of houses. Its success depends on the correct proportions of the framing, the use of good-quality well-seasoned timber and accurately made joints. The proportions of the panels and detailing generally has to be carefully considered in terms of the overall nature of the architecture. The door illustrated in fig. 7.11 has a composite elevation to illustrate the various forms of panel to be found in doors of this type.

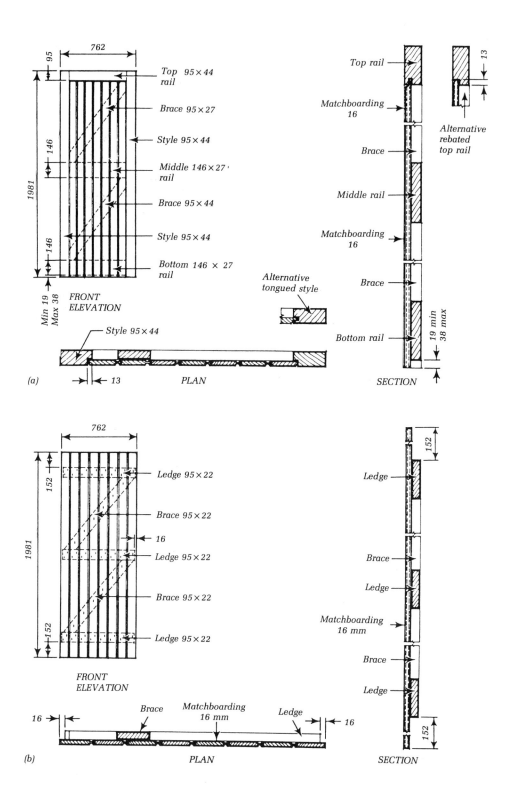

FRONT ELEVATION

762

95

1981

146

146

Min 19 Max 38

Top 95 × 44 rail

Brace 95 × 27

Style 95 × 44

Middle 146 × 27 rail

Brace 95 × 44

Style 95 × 44

Bottom 146 × 27 rail

Style 95 × 44

(a)

13

PLAN

Top rail

Matchboarding 16

Brace

Middle rail

Matchboarding 16

Brace

Bottom rail

SECTION

19 min 38 max

13

Alternative rebated top rail

Alternative tongued style

762

1981

152

152

FRONT ELEVATION

Ledge 95 × 22

Brace 95 × 22

16

Ledge 95 × 22

Brace 95 × 22

Ledge 95 × 22

16

Brace

Matchboarding 16 mm

Ledge

16

(b)

PLAN

Ledge

Brace

Ledge

Matchboarding 16 mm

Brace

Ledge

SECTION

152

152

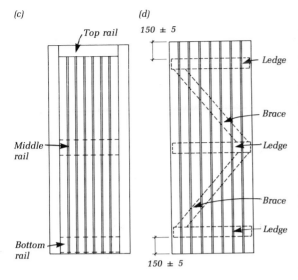

Fig. 7.10 (a) Framed, ledged and braced door; (b) Ledged and braced door; (c) Framed and ledged door; (d) Ledged and braced door with braces fitted arrow headed (All to BS 459:1988)

The horizontal rails are framed into the styles using various types of mortice and tenon joints (see *MBS: Internal Components* chapter 5 *Joinery*). The vertical middle rail or muntin is stub-tenoned into the horizontal rails. To prevent any tendency for the rails to deform, the styles are grooved and the tenons haunched into them.

The styles should not be too narrow, or there may be difficulty in fitting suitable door furniture; the bottom rail must be deep enough to allow adequate jointing. Excessive cutting away of the framing, particularly at the joints, in order to fit bolts and locks, will seriously weaken the construction. The location of door furniture and the method of framing should be designed together.

7.4.4 Standard panelled and glazed doors

Timber for framing and the plywood for panels should conform with BS 1186: *Timber for and workmanship in joinery*, Part 2 1988: *Quality of workmanship*. Exterior quality plywood to be used for external doors should be in accordance with BS 6566: 1985 *Plywood*.

Sizes The guide to the types and sizes of standard doorsets given in BS 4787: *Internal and external wood doorsets, door leaves and frames* Part 1: 1980 (1985) *Dimensional requirements* is shown in table 7.2. One of the advantages of timber doors is that they can be readily supplied to non-standard sizes particularly if required in quantity.

The framing of panelled doors can be either dowelled together or fixed by the use of mortice and tenon joints. Where dowels are used they should be of hardwood, minimum 16 mm diameter, equally spaced at not more than 57 mm centre to centre, with a minimum of three dowels in a bottom or lock rail and a minimum of two for the top rail. Where the framing is morticed and tenoned the doors

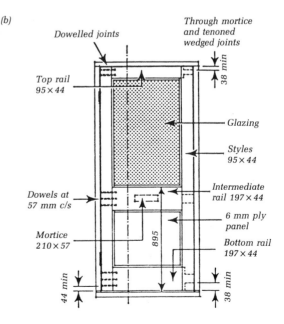

Fig. 7.11 (a) Panelled door; (b) Exterior glazed door

Table 7.2 Sizes of external doorsets and their component parts

Description	Size (mm)	Permissible deviation (mm)
Co-ordinating dimension: height of door leaf height sets	2100	
Co-ordinating dimension: height of ceiling height set	2300 2350 2400 2700 3000	
Co-ordinating dimension: width of all doorsets (S: single leaf set, D: double leaf set)	900 S 1000 S 1200 D 1500 D 1800 D 2100 D	
Work size: height of door leaf height set	2095	± 2.0
Work size: height of ceiling height set	2295 2345 2395 2695 2995	±2.0

Source: BS 4787: Part 1 1980, chapter 7, table 2

Photo 7.2 Glazed doors at the Bracknell Forest Heritage Centre. Architects: Andris Berzins and Associates, 1989. The details of these glazed external doors are shown in fig. 7.12.

must have through haunched and wedged tenons to top and bottom and one other (middle) rail. If there are more intermediate rails these are stub-tenoned (minimum 25 mm) into the styles.

Solid panels are commonly made from plywood framed into grooves to fit tightly, the panels being cut to fit, 2 mm less in width than the grooved opening. Glazed panels in exterior doors require the opening to be rebated out of the solid one side; for glazing on site mitred glazing beads are loosely pinned in position for delivery (see photo 7.2). If external doors are to be double-glazed this may require them to be thicker than might otherwise be necessary. Internal glazing beads inherently provide greater security than external beads but reference should be made to section 3.3 for a discussion of their relative merits in terms of weatherproofing. Glazed openings in doors and glazed side-panels may be subject to the requirements of Part N of the *Building Regulations 1991* depending on the height to the bottom edge of the glass (see fig. 3.18).

The adhesive used must comply with BS 5442: Part 3 1979 (1990) *Adhesives for use with wood*. Allowance should be made for a manufacturing tolerance of 2 mm on the heights and widths of the finished sizes of component parts. The diagrammatic form of each of the joints mentioned is discussed in *MBS: Internal Components* chapter 5 *Joinery*.

Furniture External doors should be fitted with one and a half pairs of butt hinges to BS 1227 *Hinges* Part 1A: 1967 *Hinges for general building purposes*; locks and/or latches to BS 5870: 1980 *Specification for locks and latches for doors in building*, and BS 3621: 1980 *Specification for thief resistant locks*; and letter plates to BS 2911: 1974.

Finish If the doors are delivered unprimed they should be handled carefully and stored in dry conditions. Knotting and priming should be carried out as soon as possible and before fixing in position.

Doors supplied with a primer of paint or stain should have the base coat made good and the finish coat applied as soon as possible after exposure to the weather. To increase protection against the weather during construction it may be advantageous to apply at least one of the finishing coats before the door is hung. The choice of dark-coloured paints and stains, depending on the orientation, results in surfaces absorbing much more of the sun's heat than if finished in a lighter colour. This can shorten the life of finishes and result in the drying out and shrinkage of timber

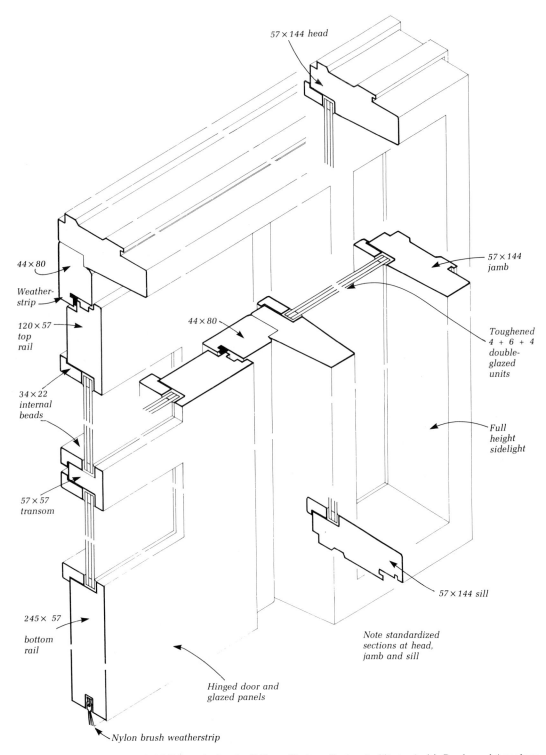

57 × 144 head

57 × 144
jamb

44 × 80

Weather-
strip

120 × 57
top
rail

44 × 80

Toughened
4 + 6 + 4
double-
glazed
units

34 × 22
internal
beads

Full
height
sidelight

57 × 57
transom

57 × 144 sill

245 × 57

bottom
rail

Note standardized
sections at head,
jamb and sill

Hinged door and
glazed panels

Nylon brush weatherstrip

Fig. 7.12 Details of glazed doors and sidelight at the Bracknell Forest Heritage Centre, Architects: Andris Berzins and Associates 1989. See photo 7.2

doors. Clear finishes on timber have limited durability if directly exposed to the sun and weather; translucent finishes containing some pigment have a longer life. Figure 7.11 illustrates typical elements for the construction of a standard exterior glazed door and fig. 7.12 and photo 7.2 show an example of the detailing of purpose-made timber glazed entrance doors.

Installation   It has been usual practice for door frames to be built into a masonry wall to ensure a good fit at the jambs and the right size for the doors which are hung later. In this case the frame must be temporarily braced to ensure that it remains plumb and square. The time delay between installing the frame and applying the final paint or stain finishes should be kept as short as possible. This method is satisfactory only for general-purpose joinery and not for frames which are to be clear finished or assemblies where a higher standard of performance is required.

The practice in the UK of specifying doors, frames, ironmongery, etc. separately for assembly on site has mitigated against the production of combined components able to achieve prescribed standards of performance. To overcome this, manufacturers are now producing doorsets comprising factory-hung doors complete with furniture, the whole component having been tested for weather resistance, security, etc. Rather than the traditional practice of building frames in as work proceeds, doorsets have to be installed as second-fix components with the size of structural opening being determined by the use of templates. As in the case of windows (see fig. 4.21), some makers provide a template that is built in as a permanent element of the construction and forms a lining to the frame when it is fixed into the pre-finished opening.

7.5 Flush doors

Although more usually associated with internal applications, flush doors are made as an economical form of external door usually faced with external grade plywood to BS 6566: *Plywood*, Parts 1 to 8: 1985. A requirement for security

Table 7.3 Manufacturer's range of external flush doors (John Carr Ltd)

Thickness	Dimensions
$1\frac{3}{4}''$ (44 mm)	$6'6'' \times 2'3''$ (1981 $\times$ 686 mm)
	$6'6'' \times 2'6''$ (1981 $\times$ 762 mm)
	$6'6'' \times 2'9''$ (1981 $\times$ 838 mm)
	$6'8'' \times 2'8''$ (2032 $\times$ 813 mm)
44 mm (Metric range)	2040 $\times$ 762 mm
	2040 $\times$ 826 mm
	2000 $\times$ 807 mm

will necessitate the use of a laminated solid core door as noted above (section 7.2.6). A typical range of standard exterior flush doors is shown in table 7.3. For the construction of flush doors see *MBS: Internal Components*, chapter 6.

7.6 Glazed doors

In addition to the standard panelled glazed door previously described, purpose-made glazed doors are much used at entrances to public buildings. Part M of the *Building Regulations 1991* specifies widths of doors, sizes of lobbies and the extent of vision panels in entrance doors that will make them suitable for use by disabled people.

Entrance doors are likely to be heavily used so appropriately robust construction is essential. Figure 7.13 shows, as an example, one of a pair of aluminium-framed glazed doors forming part of a glazed entrance screen. Aluminium doors often have steel reinforced welded corners to resist the twisting and racking to which the doors

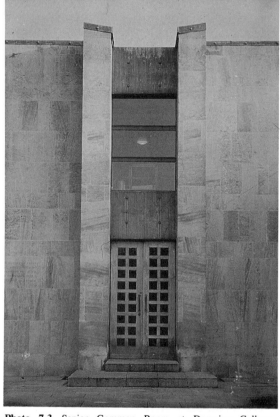

Photo 7.3 Senior Common Room at Downing College, Cambridge. Architects: Howell Killick Partridge and Amis, 1970.
The standard method of timber panelled door construction can readily be detailed to accord with the formal intentions of the design of the building (see also photo 7.4).

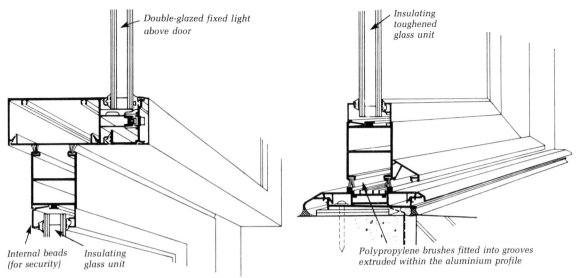

Double-glazed fixed light above door

Insulating toughened glass unit

Internal beads (for security)   Insulating glass unit

Polypropylene brushes fitted into grooves extruded within the aluminium profile

Fig. 7.13 Glazed external doors

will be subject in use and are tested to withstand repeated slamming. The hollow aluminium sections readily accommodate a variety of security locking devices and panic release bolts.

Since this type of door will be in constant use it is sensible to use an overhead or floor spring as a means of controlling the movement. Springs can be fitted with a device which checks the doors open at 90 degrees (reference should be made to *MBS: Internal Components* chapter 7).

Figure 4.44 shows an example of a glazed external door and frame fabricated from welded Jansen hollow steel section. Stainless steel rolled or pressed sections and stainless steel clad frames are also commonly used.

7.7 Glass doors

Toughened plate glass doors are often incorporated into fully glazed frontages to shops and showrooms, the glass panels being attached to one another by patch fittings (see fig. 6.4). Special patches are used to pivot the doors at their top corners, and the speed at which the doors close is controlled by an adjustable double-action floor spring. A special patch fitting incorporating a lock is used at the top centre of the doors. A continuous rail at the base of the door forms a kickplate and incorporates a double cylinder lock engaging with a keeper set into the floor, as well as a spigot recess to engage the floor spring spindle.

The obvious feature of this type of construction is its transparency but that can entail problems of safety, for example, it is sometimes difficult to know whether a door is open. For this reason a requirement has been introduced into Part N of the *Building Regulations 1991* for fully glazed

constructions of this type to be made evident by a 'manifestation' attached to the glass. This can take the form of broken or solid lines, patterns or a company logo stuck to the glass and centred at 1500 mm above floor level. As an alternative, in this instance, the use of large and obvious handles or push plates to the doors would obviate the need for any other marking.

Photo 7.4 St. Hilda's Oxford. Architects: Alison and Peter Smithson, 1969.

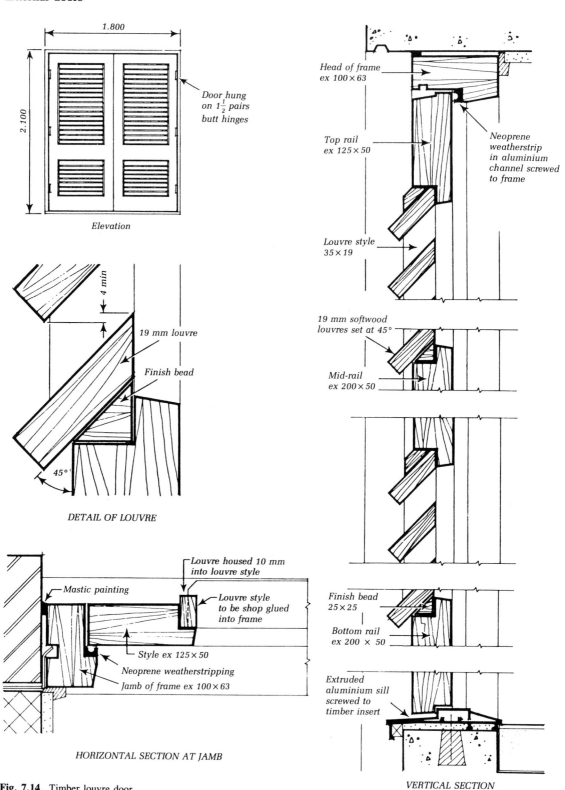

1.800

2.100

Door hung
on 1½ pairs
butt hinges

Elevation

Head of frame
ex 100×63

Top rail
ex 125×50

Neoprene
weatherstrip
in aluminium
channel screwed
to frame

Louvre style
35×19

4 min

19 mm louvre

Finish bead

45°

DETAIL OF LOUVRE

19 mm softwood
louvres set at 45°

Mid-rail
ex 200×50

Louvre housed 10 mm
into louvre style

Mastic painting

Louvre style
to be shop glued
into frame

Style ex 125×50

Neoprene weatherstripping

Jamb of frame ex 100×63

HORIZONTAL SECTION AT JAMB

Finish bead
25×25

Bottom rail
ex 200 × 50

Extruded
aluminium sill
screwed to
timber insert

Fig. 7.14   Timber louvre door

VERTICAL SECTION

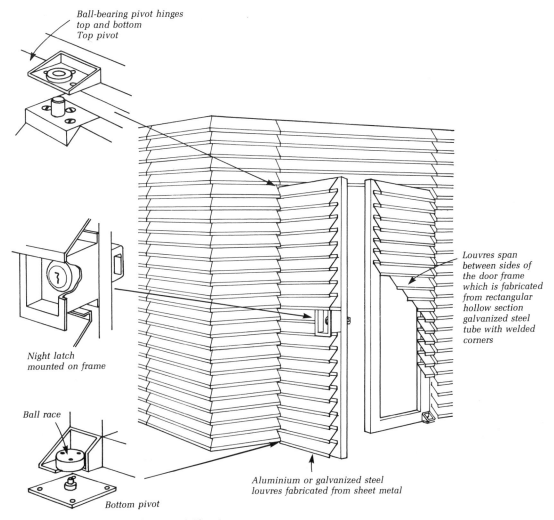

*Ball-bearing pivot hinges
top and bottom
Top pivot*

*Louvres span
between sides of
the door frame
which is fabricated
from rectangular
hollow section
galvanized steel
tube with welded
corners*

*Night latch
mounted on frame*

Ball race

*Aluminium or galvanized steel
louvres fabricated from sheet metal*

Bottom pivot

Fig. 7.15   Metal louvred door (Greenwood Airvac)

7.8 Louvred doors

This is a form of door which is used externally for the ventilation of plant and storage rooms; they are available made from both pressed metal and timber. In the timber example shown in fig. 7.14 the louvres are housed into a vertical louvre style which is then shop glued within the panel formed by the main framing of the door. The edges of the louvres, because of their projection in front of the face of the door, are chamfered off at the corners. The louvre slats are set at 45 degrees and fixed so that there is a minimum overlap of 3 mm. The door is shown closing against an extruded aluminium threshold. An alternative made from pressed metal sections is shown in fig. 7.15.

7.9 Flexible doors

Flexible doors are used in industrial buildings, warehouses and hospitals. They are designed to be pushed open (i.e. not operated by hand), where for example the user is driving a fork-lift truck or trolley or carrying bulky packages. The doors are made from canvas- or hessian-reinforced rubber with clear plastic or toughened glass vision panels clamped into position by steel flats bolted together on either side of the door. Alternatively, for less heavy duty applications, the doors can be completely transparent if made from 6 or 9 mm thick clear PVC. Where the goods to be moved through a building are fragile, flexible doors can be fitted with a pneumatic drive to open the doors automatically,

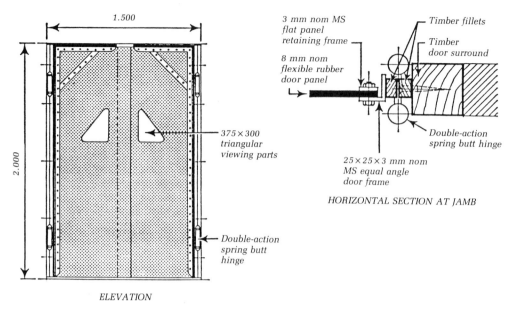

ELEVATION

HORIZONTAL SECTION AT JAMB

Fig. 7.16 Lightweight flexible rubber doors

but generally they are used as a comparatively inexpensive alternative to automatic doors.

They are designed to open on impact, the flexibility of the sheeting taking the force out the 'collision', and allowing traffic to pass through. The doors close automatically; they are controlled by double-action vertical springs which may be concealed within the tubular steel frame. The door illustrated in fig. 7.16 is a lightweight type, sturdier doors are available in sizes up to 6.0 m in height by 4.5 m in width. The heaviest doors are suitable for openings which are used by lorries and other industrial transport. The size of the welded steel tube frame varies according to the size and duty of the door, up to 90 mm OD for the largest and heaviest applications. The frame supports the top edge and side of the doors, the corner angle having a reinforcing gusset welded into the junction. The flexible sheet is clamped into position between steel flats welded to the tube frame and bolted together through holes in the doors.

7.10 Cast doors

Cast bronze or aluminium doors are still occasionally used at the entrance to the most important public buildings. These are not available as stock items; their design requires initial discussion with a foundry to determine the feasibility of the sizes, geometry and sections to be employed. The accompanying illustration (photo 7.5) shows an example of double swing doors constructed in cast aluminium.

Photo 7.5 Cast aluminium doors at the South Bank Arts Centre, London. Architects: GLC Architects' Dept., 1967.

These and the aluminium windows throughout the South Bank Centre are hollow castings that are joined together by internal bolts — access holes were formed in the cast sections so these connections could be made. The access holes are covered over by screw-fixed cover plates.

7.11 Sliding and sliding/folding doors

Sliding door gear can be used to close off sizeable openings. They are consequently much used for industrial applications. Sliding doors will work satisfactorily throughout their life provided that the moving parts of the doors are regarded as machinery and are regularly maintained. The more complex the mechanism the greater the chance of failure, a straight sliding single-leaf door is

the simplest and cheapest but takes up most space when open. Increasing the number of leaves saves space but complicates the mechanism. Folding systems take up the least space of all, but require the most complex hanging gear. The weight of the doors can be taken on wheels or rollers at the base or alternatively suspended from hangers on a track at the head of a door. The choice is made according to the weight and construction of the doors and the load-bearing capacity of the structure across the opening.

7.11.1 Straight track sliding: single leaf

The example shown (fig. 7.17) is a heavy sliding door, framed and diagonally braced, using welded steel angle and faced either in timber or steel sheet. It is mounted on bottom rollers, one at each end of the door, which is a suitable arrangement where hanging the weight of the door is not structurally feasible. This method can consequently be used for the most substantial of openings, up to 20 m in height and for doors up to 8000 kg in weight. Large heavy doors can be motor driven or operated by a geared chain wheel and winding handle. The rollers are zinc plated machined-steel wheels fitted with ball-bearings, they run on a top hat section, cold-rolled galvanized steel rail set into the concrete floor slab. The mounting brackets for the rollers are welded or bolted into the angle frame of the door.

Bolted to each end of the head of the door is a nylon or steel roller, adjustable in height to locate within the galvanized steel guide channel that is bracketed back to the

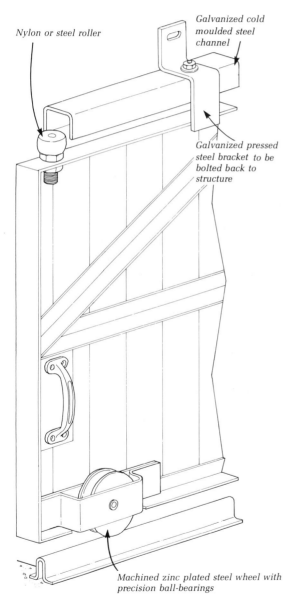

Nylon or steel roller

Galvanized cold moulded steel channel

Galvanized pressed steel bracket to be bolted back to structure

Machined zinc plated steel wheel with precision ball-bearings

Fig. 7.17   Straight track sliding, single-leaf doors

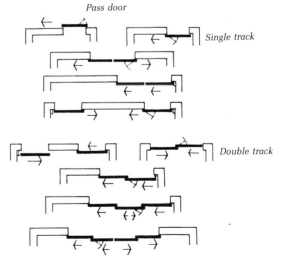

ALTERNATIVE LAYOUTS

Pass door

Single track

Double track

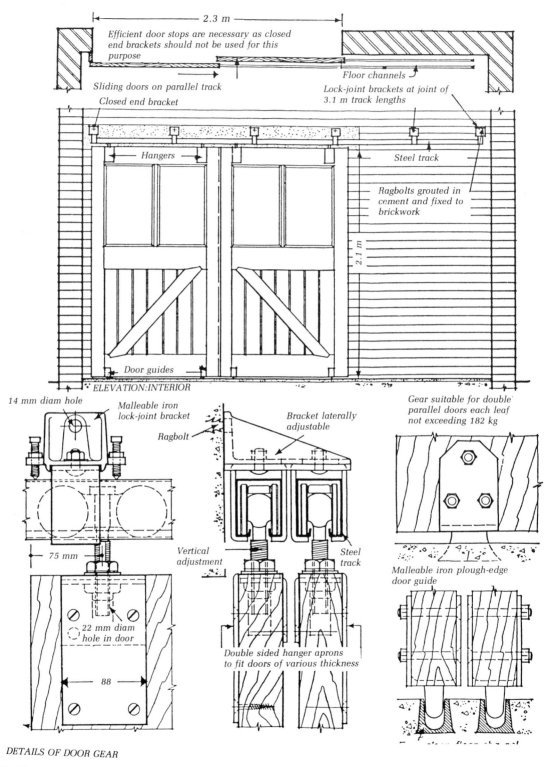

DETAILS OF DOOR GEAR

Fig. 7.18 Straight track sliding, double-leaf doors

beam spanning the opening. This gear can also be used for doors with leaves sliding past one another, a double guide channel and larger brackets being used above the doorway. The details at the bottom and sides of the door are vulnerable to the weather and, where used externally, a canopy or overhang should be provided as protection against driving rain.

7.11.2 Straight track sliding: double leaf

Figure 7.18 shows a simple two-leaf arrangement for garage-type doors suspended from a double top track. Each leaf requires two hangers each having four wheels and running in a box-section cold-formed galvanized steel guide that has to be kept greased. This track is supported on brackets of cast aluminium alloy or pressed steel. The heavier the door the larger the track, the biggest being 104 × 73 mm and capable of supporting doors weighing up to about 2000 kg. There should be about 13 mm clearance between the doors and between the door and the wall. The bottom of the doors is held in place by a brass roller mounted on a spindle running in a steel channel let into the floor. The design of the track is to prevent outward movement and so avoid any tendency for the doors to jam. The bottom rollers shown here incorporate curved guards

that 'plough' out any dirt that gets into the channel rather than letting it be pressed down, but there are many alternative designs for track and channel guides. The leaves of the door may be locked when shut by the use of a bolt at the floor and an outside fastening such as a locking bar or jamb bolt used with a padlock; special cylinder locks are also available. In the example shown, a wooden fastening post is screwed to the wall to close the gap between wall and inner door when in the closed position. The clearance required between the faces of the straight

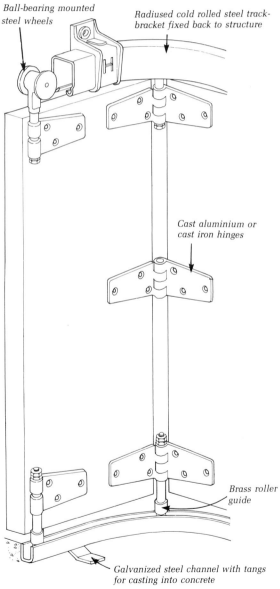

Ball-bearing mounted steel wheels

Radiused cold rolled steel track-bracket fixed back to structure

Cast aluminium or cast iron hinges

Brass roller guide

Galvanized steel channel with tangs for casting into concrete

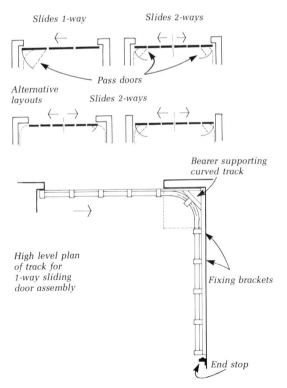

Slides 1-way

Slides 2-ways

Pass doors

Alternative layouts

Slides 2-ways

Bearer supporting curved track

High level plan of track for 1-way sliding door assembly

Fixing brackets

End stop

Fig. 7.19 Curved track sliding, sectional doors

track doors can cause problems of draught-proofing and weathering; an external canopy provides a useful degree of protection.

7.11.3 Curved track

If a curved section of track is introduced at one end of a door opening (see fig. 7.19) the doors can be pushed round

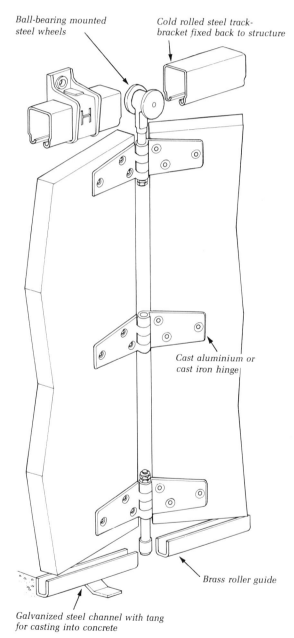

Ball-bearing mounted steel wheels

Cold rolled steel track-bracket fixed back to structure

Cast aluminium or cast iron hinge

Brass roller guide

Galvanized steel channel with tang for casting into concrete

Fig. 7.20 Straight track folding doors

a 90 degree corner to be stored against a side wall (provided the wall is of sufficient length). The doors are made in short hinged sections which fit closely to each other and can consequently have rebated styles for weathering. The hangers and guides are similar to those described previously but are combined with back-flap hinges joining the sections of door together. The hangers fixed at the ends of each section have only single pairs of steel wheels in order to traverse the curved section of track, only single track being possible. The width of the doors should be restricted to a maximum of 800 or 900 mm; the minimum curve of track for light doors is in the region of 600 mm. For smooth operation no more than five door leaves should be used and no fewer than three. The end leaf can be free swinging for use as a pass door and can be secured with an ordinary cylinder lock to a rebated door post (which will also lock the whole sliding door assembly). This makes a convenient arrangement where the doors are used mainly for pedestrian traffic and only occasionally opened for vehicular access. The curved length of track must be supported across the corner, and this is most simply done by packing out a short straight length of timber bearer at the correct angle. A finished door thickness of approximately 45 mm is the most suitable in either timber or metal construction. The doors are located at floor level in the bottom channel by means of adjustable brass rollers.

7.11.4 Folding

Sliding/folding doors (fig. 7.20) operate in much the same way as the curved track system described above. The track, hangers and guides are all similar in principle. Not more than six leaves should be used in one door, but a door

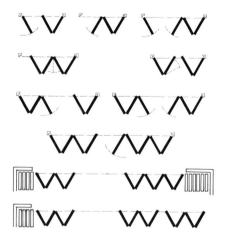

ALTERNATIVE LAYOUTS/COMBINATIONS

opening can be divided in the centre, with the folded doors bunching at both ends. The length of wall available for the doors to stack has to be calculated as sufficient in order not to obstruct the opening. The doors should be of robust construction and not less than 45 mm finished thickness. If they are made from timber, the top rail should be at least 150 mm deep to allow adequate fixing for the hangers. In the case of very large garage or warehouse doors and doors for industrial use, greater thickness and strength are often necessary. The maximum possible height using this method is 4.5 m, each leaf weighing not more than 80 kg. If the range of sliding/folding doors is designed with an odd number of leaves, the free end of the last leaf can be allowed to swing and be used as a pass door. Alternatively a single door can be hung independently on hinges fixed to a rebated timber jamb at one end of the door opening. Back-flap hinges are usual but will be seen at alternate joints on the outside elevation, and if this is to be avoided then 100 mm butt hinges can be used instead.

The systems described above for horizontal sliding/ folding doors are components designed for attachment on site to doors that have been supplied separately. Illustrated in figs 7.21 and 7.22 is a high-performance aluminium sliding/folding door system intended for industrial use. Individual door segments are made with hollow aluminium extrusions, forming styles and rails defining panels that may be insulated sandwich panels, louvres, polycarbonate or glass, single- or double-glazed. The extrusions can be thermally broken by a GRP spacer that is rolled at the factory, into the section. The meeting edges and the top horizontal edge of each door leaf is fitted with a hollow EPDM profile that weatherseals the joint when closed, but safeguards against fingers being trapped between the doors. The junction at the floor is sealed with either an EPDM compression gasket or a brush-type weatherstrip. Zinc die-cast hinges are screw-fixed into steel reinforcing plates within the aluminium extrusions. The doors are hung from a steel track attached by bolted brackets to the lintel above the opening, the doors run on rollers mounted on ball-bearings. The door leaves are secured at top and bottom with espagnolette bolts.

7.12 Folding shutters

Folding shutters are made from narrow, vertically hinged steel slats, suspended above each hinge from a roller that slides within a horizontal track above the door opening. To maintain the verticality, so that the doors do not distort when being operated, the internal hinged edges are braced by a lattice of diagonal steel hinged braces. Glazed panels can be incorporated within the width of the slats. Folding shutters are also used to form fire separations within buildings, and can achieve up to 4 hours' fire resistance (see *MBS: Internal Components* chapter 6 *Internal doors*).

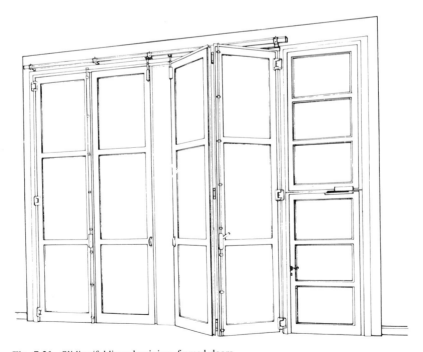

Fig. 7.21 Sliding/folding aluminium framed doors

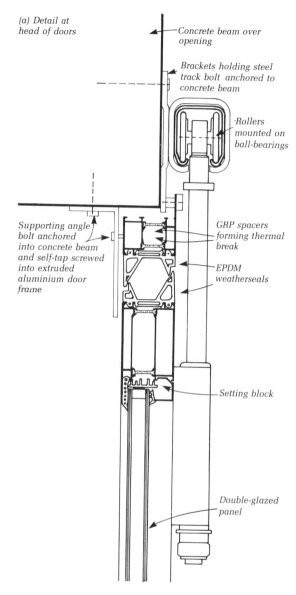

(a) Detail at head of doors

— Concrete beam over opening

— Brackets holding steel track bolt anchored to concrete beam

— Rollers mounted on ball-bearings

Supporting angle bolt anchored into concrete beam and self-tap screwed into extruded aluminium door frame

GRP spacers forming thermal break

EPDM weatherseals

— Setting block

Double-glazed panel

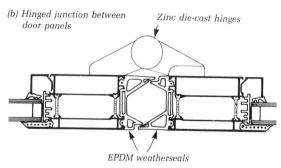

(b) Hinged junction between door panels

Zinc die-cast hinges

EPDM weatherseals

Fig. 7.22   Details of sliding/folding aluminium framed doors

7.13 Overhead doors

This type of door (see photo 7.6) commonly known as an 'up and over door' swings to open at a horizontal position overhead (see fig. 7.23); they are particularly useful for garages and wherever floor space is restricted. The doors can be constructed from a steel or aluminium frame clad with timber weatherboarding or metal cladding or framed and panelled traditional timber construction or GRP; they can incorporate windows and pass doors. The maximum size opening for a door of this type is in the region of 5.0 m wide and 2.2 m high. As the door moves from the vertical to the horizontal position, nylon wheels which are fitted to the top corners of the door run horizontally along the overhead cold-formed steel guide track, the action of the door being balanced by special torsion springs (as shown in fig. 7.24). For convenience, overhead doors are often driven by an electric motor that can be activated by remote control.

7.14 Sectional overhead doors

These operate in the same way, but rather than moving in one piece the doors are divided into horizontal hinged panels. They slide vertically and have the advantage of not projecting beyond the face of the building when in operation. Although there are no *Building Regulations*

Photo 7.6   House at Long Wittenham, Oxfordshire. Architect: Michael McEvoy, 1988.
Steel framed up and over doors can be faced in a variety of materials, in this case vertical tongued and grooved pine boarding.

requirements for conservation of energy from unheated spaces such as garages, where the garage is built into a dwelling, thermal performance is improved by using insulated doors. Sectional overhead doors are commonly made in composite construction with inner and outer profiled steel or pressed aluminium sheet skins separated by an insulating foam core of injected polyurethane foam insulation ($U = 0.3$ W/m^{2}K for a door 42 mm in thickness, see chapter 1 for specification of insulation materials in contemporary construction). Windows, if required, are cut into the panels and gasket glazed.

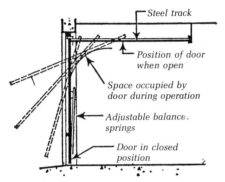

Fig. 7.23 Overhead door — space requirements for operation

In the example shown in fig. 7.25 the inner and outer metal surfaces are separated at the edges of the panels by a thermal break gasket that also forms a weatherseal between the joints when the door is closed. The top and bottom edges of the doors are fitted with PVC weatherstrips, retained within aluminium extrusions, to seal around the opening and provide a variable joint with the floor (which in garages, etc. is likely to be uneven). Galvanized steel sections are welded within the composite panels to reinforce them at locks and hinges.

Doors of this sort are very widely used in industry, for example as loading-bay doors providing access for loading or unloading into or out of a building. They need to be easily operable from both inside and outside but must also provide security and weather control as well as a certain amount of thermal insulation. Many modern buildings used for industrial processes or storage have the outside wall flush with the face of a loading platform raised above the level of the ground outside, to allow ease of unloading. The necessity for loading-bay doors to be of sufficient width to coincide with the sliding doors of a lorry often makes it necessary for a conventional size pass door to be incorporated within the large access doors into the building. Also it is sometimes desirable to provide a compressible docking pad fixed to the external face of the building around

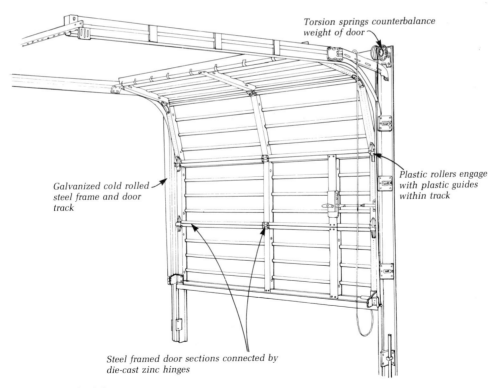

Fig. 7.24 Sectional overhead door

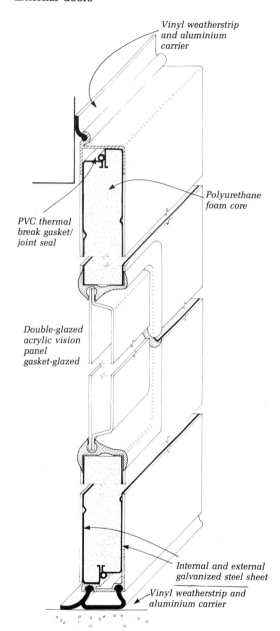

Vinyl weatherstrip
and aluminium
carrier

Polyurethane
foam core

PVC thermal
break gasket/
joint seal

Double-glazed
acrylic vision
panel
gasket-glazed

Internal and external
galvanized steel sheet

Vinyl weatherstrip and
aluminium carrier

Fig. 7.25 Sectional overhead door constructed from insulated composite panels

Photo 7.7   Translucent GRP industrial doors (see section 7.14). Photo courtesy of Envirodoor Markus Ltd.

Photo 7.8   Bi-part folding sectional overhead doors (see section 7.14). Photo courtesy of Senior Construction Services Ltd.

the loading-bay doors so that a lorry can back on to it and provide a temporary weather-proof seal when both the lorry and the building doors are open (fig. 7.26).

A variety of materials and types of sectional overhead doors are manufactured. They can, for example, be constructed from aluminium framing infilled with transparent or insulating panels similar to the horizontal sliding/folding doors described above. The system shown

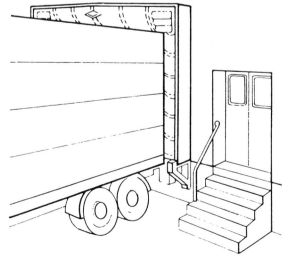

Fig. 7.26   Loading bay door

(photo 7.7) comprises translucent double-skin GRP panels glazed into aluminium frames providing a high level of illumination and a moderate degree of thermal insulation. The panels are formed in thicknesses up to 150 mm which achieve a U-value of 0.5 W/m^{2}K and can be made into doors up to 15 m in length. Clear vision windows can be cut into the double-skin GRP and held in place by gaskets. These electrically operated doors lift each panel individually on chains. As the door opens the panels are compactly stacked behind one another above the head of the structural opening.

Bi-part folding doors (see photo 7.8) are divided horizontally into two hinged panels. When the doors are lifted, ball-raced steel rollers on spindles, fixed at the centre line of each panel, slide within a vertical steel track. When the door is fully open the panels are horizontal, projecting half in and half out of the door opening, no overhead track being required. The doors are held in position by cast-iron counterbalance weights suspended alongside the vertical tracks and concealed within a cover panel. These doors can be power operated and are available in widths up to 7.3 m; pass doors can be incorporated within the height of the lower panel.

7.15 Roller shutters

Roller shutters are made from interlocking hinged steel slats that can be plain hot-dip galvanized or plastisol coated, they are available in sizes up to 5.0 × 5.0 m. Large shutters are normally operated by a chain hoist. The slats roll up into a drum above the head of the door opening, and windows can be incorporated to sizes modular with the size of the slats. Pass doors, if required, are hung independently within a steel frame at one side of the door opening, the roller shutter is cut to fit round the door, and both the pass door and its frame are hinged so they can be swung out of the way leaving a clear opening.

The roller slats can be insulated by bonding polyurethane foam with an internal lining of aluminized plastic film to the inner face of the doors. Steel roller shutters are also extensively used to close off openings in fire separation walls within buildings, they can achieve a fire rating of up to 4 hours' fire resistance when tested to BS 476: Part 8: 1972.

7.16 Fabric doors

Vertical sliding doors are also available, made from flexible fabric such as nylon-reinforced neoprene rather than solid panels. These doors are power operated, and either roll the fabric similarly to domestic roller blinds or lift horizontal sections of the fabric furled in pleats (fig. 7.27). Some types are made with a double thickness of fabric separated by

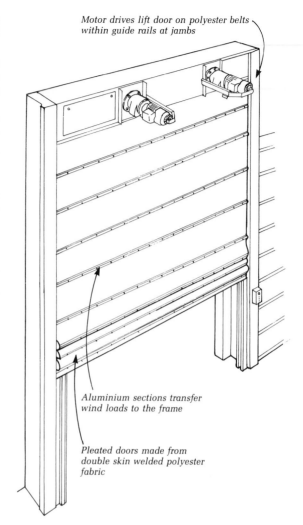

Motor drives lift door on polyester belts within guide rails at jambs

Aluminium sections transfer wind loads to the frame

Pleated doors made from double skin welded polyester fabric

Fig. 7.27 Pleated fabric, vertical sliding doors (Crawford Megadoor)

an air gap to give some thermal insulation. Being light in weight, fabric doors can be to a considerable size and are relatively quick in operation. The material is kept taut either by horizontal metal sections jointed within the fabric which span between the vertical guides at the jambs, or by the the fabric being held in tension by the guides (see photo 7.9).

7.17 Revolving doors

Since this type of door requires very particular fittings and mechanism, it is manufactured by specialist firms. Revolving doors are manufactured in a variety of different formats and sizes (fig. 7.30 and table 7.4). The curved casing to the doors can be made from a steel or aluminium

Photo 7.9 A fabric industrial door, Crawford's Megadoor (see section 7.16). Photo courtesy of Crawford Door Ltd.

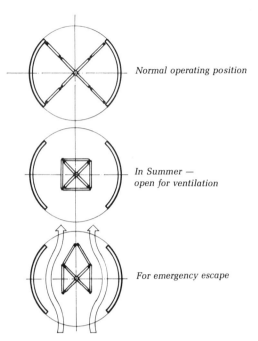

Normal operating position

*In Summer —
open for ventilation*

For emergency escape

Fig. 7.28 Revolving doors — alternative modes of operation (Grothkarst Ltd)

framework faced on both sides with sheet metal, or alternatively can be made from curved glass with a minimum of framing. The roof of the cylinder is constructed in the same way. There are various patent methods of collapsing the doors so that direct access is possible in emergency or for summer ventilation (fig. 7.28). The leaves are always made to collapse in an outward direction in an emergency. The door wings are made from security glass glazed into a steel or aluminium frame with dust-seal brush strip fixed at the edges (fig. 7.29). Four compartment doors usually have a minimum internal diameter of 1.8 m. Where space is limited a three-compartment revolving door can be used with a minimum diameter of 1.5 m. Small revolving doors do, however, create a hazard for disabled people and those with sight impairments, so the provision of revolving doors should be augmented by conventional swing doors as required by Part M of the *Building Regulations 1991*. Revolving doors can be operated by card locks and other security systems.

Much larger revolving doors are made for buildings such as supermarkets that have a constant flow of traffic. They have a central revolving drum and some are made oval rather than circular to increase the throughput of people. Diameters up to 4.8 m are possible, these larger doors being driven by a motor housed above the roof or below the floor of the revolving door. Motor-driven revolving doors may be intended for wheelchair access, so as required by section M of the *Building Regulations 1991*, they should be fitted with a sensor which stops the door if it encounters any resistance.

Of a related type and often made by the same manufacturers are circular sliding doors. In this case the cylindrical drum acts as a draught lobby between two sets of radiused glass doors that slide into the side walls. The doors are switched on and off by sensors and are motor driven to open automatically.

7.18 Automatic control of doors

Various types of pneumatic and hydraulic equipment are used to open and close sliding doors automatically. Fully automated systems are activated by a sensing device such as a push-button, sensitized mat, radar, infra-red or photoelectric beam. This initial sensor activates a timing device connected to the motorized system which causes the doors to move. Alternatively the motor can be operated by remote control from a central point such as a security cabin. The timing apparatus can vary from a simple cut-out to a complex electronic control with programmed instructions to vary the time delays for closing and opening doors as circumstances or security require. It is essential that all automatic control devices allow the doors to be moved by hand in the event of a power failure. The motor gear usually takes its initial power from electricity which is then used to generate hydraulic or pneumatic pressure to move the doors as signalled. The equipment should slow the doors down as they close to avoid the doors jarring and causing 'rebound shock' and also repeat the cycle of opening and closing should the doors meet an obstruction.

Automatic doors, whether swing doors or sliding, may be required in hospitals, hotels, shops or offices. Swing

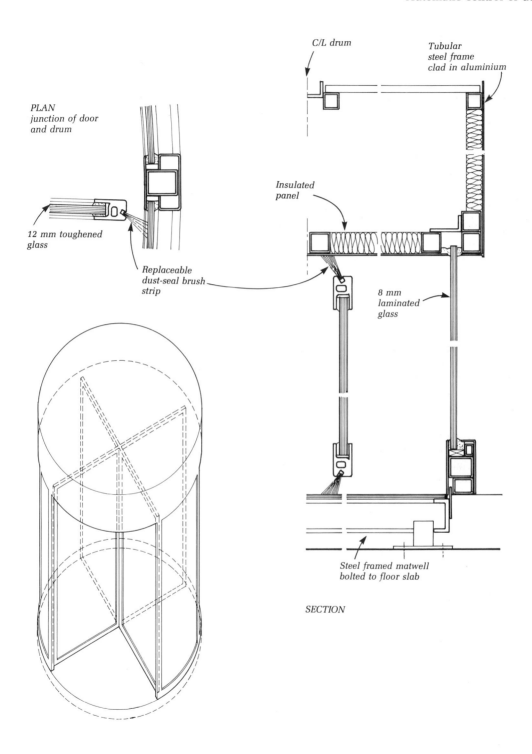

PLAN
junction of door
and drum

12 mm toughened
glass

Replaceable
dust-seal brush
strip

C/L drum

Tubular
steel frame
clad in aluminium

Insulated
panel

8 mm
laminated
glass

Steel framed matwell
bolted to floor slab

SECTION

Fig. 7.29 Typical revolving door construction (Grothkarst Ltd)

Table 7.4

Four-wing revolving doors

Interior diam (mm)	Entrance width (mm)
1800	1220
1850	1260
1900	1295
1950	1320
2000	1350
2050	1380
2100	1420
2150	1460
2200	1500
2250	1520
2300	1550
2350	1580
2400	1620
2450	1660
2500	1690
2600	1765
and so on	

Three-wing revolving doors

Interior diam (mm)	Entrance width (mm)
1500	695
1550	750
1600	775
1650	800
1700	825
1750	850
1800	875

doors have a master control with a delay to set the time that the door is to be held open and a regulating resistance for setting the speed of movement of the sliding doors. The leaves are driven electromechanically by a driving wheel attached to a moving rail at the top of each leaf. The rail regulates the width of opening, brakes the door before the end of its travel and disconnects the supply of electricity.

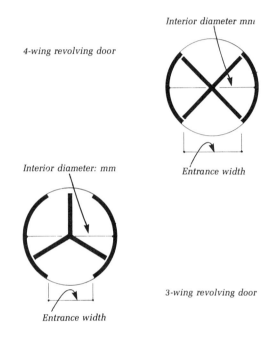

Fig. 7.30   Manufacturer's range of revolving doors (Grothkarst Ltd)

8 Roofings

8.1 Introduction

Resolution of the aesthetic and technical considerations impinging of the design of roofs (i.e. both the supporting structure and the roof finishes), is an obviously essential aspect in achieving the satisfactory overall performance of buildings (see photo 8.1). Structural behaviour and roof structure are considered in *MBS: Structure and Fabric*, Parts 1 and 2.

The configuration of the roof is a principal determinant of the form and silhouette of a building. Selection of a particular roof construction will be determined by the interrelationship between the overall shape or massing desired and the adoption of an appropriate structural system. These two factors allow a range of materials to be identified from which a choice of roof finish can be made.

The basic forms for a roof are:

- *Flat* (which for the purposes of definition means a roof sloping less than 10 degrees). Roof finishes consist of either waterproof membranes that are continuously supported by decking or self-supporting profiled sheets that interlock to form waterproof joints.
- *Pitched*. Roof finishes are either small dry jointed overlapping units that are supported by a sloping framework or decking, or waterproof membranes supported by a continuous sloping decking, or profiled interlocking sheets.
- *Curved*: Waterproof membranes are continuously supported by decking, or profiled sheets are curved to shape, or if large three-dimensional curves are required the membrane is supported by cables or another method of suspension.

In areas where the local planning authority consider it important for new construction to conform to the character of existing buildings, precedent may have an overriding influence on the choice of roof shape and materials (for

Photo 8.1 House in Ontario, Canada.
Frank Lloyd Wright said of roofs that 'you can do with a roof almost anything you like. But the type of roof you choose must not only deal with the elements in your region but be appropriate to the circumstances'. In eastern Ontario, snow can drift to the height of first floor windows. Steeply sloping roofs minimize the accumulated weight of snow during the winter.

example in conservation areas). The requirements of the local town planning officer may place considerable limitations on choice, and in some places very precise stipulations are made.

If a flat roof is the chosen alternative, the roof may have projecting eaves, forming a strong horizontal line parallel to the ground plane, a device that has been used to effect in many modern buildings. Alternatively, the roof may be concealed behind a parapet so that the façade can have the appearance of an uninterrupted planar surface. In the latter case, it is important that provision for the disposal of rainwater allows an alternative route for water to leave the roof should an outlet become blocked. In the past, flat roofs not intended for general access (which are consequently seldom seen), have often been visually unattractive. A flat roof can, however, be a usable surface and provide a building with extra amenity; in recent years landscaped 'green' roofs have become more widespread and several systems have been introduced for their construction (described later in this chapter).

The form of a pitched roof can be used to articulate each individual element of a complex of buildings or the parts of a single building. The degree of pitch is a primary design decision since it determines the area of roof which will be visible which, in turn, will influence the choice of surface finish (i.e. shape, pattern, texture and colour of the different roof coverings available). The process of selection may also involve analysis of the position and specific shape of the roof pitch, as well as the viewing distance and angle of vision. Many substitute materials are now used as alternatives to traditional pitched roof finishes. These include concrete tiles which imitate clay tiles or stone slates, fibre composite slates substituting for natural and asbestos-cement slates and metal panels profiled to look like pantiles. Pitched roofs may have projecting eaves so the roof space can be ventilated through openings in the soffit, or the edge of the roof may be recessed to form a parapet gutter. Parapet gutters require overflows in the event of rainwater outlets blocking and air bricks or an alternative method of ventilating the roof space has to be found. Some of the more common terms used in connection with roofing are illustrated in fig. 8.1.

Table 8.1 lists the more commonly used roof finishes in relation to the minimum falls and pitches for flat and pitched roofs. A roof finish which is satisfactory on a flat roof or shallow pitch may also be adequate on a steeper pitch. Also, flat and sloping roof coverings can be used in a modified form as a cladding for vertical surfaces, e.g. metal sheeting, tiling or slating. The fall of a roof is usually expressed as the rise in a stated horizontal distance or run — see fig. 8.2. For example, a flat roof finish laid at an angle of $\frac{3}{4}$ degree would have a fall expressed as 1:80.

The roof pitch is the angle of slope to the horizontal. For

Table 8.1 Minimum pitches for various types of roof finishes

Roof finish	Minimum pitch in degrees
FLAT AND SINGLE CURVED	
Asphalt; lead with drips; multi-layer bitumen felt; single-layer plastics	3/4 (1:80)
Aluminium, copper, zinc with drips	1 (1:60)
Steel, aluminium, fibre cement, or plastics profiled sheets with sealed laps	1 to 10
PITCHED	
Glass fibre reinforced bitumen slates	12
Metal multi-tiles	12
Copper and zinc with welted end seams	13.75
Timber shingles	14
Concrete interlocking tiles; flats	17.5
Concrete interlocking tiles: troughed and Roman	30
Fibre cement slates	20
Fibre cement and plastics profiled sheeting with 150 mm end laps	22.5
Slates: min 300 mm wide	30
Slates: min 225 mm wide	33.3
Clay tile: interlocking on four edges	25
Clay tile: interlocking on two edges	30
Clay tile: concrete	35
Clay tile: clay	40
Thatch	45

Note: Variation will occur according to precise exposure conditions, methods of fixing and the type of access required to the roof: manufacturers' recommendations must be taken.

the form of pitch shown in fig. 8.2 (symmetrical, sloping from a central ridge), the pitch used to be expressed as the fraction rise/span because this relationship of the rise to span was more expressive of the setting out techniques of traditional pitched roof construction (when using site-cut timber or metal structural sections). For example, a roof finish laid at an angle of $33\frac{1}{3}$ degrees would have a rise/span $= \frac{1}{3}$. However, since not all roofs have symmetrical slopes and since prefabricated roof trusses are more usual today, most manufacturers of pitched roof finishes indicate the suitability of their product according to the angle of slope.

8.2 Performance requirements

BS 6229: 1982 *Code of Practice for flat roofs with continuously supported coverings* deals with the design,

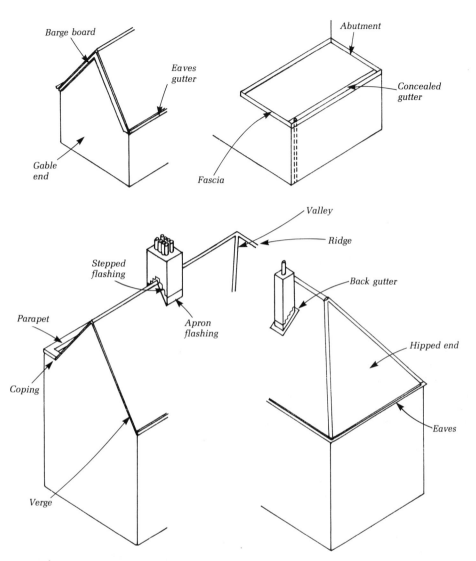

Fig. 8.1 Terms used in connection with roofing

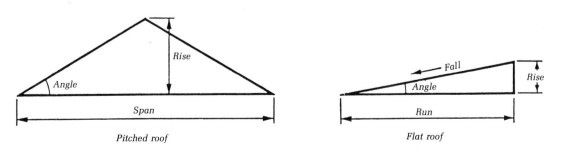

Fig. 8.2 'Pitch' and 'fall'

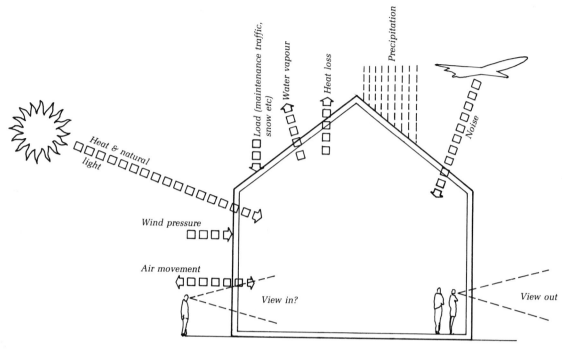

Fig. 8.3 Performance requirements for building fabric

performance requirements and construction of the flat roof as a whole (see fig. 8.3). This, together with other design guides, such as those produced by the BRE, the Property Services Agency (PSA) and the British Flat Roofing Council (BFRC), as well as other associations concerned with roofing, provide greater understanding of flat roofing design. Nevertheless, it is noteworthy that this area of construction still provides a major source of building failure and is the subject of much research and debate. A major recent reference is *Flat Roofing: Design and Good Practice*, published by the CIRIA and the BFRC. Recommended design and installation procedures should always be carefully followed.

BS 5534: *Code of practice for slating and tiling* Part 1: 1990 *Design* and Part 2: 1986 *Design charts for fixing roof slating and tiling against wind uplift* and BS 8000 *Workmanship on building sites* Part 6: *Code of practice for slating and tiling of roofs and claddings* deal with design factors and installation methods for pitched roofs using dry jointed lapped units as a covering. The construction methods adopted for pitched roofs have, in the past, given less cause for concern than those for flat roofs because they shed water more rapidly and most types rely on a multi-defence system against the penetration of moisture. In recent years, however, pitched roofs have themselves been a source of concern because of the incidence of wind damage and problems of condensation within the structure. Other British Standards are also available for specific materials used as

roof finishes, and these will be referred to at appropriate points in this chapter.

8.2.1 Weather exclusion

Roof coverings are required to prevent the entry of rain, snow and dust, as well as resist the effects of wind — both wind pressure and wind suction.

Water and dust penetration Figure 8.4 illustrates the the pattern of movement of water in relation to the two basic forms of roof construction. Flat roofs and those pitched roofs consisting of interlocking sheets, provide an *impermeable barrier* which if correctly constructed to give adequate falls, sheds water directly off the top surface. Whereas small unit pitched roof coverings form only a *semi-permeable barrier* which permits a certain amount of rainwater to penetrate before being collected at an internal impermeable layer. In this case, the construction consists of an outer surface of lapped tiles or slates which provide the initial water check, backed by an impervious water barrier of sarking felt. The minimum pitch at which the outer roof covering can be laid depends on many factors, such as exposure to wind and weather, workmanship, design and type of joints in the roofing, porosity of the material and its tendency to delaminate in frost, and the size of the unit.

For details of rainwater collection, including rainwater

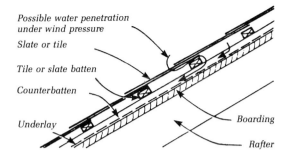

Possible water penetration
under wind pressure

Slate or tile

Tile or slate batten

Counterbatten

Underlay

Boarding

Rafter

Semi-permeable roof

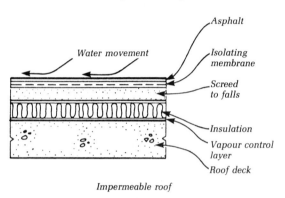

Water movement

Asphalt

Isolating
membrane

Screed
to falls

Insulation

Vapour control
layer

Roof deck

Impermeable roof

Fig. 8.4 Appropriate constructions for pitched and flat roofs

gutters, pipes and gullies, see *MBS: Environment and Services*. In the construction of flat roofs, it is vitally important that adequate falls are provided so that rainwater is not allowed to collect on the surface and cause 'ponding'. Isolated areas of water lying on a flat roof can cause excessive differential movements in the waterproof membrane because the surface below the water is likely to be at a much cooler temperature than the surrounding dry surfaces, particularly on a sunny day. Differences in expansion and contraction in the roof covering material are liable to cause stresses resulting in cracking, and the penetration of water. For this reason, it is a wise precaution not to lay flat roof coverings at the recommended minimum fall of 1:80 where workmanship or the constructional accuracy of the support decking will not guarantee against backfalls (which would cause pockets on the surface where water will collect). Either the roof should be designed to a minimum fall of 1:40, or if the roofing is to designed to a 1:80 fall, specific allowance should be made for constructional tolerances in the roof structure and its deflection under dead and imposed loads (figs 8.11 and 8.12).

It has been shown that rain penetration through the pitched roof outer coverings of slate or tiles depends both on amount of rain and wind speed, and is governed more by maximum rain intensity rather than total duration or total quantity. The effectiveness of this type of construction depends upon the adequacy of the overlap of the units, and the pitch of the roof which sheds the water by gravity. Careful selection procedures must be adopted to ensure that an appropriate roof covering is provided which is capable of providing maximum resistance to the prevailing climatic conditions, including wind-driven rainfall.

In the UK, initial assessment of climatic conditions can be obtained by reference to the driving rain index (DRI) for a particular location, as described in the maps published by the BRE. Values are obtained by taking the mean annual wind speed in metres per second (m/s) and multiplying by the mean annual rainfall in millimetres. The product is then divided by 1000, and the result used to produce contour lines linking areas of similar annual DRI in m^2/s throughout the country. Figures 8.5(a) and (b) show a simplified diagrammatic analysis:

- Sheltered exposure zone refers to districts where the DRI is 3 or less.
- Moderate exposure zone refers to districts where the DRI is between 3 and 7.
- Severe exposure zone refers to districts where the DRI is 7 or more.

The value for a particular location within 8 km of the sea coast, or a large estuary, must be modified to the next zone above (sheltered to moderate, and moderate to severe), to take account of unusual exposure conditions. Also, modifications may be necessary to allow for local topography, special features which shelter the site or make it more exposed — roughness of terrain, height of proposed building, and altitude of the site above sea level. The proportion of driving rain from various directions within one particular location can be obtained by reference to driving rain rose diagrams, an example of which is also shown in fig. 8.5(b). In conjunction with the DRI, the use of these 'roses' permits the approximate amount of driving rain to be calculated which can be expected to affect the exposed face(s) of a roof construction. DD 93: 1984 *Method for assessing exposure to wind driven rain*, describes two uses of the DRI: the *local spell index method* which measures the maximum intensity in a given period, and the *local annual index method* which measures the total rainfall in a year.

One of the most important lessons to be learnt from driving rain indexes and roses is that roof design and construction details, of necessity, vary from one exposure zone to another: details applicable in sheltered conditions may leak if simply transferred to areas where severe or even moderate conditions prevail.

There is, however, a danger in using only meteorological climatic data of this type for the final selection of appropriate materials and construction detailing since it

(a)

(b)

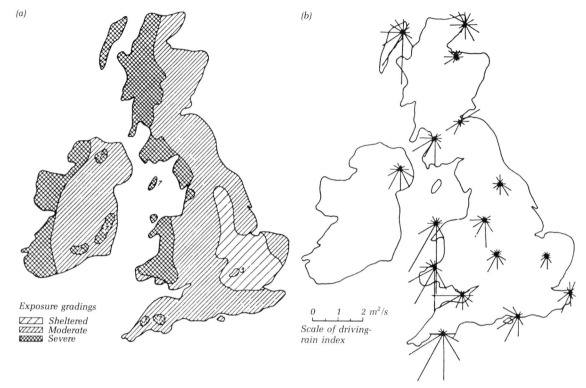

Exposure gradings

▨▨▨ Sheltered
▨▨▨ Moderate
▨▨▨ Severe

0   1   2 m²/s

Scale of driving-rain index

Fig. 8.5 (a) Driving Rain Index; (b) driving rain rose diagram

reveals only the general or macro-climatic conditions. In addition there is a micro-climate surrounding the immediate outer face of a building, not more than 1 m from the surface, which arises from its precise form, location, juxtaposition with other structures and surface geometry. Where there is no past experience to rely upon, it may be necessary to produce models of a building and its surroundings so that simulated environmental testing can be carried out.

Wind pressure and suction   The design wind loadings for wind pressure and suction can be determined from CP 3 *Code of basic data for the design of buildings*, Chapter V Part 2: 1972 *Wind loads*. The principles underlying the procedures in this document are explained in section 3.2.2.

The effect of wind pressure upon a roof depends upon the angle of pitch and the degree of exposure of the roof slope. Detrimental effects can be avoided by use of roof finishes of appropriate strength and correct fixing details.

For a symmetrically pitched roof of between 20 and 30 degrees, the *wind suction* will more or less be equally balanced on both roof slopes (fig 8.6). On steeper pitches the leeward side only is subject to negative pressure (suction), and for shallower pitches (including 'flat') greater suction occurs on the windward side. Shallow pitched roofs

may therefore present a serious problem, particularly when lightweight sheet coverings are used on lightweight timber decking. When employing this form of construction, the roof deck should be securely anchored to the roof framing and then fixed to a wall plate by means of galvanized steel straps built into the wall below — see *MBS: Structure and Fabric*, Parts 1 and 2, and *MBS: Introduction to Building*. Severe wind pressure tends to occur at the eaves of a roof, particularly in the case of projecting eaves used in conjunction with a low pitch. It is also especially important to seal the edges of flat roof coverings against their structural support to prevent wind entering beneath and lifting them. This sealing must be done in such a way so as not to restrict differential movements which may damage the flat roof coverings.

8.2.2 Sub-structure

Whereas slates or tiles used on pitched roofs need to be supported at intervals by nailing to battens, sheet materials like bituminous felts and copper, or jointless ones such as asphalt (normally used on flat roofs), are laid on a continuous supporting surface. The sub-structure which provides this support is generally referred to as decking, and there are two basic forms:

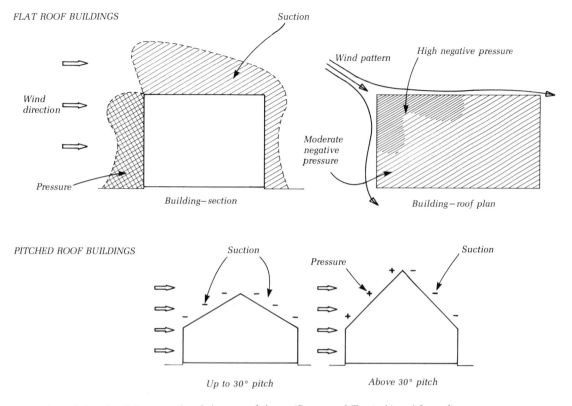

FLAT ROOF BUILDINGS

Suction

High negative pressure

Wind pattern

Wind
direction

Moderate
negative
pressure

Pressure

Building—section

Building—roof plan

PITCHED ROOF BUILDINGS

Suction

Suction

Pressure

Up to 30° pitch

Above 30° pitch

Fig. 8.6 Variations in wind pressure in relation to roof shape. (Courtesy of *The Architects' Journal*)

1. *In situ monolithic*, such as a reinforced concrete slab.
2. *Prefabricated units*, ranging from thin precast concrete slabs and metal panels, to timber and preformed boards, e.g. plywood, particle board or wood-wool.

***In situ* monolithic decking**   An *in situ* monolithic deck should be designed and constructed so that it does not deflect sufficiently to reduce falls to below the minimum acceptable slope or to adversely affect the roof covering. Suitable joints will also be required to accommodate thermal movement both within the deck and between it and the build-up of the roof. Precise specification and careful site supervision are essential. Screeds which are used to provide falls or insulation should have adequate strength so that they do not induce stresses which will affect the roof covering. It is advisable not to place a vapour control layer underneath water-based insulating screeds as this will seal any residual moisture content against the impermeable roof covering. The provision of temporary weep holes through the supporting slab will allow this residual moisture to drain away during the remaining construction period, and the provision of strategically placed roof ventilators will avoid the dangers of subsequent interstitial condensation which would otherwise be avoided by the vapour control layer.

As an alternative, bitumen-coated insulating screed can be used, above a vapour control layer, as it does not contain water.

Generally, however, falls are better built into the structure itself, for example by pre-cambering or sloping supporting beams. Screed and other wet applied materials that are liable to trap moisture within the roof construction are not then required. This is a particularly important consideration when metal decking is used as permanent shuttering to a concrete roof slab. (The ability of the construction to dry out is limited by the impermeable underlayer formed by the metal decking.)

In situ or reinforced concrete must be well cured and dry. The surface should be hard and smooth, clean without irregularities and *wood-floated* (a concrete slab to receive a vapour barrier should not be *tamped*). Where it would otherwise be of open texture, it must be floated or screeded. Adequate crack control methods including movement joints must be provided in the sub-structure.

Lightweight aggregates may be used in *in situ* concrete. The types generally in use are either prepared from manufactured materials such as expanded vermiculite, or expanded clay. Lightweight concrete should comply with requirements described above for ordinary concrete, i.e.

specification of surface finish. The water content should be kept to a minimum and sufficient time must be allowed for any water to dry out or disperse. The porous nature and high residual water content of lightweight concrete make it essential to take special precautions to disperse this trapped water so that it does not damage the roof finish.

Precast (including pre-stressed) concrete beams or slabs Where this type of structural unit is used for a flat roof deck, a 1:4 cement and sand screed is required of at least 25 mm thickness. Drainage holes should be provided to prevent water being trapped in the construction after laying. It is important to establish the maximum amount of deflection likely to take place in the units after all dead loads have been applied, including that provided by the screed (the weight of which can be significant). The units can then be laid at a slight incline to provide the necessary falls for the roof covering.

Lightweight concrete slabs Where this type of deck is laid carefully, a cement and sand screed may not be needed and the construction is totally 'dry' (except for the grouting which may be required between units). A bitumen emulsion primer should be applied to the top surface of the slabs as soon as possible after installation to prevent absorption of rainwater.

Profiled metal panels Profiled metal sheets may be used as decking for both flat and pitched roofs. It is usually laid complete with thermal insulation and built-up felt waterproofing or a single ply membrane as a composite roof construction installed by specialist sub-contractors. The decking may be of galvanized steel or of aluminium, both of which are made by rolling sheet metal to a variety of profiles and gauges to suit varying spans (fig. 8.7). Mineral fibre insulation board, polystyrene, cork or other flexible materials are used in sheet form to provide thermal insulation. The insulation is protected by a bituminous felt or plastic film vapour control layer where necessary. The decking units are fixed by self-tapping screws or by shot-firing to the steel supporting structure. The vapour control layer is bonded to the top surface of the metal decking followed by the insulation board and the outer roof finish. Where the decking is to be exposed to the space below and acoustic absorption is required, perforated decking is available on to which acoustically absorbent board is laid, the vapour control layer is then sandwiched between this and the thermal insulation above.

Alternatively, the insulation can be contained between outer and inner skins of metal sheet to form panels (for both flat and pitched roof applications). The outer sheet can have a plastic coating, in a wide range of colours, which forms the outer waterproof covering. This sort of roof is

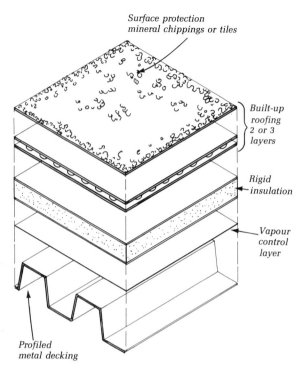

Fig. 8.7 Profiled metal decking and built-up roof construction

constructed by mechanically fixing the three layers together. In some systems the insulation is preformed to the shape of the profiled sheet (to exclude air gaps within the corrugations of the sheets that can trap condensation).

Fully composite panels are also available (fig. 8.8), the insulation being bonded to the profiled metal sheets so that the whole depth of the panel is used structurally, the insulation contributing to the structural strength. The edges of the panels upstand to interlock and seal against the adjoining panels to form an impermeable joint.

Timber boarding The whole of the timber sub-structure should be constructed in accordance with BS 5268 *Structural use of timber*, Part 2: 1988 *Code of practice for permissible stress design, materials and workmanship*, which provides guidance on the structural use of timber, plywood, glue-laminated timber and tempered hardboard in load-bearing members. See also *Building Regulations 1991*: Approved Document A, and *MBS: Materials* chapter 2.

The construction should minimize the effects of shrinkage, warping and displacement, or relative movement of the timber. All timber used in roof construction must have a preservative treatment in accordance with BS 1282: 1975 *Guide to choice, use and application of wood preservatives*, and the *Building Regulations 1991*: 'Approved Document to support Regulation 7 Materials and

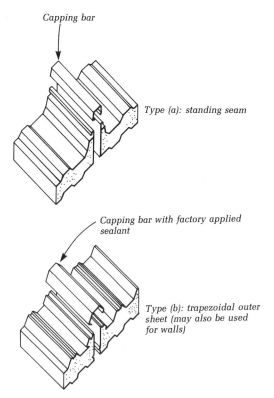

Capping bar

Type (a): standing seam

Capping bar with factory applied sealant

Type (b): trapezoidal outer sheet (may also be used for walls)

Fig. 8.8 Composite metal panel roofing types (British Steel Strip Products)

workmanship' which require that in certain geographical areas, softwood used for roof construction or fixed in the roof space should be adequately treated with a suitable preservative to prevent infestation by the house longhorn beetle.

Care should be taken to guard against all conditions which might allow decay through moisture already present in unseasoned timber or resulting from the ingress of water from other parts of the structure, or from condensation. To avoid dry rot in cold deck flat roof construction, ventilation should also be provided between the roof boarding and the ceiling (see p. 201). Roof boarding should be well seasoned to avoid the tendency to shrink and cup. It should not be less than 25 mm thick and 100 mm wide and should be tongued and grooved. Arrises should be rounded, upstanding edges planed flat and nail heads well sunk. The supporting roof construction should be rigid with joists at maximum 450 mm centres and the board fixed by nails at each edge to minimize the danger of curling. Boarding should be laid either in the direction of the fall of the roof or diagonally, and the roof surface should be protected as far as possible from rain during the course of construction.

Plywood BS 5268 also deals with plywood and BS 6566 *Plywood*, Parts 1 to 8 1985, specifies characteristics in greater detail. Tongued and grooved sheets should be used for roof decking to provide perfect alignment at joints and it is essential to supply sufficient support between the supporting joists at sheet edges. Nailing, or preferably screwing, should be at 150 mm centres and at all edges as well as at intermediate supports. Rafter centres and nail lengths, according to plywood thickness, are as follows:

- *Plywood thickness*: 8, 12, 16 and 19 mm.
- *Rafter centres*: 400, 600, 800 and 1200 mm.
- *Nail length*: 38, 38, 50 and 50 mm.

Particle boards These are described in BS 5669: 1989 Part 2: *Specification for wood chipboard*. Because CP 144 *Roof coverings* Part 3: 1970 *Built-up bitumen felt* is currently under review the British Flat Roofing Council do not at present recommend the use of particle board for flat roofs other than those with an open soffit (e.g. garages) in which case the most moisture-resistant grade should be used. Boards should be tongued and grooved used for the reasons stated above, and must be a type whose binders are either MF/UF (melamine formaldehyde/urea formaldehyde) or PF (phenol formaldehyde) in order to comply with the recommendations of BS 5669, Types II/III; they should also incorporate a fungicide.

For pitched roof construction, pre-felted and wax or bitumen-coated boards have in the past been used under tiles or slates to act in conjunction with counterbattens to provide the final water barrier in a similar fashion to underlay. However, higher standards of loft insulation are associated with an increase in the moisture content within boards used in this way.

Wood-wool slabs Wood-wool slabs are made from long wood fibres, chemically impregnated and bonded together under pressure. For roof decking the heavy duty type of slab should be used, fixed according to the manufacturer's instructions by clips or screwing to timber joists or steel purlins. The joints between the slabs should be filled with sand and cement and the slab should be topped with a 1:4 sand and cement screed 12 mm thick in bays not exceeding 9 m². On roof slopes of more than 20 degrees the screed may be omitted and a sand and cement slurry used instead. Where wood-wool units are fixed directly to the roof supports, the span should not exceed 600 mm and the bearing surface at the edge of each unit should not be less than 50 mm and fixed securely at every bearing point. Wood-wool slab units of this type are usually 50 mm thick and it is important that, once fixed, they are temporarily protected by crawling boards to avoid the danger of fracture or collapse under point-loads which could occur when completing the roof. To avoid the use of crawling boards, a wood-wool slab unit is available which incorporates a plastic 'safety-net' reinforcement.

Galvanized mild steel channel reinforced wood-wool units must be used for increased loading and/or span requirements. These vary in thickness from 50 mm to 100 mm and span up to 4 metres; trough section reinforced units capable of spanning 6 metres are also available.

There is a risk of condensation due to the cold bridge made by the steel reinforcing channels. In order to counteract this, thick slabs are available rebated at the joints to receive inserts which provide continuity of insulation at each joint. Pre-screeded and pre-felted wood-wool slabs are also available. The felt is a protective layer only and does not form part of a built-up felt system. The joints of this type of wood-wool slab should be taped immediately the slabs are laid to prevent the infiltration of moisture. In general terms, it is advisable that where moisture-sensitive materials are used for roof decking they should be sealed on all surfaces if they are to maintain stability in cases where condensation is likely.

Most forms of structural decking require an intermediate layer of material between its top surface and the outer roof covering to allow for differential movements; a fall for drainage purposes; additional thermal insulation; sound control; or to provide any combination of these factors. The use of appropriate materials to fulfil the requirement for thermal insulation is covered under section 8.2.3. For isolation purposes only, vegetable fibre boards can be used, including those made from wood or cane fibres of natural or regranulated cork. Care should be taken to guard against decay of the boards, particularly through effects of moisture. Where this type of board is laid on concrete or similar materials, and where there will be moisture vapour diffusion from within the building, a vapour control layer is essential as described under section 8.2.4. Alternatively, mineral fibre and granular boards can be used which are made from glass, mica or similar granules compressed and bonded with bitumen, synthetic resin, or other material. In theory, a vapour control layer might be omitted when using these materials except in conditions of very high humidity, e.g. as found in laundries. Extruded polystyrene is also a commonly used board (which has the advantage of being free of CFCs, see chapter 1), and for most practical purposes is vapour proof when the joints are sealed. When placed on metal decking, joints in the insulating boards must not be made over the trough of the deck or the board may collapse under normal maintenance traffic.

The provision of a *screed* may be necessary on certain flat roof decks, including those of *in situ* or precast concrete and wood-wool slabs. This screed provides the falls and a smooth surface upon which to lay the roof waterproofing, and the selection of a suitable mix and thickness is very

important. Whilst the mix must be related to that of the sub-structure, a cement/sand screed should not be richer than 1 part cement to 4 parts sand by volume. There should be a low water to cement ratio and the screed should be laid in areas not exceeding 9 m², even if they are reinforced. The minimum thickness should be 25 mm except where they are laid as topping on wood-wool slabs, when the minimum thickness can be reduced to 13 mm. Additional thermal insulation to that already provided can be achieved when using a *lightweight concrete screed*. In such cases entrapped water is always a problem, and if allowed to remain it will seriously reduce the thermal insulation value of the screed and may cause blistering of the roof finish where built-up felt, asphalt or single-layer roofing is used. It is customary to provide a topping of 12 mm sand/cement screed immediately after the lightweight concrete screed has been laid. This topping provides a surface which will shed water to prearranged temporary drainage holes passing through the screed and roof decking at the lowest part of the roof. The problems that may arise where water is trapped in a screed are more fully discussed on p. 193. As an alternative to a water-mixed lightweight concrete screed, a *bitumen-mixed screed* can be used which has a dry exfoliated vermiculite aggregate. This screed does not present the same problems as the sand/cement screed as water is not used in the mix. However, should bitumen screed get wet before the roof is complete, the screed must be allowed to dry out before the roof finish is applied (otherwise temporary ventilators may have to be incorporated and removed once the moisture has dissipated).

Whatever type of screed is employed to provide falls and a smooth surface, the roof covering will still require to be isolated from it in order to allow for differential movements. A layer of underfelt may be all that is necessary where thermal insulation is provided by other means.

8.2.3 Thermal control

The *Building Regulations 1991*: Approved Document L: 'Conservation of fuel and power' requires that the maximum thermal transmittance value (*U*-value) for the roofs of dwellings should be 0.25 W/m²K (excepting loft conversions in existing houses where 0.35 W/m²K is required); for other residential buildings, shops, offices or assembly buildings, 0.45 W/m²K; and for buildings used for industrial, storage or other purposes so far not described, 0.45 W/m²K. These values are applicable to a roof of less than 70 degrees pitch. Above this angle, a roof is classified as a wall and the required *U*-value is 0.45 W/m²K, for all purpose groups.

It always used to be assumed that heat loss to the ground through the base floor slab of a building was relatively unimportant (other than at the perimeter), because the slab and subsoil reach a state of thermal equilibrium. It has since been realized, however, that movement of ground water, particularly through clay soils, can substantially cool the slab and thereby transmit heat from the building. Consequently there is now a building regulations requirement for ground floors to be insulated to 0.45 W/m²K applicable to all purpose groups, according to the building's plan dimensions.

Some elements can be insulated to a higher standard to reduce the requirements elsewhere. For example, if some of a building's windows are to be double-glazed then the insulation value of the walls may be reduced. If all the windows are to be double-glazed, the exposed walls must achieve only 0.6 W/m²K and the roof 0.35 W/m²K and the ground floor can be left uninsulated. Similarly, a ground floor may be insulated to give a U-value of 0.35 W/m²K, so the roof insulation can be reduced to 0.35 W/m²K.

Roofs, walls and floors having the U-values stated in the *Regulations* are considered to provide an acceptable limit on the overall heat loss from a building (although UK standards are considerably lower than in some other northern European countries), and to help in the overall conservation of energy. As rooflights and windows (which lose more heat than the solid parts of roofs and walls) are liable to cause greater overall heat loss from a building, their combined areas are limited in proportion to the amount of solid construction comprising the roof and walls. These permissible areas increase if double or triple glazing or low-emissivity glass is used. Where the permitted area of rooflights and windows is required to be increased for other reasons, the permitted overall heat loss from a building can be maintained by a corresponding increase in the thermal insulation properties of the rest of the building envelope. This is established by *calculation* rather than the *elemental* approach outlined so far. Further information about the procedures to be followed when determining the areas of rooflights and windows relative to the overall heat loss from a building is given in sections 4.2.5 and 5.2.

Thermal control both in terms of heat loss and heat gain, may be achieved by either a *cold roof* construction, or a *warm roof* construction.

Cold roof construction   A layer of thermal insulation is built within or at the lowest level of the roof structure (figs 8.9(a) and (d). This is an economical method in terms of energy costs since the space contained by the roof construction above the insulation is not heated. It is also an advantageous one, if a building is to be intermittently occupied resulting in frequent variations in internal temperature, because in a cold roof no heat is expended to warm the structure. In domestic construction (flat and pitched roofs), the insulation is placed between joists and

immediately above the ceiling. It should be noted however, that timber joisted cold flat roofs have inherent problems of condensation due to the difficulty of providing an effective vapour control membrane and ensuring adequate ventilation of the roof void. Similarly, when building a structure in concrete, the insulation can be fixed on the underside of the structural slab or form part of the ceiling construction.

Warm roof construction   A layer of thermal insulation is located immediately below the roof covering but external to the roof structure (see figs 8.9(b) and (e)). This method insulates the roof structure and as a result reduces stress due to changes in ambient temperature. The insulation can be fully continuous to eliminate cold spots in the external envelope and the insulation is conveniently supported by the structure. This is a 'slow-response' construction because the structure has to be heated to attain comfort conditions, consequently warm roofs are well suited to buildings that are in continuous occupancy. In flat roof construction, the insulation can be placed immediately beneath the waterproof covering. In pitched roofs it can be placed over the rafters.

Inverted warm roof construction for flat roofs (fig. 8.9(c))   Slabs of thermal insulation are laid loose external to the whole roof construction (including the roof covering) and are weighted down with gravel or paving. This is known as an 'inverted roof' construction, and has the advantage of ensuring that minimal thermal stresses develop in the roof covering and the structure below. A vapour control layer is not required in this configuration as the waterproofing itself acts as a vapour barrier, being on the warm side of the insulation.

There are numerous materials which can be used to provide thermal insulation (see *MBS: Materials*); their relative merits will vary according to their location within the roof construction and to any other performance requirements which have to be fulfilled. They are available as quilts, foams and rigid batts, as well as loose granules. Rigid urethane foams in current use contain varying amounts of polyurethane and polyisocyanurate to meet different performance requirements; they and mineral fibre boards, foamed glass and cork are used under built-up bituminous felt, single ply roofing and asphalt. Extruded cellular polystyrene slabs, weighted down with paving stones, are used over the roof covering in inverted roofs including those having to withstand the weight of vehicles. High-density mineral wool boards are also suitable for inverted roof construction and, if access is to be limited to light maintenance foot traffic, have the advantage of not requiring ballasting other than at the roof's perimeter.

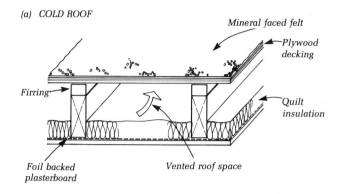

(a) COLD ROOF

Mineral faced felt

Plywood decking

Firring

Quilt insulation

Foil backed plasterboard

Vented roof space

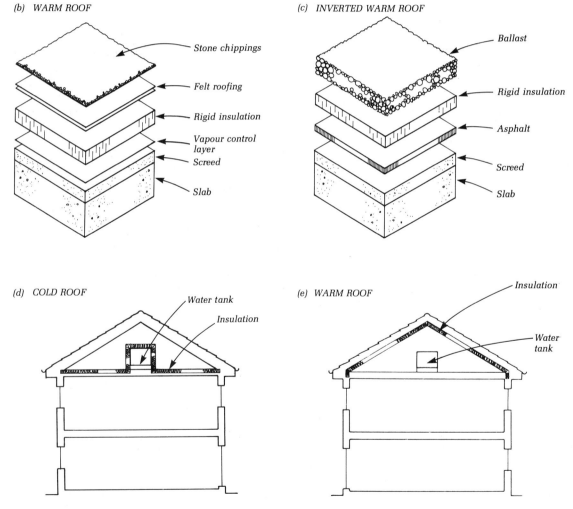

(b) WARM ROOF

Stone chippings

Felt roofing

Rigid insulation

Vapour control layer

Screed

Slab

(c) INVERTED WARM ROOF

Ballast

Rigid insulation

Asphalt

Screed

Slab

(d) COLD ROOF

Water tank

Insulation

(e) WARM ROOF

Insulation

Water tank

Fig. 8.9 Examples of flat roof construction: (a) cold roof; (b) warm roof; (c) inverted warm roof. Pitched roofs: (d) cold roof; (e) warm roof

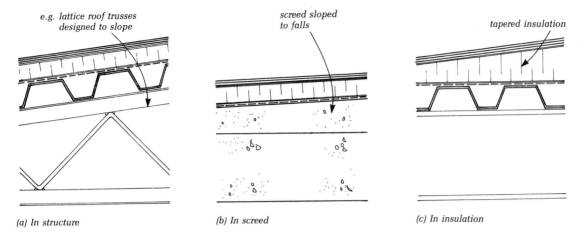

e.g. lattice roof trusses designed to slope

screed sloped to falls

tapered insulation

(a) In structure   (b) In screed   (c) In insulation

Fig. 8.10 Methods of obtaining falls

Mineral fibre or glass fibre are used in friction-fit batt form between joists and rafters, or in quilt form over rafters.

Polyurethane and polyisocyanurate insulants are foamed using CFC gases that, by international agreement, are to be discontinued from use during the 1990s. Some manufacturers are now producing polyurethane foamed without the use of CFC gases or halogens (see chapter 1). Other materials such as mineral fibre or cork that may be suitable substitutes for some applications require a considerably greater thickness to achieve the same thermal resistance.

A large proportion of insulating materials are combustible, so the type included in a roof construction can affect the spread of fire from one building to another. Some cannot be used at all when roof coverings are heat applied to the decking. Only mineral wool and foamed glass can be said to be truly non-combustible.

In roofs where there is no air space above the insulation and therefore no possibility of ventilation, insulation of high vapour resistance is necessary to minimize moisture transfer to the cold side of the insulation. Most insulation materials will require some form of vapour control on their 'warm' side (see section 8.2.5) to prevent vapour diffusion, the exceptions perhaps being mineral fibre, polystyrene and foamed glass when not used in highly humid conditions. Nevertheless, under many circumstances, it is wise to include a vapour control layer in order to prevent the formation of condensation, and essential in timber cold deck flat roof construction where there is limited space above the insulation which is difficult to ventilate adequately. In roofs where there is a large readily ventilated space within the roof construction, above the insulation, a vapour control membrane is not normally required, for example a typical domestic pitched roof insulated at ceiling level (other than possibly above kitchens and bathrooms).

8.2.4 Vapour control

The atmosphere contains water in the form of water vapour but the amount varies according to the temperature and the humidity. The temperature at which air is saturated is called *dew point*; warm air is able to contain more water vapour than cold air before it becomes saturated. The measure of the level of humidity within air is known as its *vapour pressure*. If air containing a given amount of water vapour is cooled there will be a temperature at which condensation of the water vapour into water droplets will occur, either as *surface condensation* or as *interstitial condensation* (see fig. 8.11).

Surface condensation occurs when moisture-laden air comes into contact with a surface which is at a temperature below the dew point of the air. In roof construction, surface condensation can be avoided quite simply by keeping the internal surface at a temperature above the dew point of the internal air. This is achieved by provision of adequate thermal insulation.

Due to people and activities in buildings that generate heat and humidity, internal environments tend to be warm and humid relative to outside air, for much of the year. The difference in the vapour pressure of the internal atmosphere and the cold external air may result in the migration of water vapour into the roof towards the colder side which then condenses within the roof construction. This is known as interstitial condensation. Where this type of condensation occurs, the water will most likely cause the roof structure and insulation to deteriorate and some of the moisture may drip back into the building. A vapour control layer may be necessary to prevent this moisture movement, depending on the anticipated internal conditions.

Even if materials impermeable to water vapour are used, joints can cause a breach in the membrane as can holes for

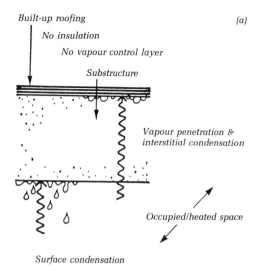

Built-up roofing   *(a)*

No insulation

No vapour control layer

Substructure

*Vapour penetration &
interstitial condensation*

Occupied/heated space

Surface condensation

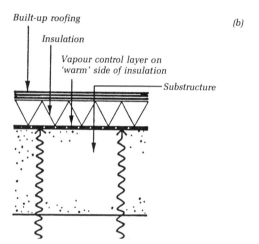

Built-up roofing   *(b)*

Insulation

*Vapour control layer on
'warm' side of insulation*

Substructure

*No vapour penetration into cold part of construction
No surface condensation*

Fig. 8.11 (a) Condensation types; (b) Provision of vapour control layer related correctly to insulation

fixings and services. Within timber cold roof construction for example, the vapour control layer (which is positioned at a point where the temperature remains above the dew point of the vapour, on the warmer side of the insulation), is likely to be a foil-backed plasterboard or other sheet material that will be nailed to the joists and have within it holes for light fittings, etc. This will not provide a full vapour barrier but only a vapour check which must be supplemented by substantial ventilation of the roof void to dissipate any water vapour that may bypass the vapour check (reference should be made to BS 6229: 1982 for vapour control layers incorporated into flat roofs).

On the other hand, a warm roof on a structural deck can accommodate the vapour control layer above the structure and below the insulation and generally without holes for fixings or services. This can form a vapour barrier (rather than a check), consisting of one or more layers of plastic sheet, aluminium foil or roofing felt. To avoid thermal movement within the sub-structure weakening and cracking the vapour control layer, it may be only partially bonded to the supporting deck depending on the materials used and the wind loads that are anticipated.

The *Building Regulations 1991*: Approved Document F requires that for roofs with insulation beneath or within the structure (a 'cold roof'), openings for ventilation must be provided, as shown in fig. 8.13. These requirements apply only to dwellings. Roofs with a pitch of 15 degrees or more must have ventilation openings at eaves level on opposite sides of the roof, equivalent to a 10 mm slot running the full length of the eaves. In the case of a lean-to roof, ventilation should be provided at the high point of the roof, as well as at the eaves. These rules do not apply to roofs in which the ceiling follows the pitch of the roof, as they are treated as having a pitch of less than 15 degrees.

The ventilation openings in roofs with a pitch of less than 15 degrees should be equivalent to a slot 25 mm wide, running the length of the eaves. The ventilated void between the roof deck and the top of the insulation should be at least an unobstructed 50 mm. If joists run at right angles to the flow of air, the space between the tops of the joists and the underside of the deck should be maintained by means of counterbattens.

The *Regulations* provide an alternative approach to the rules described in Approved Document F, allowing the ventilation to conform to the recommendations of BS 5250: 1989 *Code of basic data for the design of buildings: the control of condensation in dwellings*. The design recommendations are similar to those of the *Regulations* but give more detailed information including requirements for pitched roofs with inclined ceilings and roofs containing rooms. Also by using the method of condensation risk calculation described in appendix A of BS 5250, it is possible to establish whether a vapour control layer will be necessary in a warm roof construction depending on the conditions that are anticipated.

8.2.5 Fire precautions

BS 476 *Fire test on building materials and structures* Part 3: 1975 *External fire exposure roof test*, lays down procedures to measure the ability of a representative section of a roof, rooflight, dome light or similar components to resist the penetration by fire, when its external surface is exposed to heat, radiation and flame. It also provides methods of establishing the extent of surface ignition. The

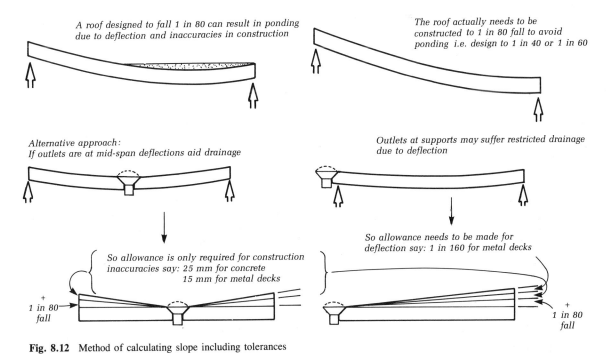

A roof designed to fall 1 in 80 can result in ponding due to deflection and inaccuracies in construction

The roof actually needs to be constructed to 1 in 80 fall to avoid ponding i.e. design to 1 in 40 or 1 in 60

Alternative approach:
If outlets are at mid-span deflections aid drainage

Outlets at supports may suffer restricted drainage due to deflection

So allowance is only required for construction inaccuracies say: 25 mm for concrete
15 mm for metal decks

So allowance needs to be made for deflection say: 1 in 160 for metal decks

+
1 in 80
fall

+
1 in 80
fall

Fig. 8.12 Method of calculating slope including tolerances

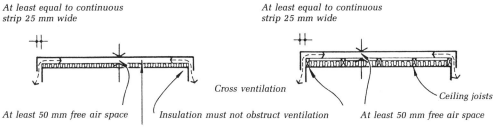

At least equal to continuous strip 10 mm wide

Roof of any pitch

At least equal to continuous strip 10 mm wide

Cross ventilation

Insulation must not obstruct ventilation

PITCHED ROOF

LEAN-TO ROOF

At least equal to continuous strip 25 mm wide

At least equal to continuous strip 25 mm wide

Cross ventilation

Ceiling joists

At least 50 mm free air space *Insulation must not obstruct ventilation*   *At least 50 mm free air space*

Gap may be created by counterbattens where joists run at right angles to flow of air

FLAT ROOF: pitch less than 15°

Fig. 8.13 Ventilation requirements for 'cold roof' coverings (*Building Regulations 1991*)

purpose of the tests described in the BS is to provide information on the behaviour of roofs when there is a fire nearby but outside the building itself. It is important to note that the tests are neither able, nor intended, to predict the performance of a roof in the event of an internal fire.

BS 476 Part 3 is under revision, but in an earlier version (1958) of the current edition, particular forms of roofing systems were designated by two letters (each A to D) for external surface spread of flame (see *MBS: Internal Components*: chapter 2 *Demountable partitions*). These two-letter combinations indicated the permissible proximity of a building with a specific roof covering to a boundary or adjoining buildings, given the necessity to limit the spread of a fire. This double-letter designation is still used for classification purposes by the *Building Regulations 1991*.

However, the current edition of BS 476 Part 3 contains different but related designations for roof coverings, firstly to avoid confusion with the surface spread of flame test characteristics associated with wall and ceiling linings, and secondly to give reference to actual performance data. Accordingly, these designations are P60, P30, P15, and P5: they relate to precise groups of the two-letter designations, and reference should be made directly to the BS for clarification as well as *MBS: Materials*.

In the test procedures laid down in the 1958 version of BS 476: Part 3, samples of the roof construction are subjected to radiant heat on the upper surface and measurements are made of the possibility of fire penetration during 1 hour. A test flame is applied after 5 minutes to simulate the fall of a burning brand and the spread of flame is observed. A preliminary test is also made in which the specimen is subjected to a flame in the absence of radiant heat to identify highly flammable coverings. The two criteria of performance are *penetration time* and distance of external *surface spread of flame*, and the performance of the total roof construction is represented as follows, with an AA designation indicating the best performance that can be obtained:

1. *Penetration classification*
   - A — specimen not penetrated within 60 minutes.
   - B — specimen penetrated in not less than 30 minutes.
   - C — specimen penetrated in less than 30 minutes.
   - D — specimen penetrated in the preliminary flame test.
2. *Spread of flame classification*:
   - A — specimen with no spread of flame.
   - B — specimen with not more than 533 mm spread of flame.
   - C — specimen with more than 533 mm spread of flame.
   - D — specimen which continues to burn for 5 minutes after the withdrawal of the test flame or spread more than 381 mm in the preliminary test.

Roof coverings are given designations such as AA, AC, BA, BB, CA, CD, or 'unclassified' according to their specimen test result. Although not stipulated as a requirement for consideration in the *Building Regulations 1991*, the British Standard also gives an additional designation (suffix 'x' to the two letters) which indicates the likelihood of dripping from the underside of the specimen during test and of mechanical failure or the development of any hole.

The two-letter designations achieved during specimen test are used in Approved Document B of the *Regulations* to define acceptable roof constructions, including coverings, relative to their distance from a possible external fire source. For example, mastic asphalt provides AA designation over deckings of timber, wood-wool, plywood, particle board, concrete, steel, or aluminium; and natural slates achieve this designation when used with timber rafters with or without underfelt, sarking, boarding, wood-wool slabs, plywood, wood or particle board. Nevertheless, the provisions for roof constructions and coverings are made by reference to the size and use of the building and the proximity of the roof to the site boundary. Reference must be made to the *Building Regulations 1991*: Approved Document B for precise requirements, which not only include required designations for roof coverings for purpose groups of building, but also state the area limitations imposed on coverings with less than an AA rating. Generally, those roof coverings designated AA, AB or AC are allowed to be placed close to a boundary irrespective of purpose group; whereas those designated BA, BB and BC need to be at least 6 m from a boundary, and DA, DB, DC or DD at least 20 m from a boundary. A boundary is defined as a line of land belonging to the building, including up the centre of an abutting street, railway, canal or river. The *Regulations* allow concessions for certain plastic materials used in roofing, see section 5.2.

8.2.6 Durability and maintenance

In general, a roof is more vulnerable to the effects of rain, snow, solar radiation and atmospheric pollution, than any other part of a building. Traditional pitched roof coverings such as tiles, slates, lead sheet and even thatch remain serviceable for many years whilst some flat roof coverings have a shorter life (see table 8.2). The question of the rate at which water will 'run off' a roof is of fundamental importance. Pitched roofing has a high rate of runoff and, provided that the detailing of overlaps or jointing is satisfactory, the materials used to cover pitched roofs can have a long life. The runoff from a truly flat roof is very slow indeed and in practice most materials used for flat roof covering do not remain perfectly level and true after laying. As a result, water is very often retained in shallow

Table 8.2 PSA expected life span of flat roofing materials (years)

	Asphalt		Felt	
	30 +	Up to 30	Up to 30	Up to 15
Concrete	A	B	A/B	A/B
Metal deck			A/B	A/B
Plywood			A/B	A/B
Tongued and grooved boards			A/B	A/B
Wood-wool			A/B	A/B

A = inverted roof; B = warm roof.

pools on the finished roof surface. This is known as *ponding*. It is a cause of deterioration because local variations in temperature between the wet and dry area of the roof result in differential thermal movement, which together with accumulations of acid left by evaporating rain, break down the roof surface. Unless ponding is extensive, it is a less significant factor in the failure of flat roofs than is accelerated ageing due to loss of solar protection (resulting in hardening and oxidation), ponding is none the less to be avoided. The most satisfactory method is to construct the roof deck so that it slopes or falls towards the roof outlet to a sufficient degree to shed the surface water. A fall of 1 in 80, say 12 mm in 1000 mm, is required for sheet metal covering and mastic asphalt although a greater fall is desirable where possible. Skirtings at all abutments must be at least 150 mm high.

Apart from the care needed when detailing roofs to ensure adequate performance, the quality of workmanship is another obviously important factor; reference should be made to BS 8000 *Workmanship on building sites* Part 5: 1990 *Code of practice for carpentry, joinery and general fixings* and Part 6: 1990 *Code of practice for slating and tiling of roofs and claddings*. For technical advice on flat roofing, the Flat Roofing Contractors' Advisory Board and the Mastic Asphalt Technical Advisory Centre provide valuable information.

The National Federation of Roofing Contractors (NFRC) has made training a major issue in order to raise standards of workmanship on site. In addition, many roofing companies participate in the BSI Quality Assurance Scheme recommended in BS 5750 (see *MBS: Internal Components* section 1.4 *Component testing and quality assurance'*). Insurance companies have played a large role in upgrading flat roofing methods, including their covering materials, by insisting on appropriate standards (the serviceable life of the roof is reflected in the amount of annual premium payable by the building owner). Designers are increasingly being required to take this factor into account when establishing performance requirements.

Following its construction, a roof requires regular maintenance inspections by a roofing contractor or other person familiar with roof coverings. Should repairs become necessary, the work should be done immediately to avoid deterioration of the roofing. These repairs should always be carried out by skilled operatives using similar materials and techniques to those used in the original roof installation. It is very important that the designers of a roofing system prepare guidelines for building users as part of a building maintenance manual (see *MBS: Introduction to Building* section 16.8).

When calculating the 'cost-in-use' of a roof covering, the cost of maintenance and occasional repair must be taken into account. Although roof coverings can be replaced or mended with less disturbance than for most other parts of a building's fabric, failure to do so will usually involve very costly damage to the structure as well as to the building's contents. From a study cost analysis on various projects, the waterproofing element of a flat roofed building represents approximately 1.5 per cent of the total cost per unit area (assuming the the use of built-up felt roofing or equivalent materials).

8.3 Flat roofing materials

Flat roofing materials form an impermeable barrier to weather penetration (see section 8.2.1) and consist of continuously supported coverings of mastic asphalt, bitumen felt, single ply plastics sheeting, liquid coatings, and dry jointed flat sheets of lead, copper, zinc, aluminium, and stainless steel; or intermittently supported profiled sheets or units. The installed weight of these flat roof coverings are as indicated in table 8.3; it is important to check with manufacturers and/or suppliers to obtain exact weights according to the precise specification of materials to be used.

8.3.1 Mastic asphalt

Mastic asphalt is used for various purposes, such as roofing (see photo 8.2), flooring and tanking, covered by a series of British Standards. Those relevant to asphalt for roofing are BS 6925: 1988 *Specification for mastic asphalt for building and civil engineering (limestone aggregate)*; and CP 144 *Roof coverings*, Part 4: 1970 *Mastic asphalt*, describing methods of application, design and craftsmanship. Where the roof is liable to be used by vehicular traffic, for example as a car park, the relevant British Standard is BS 1447: 1988 *Specification for mastic asphalt (limestone fine aggregate)* for roads, footways and paving in building. A recent development, not yet covered by British Standard, is asphalt modified by small additions of a specially formulated mix of polymers. The resulting

Table 8.3 Installed weights of flat roof coverings

Material	Approx installed weight (kg/m²)
Mastic asphalt	46.0
Bitumen felt (no chippings)	9.6−12.0
Single ply sheeting	1.9−4.7
Liquid coatings	0.5−1.5
Flat metal sheets	
lead	24.4−40.2
copper	2.82−6.3
zinc	4.32−7.2
aluminium	1.92−3.4
stainless steel	3.0
pre-bonded panels	17.0−29.3
Corrugated fibre cement sheets	
single	15.0−19.2
sandwich	22.0−32.0
Profiled metal sheets	
aluminium	
single	2.5−5.2
sandwich	6.0−10.0
steel	
single	4.9−21.9
sandwich	12.0−15.5

Photo 8.2 Laying asphalt. Photo courtesy of Permanite Asphalt Ltd.

Asphalt is delivered to the site as solid blocks that are broken up and then heated in a boiler to about 230 °C. For horizontal roofing, the asphalt is applied with a float at its maximum working temperature, usually in two coats each 10 mm thick. Because the material is applied from a kneeling position, each bay is 2.5−3 m in width. Kerbs and details are formed with the asphalt at a lower temperature and for vertical applications the coats are thinner.

material has properties of high temperature stability (reducing its tendency to flow when used on slopes and improving its long-term durability when used above insulation in warm roof construction), as well as increasing its low-temperature flexibility (see p. 212).

Good general references on the subject of asphalt flat roofing include *Flat roofing: a guide to good practice*, which is a trade publication sponsored by Tarmac Building Products Limited (1982); the *technical information sheets* published by the British Flat Roofing Council (BFRC); and the Mastic Asphalt Council and Employer's Federation's *Roofing Handbook*.

As will be seen from the British Standards, asphalt is an amalgam of asphaltic cement to which is added fine and coarse aggregate. The asphaltic cement consists of bitumen or a mixture containing bitumen and refined lake asphalt according to the grade specified. The fine aggregate is naturally occurring limestone rock and the coarse aggregate is clean igneous or calcareous rock or siliceous material obtained from natural deposits either directly or by screening, crushing or other mechanical process.

The British Standard permits two different percentages of Trinidad Lake asphalt to be incorporated in the asphaltic cement. BS 6925 type R988 may contain either 50 or 25 per cent refined lake asphalt; specifiers should indicate which composition of asphaltic cement is required. Asphalt

with a higher percentage of bitumen may be adequate for use over concrete decks but may be suspect when thermal insulation is incorporated directly below the asphalt in warm roof construction. The higher the percentage of Trinidad asphalt the better from this point of view, although BRE tests suggest that the performance of polymer-modified asphalt is likely to exceed either of these mixes. This is an important consideration because the increasing amounts of thermal insulation being incorporated into roofs is subjecting mastic asphalt waterproofing to a wider range of temperature variations.

CP 144 requires the use of black sheathing felt to BS 747: 1977 *Specification for roofing felts*, type 4A (i) as an isolating membrane under the asphalt. The purpose of the sheathing felt is to isolate the asphalt from movements of the roof decking whilst developing sufficient friction with the deck surface to help restrain the asphalt from contracting in cold weather. Under horizontal asphalt it is laid with 50 mm lapped joints and on vertical timber, or lightweight concrete surfaces, it is overlaid with expanded metal lath to form a key to the asphalt.

Table 8.4 Typical examples of materials used for thermal insulation under mastic asphalt (warm roof) relative to sub-structure

Sub-structure	Insulation	Thickness (mm)	
		0.45 W/m²K	0.25 W/m²K
150 mm *in*	Polyisocyanurate	41	80
situ	Glass fibre board	64	124
concrete	Mineral wool slab	64	124
slab and	Cork board	79	154
screed	Foamed glass	85	164
100 mm	Polyisocyanurate	30	69
precast	Glass fibre board	46	107
concrete	Mineral wool slab	46	107
lightweight	Cork board	57	132
concrete	Foamed glass	61	142
Plywood or	Polyisocyanurate	40	79
particle	Glass fibre board	62	122
board deck	Mineral wool slab	62	122
	Cork board	76	151
	Foamed glass	82	162
Metal deck	Polyisocyanurate	44	83
	Glass fibre board	68	129
	Mineral wool slab	68	129
	Cork board	84	158
	Foamed glass	90	170

Notes: The insulation materials must be overlaid with loose-laid separating felt: mineral wool and glass fibre must also be overlaid with firm heat-resistant board (cork, perlite or woodfibre) before applying separating felt. Foamed glass requires mopping with hot bitumen and then two layers of non-bituminized paper to avoid adhesion to separating felt (not included in these figures).
Source: *Flat roofing — guide to good practice*, produced by Tarmac Building Products Limited.

Roofing asphalt can be used to form a continuous waterproof covering over either flat, pitched or curved surfaces and can be formed more easily than sheet materials into complex shapes and worked round pipes, rooflights and other roof projections.

It can be laid on most types of rigid sub-structure such as concrete (either screeded, precast or *in situ*), plywood decking or a variety of proprietary structural deck units. Because asphalt is, however, a homogeneous, seamless material that is relatively brittle at low temperatures, it is most durable and has its greatest life expectancy when laid upon a heavyweight stable structure made of concrete. The Property Services Agency accord the longest anticipated life of any flat roofing material to asphalt but only when laid on concrete as shown in fig. 8.14. Timber boarding should be avoided for use as decking, drying of the timber results in distortion of the face of the boards and the surface of the asphalt. Plywood is satisfactory but preferably for roofs of limited size so that the number of joints in the decking are minimized.

Durability   Mastic asphalt when laid by a good specialist roofing sub-contractor on a sound base will not require major repairs for at least 60 years. When repairs are required they should always be carried out by a specialist. Eventually, exposed mastic asphalt is broken down by acids in the atmosphere and by UV radiation. So a surface finish such as stone chippings greatly increases the durability of the waterproofing. Where there will be pedestrian or vehicular traffic the roofing asphalt will require a protective layer of, for example, paving grade asphalt.

Mastic asphalt is a dense material and being very dark in colour absorbs solar heat very readily and can soften, especially where insulation is laid below the asphalt (from which point of view, the use of an inverted warm roof construction is particularly advantageous). A wide range of solar reflective chippings is available to cover exposed asphalt roofing. Various coloured granites, white limestone, calcined flint and white spar, usually in sizes up to 13 mm, are the most widely used. Reflective chippings are suitable for use on roofs up to 10 degree pitch. They are embedded in a layer of bitumen dressing compound to form a textured surface. Vertical and inclined surfaces should be coated with solar reflective paint.

Thermal insulation   For general comments on thermal insulation standards for roofs see p. 196. Table 8.4 indicates forms of construction and thicknesses of various insulation boards which each give a *U*-value within the *Building Regulations 1991*, Part L. Included in the specification for table 8.4 are surface chippings on 20 mm mastic asphalt and black sheathing felt below; the effect of the vapour control layer beneath the insulation has also been assumed but not an internal ceiling finish.

Vapour control   It is frequently necessary to provide a vapour control layer on the warm side of the insulation. A vapour control layer should consist of not less than BS 747 Type 3B roofing felt lapped and bedded in hot bitumen, though the use of a more substantial membrane is desirable. A vapour control layer should consist of one layer of felt to BS 747 type 3B and one layer of type 5. The best types incorporate an impermeable metal foil or thick plastic sheet. A good example of a proprietary vapour control layer as supplied by specialist asphalt contractors consists of a sheet of aluminium foil protected by a coating of bitumen on both sides and reinforced with a sheet of glassfibre tissue. The vapour control layer should be folded back at least 225 mm over the outer edges of the insulating layer and the asphalt roofing bonded to the overlap as shown in fig. 8.14. A vapour control layer should not be used beneath a wet applied, cement-based screed; separation of the screed from the concrete deck is liable to cause the screed to crack or curl and damage the waterproofing (see

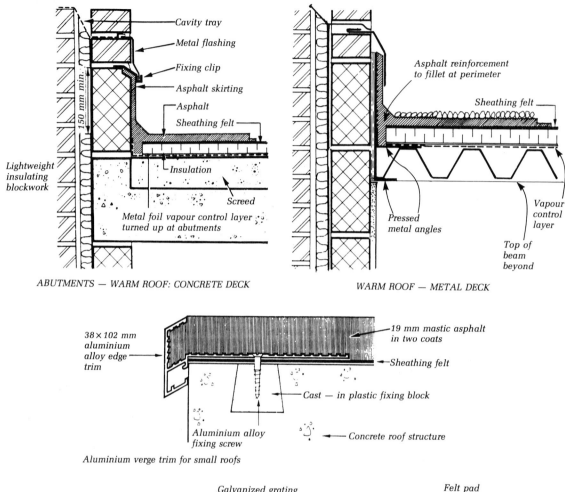

Cavity tray

Metal flashing

Fixing clip

Asphalt skirting

Asphalt

Sheathing felt

150 mm min.

Lightweight insulating blockwork

Insulation

Screed

Metal foil vapour control layer turned up at abutments

ABUTMENTS — WARM ROOF: CONCRETE DECK

Asphalt reinforcement to fillet at perimeter

Sheathing felt

Pressed metal angles

Vapour control layer

Top of beam beyond

WARM ROOF — METAL DECK

38 × 102 mm aluminium alloy edge trim

19 mm mastic asphalt in two coats

Sheathing felt

Cast — in plastic fixing block

Aluminium alloy fixing screw

Concrete roof structure

Aluminium verge trim for small roofs

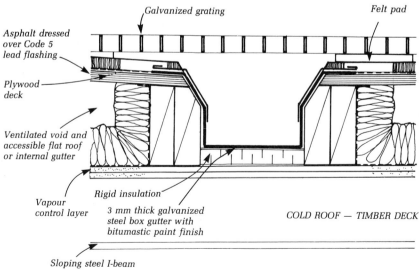

Galvanized grating

Felt pad

Asphalt dressed over Code 5 lead flashing

Plywood deck

Ventilated void and accessible flat roof or internal gutter

Vapour control layer

Rigid insulation

3 mm thick galvanized steel box gutter with bitumastic paint finish

COLD ROOF — TIMBER DECK

Sloping steel I-beam

Fig. 8.14 Asphalt details: abutments, eaves and verge

DOE advisory leaflet no. 79 *Vapour barriers and vapour checks*). In an inverted warm roof the asphalt will act as both the waterproofing membrane and the vapour barrier. A vapour control layer if provided at ceiling level (such as foil-backed plasterboard), will be punctured by frequent fixings and will only form a vapour check not a barrier. See p. 199 for general comments.

Fire precautions Mastic asphalt fulfils all the requirements for a roof covering as described in BS 5588 *Fire precautions in the design and construction of buildings* Part 1, section 1.1: 1984. Also asphalt achieves the designation P60/AA under the test requirements of BS 476 Part 3 *External fire exposure roof tests*. See also p. 203 for general comments.

Application *In situ* concrete, screeded and precast concrete beams and slabs, wood-wool slabs, plywood decking, and metal decking are all suitable methods of construction for the sub-structure upon which asphalt may be laid (subject to the provisos mentioned above). In all cases the sub-structure must be strong enough to prevent excessive deflection. And in particular, where metal decking is used, the deflection limit must be reduced to 1/325 of the span instead of the more normal 1/240. For plywood or metal decking, at the upstand walls around the roof, a timber or pressed metal kerb on which expanded metal is fixed forms a key for the asphalt. A space between the wall and kerb allows for movement as shown in fig. 8.14.

The asphalt is laid over a timber roll or solid mastic asphalt water check is formed to prevent water being blown over verges. If the sub-structure of a 'cold' roof is to be formed from wood-wool, timber or plywood, provision must be made for adequate ventilation between the roof deck and the insulated ceiling (though this form of construction should be avoided if possible — see section 8.2.4).

The sub-structure on which the asphalt is to be laid should ensure the rapid dispersal of rainwater. Falls should be designed to be not less than 1:50 in order to attain actual falls when constructed of not less than 1:80 (see fig. 8.12). Any change in the direction of the roof surface in buildings shaped as letter T or L indicates the possible need for movement joints. They should be continuous through the entire structure, including roof, walls and upstands, as shown in fig. 8.16(d), which is similar in one of those illustrated for built-up felt roofing in fig. 8.27(c).

To assist in drying out wet laid cement screeds on concrete decks and to release trapped moisture, it may be considered necessary to install drying vents. Either a proprietary ventilation system or large-diameter ventilators should be located at a spacing of one ventilator per 25 to 40 m². Small pressure relief ventilators are not generally

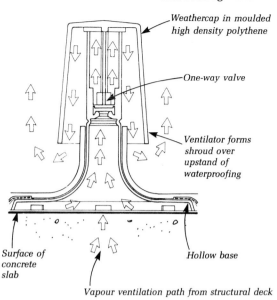

Weathercap in moulded high density polythene

One-way valve

Ventilator forms shroud over upstand of waterproofing

Surface of concrete slab

Hollow base

Vapour ventilation path from structural deck

Fig. 8.15 Proprietary roof ventilator (Euroroof Ltd)

considered useful in conjunction with mastic asphalt. A proprietary example is shown in fig. 8.15.

On flat roofs and roofs up to 30 degrees pitch the roofing asphalt is applied with a wooden float in two coats laid to a minimum thickness of 20 mm on an underlay of black sheathing felt laid loose with 50 mm lapped joints. The asphalt if exposed, should be dressed with reflective mineral chippings to reduce the temperature induced by solar heat and to protect it from UV radiation and fire. On flat roofs to be used for foot traffic, the asphalt should be laid in two coats to a minimum total thickness of 25 mm, the second coat being 15 mm thick and having 5−10 per cent additional grit added.

Because asphalt 'flows' in hot weather, its suitability as a material for use on sloping surfaces can be questioned, if this situation is encountered special precautions have to be taken. On slopes of over 30 degrees the asphalt is applied without sheathing felt in three coats, the first coat being applied very thinly with a steel trowel. The second and third coats are then applied to the breaking joint to give a total thickness of not less than 20 mm. Where the asphalt is laid on vertical or sloping surfaces of more than 30 degrees a positive key is required. In the case of sloping surfaces over 10 degrees formed in plywood boarding a layer of black sheathing felt is nailed to the plywood boards and bitumen-coated expanded metal lathing is then fixed at 150 mm centres with galvanized clout nails or staples to form the key for the asphalt which is then applied in three coats. In all cases where the asphalt is laid on flat or slightly sloping roofs, clean sharp sand is rubbed evenly over the surface of the asphalt whilst it is still hot. This breaks up

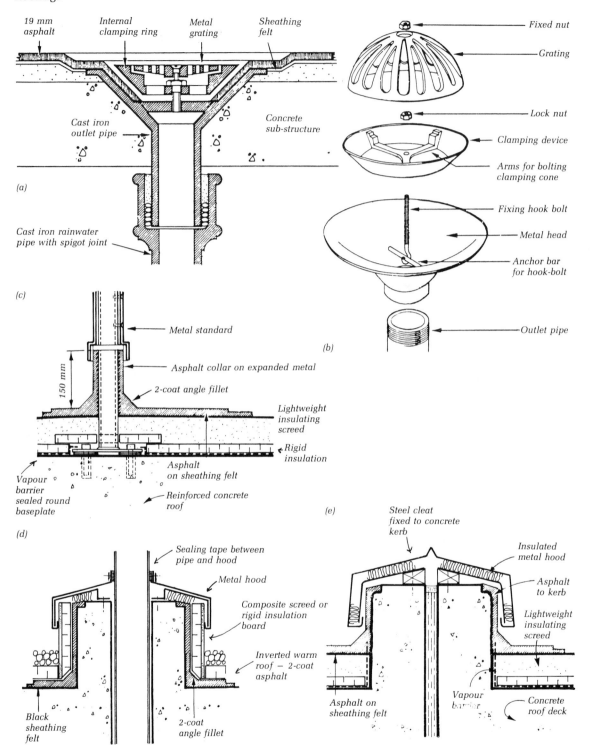

Fig. 8.16 Details of asphalt roof penetrations: (a) uninsulated roof − rainwater outlet; (b) alternative rainwater outlet; (c) warm roof − base fixing for handrails, asphalt finish to metal standard; (d) projecting vent pipe through roof; (e) warm roof − twin kerb expansion joint

the skin of the bitumen brought to the surface by the wooden float at the time of application. The object of this is to minimize the gradual crazing of the surface due to the action of the sun.

Details at abutments and edges are shown in the drawings as follows: fig. 8.14 abutments, eaves and internal gutter; fig. 8.16(a) rainwater outlet with grille; fig. 8.16(b) an alternative type of outlet made from gunmetal or spun steel which has a domed grating and a clamping device which allows the grating to be tightened against the waterproofing; balustrade detail 8.18(c); fig. 8.16(d) projections through roof; fig. 8.16(e) movement joint. Where there is continuous foot traffic, mastic asphalt can be protected with concrete tiles or with a jointed screed.

Tiles Concrete (GRC) tiles are approximately 300 × 300 × 25 mm thick. The tiles are laid in bays of maximum 9 m² with 25 mm joints between the bays which are filled with bitumen compound. The tiles are set 25 mm back from the base of angle fillets and the margin is completed with bitumen compound. A bitumen primer is applied to the surface of the mastic asphalt roof covering and the backs of the tiles and allowed to dry (though priming is not necessary if porous concrete tiles are used). The tiles are then bedded in hot bitumen bonding compound, taking care not to squeeze the compound upwards between the individual tiles.

Jointed screed A cement and sand screed 25 mm thick is laid on a separating membrane of building paper and grooved into a 600 mm square tiled pattern. The screed should be laid in bays of not more than 9 m² with a 25 mm joint between the bays. The grooves and joints can be filled with hot bitumen compound on completion.

The following are examples of typical mastic asphalt roofing specifications:

Plywood decking (cold roof) A construction to be used with caution, see section 8.2.3.
- Decking of 19 mm tongued and grooved exterior grade WBP plywood to BS 6566, well nailed to timber joists and noggings, laid to falls. Joints in the decking should be minimized; those not closed off by the support system should be taped. Mineral wool boards above 9 mm foil-backed (vapour check) plasterboard ceiling finished with skim coat plaster (to achieve a combined *U*-value complying with the *Building Regulations*, see section 8.2.3). Cavity above insulation vented to outside (see section 8.2.4). Separating layer of loose-laid BS 747 type 4A(i) sheathing felt with non-bitumenized paper underlay to prevent adhesion to the decking.
- Roof covering of 20 mm two-coat mastic asphalt to BS 6925 type R988 T25 (25 per cent lake asphalt, 75 per cent bitumen) finished with 10−13′m stone chippings bedded in bitumen based adhesive compound.

Concrete decking (warm roof)
- *In situ* reinforced concrete deck incorporating temporary drainage holes.
- Screed of 25 mm minimum thickness to falls 1:50 with top surface finished smooth using wood float; apply a bitumen-based primer to bind damp or dusty surfaces.
- Vapour check of one layer of BS 747 type 3B glass fibre base felt bonded in bitumen. Assuming that calculations have shown a full vapour barrier to be unnecessary in this instance, the single layer of felt is acting as a temporary 'damp proof course' until the concrete and screed dry out (to prevent dampness rising into the insulation).
- Thermal insulation of insulation board (to achieve a combined *U*-value complying with the *Building Regulations*, see section 8.2.3). Heat-sensitive boards such as expanded polystyrene, polyisocyanurate rigid urethane foam, mineral wool or glass fibre must be overlaid with a firm heat-resistant board, such as wood fibre board, perlite board or cork.
- Separating layer of loose-laid BS 747 type 4A(i) sheathing felt to allow for differential movements.
- Two-coat polymer modified mastic asphalt (20 mm) finished with 10−13 mm stone chippings bedded in bitumen-based adhesive compound.

Figure 8.17 shows an alternative and increasingly used form of flat roof construction where the thermal insulation is placed external to the mastic asphalt (or built-up bituminous felt — see p. 211). This method creates a 'warm roof' (see section 8.2.3), and is referred to as a 'protected membrane roof', 'inverted roof,' 'upside-down roof' or simply as an 'externally insulated roof'. Non-absorbent, usually foamed plastic insulation, is placed over the mastic asphalt (or felt) and is held down by gravel, by paving slabs or by the use of pre-screeded insulation board. A recent, relatively unproven, alternative is mineral fibre insulation which is produced in dense slabs that are sufficiently heavy not to require ballast, although additional fixing may be required at the edges of roofs or in particularly exposed locations.

This form of construction has several advantages over the conventional warm roof:

- The insulation and paving or ballast protect the asphalt.
- Temperature variations in the asphalt are reduced, an important consideration given the increasing thermal requirements with which roofs now have to comply. Also the waterproof layer is protected from UV light and impact damage.
- The layers of the construction are laid loose one above the other and so differential thermal movements between the materials are not transmitted between them.
- The waterproofing also acts as a vapour barrier, being on the warm side of the insulation, thus avoiding the problems of entrapped moisture associated with the

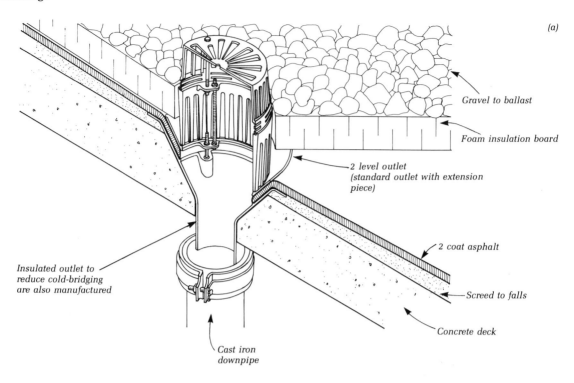

(a)

Gravel to ballast

Foam insulation board

2 level outlet
(standard outlet with extension
piece)

2 coat asphalt

Screed to falls

Concrete deck

Insulated outlet to
reduce cold-bridging
are also manufactured

Cast iron
downpipe

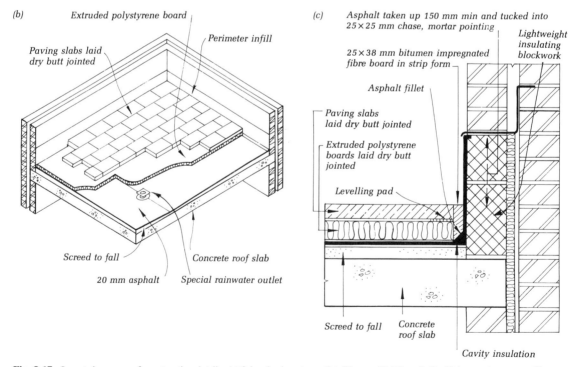

(b)

Extruded polystyrene board

Perimeter infill

Paving slabs laid
dry butt jointed

Screed to fall

20 mm asphalt

Special rainwater outlet

Concrete roof slab

(c)

Asphalt taken up 150 mm min and tucked into
25 × 25 mm chase, mortar pointing

Lightweight
insulating
blockwork

25 × 38 mm bitumen impregnated
fibre board in strip form

Asphalt fillet

Paving slabs
laid dry butt jointed

Extruded polystyrene
boards laid dry butt
jointed

Levelling pad

Screed to fall

Concrete
roof slab

Cavity insulation

Fig. 8.17 Inverted warm roof construction details: (a) 2-level rainwater outlet (Harmer Holdings Ltd); (b) inverted warm roof layout; (c) abutment detail − 2-coat asphalt roofing with felt underlay; (d) abutment detail −bitumen felt roofing

(d)

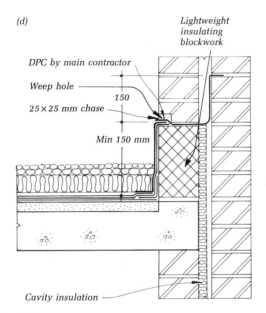

DPC by main contractor

Weep hole

150

25×25 mm chase

Min 150 mm

Lightweight insulating blockwork

Cavity insulation

Fig. 8.17　*cont.*

'sandwich' roof constructions previously described.

- Inverted warm roof construction is relatively heavy and therefore best suited for application to a concrete structural deck (as indeed is asphalt).

There are, however, some disadvantages:

- Because the waterproofing is not visible, leaks may be difficult to find.
- The waterproofing will be wet for longer periods than in a traditional flat roof construction.

In the example shown the waterproof covering is applied in the normal manner to a dry screed which has been laid to a fall of 1:50. In the case of mastic asphalt, it is laid on sheathing felt and the upstand at the wall is taken up to at least 150 mm above the level of the finished roof surface. Extruded polystyrene foam boards are placed on the asphalt: they are laid loose, with tight butt joints and all joints staggered. The polystyrene must be 20 per cent thicker than when used in warm roof construction in order to cater for the sudden cooling of the membrane that will occur during rain and to overcome the external exposure conditions (snow and frost). It is considered a suitable material for this application because it is water and vapour resistant, dimensionally stable, is readily available, CFC free, tough and resistant to temperature cycles, and is available in a flame-retardant, self-extinguishing grade. The polystyrene is protected by hydraulically pressed precast concrete paving slabs (their weight and thickness need to be calculated relative to the anticipated wind loads on the roof), laid dry with butt joints. No bedding is

necessary, but they must be laid on small blocks or spacers at corners (e.g. 150 × 150 mm inorganic felt pads) to avoid water retention between the underside of the slabs and the insulation, as well as to facilitate drainage. A 38 mm minimum gap must be left between perimeter abutments and the insulation/paving to allow for differential movements, and this gap can be filled with bitumen-impregnated fibre board, a compressible plastics strip or loose gravel.

No gutter is formed in the roof as the screed falls to a lowest line and the rainwater runs to this line before being collected by outlets located to ensure each serves about 35 m² of roofing. The outlets are required to be at two levels: some to collect water from the waterproof membrane and some for the water flowing at paving level, and specially designed rainwater goods are manufactured for this purpose (see fig. 8.17(a)).

8.3.2 Bitumen felt

The following terms are commonly used in connection with bitumen felt roofing (see photos 8.3 and 8.4):

- *Built-up roofing* — two or three layers of bituminous sheet roofing fused together on site with bitumen compound and consisting of a first layer, an intermediate layer (in three-layer systems) and a top layer or cap-sheet.
- *Layer* — a single thickness of membrane roofing. The word 'ply' is often synonymous with layer, but is sometimes used to denote the thickness of bitumen sheet.
- *Underlayer* — an unexposed layer in built-up roofing.
- *Cap-sheet* — an exposed or final layer in built-up roofing.

BS 747: 1977 (1986) *Specification for roofing felts* describes different types of bituminous-based roof felts, and CP 144 *Roof coverings*, part 3: 1970 *Built-up bitumen felt* gives suitable methods for their application.

The manufacture of standard roofing felt is a continuous process involving the impregnation of a base material with a penetration grade bitumen, and then coating the product with a filled oxidized bitumen to provide the waterproofing. Finally, a sand surfacing is applied to the roofing felt to prevent sticking within the roll form in which they are supplied. Table 8.5 shows the BS 747 classification of felts according to base and surface finish. Each type of felt is further subdivided according to its weight per length and width of roll. The heavier felts are normally used as a top layer and taking into account the different types, finishes and weights available, the specifier has a wide choice.

The ageing and weathering characteristics of bitumen over a period of time approximating to 15 years are well known. The addition of a polymer to the bitumen improves

Photo 8.3 Laying a felt roof. Photo courtesy of D. Anderson and Son Ltd.

Photo 8.4 Laying a felt roof. Photo courtesy of D. Anderson and Son Ltd.

Bitumen which has been heated in a boiler to between 200 and 250 °C is poured in front of the roofing felt which is unrolled into it. For detail work, the bitumen may be applied by mop, or the bitumen may be poured from jugs rather than buckets. Most felt roofing is applied with 100 mm end laps and 50 mm laps at the sides.

its properties as a roofing material in almost all respects, and in particular, flexibility, strength and fatigue resistance. The most commonly used modifying additives are styrene butadiene styrene (SBS) and atactic polypropylene (APP). Roofings with SBS additives have the greatest elasticity and elongation and generally involve conventional hot bitumen bonding techniques. Those with APP additives have improved high-temperature performance, and can have better weathering characteristics. But they are not suitable for bonding with ordinary oxidized bitumens and so are generally bonded by torching.

Polymer-modified bitumens are usually applied to a base of polyester or glass and form part of a range of *high-performance roofings* recommended by the BFRC. Others in the range are *calendered polymeric roofings* which do not contain a base as they are formed from proprietary compounds of polymers and bitumen which are calendered to form sheets of high flexibility and elasticity; and *metal foil surfaced felts* which achieve high strength with a polyester or woven glass base and have a facing of aluminium or copper. The metal facing provides an effective protection to the membrane as it excludes UV light, oxygen and ozone which are the chief factors causing ageing and hardening of bitumen. Aluminium also has good reflective properties and effectively reduces the temperature of the bitumen, but should not be used where there is an alkaline atmosphere.

It is usual to provide a protective finish to the top layer of built-up felt roofing to give a reflective finish and protect the bitumen from sunlight or to provide a wearing surface. Felts are made with a factory applied reflective mineral surface. Alternatively, stone chippings 10–13 mm in size are often used bedded in bitumen and there is a wide range of types, Derbyshire, white spar and Leicester red granite being examples. If the roof is to be accessed (for

Table 8.5 Bitumen felt classifications

Class 1	Fibre based	Weight/roll
1B	fine granule surfaced	36 kg/20 × 1 m
1E	mineral surfaced	38 kg/10 × 1 m
1F	hessian reinforced base with fine granule surfaced	22 kg/15 × 1 m

These are the original felts used in industry, the cheapest, and have failings which suggest they should no longer be used in built-up roofing.

Class 3	Glass based	Weight/roll
3B	fine granule surfaced	36 kg/20 × 1 m
3E	mineral surfaced	28 kg/10 × 1 m
3G	(perforated) grit finished underside, fine granule surfaced topside	32 kg/15 × 1 m

These give the best waterproof performance and should give many years of service. They are not suitable for nailing as they do not have enough strength. Type 3G is used for the first layer in partially bonded systems.

Class 4	Sheathing felts	Weight/roll
4A (i)	black sheathing felt (bitumen)	17 kg/25 × 0.81 m
4A (ii)	brown sheathing felt	17−21 kg/25 × 0.81 m
4B (i)	black hair felt	41 kg/25 × 0.81 m
4B (ii)	brown hair felt	41 kg/25 × 0.81 m

These are used under mastic asphalt roofing because they are dimensionally stable, and as sarking felts under metal roofing. Hair felts are used for heat insulation, sound absorption and other purposes.

Class 5	Polyester based	Weight/roll
5B	fine granule surfaced bitumen polyester top layer	34 kg/8 × 1 m
5E	mineral surfaced bitumen polyester cap-sheet	38 kg/8 × 1 m
5U	fine granule surfaced bitumen polyester underlayer	14 kg/8 × 1 m

Development work on built-up roofing felts has led to the recent inclusion of this type in the BS. Two main areas of deficiency in the other types had to be addressed: age hardening of bitumen, and long-term strength of elongation properties of the felt base. Polyester bases appear to be stable and strong for a long life, and to be compatible with bitumen. It provides a non-rotting base, with greater strength than glass and greater elongation before breaking. See also main text regarding the use of polymer-modified bitumen.

Source: BS 747: 1977 (1986) *Classification of bitumen felts.*

maintenance, etc). the spar finish will not be suitable because of the danger of the sharp granules cutting the felt. Roofs to be used for foot traffic may be finished with concrete, GRC or fibre cement tiles.

Durability Successful built-up felt roofing is dependent on the correct specification for each situation. Good materials, careful detailing and correct application are all necessary and so, since supervision is difficult, the contract should be carried out by a specialist contractor experienced in this type of work.

Thermal insulation For general comments on thermal insulation standards for roofs see p. 196. Table 8.6 indicates forms of construction and thicknesses of various insulation boards which each give a *U*-value within the *Building Regulations 1991* requirements. The figures assume solar control chippings on three layers of built-up felt; a vapour barrier beneath the insulation quoted has also been included but a ceiling finish has not.

Table 8.6 Typical examples of materials used for thermal insulation under built-up felt (warm roof) relative to sub-structure

Sub-structure	Insulation	Thickness (mm)	
		0.45 W/m²K	0.25 W/m²K
150 mm *in situ* concrete slab and screed	Expanded polystyrene*	55	116
	Glass fibre board	64	124
	Mineral wool slab	64	124
	Cork board	79	154
	Foamed glass	85	164
100 mm precast concrete lightweight concrete	Expanded polystyrene*	37	98
	Glass fibre board	46	107
	Mineral wool slab	46	107
	Cork board	57	132
	Foamed glass	61	142
Plywood or particle board deck	Expanded polystyrene*	53	113
	Glass fibre board	62	122
	Mineral wool slab	62	122
	Cork board	76	151
	Foamed glass	82	162
Metal deck	Expanded polystyrene*	59	119
	Glass fibre board	68	129
	Mineral wool slab	68	129
	Cork board	84	158
	Foamed glass	90	170

* The *U*-value for expanded polystyrene allows for additional 13 mm wood fibreboard overlay.
Source: *Flat roofing — guide to good practice*, produced by Tarmac Building Products Limited.

Vapour control Condensation is liable to occur on the internal surfaces of a roof construction within a building if the temperature and humidity of the air inside are appreciably higher than the outside temperature and humidity. Thus a vapour barrier (two layers of felt or a proprietary aluminium cored felt) will be required or, in conditions of less risk, a vapour check (a single layer of felt) will suffice. The choice of method can be decided by *moisture gain analysis* (see section 8.2.4). The vapour control layer will be on the underside (the warm side) of the insulation below the built-up roofing.

Fire precautions Where stone chippings are used as topping all felt flat roofs have the highest AA BS 476 Part 3 fire rating. On pitched roofs the rating varies according to the type of felt in each layer and the combustibility of the roof deck.

Application Hollow precast concrete beams or slabs, *in situ* concrete, aerated concrete, wood-wool slabs, timber construction or profiled metal, all provide suitable roof decking for built-up felt roofing. The use of these materials is more fully discussed on p. 192. All built-up felt roofing should be carried out in accordance with the requirements of CP 144 Part 3 1970 *Built-up bitumen felt*. It is necessary to provide falls to clear the water from a flat roof and for built-up roofing a minimum of 17 mm in 1000 mm is recommended (1 degree slope). The first layer of roofing felt is fixed by nailing, or by full bonding or by partial bonding according to the nature of the sub-structure. Partial bonding to the deck prevents the formation of blisters in the waterproofing due to vapour pressure, and gives a measure of freedom of movement between the roof deck and covering.

BRE digest 51 *Developments in roofing* discusses at length the problem of water vapour in the roof deck, particularly in connection with lightweight screeds. The sources of screed moisture are as follows:

- Mixing water.
- Rainwater during the drying out period.
- Condensation formed within the building.

The effects of a saturated screed are as follows:

- Diminished thermal insulation.
- Blistering and damage to waterproofing by vapour pressure.
- Staining of internal decoration.

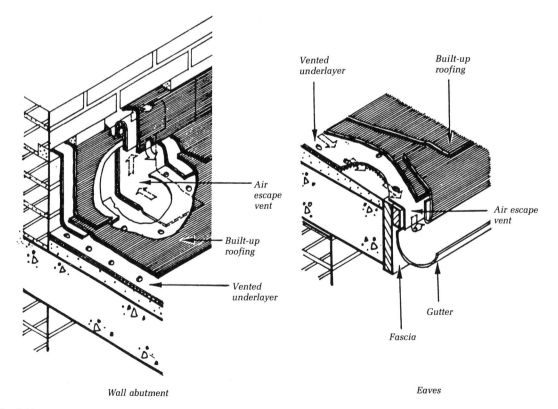

Wall abutment

Eaves

Fig. 8.18 Bitumen felt details: water vapour release by vented underlay

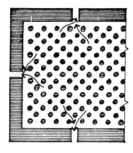

Spot sticking

Fig. 8.19 Bitumen felt details: water vapour release by spot sticking

Experiments at the BRE on exposed aerated screed found that after 4 wet days and with subsequent shielding from further rain, complete drying out took 36 good summer days or 180 winter days. Research and long observation has revealed that nearly all roof blisters are caused by the entrapping of constructional moisture in the roof deck. Solar heat vaporizes this moisture and causes considerable pressure which weakens the bond between the roofing layers and the roof sub-structure. Ventilation of the roof deck, below built-up felt roofing, can be obtained in several ways as follows.

Where built-up roofing is applied to a substrate liable to have retained moisture, for example screed, concrete, plywood or wood-wool, the base layer should be only partially bonded to the decking. The first layer used is a type 3G felt (to BS 747) which is perforated with a regular pattern of holes. This *vented underlay* is laid loose but the bitumen poured across it to fix the second layer also seeps through and around the holes to partially bond it to the decking. The under-surface of these perforated felts is impregnated with coarse granules which allows a sufficient spread of bitumen and prevents the bitumen within the felt from adhering to the deck in warm weather. A substantial area of roofing is left partially bonded so water vapour can migrate to a vent detailed at the edge of the roof (fig. 8.18) or to roof ventilators (trapped vapour would otherwise cause the weatherproofing to blister and tear). Proprietary plastics or metal breather vents are made in a variety of types specially designed to dry out wet screeds and/or to act as pressure release vents. Figure 8.19 shows a typical pattern of partial bonding; additional attachment to resist wind uplift is provided by fully bonding a strip of the underlay around the perimeter and at penetrations such as rainwater outlets. Vented underlay is in fact recommended for most roof decks other than timber boarding (to which the first layer should be nailed to provide positive attachment whilst allowing

for movement of the boards). Perforated felts are also used in warm roof constructions containing polyurethane and polyisocyanurate insulation boards which release gas on the application of hot bitumen.

BS 6229: 1982 *Flat roofs with continuously supported coverings* recommends that all timbers used for the structure of a flat roof should have preservative treatment to BS 5268: Part 5 (see section 8.2.4). Typical examples of built-up felt roofing specifications are described under the headings below.

Timber decking (cold roof) This is a construction to be used with care (see section 8.2.4).

- Roof deck of 19 mm nominal thickness tongued and grooved boards, closely clamped and securely nailed to 200 × 50 mm timber joists at 450 mm centres, all timbers treated with preservative to BS 5268: Part 5. Mineral wool boards between joists above 9 mm foil-backed plasterboard ceiling, finished with skim coat plaster. Cavity above insulation vented to outside (see section 8.2.4).
- Roof covering of: first layer — type 5U polyester-based underlay nailed to the decking; second layer — type 5B polyester-based felt fully bonded to the first layer; and cap-sheet — sanded type 5U polyester-based felt fully bonded to the intermediate layer and finished with 10 mm stone chippings bedded in bitumen based adhesive compound.

Timber decking (warm roof)

- Warm roof construction is recommended by the BFRC in all situations — see section 8.2.4.
- Decking of exterior grade WBP 19 mm tongued and grooved plywood panels to BS 6566 well nailed to 200 × 50 mm wide timber joists and noggings, laid to falls. Any deck joints not closed off by the support system should be taped. Single layer vapour check of type 5U polyester base or type 3B glass fibre base felt bonded in bitumen (suitable if there is a low condensation risk).
- Mineral wool insulation slab bonded in hot bitumen (giving a combined *U*-value sufficient to comply with *Building Regulations* Part L, see section 8.2.3).
- Roof covering of: first layer — type 3B glass fibre base felt, bonded in bitumen; second layer — type 5U polyester base felt bonded in bitumen, and cap-sheet — type 5B polyester base felt or similar, bonded in bitumen and finished with 10 mm stone chippings bedded in bitumen-based adhesive compound.

Concrete decking See fig. 8.17 — protected membrane warm roof.

- *In situ* cast dense concrete slab deck (175 mm) with wood float finish to falls and incorporating temporary

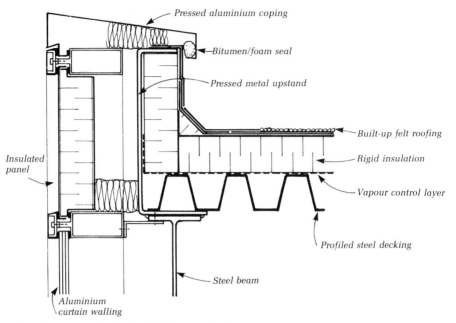

Fig. 8.20   Bitumen felt details: flashing at roof edge

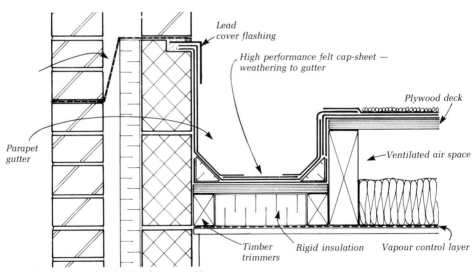

Fig. 8.21   Bitumen felt details: abutment with parapet

drainage holes. Apply a bitumen-based primer to bind damp or dusty surfaces.

- High-performance roof covering of: first layer — type 3G glass fibre base perforated felt, partially bonded in bitumen; second layer — type 3B glass fibre base felt, bonded in bitumen; and cap-sheet of bitumen polymer or pitch polymer roofing bonded in bitumen (pour and roll method), or an APP modified bitumen roofing (torch applied).

- Underlay sheet (if required) to manufacturer's specification laid loose to even out cap layer surface irregularities.

- Thermal insulation of extruded polystyrene boards laid butt-jointed and staggered (giving combined U-value to comply with *Building Regulations* Part L, see section 8.2.3).

- Gravel of 50 mm minimum thickness (20–30 mm nominal diameter). If significant quantities of fine gravel

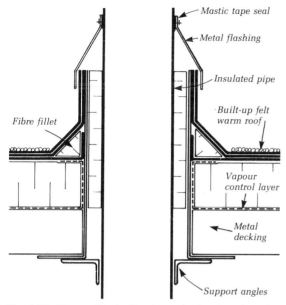

Fibre fillet

Mastic tape seal

Metal flashing

Insulated pipe

Built-up felt warm roof

Vapour control layer

Metal decking

Support angles

Fig. 8.22 Bitumen felt details: pipe projection through roof

are present, it will be necessary to add a filter layer above the insulation to prevent fine material working through the joints and accumulating on the underside of the boards. A 50 mm depth of gravel will prevent flotation of the insulation, providing there is efficient drainage (see p. 211).

In view of the very many alternative specifications possible in built-up roofing and the various weights and types of felt available, it is advisable to take advice from the specialist contractor regarding the intended specification with regard to suitability and particularly with regard to cost. Typical details of three-layer built-up felt roofing are shown in the following figures: fig. 8.20 detail at roof edge; fig. 8.21 detail at abutment with parapet gutter; fig. 8.22 detail showing weathering to a pipe projecting through the roof; fig. 8.23 detail at a rainwater outlet and an eaves gutter; fig. 8.24 detail at verge.

As an alternative to the welted drip shown at the verge, aluminium trim is available in various depths and profiles to receive the built-up felt and asphalt. A typical profile is shown in fig. 8.25. The welted drip (fig. 8.24) is formed by nailing the felt over the roof surface lapping with the roof covering according to the direction of the fall. The depth of the apron can be varied but will not be satisfactory if it is less than 50 mm. Aluminium trim is 'built-in' to the three-layer felt system as shown in fig. 8.25. Because of the possibility of electrolytic action between steel and aluminium, it is fixed with stainless steel or aluminium alloy screws at 450 mm centres. The trim is produced in standard lengths of 3050 mm, long pieces are joined by a spigot,

a 3 mm gap being left between each length to allow for expansion. A glass fibre reinforced polyester resin verge trim is shown as an alternative detail in fig. 8.26. Since there is no possibility of electrolytic action galvanized steel screws can be used for fixing. Because of the lower thermal expansion of GRP, it is not necessary to leave a gap between the lengths of trim. Built-up felt roofing will often be detailed to incorporate an internal or secret gutter. A typical detail incorporating a plastic outlet is shown in fig. 8.26.

In order to prevent the failure of the roofing due to movement in the structure it is often necessary to incorporate joints in the roof finish which will accommodate relative movement; figs 8.27(a), (b) and (c) illustrate suitable joints for minor, moderate and major movement respectively.

8.3.3 Single ply sheeting

A single ply membrane laid over an insulated roof deck is often seen by designers as an ideal flat roof covering: it forms a seamless umbrella that protects the building from rainfall without the risk of water penetration through joints. However, for satisfactory results to permit acceptance by insurance companies, they must be installed by specialists. The recommendations of the members of the Single Ply Roofing Association (SPRA) should be followed and the system adopted must conform to the requirements of published standards, or have been independently assessed for performance by the British or European Boards of Agrément. Materials currently used for single ply flat roofing are:

- *Thermoplastic*: including black-coloured polyiso-butylene (PIB) and ethylene copolymerized bitumen (ECB), light-coloured, white or grey, chloropolythene (CPE), polyvinylchloride (PVC), and chlorosulphonated polyethylene (CSPE) 'Hypalon'.
- *Thermosetting/elastomeric*: including normally black coloured, but also available in white, ethylene propylene dimonomer (EPDM); white-coloured silicone; neoprene and butyl rubber, both of which are susceptible to ozone degradation and are now little used for roofing; and light-coloured glass fibre reinforced plastics (GRP).
- *Modified bitumen* (see p. 211): including metal foil finished or sand-faced self-adhesive bitumen polymer felts. Generally, bitumen-containing products are less resistant to UV radiation than other alternatives.

EPDM, ECB, PIB, CSPE and CPE are compatible with bitumen, and ECB is also resistant to UV light, ozone, most chemicals as well as biological attack. PVC is not compatible with bitumen or polystyrene and is vulnerable to plasticizer migration. PIB is comparatively soft and easily punctured and needs protection, even for normal foot

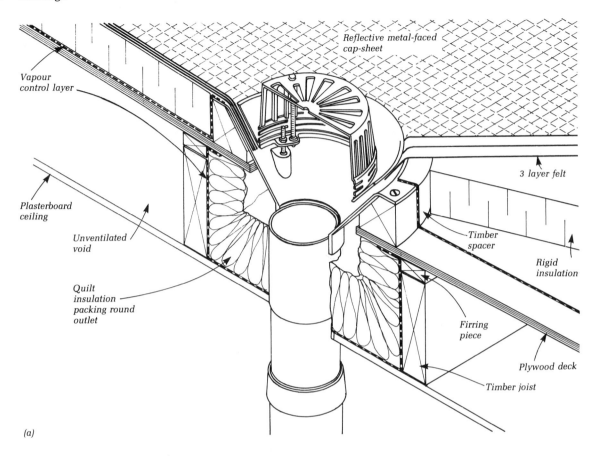

Reflective metal-faced cap-sheet

Vapour control layer

Plasterboard ceiling

Unventilated void

Quilt insulation packing round outlet

3 layer felt

Timber spacer

Rigid insulation

Firring piece

Plywood deck

Timber joist

(a)

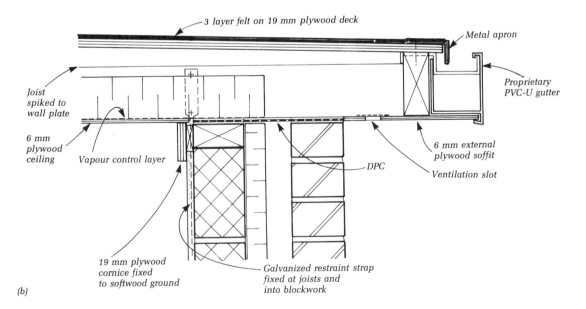

3 layer felt on 19 mm plywood deck

Metal apron

Joist spiked to wall plate

6 mm plywood ceiling

Vapour control layer

Proprietary PVC-U gutter

6 mm external plywood soffit

Ventilation slot

DPC

19 mm plywood cornice fixed to softwood ground

Galvanized restraint strap fixed at joists and into blockwork

(b)

Fig. 8.23 Built-up warm roof details: (a) rainwater outlet (Harmer Holdings Ltd); (b) bitumen felt at eaves and gutter

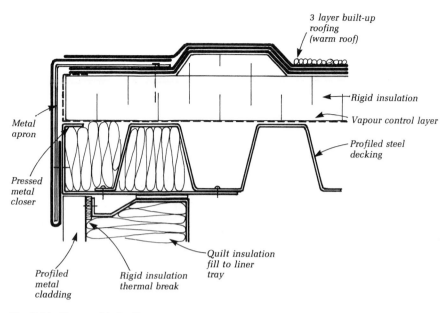

Fig. 8.24 Bitumen felt detail at verge

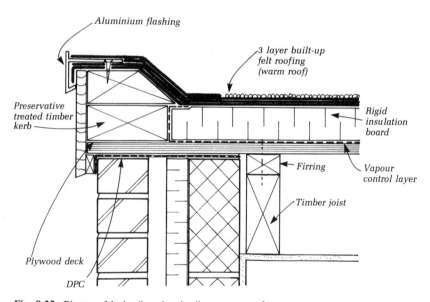

Fig. 8.25 Bitumen felt details: edge detail — warm roof

traffic. It is also susceptible to attack by hydrocarbon solvents and intolerant of local flexing caused by excessive differential movements. The use of EPDM allows rapid site installation and is similar to neoprene in providing a highly elastic but ozone resistant covering. EPDM also provides excellent resistance to UV radiation, but is vulnerable to solvents, oils and grease (though these can have a deleterious effect on possibly all roofing membranes, a

consideration that will influence choice only in environments with a high concentration of chemicals). Most roofing products have a susceptibility to biological attack but the extent can be changed by compounding materials together.

Each of the membranes has different *fire control* properties: PVC has good fire resistance due to its chloride content but CPE will vary according to the amount of this

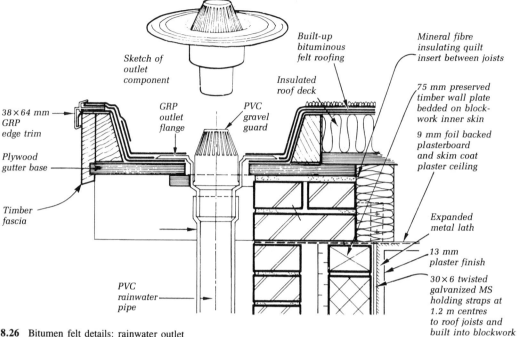

Sketch of outlet component

Built-up bituminous felt roofing

Mineral fibre insulating quilt insert between joists

38 × 64 mm GRP edge trim

GRP outlet flange

PVC gravel guard

Insulated roof deck

75 mm preserved timber wall plate bedded on block-work inner skin

9 mm foil backed plasterboard and skim coat plaster ceiling

Plywood gutter base

Timber fascia

PVC rainwater pipe

Expanded metal lath

13 mm plaster finish

30 × 6 twisted galvanized MS holding straps at 1.2 m centres to roof joists and built into blockwork

Fig. 8.26 Bitumen felt details: rainwater outlet

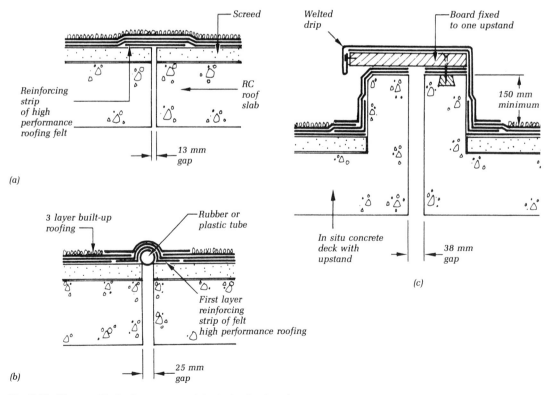

Screed

Welted drip

Board fixed to one upstand

Reinforcing strip of high performance roofing felt

RC roof slab

150 mm minimum

13 mm gap

(a)

3 layer built-up roofing

Rubber or plastic tube

First layer reinforcing strip of felt high performance roofing

In situ concrete deck with upstand

38 mm gap

(c)

25 mm gap

(b)

Fig. 8.27 Bitumen felt details: movement joint (uninsulated roof)

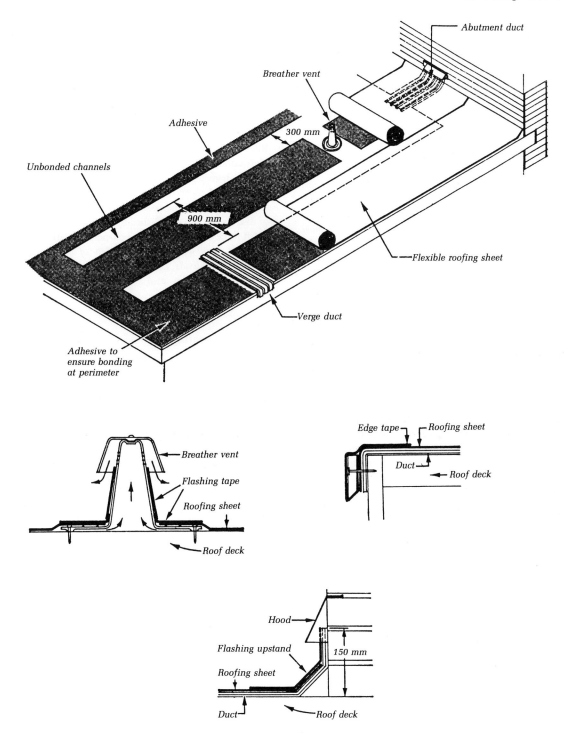

Fig. 8.28 Single ply sheet roofing

constituent. Although chloride content is an important factor so are the compounding ingredients employed in single ply materials both with respect to inflammability and fire spread. PIB, ECB, EPDM and neoprene are flammable and silicone is inherently fire resistant.

Although most roofing systems using these materials have British Board of Agrément (BBA) certification, at present there is no British Standard code of practice. Single ply roofing may be laid loose, with or without ballast, or partially adhered, fully adhered or mechanically attached. The choice of system depends upon exposure rating, substrate, and whether the roof is 'cold', 'warm' or has a protected membrane (but the roof design should conform with BS 6229). Figure 8.28 shows typical details of a partially adhered PVC single ply roofing. As a typical example of single ply roofing, the PVC is 1.5 mm thick, and is supplied in 0.6 × 25 m, 1.1 × 20 or 1.8 × 15 m rolls. Joints between sheets are normally achieved with a contact adhesive and double-sided tape, or often by solvent welding in the case of PVC sheets.

Most single ply roofs are supplied as complete systems. Within each type, different specifications of weather-proof sheet are used for warm roof or inverted roofs, for horizontal and vertical applications. Manufacturers also supply junction pieces for intersections and metal fittings for mechanical fixing or sheets, copings, etc. Since features such as membrane composition, thickness and design of manufacturer's components are based on optimum performance of the completed system, manufacturer's recommendations have to be adhered to. As with all other forms of flat roofing however, the conditions under which the material is applied and the standard of workmanship achieved are of principal importance to the long-term durability of the installation.

Despite their greater cost, single ply polymeric roofing has taken an increasing share of the roofing market in recent years. This is because they are made from generally inert materials which are jointed in different ways from traditional materials and using new technologies. They are designed to have wide temperature stability which, coupled with the methods of mechanical attachment employed, isolates the waterproofing from localized and general structural movements which should give enhanced life and performance. Because they can be laid relatively quickly, they are well suited to contemporary methods of fast-track construction (see chapter 2). There have been some recent failures of single ply materials used in the USA. Ballasted, unreinforced PVC sheets have proved subject to shattering in cold weather conditions; the use of this material in inverted roof constructions is consequently less suitable than alternative coverings.

Metal roof deck (warm roof)   This is a typical example

of a single ply roofing specification (see fig. 8.29):

- Metal trough roof decking 150 mm deep fixed by specialist contractor incorporating a vapour check such as a single-layer high-performance felt bonded in bitumen to top flats of deck; glass tissue faced mineral fibre thermal insulation (thickness as required to achieve a combined *U*-value complying with the *Building Regulations*, see section 8.2.3).
- Roof covering of 1.5 mm thick 'Trocal S' light grey plasticized PVC held against wind uplift with 1 × 80 mm diameter metal discs mechanically fixed first to decking and incorporating factory-laminated face of PVC on to which the roofing is welded. The number of disc fixings are established by reference to the wind uplift force calculated to CP 3 chapter V: Part 2: 1972 *Code of basic data for the design of buildings — Loading: wind loads*. Laps in sheets must be a minimum of 50 mm wide and solvent welded using tetrahydrofurane (THF).

8.3.4 Liquid coatings

Although there are a vast number of liquid coating materials suitable for roof covering, most are intended for short-term repairs. Formulations may be spirit based or water based, and the choice relies on the prevailing climatic conditions when application is required. Many are liquid versions of thermoset sheet material, such as neoprene and silicone, and others include polyurethane, acrylic resin, bitumen, modified bitumen, and asphalt. There are no British Standards for the application of these coatings, although many have BBA certification. Some are marketed as complete systems including for example, flexible plastic sheet that is laminated within the liquid coats to reinforce the waterproof covering at joints in the structure and at perimeter junctions. Liquid coatings require exacting preparation, surfaces must be dust free to ensure full adhesion as well as providing a degree of roughness to provide adequate mechanical bonding. Achieving a consistent and adequate film thickness is a particular difficulty with this type of construction.

8.3.5 'Pour and spread' reinforced hot asphalt

A material related to bitumen-based liquid coatings but derived from traditional asphalt practice is 'pour and spread' reinforced hot asphalt (fig. 8.30). This aims to combine the advantageous characteristics of mastic asphalt with those of high performance roofing felt. A separating layer of BS 747 felt is required fully, partially or mechanically bonded according to the substrate (see under built-up roofing above). The waterproofing is a polymer-modified

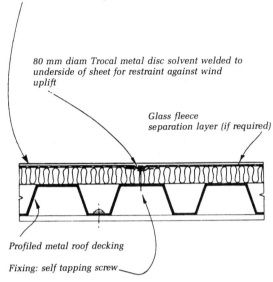

1.5 m thick Trocal 'S' light grey plasticized PVC sheet roof membrane, high flexibility and extendibility, high tensile strength, and excellent resistance to air-borne pollutants

80 mm diam Trocal metal disc solvent welded to underside of sheet for restraint against wind uplift

Glass fleece separation layer (if required)

Profiled metal roof decking

Fixing: self tapping screw

Fig. 8.29 Single ply sheeting details: Trocal 'S' PVC roofing system

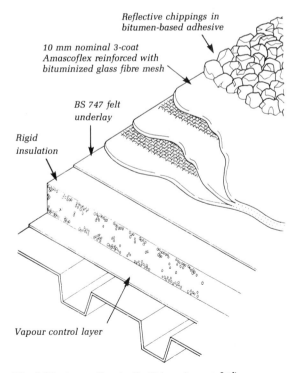

Reflective chippings in bitumen-based adhesive

10 mm nominal 3-coat Amascoflex reinforced with bituminized glass fibre mesh

BS 747 felt underlay

Rigid insulation

Vapour control layer

Fig. 8.30 Amascoflex details (Briggs Amasco Ltd)

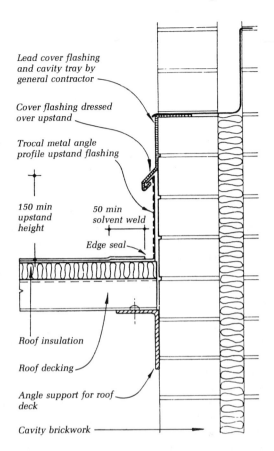

Lead cover flashing and cavity tray by general contractor

Cover flashing dressed over upstand

Trocal metal angle profile upstand flashing

150 min upstand height

50 min solvent weld

Edge seal

Roof insulation

Roof decking

Angle support for roof deck

Cavity brickwork

asphaltic compound comprising oxidized bitumen blended with fine graded limestone filler. It is applied in three coats with two intermediate reinforcing layers of bituminized glass fibre mesh. The bitumen coating of the mesh combines with the hot asphalt to form a homogeneous membrane. The site application of the reinforcement requires a wider spread mesh to be used than is used for roofing felt. The overall thickness of the three coats is 10 mm, forming a seamless lightweight finish that by virtue of its reinforcement is strong and flexible. Reinforced asphalt may be used either above or below insulation; a solar reflective top surfacing of chippings is applied with bitumen-based adhesive. Its manufacturers anticipate its lifespan to be similar to that of traditional asphalt, it also achieves an AA fire rating when tested to BS 476.

8.3.6 Metal sheets

Metal roofing using traditional techniques can be laid to a shallow pitch but not dead flat. Materials fall into two categories, *soft*, i.e. lead and *hard*, i.e. copper, zinc, stainless steel and aluminium. Fully supported sheet metal roofing has standing seams, rolls, drips and welted joints to connect the sheets. These roof coverings are suitable provided foot traffic is restricted to maintenance personnel. For purposes such as escape in case of fire and for maintenance inspections duck boards should be provided to distribute weight evenly without restricting the flow of rainwater.

With the exception of lead, sheet metal roof coverings are much lighter than tiles and slates and differences in weight between copper, zinc and aluminium are not significant, though the relatively recently introduced terne-coated stainless steel is the lightest of these alternatives.

A cause of failure in both hard and soft metal roofing is *thermal pumping*. Showers during a hot summer's day or frost at night can rapidly contract air within the roof drawing in air and water at joints. A similar effect can occur in 'hard' metal roofs, wind uplift on the roof covering creates a vacuum and suction through joints. If due to poor workmanship the metal has not been laid flat, this problem is exacerbated particularly at flattened, welted or lapped junctions.

Correctly laid lead and copper roof coverings have given trouble-free protection for buildings for centuries but premature decay can result from bimetallic electrolytic action or in the presence of corrosive agents. For instance, timber such as western red cedar and those treated with corrosive preservatives should not be used for the duck boards. Lead can be perforated by constant concentrated dripping of water from roofs upon which algae are growing. See MBS: *Materials* chapter 9. Durability increases with the pitch of roofs and all the metal sheets can be fixed at any pitch or vertical.

Although lead has been used for longer than any of the hard metals, and many lead roofs hundreds of years old are in existence, eventually lead roofs fail from metal fatigue. Many more problems have, however, been encountered in recently constructed lead roofs. One probable reason is that for reasons of economy, lighter codes of lead sheet have been used in bay sizes that are too large. Also, an increasingly frequent cause of failure is corrosion from the underside of the metal. There seem to be two main reasons for this:

1. Greater internal humidity due to changes in heating, ventilation, insulation and the pattern of use of buildings.
2. Contemporary decking materials that retain moisture during construction and obstruct drying out.

Condensation on the inner face of the metal in the relative absence of air (and the pollutants — carbon and sulphur dioxide — that normally combine with the metal to form a long life passive surface skin), causes corrosion. These problems are not limited to lead roofing but are also a potential difficulty for 'hard' metal roofs.

The currently advocated form of construction is the 'ventilated warm roof' (fig. 8.31). The air space under the outer boarding helps to disperse moisture which has succeeded in bypassing the vapour barrier. It also ensures pressure equalization that guards against 'thermal pumping' and the hazards outlined above. The roof void should be adequately ventilated as described in BS 6229 and 5250. Sheet metal gauges are compared in table 78 MBS: *Materials* p. 195.

Method of fixing The underlying methods of fixing lead are slightly different from those for hard metals. Stresses which could arise from constant thermal movements will cause fatigue but can be minimized by reducing friction between the metal and the decking through the provision, at suitable centres, of joints designed to absorb movement. Sheet metals are laid in bays with their lengths in alignment with the fall of the roof. The sheets are turned up to form upstands against abutments which are protected by a cover flashing taken into a raked joint in brickwork or a raglet groove in masonry or concrete. The cover flashing is retained by wedges and then afterwards pointed. Joints in the direction of the fall are formed into rolls. Rolls with solid cores are preferable where there may be foot traffic and also on flat roofs since their greater height is an advantage. An alternative to the roll is a standing seam (although not suitable for lead-covered flat roofs or pitched roofs up to 45 degrees). Differences between the properties of the metals determine the techniques for laying, thus lead is malleable but roofing grades of copper and aluminium are less so and zinc and stainless steel are relatively stiff. Consequently, lead can be formed into complex shapes by bossing, copper and aluminium can be made into standing seams and welts while details in zinc roof covering are generally restricted to simple folds, the sheets being preformed before being placed in position. Joints across the fall are made as follows:

- In pitches up to 5 degrees, as steps called drips — to be at least 50 mm high.
- Pitches over 5 degrees, as welts (cross welts are not suitable for lead roofs as they do not provide adequate support at the top of each sheet. Welts in the directions of the fall are only suitable for small roof areas or vertical surfaces where the joints are unlikely to be submerged in water).
- Lap joints across the fall of the roof are suitable for lead sheet laid on roofs 11 degrees and over, drips should be used on roofs shallower than 11 degrees.

Where the longitudinal joints are standing seams, welts across the fall must be staggered to avoid the problem which arises if the corners of four sheets coincide.

Underlays Underlays are required: to allow free

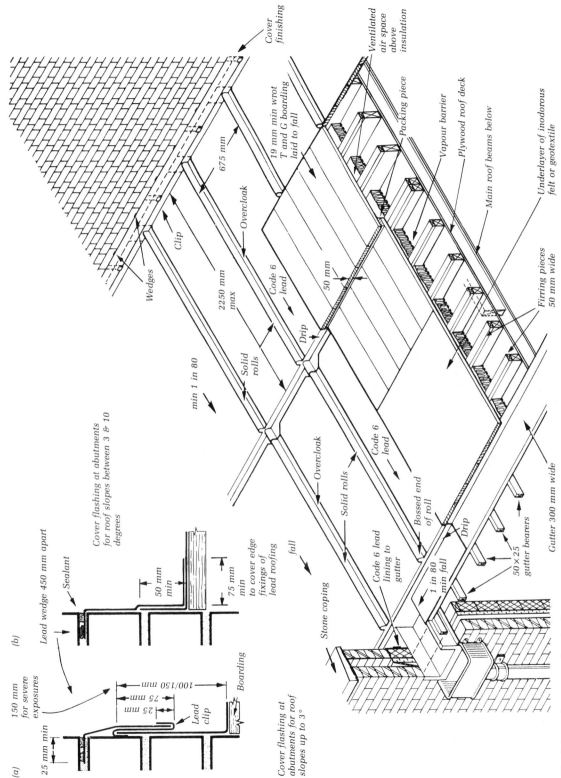

Cover finishing

Ventilated air space above insulation

Packing piece

Vapour barrier

Plywood roof deck

Main roof beams below

Underlayer of inodorous felt or geotextile

19 mm min wrot T and G boarding laid to fall

Overcloak

675 mm

Clip

Code 6 lead

50 mm

Firring pieces 50 mm wide

Wedges

2250 mm max

Drip

Solid rolls

min 1 in 80

Overcloak

Code 6 lead

Solid rolls

Bossed end of roll

Drip

Cover flashing at abutments for roof slopes between 3 & 10 degrees

Stone coping

Code 6 lead lining to gutter

1 in 80 min fall

50 × 25 gutter bearers

Gutter 300 mm wide

fall

Cover flashing at abutments for roof slopes up to 3°

150 mm for severe exposures

Lead wedge 450 mm apart

Sealant

(b)

50 mm min

75 mm min

50 mm min to cover edge fixings of lead roofing

25 mm min

(a)

100/150 mm

75 mm

25 mm

Lead clip

Boarding

Fig. 8.31 Lead sheet roofing details

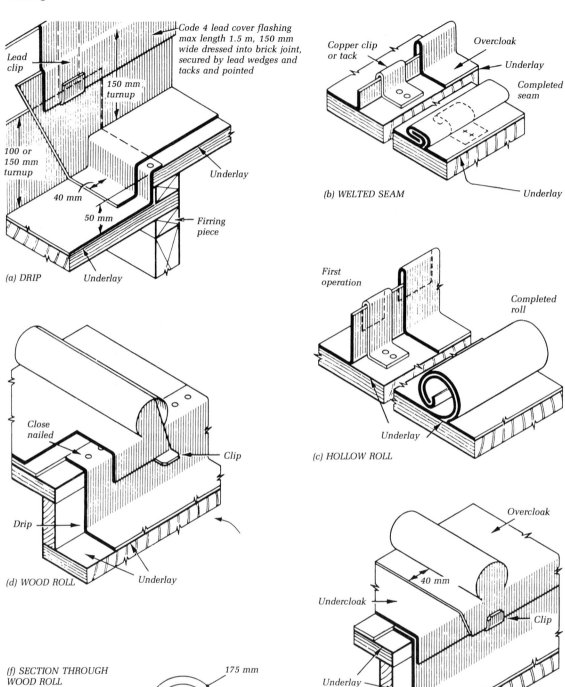

Code 4 lead cover flashing
max length 1.5 m, 150 mm
wide dressed into brick joint,
secured by lead wedges and
tacks and pointed

Lead
clip

150 mm
turnup

100 or
150 mm
turnup

40 mm

50 mm

Underlay

Firring
piece

(a) DRIP   Underlay

Copper clip
or tack

Overcloak

Underlay

Completed
seam

(b) WELTED SEAM   Underlay

First
operation

Completed
roll

Underlay

(c) HOLLOW ROLL

Close
nailed

Clip

Drip

(d) WOOD ROLL   Underlay

Overcloak

40 mm

Undercloak

Clip

Underlay   Underlay

(e) WOOD ROLL

(f) SECTION THROUGH
WOOD ROLL

175 mm

Open
nailed

Underlay

Fig. 8.32 Lead sheeting details

Table 8.7 Use of lead sheet

Code No.	Thickness (mm)	Use
5	2.24	Roofing and gutter lining
6	2.64	
7	3.15	
4	1.80	Flashings and lead 'slates'
5	2.24	
3	1.32	Soakers
4	1.80	

movement of metal; to prevent corrosion by screeds or timbers; and for sound deadening. Recently introduced heavy gauge *geotextile felt* underlays are good from this point of view; they also reduce the extent of friction between the metal and its substrate, but are otherwise relatively unproven. Rain and hail can be very noisy on copper, aluminium and zinc sheets. The noise of rain 'drumming' on hard metal roofs is made worse if the sheets have not been laid flat.

Lead, copper, stainless steel and zinc sheets should be laid on an inodorous felt (to BS 747 Type 4a(ii) *Brown sheathing felt (no. 2) inodorous*) butt jointed and fixed with clout nails. The same felt is suitable for aluminium which is laid over timber boarding but on other bases, 200 gauge polythene sheets (0.508 mm thick) should be used. The underlay in this case should be laid with 50 mm sealed laps. Underlays should be dry when the roof coverings are laid and this is particularly important in the case of inodorous felt. In recent years inodorous felt has caused problems due to the impregnating material becoming too sticky in hot weather leading to failure of lead sheet. Consequently the Lead Sheet Association now recommend the use of polyester geotextile felts.

Lead   This form of roofing is described in BS 6915: 1988 *Specification for design and construction of fully supported lead sheet roof and wall coverings*. The metal is discussed in MBS: *Materials* chapter 9, and should conform to BS 1178: 1982 *Specification for milled lead sheet and strip for building purposes*. Sand-cast lead is still produced, mainly for use in conservation work.

The use of this extremely durable roof covering is limited by its weight, high initial cost and environmental concerns about its toxicity. It is, however, still widely employed because of its malleability and consequent suitability for items such as flashings which require to be bossed into complex shapes. The use of lead for vertical cladding has enjoyed a revival and its use preformed on panels is likely to increase. Fully supported lead sheet for roofing has an AA fire rating in respect of BS 476 Part 3 (see p. 200)

except where laid on plain edge boards when the rating is reduced to BA. Table 8.7 is a guide to the thicknesses of lead sheet suitable for various uses, and the appropriate code number.

A typical small lead flat roof is shown in fig. 8.31. The upstand at the abutment is protected by a cover flashing secured by means of lead tacks and wedges, illustrated at (a) and (b). The correct spacing of lead tacks is a function of the exposure conditions of the roof; reference should be made to *Lead Sheet Roofing and Cladding*, volume two of *The Lead Sheet Manual* published by the Lead Sheet Association. The cover flashing is tucked at least 25 mm into the brickwork joint. The object of the cover flashing is to seal the joint between upstand and wall, and at the same time allow the covered sheet freedom to contract or expand.

The joints shown in fig. 8.32 are:

- (a) — the junction of a drip with an abutment. The flat roof consists of plane surfaces slightly inclined and separated by low steps or drips to facilitate water runoff at the joints where the ends of the lead sheets overlap.
- (b) — a welted seam for a joint running with the fall on a small steeply pitched roof or on vertical surfaces. The seam is made by fixing copper clips or tacks at about 600 mm centres at the junction of the sheets. The clips should be 'dead soft' temper and cut from 24 SWG (0.559 mm sheet). The edges of the sheets are then turned up and dressed flat as shown.
- (c) — an alternative to the welted seam is a hollow roll of lead. It was extensively used on steep pitches in old buildings.
- (d), (e) and (f) — these show a solid roll, made over a wood former. This joint is used on flat roofs as shown in fig. 8.31 or as a ridge joint. Wooden rolls of 50 mm diameter are fixed at the joint either by screwing through the roll or by using a double-headed nail. The lead is then dressed as shown in (f) being formed well into the angles to obtain a firm joint.

Further reference should be made to *The Lead Sheet Manuals* published by the Lead Sheet Association.

Copper   Applications for copper roofing are described in CP 143 *Sheet roof and wall coverings* Part 12: 1970 (1988) *Copper*, and the material in BS 2870: 1980 *Rolled copper and copper alloys: sheet strip and foil*.

Copper is strong in tension, tough, ductile and in suitable tempers it is malleable, but has negligible creep. See MBS: *Materials*, chapter 9. Copper sheets for traditional roofing and flashings should be in soft or $\frac{1}{4}$ hard temper. Welts and folds should be made with a minimum number of blows rather than the succession of taps with which the plumber works with lead sheet. Half-hard temper metal is sometimes required for weatherings to window frames and copings. Like lead, copper is extremely durable.

When exposed to most atmospheres a thin coating of basic sulphate of copper forms which in a number of years becomes green after passing through darkening shades of brown. The presence of water is necessary for patination to take place and vertical surfaces, which shed water too quickly, are likely not to turn green at all. The length of time required for a patina to form on a roof depends on the atmospheric conditions:

- In marine locations   4−6 years
- City/industrial locations   5−8 years
- General urban conditions   8−12 years
- Clean rural atmosphere   30 years

This coating protects the underlying metal from continuing corrosion — even in industrial areas. Continental manufacturers produce copper sheet which has been artificially weathered so the roofing can be installed already with a patina. It can normally be bent, welted and seamed without damaging the surface which stabilizes and builds up an even thicker layer when exposed to the atmosphere.

The electrical potential of copper is high compared with other metals used in building so copper is not usually attacked by them. Copper can, however, cause bimetallic corrosion to steel, aluminium or zinc if there is direct contact in the presence of water. Consequently, fixings for copper roofing should also be made from copper. The coefficient of linear expansion of copper is less than that for lead, aluminium and zinc.

Traditional copper roofing relies upon the availability of skilled craftsmen. There are two traditional methods of forming the longitudinal joints in copper roofing: the standing seam, and the batten or wood roll.

1. *Standing seam*. The three stages in the formation of a standing seam in copper are shown in fig. 8.33.
2. *Wood roll*. This method uses timber battens to form a shaped wooden core against which the edges of the sheet are turned up. A prepared capping strip is then welted to the flanges. The timber battens are screwed to the decking and the roof sheeting is secured to the battens by means of 50 mm wide copper strips. The four stages in the formation of a batten roll are shown in fig. 8.34.

Transverse joints in each case are formed by double lock cross welts (or for very flat roofs — drips). The formation of a double-lock cross welt is shown in fig. 8.35 and the application of the standing seam method is shown in fig. 8.36. The minimum fall for copper roofing is 12 mm in 1 m (1 in 60), and drips 65 mm deep, spaced not more than 3 m apart, should be used in roofs of 5 degree pitch or lower.

Long strip roofing employs harder temper metal ($\frac{1}{4}$, $\frac{1}{2}$

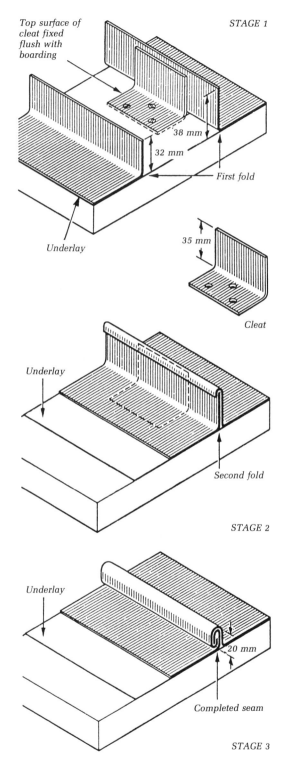

Fig. 8.33 Copper sheeting details: formation of a standing seam

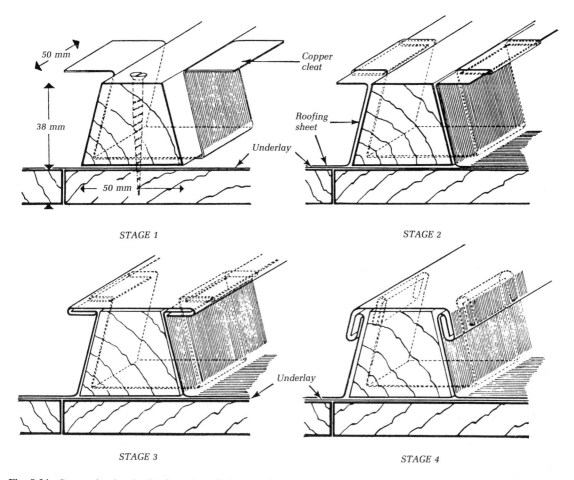

STAGE 1

STAGE 2

STAGE 3

STAGE 4

Fig. 8.34 Copper sheeting details: formation of a batten roll

hard) than does traditional roofing; long strips are formed into profiled panels and seamed together by machine. The many cross welts previously required are eliminated, resulting in a less labour intensive and cheaper form of roofing which is smoother and more mechanical in appearance than the traditional form. Long strip roofing can be used where the total distance between eaves and ridge does not exceed 10.0 m. For runs over 3 m sliding clips are used in the standing seam joint to allow longitudinal movement. The minimum recommended roof slope is 3 degrees but 5 degrees is a better specification.

Fully supported copper sheet for roofing has a fire rating of AA in respect of BS 476. Sheets are usually supplied 1.8 × 0.6 m for traditional standing seam applications and up to 1.8 × 0.6 m for batten roll roofing; in both cases the material is 0.6 m in thickness. For long strip roofing, 10 m maximum length panels are made from copper coils 450, 600 or 670 mm in width and 0.6/0.7 mm in thickness.

Fully supported metal roofing needs to be laid on continuous decking, most commonly timber. This has to be thick enough to fix the 25 mm long nails used to attach the roofing clips. Other materials can be used provided that they are dimensionally stable and that the clips can be fixed securely subject to the provisions of CP 143: Part 12: 1970. Panels and boarding should be tongued and grooved, the latter being laid either diagonally or in the direction of fall. Concrete decks should be screeded, in all cases the roof should incorporate a suitable vapour control layer (see section 8.2.4). The construction needs ventilating as described earlier in this section, either by openings at the ridge and eaves or by building prefabricated copper air vents into the roof covering; further information should be sought from the Copper Development Association.

It is necessary to use an underlay of BS 747 Type 4a(ii) *Brown sheathing felt (no. 2) inodorous*, or a geotextile felt, whatever the decking material. On timber boarding it serves to separate the copper from wood preservatives which can corrode the metal and serves as a temporary weathering

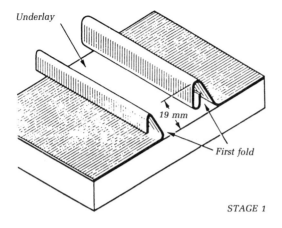

Underlay

19 mm

First fold

STAGE 1

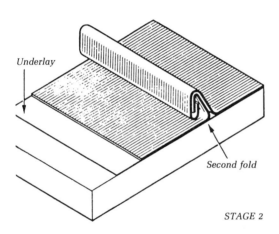

Underlay

Second fold

STAGE 2

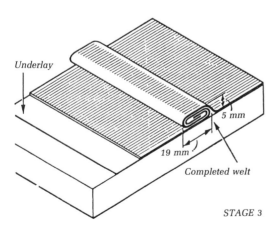

Underlay

5 mm

19 mm

Completed welt

STAGE 3

Fig. 8.35 Copper sheeting details: formation of a welted joint

for the roof during the course of construction. The underlay is secured to the timber deck by copper nails and laid butt jointed. The underlay lessens the possibility of 'wearing' the copper as it expands and contracts, and deadens the drumming sound of rain.

A proprietary roofing material utilizes an indented copper sheet (42 gauge) backed with bitumen and laid as a top layer of built-up felt roofing on an underlayer of glass fibre based bitumen felt. This copper/bitumen roofing gives approximately the appearance of traditional copper at less cost.

Zinc Applications for zinc roofing are described in CP 143 *Roof and wall coverings* Part 5: 1988 (draft) *Zinc sheet*. The metal is considered in MBS: *Materials* chapter 9 and is described in BS 6561: 1985 (1991) *Specification for zinc alloy sheet and strip for building. Working zinc sheet and strip* and *Zinc in building design* by the Zinc Development Association are further useful sources of information.

Zinc–copper–titanium alloy (generally 0.08–1.0 per cent copper and 0.07–0.2 per cent titanium, remainder zinc) is, in contemporary applications, laid in long strip form, which is a relatively economical and efficient method of roofing. The durability of zinc alloy is proportional to its thickness, also the steeper the pitch to which the roof is laid, the longer will be its life. When using the usual and recommended 0.8 mm zinc sheeting, under normal urban conditions, a period of 60–80 years without first maintenance and an overall life expectancy of more than 100 years can be anticipated.

Fixings for zinc roofing are generally made from galvanized steel or stainless steel; water running off unprotected steel or iron stains zinc but without any apparent effect on its performance. Contact with aluminium or lead does not risk significant electrolytic corrosion but zinc should not be situated close to copper, and water draining from copper and copper alloys should not be discharged on to zinc. Zinc can be located adjacent to seasoned softwoods and most other timber but it should not be laid in direct contact with, nor receive drainage from, western red cedar, oak or sweet chestnut (which contain organic acids that accelerate corrosion of non-ferrous metals).

The coefficient of thermal expansion for zinc–copper–titanium is 0.022 mm/m°C, allowing zinc to be used in continuous lengths of up to 9 m without need for expansion joints, depending on the width of the bay. Zinc sheet for roofing has a fire rating of AA in respect of BS 476. Typical details for zinc are shown in fig. 8.37. Roll cap roofing is used to a minimum slope of 3 degrees and standing seam roofing above 7 degrees. The welting of a corner piece to the upper sheet is shown in (a) and the forming of saddle pieces on the ends of cappings at walls and drips is shown in (b). Detail (c) shows the formation of a 'dog ear' at an

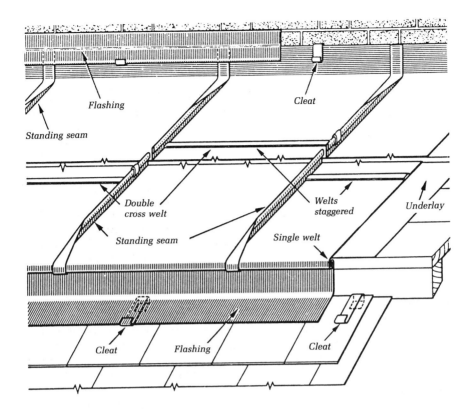

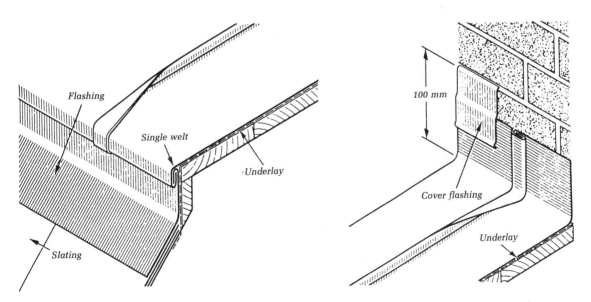

FINISH OF STANDING SEAM AT LOWER END

FINISH OF STANDING SEAM AT UPPER END

Fig. 8.36 Copper sheeting details: application of standing seam method

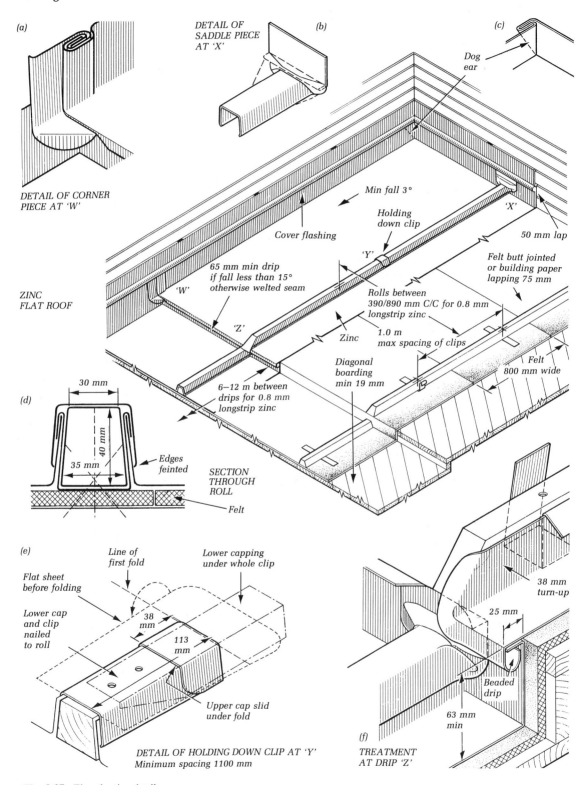

(a)

DETAIL OF CORNER
PIECE AT 'W'

DETAIL OF
SADDLE PIECE
AT 'X'

(b)

(c)

Dog
ear

ZINC
FLAT ROOF

Min fall 3°

Holding
down clip

Cover flashing

'Y'

'X'

50 mm lap

65 mm min drip
if fall less than 15°
otherwise welted seam

'W'

Felt butt jointed
or building paper
lapping 75 mm

Rolls between
390/890 mm C/C for 0.8 mm
longstrip zinc

'Z'

Zinc

1.0 m
max spacing of clips

Felt
800 mm wide

Diagonal
boarding
min 19 mm

(d)

30 mm

40 mm

35 mm

Edges
feinted

SECTION
THROUGH
ROLL

Felt

6–12 m between
drips for 0.8 mm
longstrip zinc

(e)

Line of
first fold

Lower capping
under whole clip

Flat sheet
before folding

Lower cap
and clip
nailed
to roll

38
mm

113
mm

Upper cap slid
under fold

DETAIL OF HOLDING DOWN CLIP AT 'Y'
Minimum spacing 1100 mm

38 mm
turn-up

25 mm

Beaded
drip

63 mm
min

(f)

TREATMENT
AT DRIP 'Z'

Fig. 8.37   Zinc sheeting details

internal corner. A section through the batten roll is shown in (d) and the detail of a holding down clip in (e). Where a roof abuts the wall at a drip, a corner piece is welted to the upper sheet, as in (f).

Aluminium   Applications for aluminium flat roofing are described in CP 143: *Sheet roof and wall coverings* Part 15: 1973 *Aluminium*. The metal is considered in MBS: *Materials* chapter 9, and is described in BS 1470: 1987 *Specification for wrought aluminium and aluminium alloys for general engineering purposes — plate, sheet and strip*. Grades S1, S1A, S1B, S1C and NS3 are all suitable for fully supported roof coverings.

Since aluminium forms a protective oxide when exposed to the atmosphere, the alloy used for roofing is normally extremely durable even in industrial and marine environments. Precautions must, however, be taken to avoid galvanic attack by other materials and aluminium should be protected from wet concrete and mortar. Timbers containing acid and preservatives are also dangerous to the sheeting. Fully supported aluminium sheet for roofing has a fire rating AA in respect of BS 476. The techniques of laying aluminium fully supported roof coverings are similar to those of copper.

The minimum fall for a 'traditional' aluminium flat roof system is 1:60, using 0.8 mm thickness sheets. Bay width should be between 450 and 600 mm with lengths of between 2.5 and 2.0 m respectively. For a longer bay length of 3.0 m the width should be restricted to 450 mm. Similarly to copper, aluminium may be laid as long-strip 'economy' roofing to a maximum length from ridge to eaves of 7 m, the material is available in natural finish or a wide variety of colours.

Aluminium is the lightest of roofing metals — it has ample strength and ductility and creep is not significant. Hand forming is easiest in soft temper and high purity metal. It has a high reflectivity to solar heat. The durability of high purity aluminium is good in normal atmospheres provided that it is washed by rain. Its initial bright appearance can be retained for several years in rural areas, but in highly industrial regions the surface will turn matt black. Aluminium to BS 1470 (SIC, 0-grade 99 per cent purity) can be colour coated using PVF2 paint formulations: this provides a finish which is both malleable to work and durable.

Fixings should be preferably of aluminium but where steel is used it should be galvanized; however, it must not be used in contact with copper or copper alloys. Water must not be allowed to drain on to aluminium from copper roofing and particularly not from copper pipes.

Stainless steel   BS 1449 *Steel plate, sheet and strip* Part 2: 1983 *Specification for stainless and heat resisting steel plate, sheet and strip* provides information about the chemical composition, mechanical properties and dimensional tolerances for stainless steel, and the material is described further in MBS: *Materials* chapter 9.

Of the two grades that are used for roofing: type 304 S16 for normal situations and type 316 S16, only the latter will stay reasonably stain free. The standard thicknesses are 0.38 and 0.46 mm; other thicknesses are available to order. Widths of sheets vary; for 'long-strip' roofing, lengths of up to 15 m are possible for pitches of more than 5 degrees. Sheets are laid using long-strip or traditional methods with standing seams or batten rolls at centres depending on the gauge, exposure and roll width; 375 and 435 mm are usual. Otherwise the sheets are installed on an underlay and follow the same requirements applicable to copper roofing.

Stainless steel is inherently resistant to corrosion because of its self-repairing oxide film that is only a few atoms thick and is maintained by the oxygen in the air; it does, however, have a tendency to 'oil-can', where the sheet does not lie perfectly flat against the substrate, making for an unattractive appearance. Deposits on the metal due to pollution will prevent oxygen from reaching the surface and cause pitting or crevice corrosion; stainless steel surfaces must be regularly washed clean to prevent this. Bimetallic corrosion of zinc and aluminium occurs when adjacent to stainless steel in permanently damp conditions. Also stainless steel is highly prone to contamination and rust when in contact with carbon steel so careful choice of fixings is necessary.

Terne-coated stainless steel (see photo 8.5)   Terne-coated stainless steel (TCSS) is now widely used. It consists of standard stainless steel grades used for roofing to BS 1449: Part 2: 1983 to which terne is metallurgically bonded in a hot-dip process. Terne consists of 80 per cent lead and 20 per cent tin and combines with the stainless steel to provide better performance characteristics than some other metals used for roofing. These include the reduction of 'oil-canning', increased fatigue and creep resistance, use in all atmospheres and its non-susceptibility to bimetallic corrosion from lead, copper, zinc or aluminium, and mortar or timber preservatives. It is cheaper than lead and, as it has little scrap value, is not subject to theft.

The terne coating weathers to a pewter grey after some months' exposure to the atmosphere, though the final shade may not be achieved until approximately 18 months. The thermal expansion of TCSS is lower than other metals used for roofing which makes it particularly suitable for long-strip 'economy' applications up to 15 m in length and with relatively seamless gutters. Roofs may be constructed with either batten rolls or standing seams at a maximum spacing of 500 mm, thermal movement is accommodated by sliding

Photo 8.5 Lead coated stainless steel roof and wall cladding, the theatre at Blue Boar Court, Trinity College, Cambridge. Architects: MacCormac, Jamieson, Pritchard and Wright, 1990.

cleats built into the seams.

Figure 8.38 illustrates the use of preformed flashings in conjunction with standing seam roofing to form a ventilation opening at the top of a mono-pitch roof.

8.4 Pitched roofing sheets

8.4.1 Corrugated fibre cement sheets

A comprehensive range of reinforced cement corrugated sheets, slates and accessories are available for covering pitched roofs. Asbestos fibre was formerly used as the reinforcing material because of its excellent fire resistance, strength, weatherability and durability (see MBS: *Materials* chapter 10). Loose fibres of asbestos are, however, a considerable health hazard so the material is now manufactured using safer alternatives such as cellulose fibre reinforced calcium-silicate or polymeric fibre reinforced cement. The standard profiles for roof sheeting remains are still those described in BS 690 *Asbestos cement slates and sheets* Part 3: 1973 (1989) *Corrugated sheets* which includes a wide range of symmetrical and asymmetrical sheets of straight, cranked and curved configuration. Only a few of these are still made in the UK although manufacturers might be able to supply some patterns for replacement of damaged panels on existing buildings. Also

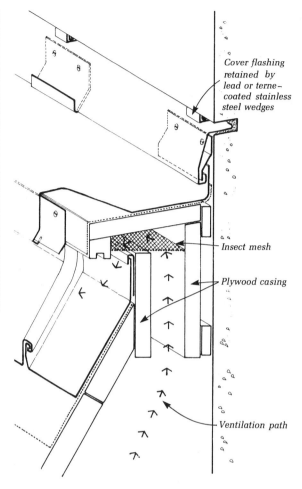

Fig. 8.38 Terne-coated stainless steel roofing details: ventilation opening to a monopitch roof (Lee Steel Strip Ltd)

some new components have been introduced to meet requirements laid down in BS 4624: 1981 *Methods of test for asbestos cement building products*, as well as BS 5427: 1976 *Code of practice for performance and loading criteria for profiled sheeting in building*.

Fibre cement sheet is available with either a textured natural grey finish, or a factory-applied acrylic surface coating in a range of colours. Although complex shapes can be covered by fibre cement sheets, maximum economy is achieved where the roof is simple in plan shape. The roof should be planned so that the purlin spacing allows the use of standard sheets without cutting. Figure 8.39 gives an idea of the range of accessories available for use with standard profile sheets, for example, projecting pipes can have integral soaker flanges to dispense with the use of separate flashings.

BS 690: Part 3 divides roofing and vertical cladding into

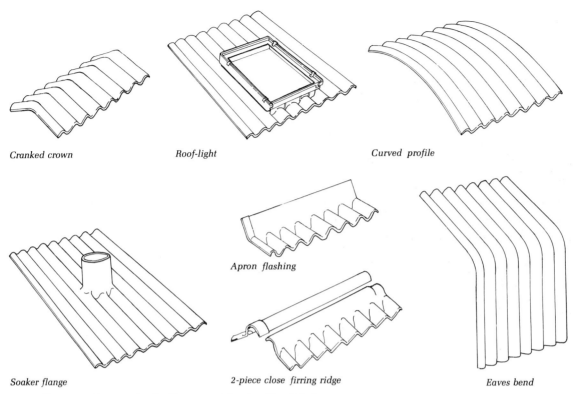

Cranked crown   *Roof-light*   *Curved profile*

Apron flashing

Soaker flange   *2-piece close firring ridge*   *Eaves bend*

Fig. 8.39 Range of accessories for fibre cement sheeting (Eternit Ltd)

five profile classes according to depth and minimum load-bearing capacity. Fittings are detailed in BS 690 Part 6: 1976. Figure 8.40 shows two of the more commonly used profiles for roof sheeting. Metal fixing accessories are covered by BS 1494 *Fixing accessories for building purposes* Part 1: 1964. *Fixings for sheet, roof and wall coverings*. Sheets may be coloured by a factory-applied process in a range of subdued colours with a high resistance to fading. Profiled translucent sheeting made from glass fibre reinforced polyester-resin, and transparent sheets from rigid PVC are manufactured to fit the various sheet profiles and provide natural daylighting. Where it is necessary to provide insulated rooflights in conjunction with insulated roofing, hermetically sealed insulated rooflights or translucent lining panels are available.

The thermal transmittance (U) through a single layer of unlined roof sheeting is only approximately 6.1 W/m² K. A method of insulating the sheets which can satisfy the insulation requirements of the *Building Regulations* and which does not require the use of additional supporting members, is to incorporate a top corrugated sheet with an internal lining panel. BS 690 Part 5: 1975 is applicable to lining sheets and panels. The space between the two sheet layers can contain glass fibre or mineral wool which

improves the insulation value of the structure and also restricts the flow of air circulating within the cavity. Consequently, this method both achieves improved insulation whilst providing a reasonably dust-tight enclosure. Two types of sandwich construction are illustrated in fig. 8.41. Insulated double cladding can have either timber or proprietary metal spacer pieces fixed between the lining panels and corrugated sheets. This avoids the insulation being compressed by the weight of the top sheet and helps to further improve the thermal characteristics of the roof. The current *Building Regulations* requirement for roofs of non-domestic buildings to have a U-value of 0.45 W/m² K (see p. 196), makes it essential to use spacers in sandwich construction. The internal lining sheet can be made either of fibre cement or profiled sheet steel (which can be finished with an acrylic coating available in a wide range of colours to BS 4904: 1978 (1985) *Specification for external cladding colours for building purposes*).

As an alternative to sandwich construction, satisfactory thermal insulation can be obtained by 'under-drawing' or lining the roof above or below the purlins with rigid sheets of fibre building board or plasterboard bonded to insulation.

Another commonly used construction employs 40 mm

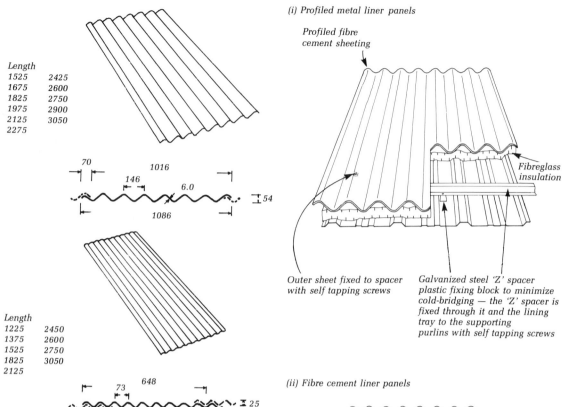

Length
1525   2425
1675   2600
1825   2750
1975   2900
2125   3050
2275

70   1016
146   6.0
1086
54

Length
1225   2450
1375   2600
1525   2750
1825   3050
2125

73   648
782
25

Fig. 8.40 Two commonly used fibre cement sheet profiles (Eternit Ltd)

(i) Profiled metal liner panels

Profiled fibre cement sheeting

Fibreglass insulation

Outer sheet fixed to spacer with self tapping screws

Galvanized steel 'Z' spacer plastic fixing block to minimize cold-bridging — the 'Z' spacer is fixed through it and the lining tray to the supporting purlins with self tapping screws

(ii) Fibre cement liner panels

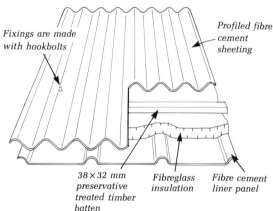

Fixings are made with hookbolts

Profiled fibre cement sheeting

38 × 32 mm preservative treated timber batten

Fibreglass insulation

Fibre cement liner panel

Fig. 8.41 Corrugated fibre cement sheeting details: insulated double cladding with; (a) profiled metal liner trays and (b) fibre cement liner panels (Eternit Ltd)

thick polyisocyanurate foam insulation slabs laid directly on the roof purlins. Longitudinal joints between adjacent insulation boards are supported on inverted 'T' sections. The joints between insulation boards can be vapour sealed with self-adhesive aluminium foil tape. A strip of fibre cement board is placed over the insulation at purlin locations to spread the fixing load, the fibre cement weathering sheet being 'top-fixed' with self-drilling/tapping screws of a type able to form an oversized hole in the crown of the fibre cement sheet as it is installed (see fig. 8.42).

Fibre-reinforced cement sheeting should not be rigidly fixed; allowance must be made for slight movement. Hook bolts which pass through the asbestos sheet and clip around steel or concrete purlins are commonly used as are drive screws into timber purlins or timber plugs in concrete purlins as shown in fig. 8.43. To accommodate movement and render the detail weathertight, a plastic washer with a separate dome-shaped cap-seal is used, as shown in fig. 8.44. Alternatively, galvanized steel or bitumen washers are available. The minimum roof pitch that is feasible will

vary according to the profile of the sheet and the degree of exposure of the site. Profiled sheet is designed to provide resistance to the penetration of rain at end and side laps provided that the roof pitch is adequate and the site is not severely exposed. Under these circumstances the base of the corrugations act as gutters and the rainwater will usually

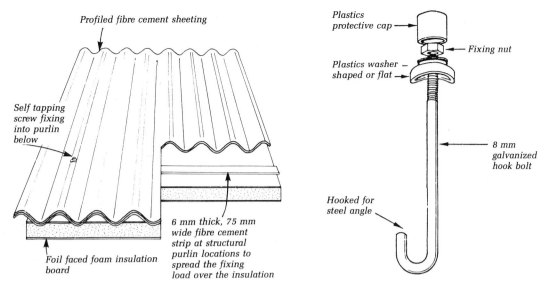

Fig. 8.42 Fibre cement sheeting: single skin with over-purlin polyisocyanurate insulation boards (Eternit Ltd)

Fig. 8.44 Corrugated fibre cement sheeting details: hook bolt

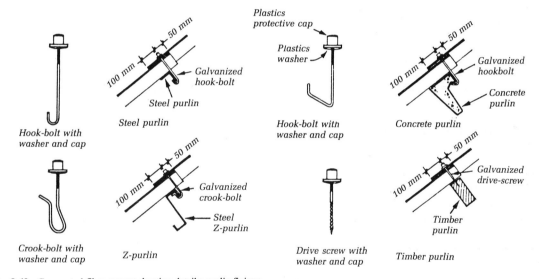

Fig. 8.43 Corrugated fibre cement sheeting details: purlin fixings

run down the roof slope without breaching the enclosure. Use of a shallow pitch or wind blowing at the slope of the roof may reduce the velocity of flow sufficient to cause a build-up of water which may then be forced under the joints in the sheets. To prevent this and to exclude dust from the building it may be necessary to seal the laps; manufacturers' advice should be sought.

A 4 degree pitch is recommended only for a limited number of profiles and the manufacturers should be consulted to check suitability. At low pitches and depending on the exposure of the roof, end and/or side laps should

be sealed with 6 mm diameter mastic strip. The type and method of laying must be as directed by individual manufacturers. The efficiency of the seal can, however, be affected by the temperature at which it is laid so routine checks should be made on the compression of mastic laid during winter months.

Laying procedure Sheets should be fixed in accordance with the recommendations of BS 5427: 1976 *Code of practice of performance and loading criteria for profiled sheeting in building*. Fixing holes should never be punched, they should always be drilled through the crown of the corrugations. Roof ladders must be used to avoid damaging the sheets and properly constructed walkways or roof boards will be necessary to give regular access to rooflights, or other places likely to need occasional attention and maintenance.

Sheets are designed to be laid smooth side to the weather with a side lap of one corrugation. They are fastened through the crown of the corrugations on each side of the side laps except at each intermediate purlin where one fixing only on the overlapping side is adequate. The laying procedure is shown in figs 8.45 and 8.46. Working upwards from the eaves, sheets may be laid either from left to right,

or right to left, but it is advisable to commence at the end away from the prevailing wind. The starter sheet and the last sheet to be fixed are laid unmitred; all other sheets require mitring where the overlap occurs as shown.

If using insulated double cladding the fixing procedure is similar. The lining panels are first laid mitred as for the roofing sheets except that they are laid smooth side to the underside. The sheets are secured with a short bolt through the intermediate corrugation. Lining panels are then overlaid with a glass fibre or mineral wool insulation mat which should have a minimum lap to all joints. The fixing of the final covering sheet then proceeds as before. Figure 8.47 shows a typical roof using single skin construction (suitable for a storage building where space heating is not required). Detail (a) shows the finish at the eaves. The sheets should not have an unsupported overhang of more than 300 mm beyond the eaves purlin and the detail is completed by an eaves filler component. Details (b) and (c) show a method of construction employing translucent rooflights. The translucent sheet is unmitred, and 10 mm diameter sealing strips are used at the end laps. Detail (d) shows a close-fitting ridge; because it is in two parts, it is adjustable to suit various angles of slope. An alternative detail is possible using a cranked crown sheet. The correct positioning of the top purlin is important; it should be located so that the hook bolt fixing is not less than 100 mm from the end of the crown sheet.

Figure 8.48 illustrates the use of insulated double cladding with a profiled metal liner tray used internally. Detail (a) shows a typical method of joining sheets at the ridge of a roof with a 'cranked crown sheet'. Detail (b) is a section through the roof eaves and illustrates the use of a curved

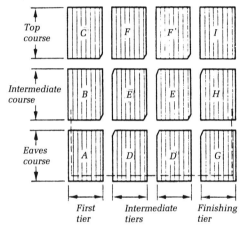

Fig. 8.45 Corrugated fibre cement sheeting details: laying procedure

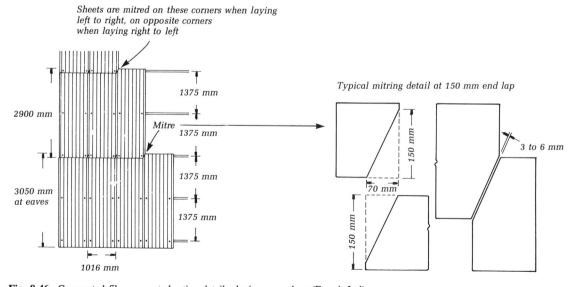

Fig. 8.46 Corrugated fibre cement sheeting details: laying procedure (Eternit Ltd)

panel providing a neat junction between the vertical cladding and roof sheeting without the use of a gutter (the disposal of rainwater consequently has to be detailed into the base of the building). Detail (c) is a method of forming the verge with a masonry wall below, special bargeboard profiles are made for this purpose. Detail (d) shows the method of detailing translucent roof sheeting; note the use of the closure pieces within the thickness of the roof.

Fibre cement sheets for roofing have P60 (Ext S AA) fire rating to BS 476: Part 3, and are non-combustible to BS 476: Part 4 (see p. 200). When an insulated sandwich roof system incorporating a metal internal lining system is used, the combined construction achieves an internal fire resistance of up to 4 hours (fibre cement itself has no fire resistance). The internal lining has a Class 1 surface spread of flame classification to BS 476: Part 7, or Class 0 classification in accordance with the *Building Regulations*.

8.4.2 Profiled aluminium sheeting

BS CP 143 *Sheet roof and wall coverings*, Part 1: 1958 *Aluminium, corrugated and troughed* describes the main types of roof sheeting and gives information on fixing accessories, contact with other materials, weathering, thermal insulation, fire resistance and condensation, as well as recommendations about minimum pitch, methods of fixing side and end laps. BS 4868: 1972 *Profiled aluminium sheet for building* specifies two suitable alloys: NS3-H8 (3103 — International Alloy Designation) to BS 1470: 1987, and NS31-H6/3105 to BS 4300/6: 1969 that has slightly higher tensile strength. These alloys are of the work-hardening type; they develop their greatest strength through being cold rolled. The 'H' suffix designates the degree of hardness that can be achieved through cold working (H8 is the hardest temper for aluminium building sheet).

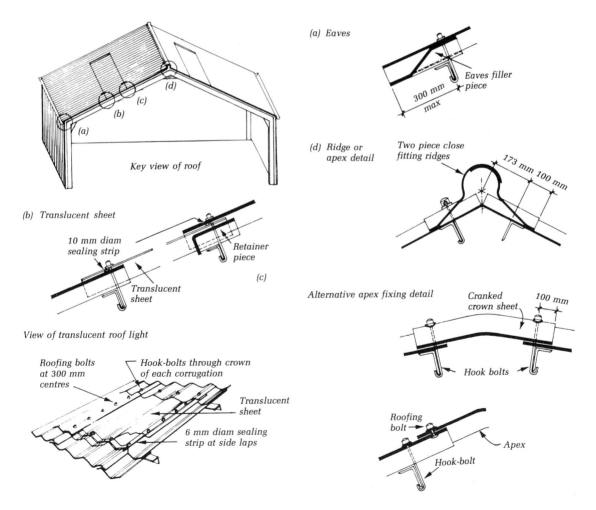

Fig. 8.47 Corrugated fibre cement sheeting details: single skin construction

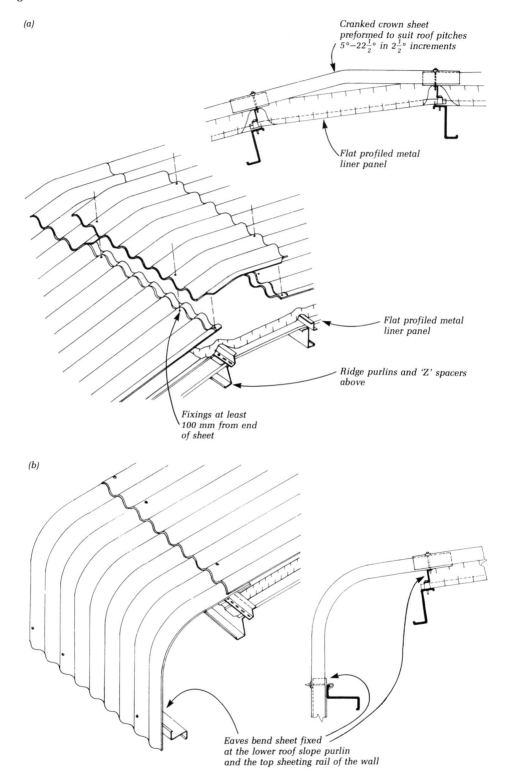

(a)

Cranked crown sheet
preformed to suit roof pitches
5°–22½° in 2½° increments

Flat profiled metal
liner panel

Flat profiled metal
liner panel

Ridge purlins and 'Z' spacers
above

Fixings at least
100 mm from end
of sheet

(b)

Eaves bend sheet fixed
at the lower roof slope purlin
and the top sheeting rail of the wall

Fig. 8.48   Corrugated fibre cement sheeting with internal profiled metal liner, insulated double cladding construction details (Eternit Ltd)

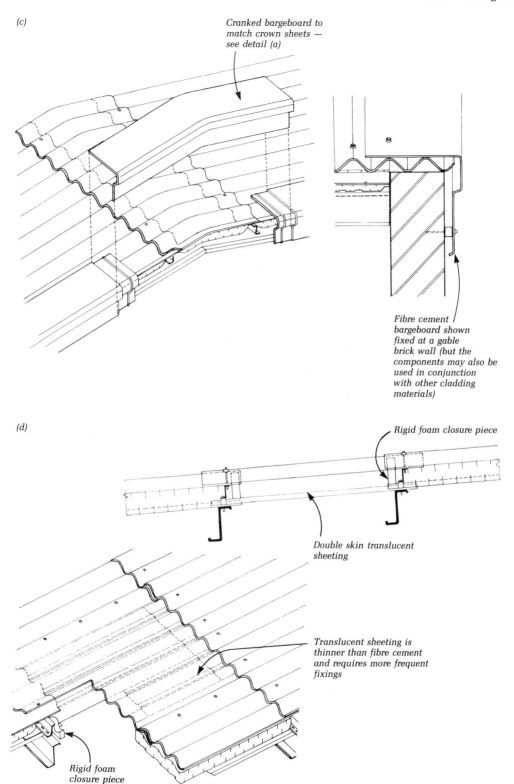

(c)

Cranked bargeboard to
match crown sheets —
see detail (a)

Fibre cement
bargeboard shown
fixed at a gable
brick wall (but the
components may also be
used in conjunction
with other cladding
materials)

(d)

Rigid foam closure piece

Double skin translucent
sheeting

Translucent sheeting is
thinner than fibre cement
and requires more frequent
fixings

Rigid foam
closure piece

Table 8.8 Profiled aluminium sheeting details: profiles

Type	Profile with nominal pitch (dimensions in mm)	Gauge (mm)	Available sizes (max and min)	
			Width (mm)	Length
Corrugated sheet BS 4868 type S	 76.2 76.2 19	1.00 to 0.5	1118 to 508	Any length to 1.22 m
Trough sheet BS 4868 type A	127 38.1	0.9 to 0.7	1187 to 579	Any length to 7.62 m
Heavy trough sheet BS 4868 type B	130.2 44.5	1.2 to 1.00	1229 to 705	Any length to 7.62 m

Self-tapping screw with sealing washer and plastic cap

Mushroom head bolt fixed through crown of profiled sheet

Plastic headed wood screw + sealing washer

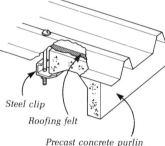
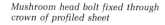
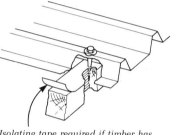
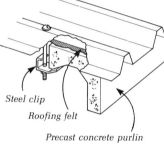

Isolating tape between profiled aluminium sheeting and galvanized cold-formed steel purlin

Steel clip

Roofing felt

Precast concrete purlin

Isolating tape required if timber has been tanalized or copper impregnation treated

Fig. 8.49 Profiled aluminium sheeting: fixings to steel, concrete and timber structure

The behaviour of aluminium when exposed to the atmosphere is discussed on p. 103. Unlike steel where corrosion is progressive, aluminium forms a protective film that is 'self-sealing'. This enables aluminium alloy to be cut, drilled and punched for fixing on site with less detriment to the long-term durability of the material. This, and its resistance to corrosion in marine and polluted atmospheres, is often the reason for its choice. It should be noted, however, that of all metals, aluminium requires the greatest imput of energy to make the conversion from ore to usable material.

Profiled sheets are available with a plain mill finish to flat or embossed sheets which darkens as it weathers, or with an organic coating made from polyester resin (see p. 103) in a range of bright colours. There are also alkyd-amino coatings which can be stoved on in the factory after forming, as well as a range of PVF/2 acrylic paints which give the best resistance to sunlight. These are available in a wide range of colours from BS 5252: 1976 and BS 4904: 1978 (1985) and non-standard colours subject to quantity.

Table 8.8 shows the BS 4868 profiles; specialist manufacturers produce a much more extended range in lengths up to 12 m. Unsealed end and side laps are only possible for slopes in excess of 15 degrees, for flatter pitches joints are made with both mechanical fastenings and mastic tape, and at lower slopes the weather-proofing on the building is solely reliant on the sealant.

For profiled roof sheeting, fixings should be through the crown into the purlins below; an isolating membrane will be necessary where the aluminium is fixed to mild steel, concrete or pressure-impregnated timber. Self-tapping screws are used when fastening to steel, hook bolts to concrete purlins and wood screws into timber (fig. 8.49). Side laps are also fixed through the crown where the sheets are overlaid.

The number and spacing of fixings depend on the loading anticipated, the centres of purlins and the permissible working load of the fastenings. The design loads comprise the imposed load (from BS CP3: chapter 5 Part 1: 1967) and the wind loads (from BS CP3: chapter 5 Part 2: 1972).

The *Building Regulations* also require roof cladding to comply with BS 6399: *Design loading for buildings*, Part 1: 1984. This states that the imposed load on sloping roofs of 30 degrees or less is to be calculated as a minimum of 0.75 kN/m² measured on plan or a 0.9 kN concentrated load. Wind loads can, however, exceed these figures and be considerably in excess of them near the edges of roofs and walls. Manufacturers publish span tables for different sheeting profiles. Sheet length may also influence the centres of purlins. Very long sheets may be impractical to use on site; usually a maximum length of 7–8 m best suits the requirements of transport and handling. The strength of a fixing is a function of the strength of the purlins in combination with the strength of the fastener itself. For aluminium sheeting the fixings are best made from aluminium alloy or stainless steel to avoid bimetallic corrosion although galvanized fittings may be acceptable in a non-polluted atmosphere.

A variety of subsidiary components have been developed for use with profiled metal, including expanded foam filler blocks to seal the open ends of corrugations at gutters, etc., self-tapping fixings and standard flashings finished to match the sheeting (fig. 8.50). Flashings for aluminium roofing are preferably preformed and of 1/22 H or 3/4 temper aluminium; these accessories should conform to BS 1470: 1987 *Wrought aluminium and aluminium alloys for general engineering purposes*. Self-tapping fasteners are made in a number of different patterns for fixing into steel or timber or with blank sections of thread or spacers for fixing through insulation. The principal fasteners, for attaching the sheeting to the purlins, are made with self-coloured plastic heads to match the finish of the sheets (fig. 8.51). They are designed to seal around the fixing holes by tightening on a neoprene or EPDM (ethylene propylene dimonomer) washer to the correct pressure against the face of the metal. To ensure that they are not over-tightened, a power screwdriver should be used with an adjustable depth-sensing clutch. In addition there are secondary fasteners such as bulb-tite rivets that are used to stitch together side laps and ensure that sealants are effectively retained.

Single and translucent plastics sheets to match the aluminium profiles are available. The minimum recommended gauges of aluminium for durability related to the use of sheeting are as shown in table 8.9.

Table 8.9 Recommended minimum aluminium gauges

Use	SWG
Heavy and marine industrial	18
Industrial	20
Light industrial	22
Agricultural	24

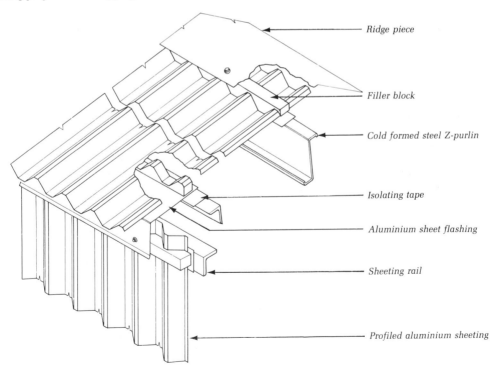

Fig. 8.50 Profiled aluminium sheeting: subsidiary components

Ridge piece

Filler block

Cold formed steel Z-purlin

Isolating tape

Aluminium sheet flashing

Sheeting rail

Profiled aluminium sheeting

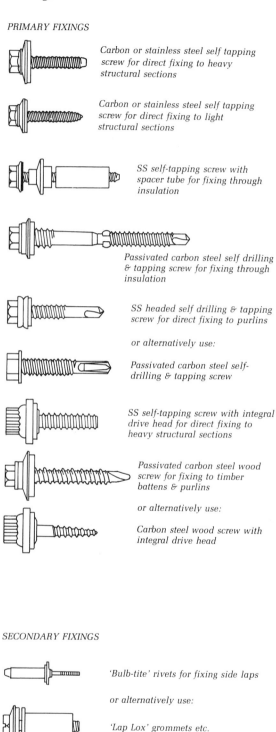

PRIMARY FIXINGS

Carbon or stainless steel self tapping screw for direct fixing to heavy structural sections

Carbon or stainless steel self tapping screw for direct fixing to light structural sections

SS self-tapping screw with spacer tube for fixing through insulation

Passivated carbon steel self drilling & tapping screw for fixing through insulation

SS headed self drilling & tapping screw for direct fixing to purlins

or alternatively use:

Passivated carbon steel self-drilling & tapping screw

SS self-tapping screw with integral drive head for direct fixing to heavy structural sections

Passivated carbon steel wood screw for fixing to timber battens & purlins

or alternatively use:

Carbon steel wood screw with integral drive head

SECONDARY FIXINGS

'Bulb-tite' rivets for fixing side laps

or alternatively use:

'Lap Lox' grommets etc.

Fig. 8.51 Profiled aluminium sheeting: fixings

Except for agricultural buildings, the aluminium roofing panels are usually combined on site with an inner shallow-profiled liner sheet containing the insulation between the two layers and separated by Z-spacers. The problem of cold-bridging the two layers is avoided in one system by the use of spacer purlins between the inner and outer sheets made from high density mineral wool insulation cut into planks (fig. 8.52) with galvanized steel U-channels forming top and bottom flanges and strapped together with stainless steel bands and clips (which form the only cold-bridge). Because aluminium is a relatively soft metal, sheets less than 0.9 mm in thickness need to be protected by crawl boards when being installed or if access is required for maintenance.

Where conditions of high internal humidity are anticipated a vapour control layer can be provided by sealing together the laps joining the liner sheets, although this is difficult to achieve. Alternatively plastic foil can be laid over the liner trays preferably with welded rather than taped joints (although the film is inevitably holed at fixings). Care is required to adequately seal the junction between the membrane in the roof and the walls, so as to reduce air leakage and as a consequence, energy consumption. A particular problem with profiled roofing is that on hot humid summer days a rapid fall in temperature after nightfall can cause condensation to form on the inside of the sheeting. This can be alleviated by installing a breather membrane above the insulation and lapping into the gutter at the eaves, and by ventilating the roof corrugations at the ridge (fig. 8.52).

A building clad in this way presents an acoustically reflective surface to the interior. To provide some compensating acoustic absorption the inner liner sheet can, in some systems, be provided with a perforated grid of small holes to part or all of the sheet. Acoustic insulation is then packed within the corrugations. If a vapour control layer is required, it is sandwiched between the acoustically absorbent material and the main thickness of thermal insulation.

Profiled sheets are also available with slab-foamed polyisocyanurate or polyurethane insulation (see section 8.2.3) bonded to the inside corrugations. A *bonded panel* with insulation thickness of 50 mm can achieve a *U*-value of 0.40 W/m°C. Bonded panels are subject to greater thermal movements (which are further accentuated if the sheets are of a dark colour), and are consequently used in shorter lengths. Expansion across the width of the panels is not usually significant as the corrugations are capable of distorting to accommodate it. Within the length, end laps can accommodate thermal movement by fixing through the upper sheet only. 'Stand-off' type fasteners are used to avoid compressing the insulation and distorting the metal at the fixings.

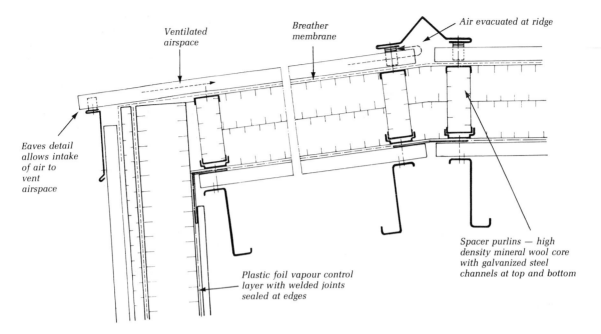

Ventilated airspace

Breather membrane

Air evacuated at ridge

Eaves detail allows intake of air to vent airspace

Plastic foil vapour control layer with welded joints sealed at edges

Spacer purlins — high density mineral wool core with galvanized steel channels at top and bottom

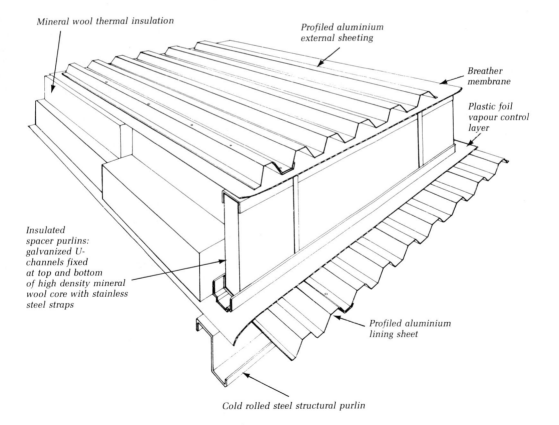

Mineral wool thermal insulation

Profiled aluminium external sheeting

Breather membrane

Plastic foil vapour control layer

Insulated spacer purlins: galvanized U-channels fixed at top and bottom of high density mineral wool core with stainless steel straps

Profiled aluminium lining sheet

Cold rolled steel structural purlin

Fig. 8.52 Profiled aluminium sheeting: insulated purlin construction

Also available are aluminium 'planks' with interlocking standing seam edges. They are attached to internal clips or secretly fixed to the supporting structure so fixing holes through the cladding are eliminated. Some systems can be laid to a pitch as low as 1 degree (without reliance on mastic). The planks, which generally have a cover width between 300 and 600 mm, are available in lengths up to in some cases 40 m (depending on transportation and site access restrictions). As a result, joints within the length of the sheets can also be avoided or the joints may be welded to maintain an uninterrupted surface; by curving the sheets a junction at the ridge can also be eliminated (see fig. 8.54 — steel profiled 'plank' systems are similar). In one current system, panels can be formed to a taper which enables roofs to be made curving in plan and section (see photo 5.4).

Composite panels are increasingly widely used. A sandwich of insulation between an outer weathering aluminium skin and an internal shallow-profiled liner is made by pouring liquid foamed thermal insulation between the layers of aluminium sheeting. Polyurethane foamed with water or carbon dioxide (and consequently free from CFCs) is now used by some manufacturers. On cooling the insulation becomes rigid and adheres to the metal; not only does it insulate the panel but it also helps provide structural rigidity. As a consequence, relatively thin gauge metal is used, around 0.7 mm for the weathering face and less for the internal lining. The panels have a high strength to weight ratio and a 50 mm thick panel can achieve a U-value less than the 0.45 W/m²°C currently required for the roofs of industrial buildings. The method of fixing composite panels is similar to conventional sheeting and insulation-bonded sheet (see fig. 8.55 which shows a steel equivalent) but some systems employ secret fixings similar to standing seam types. Because of its integral insulation and the relative rigidity of composite panels the outer skin can reach a much higher temperature than the inner, the resulting stresses would be detrimental over long distances so the maximum lengths of composite panels are kept relatively short. A great advantage is that the absence of internal voids within their construction makes them less subject to the detrimental effects of interstitial condensation.

The greatly increased use of profiled cladding in recent times has resulted in the introduction of special radiused sheet that is mainly used to form a uninterrupted junction at the ridge or eaves. Most curved sheeting is made by crimping the metal within the corrugations but completely smooth curves have recently been introduced; a minimum radius of approximately 500 mm is possible. Specialist manufacturers can also supply matching flashing profiles to suit the ends and edges of curved sheets. The ductility of aluminium makes it particularly suitable for the fabrication of these geometrically relatively complex components, although panels which curve in two directions are not made in profiled metal sheet. Junctions such as a curved corner in plan meeting a radiused panel joining wall and eaves can be fabricated from flat metal sheet or alternatively GRP (although over time the latter will weather differently).

Where boundary conditions permit a roof to be designed as an *unprotected area*, requiring no fire resistance, the use of aluminium can be advantageous. Because of its relatively low melting point, the roof will burn through relatively quickly at the centre of the fire, releasing the heat and smoke to the atmosphere. Under the test defined by BS 476 Part 3, aluminium profile roofing is given the highest rating 'ext. AA' indicating the reduced hazard of fire spreading from adjacent buildings. Aluminium is non-combustible, both mill-finished and colour-coated sheets are rated class 0 under BS 476 Part 4 and class 1 'very low spread of flame' under Part 7. It should be noted, however, that bonded and composite panels utilizing foam plastic insulation may be subject to increased fire insurance premiums.

Photo 8.6   Radiused profiled sheeting at Tottenham Hale Station. Architects: Alsop and Stormer, 1992.
Both aluminium (as used here) and steel radiused sheeting can be made by crimping the surface of the metal.

8.4.3 Profiled steel (see photo 8.6)

Profiled steel, like its predecessor corrugated iron, has become a vernacular material in many parts of the world. Its widespread use not only for wall and roof cladding but also as a decking for flat roofs and permanent shuttering to cast concrete floors explains the great variety of profiles that are available. The increasing use of 'crinkly tin' in recent years has followed the introduction of more sophisticated long-life finishes that both inhibit corrosion and are maintenance-free for a considerable period. Most of these finishes comprise a protective base coat of zinc followed by primers and a weather surface consisting of a plastic coating with integral pigment. The first of these 'plastisol' treatments was introduced in the UK in 1965.

Strip steel products are among the first for which national standards have been replaced with European CEN standards. Steel for profiling should comply with BS EN 10 142: 1991 and BS EN 10 147: 1992 *Continuously hot-dip zinc coated unalloyed structural steel sheet and strip — technical delivery conditions.*

Profiles are available in both sinusoidal and trapezoidal profiles; for longer spans further ribbed corrugations are added to stiffen the profile along the length of the panel. They are made up to 44 m maximum length and 1.20 m maximum width, thickness gauges from 0.4 mm to 1.0 mm Few manufacturers produce the sinusoidal profile and only up to 19 mm in depth, whereas trapezoidal sheeting can be obtained up to 63 mm deep (from UK sources) and is accordingly stiffer. Flashings and other subsidiary components should comply with BS 1091: (1963) 1980 *Pressed steel gutters, rainwater pipes, fittings and accessories.*

Steel is stiffer than aluminium (its Young's modulus is three times as great) and can consequently be used in thinner gauges usually in the range 0.4–0.7 mm thickness. Limiting deflection may as a result not be so important as when using aluminium which for a given load/span will have a greater deflection and its attendant problems, namely failure of the roof to drain adequately at low pitches and strain at fixings and overlaps.

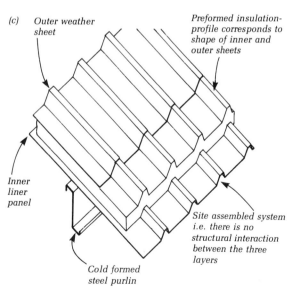

(a) *Profiled steel/aluminium sheeting*

Steel liner trays filled with glass fibre or mineral wool insulation

Batten cut from insulation board to alleviate cold-bridging mounted on steel top hat section

(b) *Plastisol coated profiled steel sheeting*

Quilt insulation

Fixing ferrule

Liner panel

Z section spacer

Cold formed steel purlin

(c) *Outer weather sheet*

Preformed insulation-profile corresponds to shape of inner and outer sheets

Inner liner panel

Site assembled system i.e. there is no structural interaction between the three layers

Cold formed steel purlin

Fig. 8.53 Profiled steel sheeting, alternative constructions: (a) liner trays; (b) over purlin, three-layer construction; (c) preformed insulation

Steel profiled sheet is used in much the same forms of construction as its aluminium equivalent, but given the wide variety of building environments that are roofed in steel profiled sheet and the alternative combinations of material, insulation and components, British Steel have isolated 31 types of system based on the following general varieties:

1. A single weather-proof sheet for industrial buildings or in combination with insulation and an internal lining of plasterboard or similar facing board.
2. This first alternative is being largely superseded by *three-layer construction* which is shown in fig. 8.53. The use of 'mini-zed' spacers is illustrated but a variety of alternative sections are available; the outer sheeting fixes directly to the spacers to make a firm metal-to-metal contact. The plastic ferrules between the spacers and the inner sheet and purlins minimize cold-bridging. Insulation is sandwiched between the outer profiled weathering sheet and the internal non-structural steel lining sheet (which is made with less pronounced corrugations) or a liner tray.

Steel *liner trays* are interlocking sections of folded steel (fig. 8.53(a)) which contain the insulation within them. By virtue of their shape, they are able to span up to about 4−5 m and eliminate the need for purlins. The width of liner trays varies between 400 and 600 mm, and their spanning capability depends on the depth of section (between 70 and 100 mm). The structural interaction of the liner trays and the cladding makes for a composite construction. Relatively good sound insulation can be achieved using liner trays (as well as fire resistance of up to 4 hours). The external profiled sheet can be either steel or aluminium and is fixed through to the flanges of the liner tray with self-tapping screws. To avoid cold-bridging, the internal and external metal surfaces are separated by a strip of rigid insulation. The metal surface that the liner trays present to the interior of the building is vapour-proof and the joints can be sealed with mastic, PVC or metallic tape to form a more successful vapour control layer than if using plastic sheeting with taped edges (as described previously), though unless curved panels are used at the ridge and the eaves, continuity of the membrane will be difficult to achieve at these changes in direction.

As described for profiled aluminium, sheets with insulation pre-bonded to the internal face are available. Some manufacturers produce insulation boards with internal and external profiles matching those of the steel panels, for use in conjunction with steel lining sheets (fig. 8.53(c)). This eliminates air voids and reduces the likelihood of interstitial condensation as explained on p. 244.

3. Of a similar nature to the systems described for aluminium, profiled steel 'planks' with upstand edges enable both low roof slopes and long spans. They have concealed fixings and either snap-fix together or the standing seams are folded by machine on site; they are produced in lengths up to 16 m to avoid the difficulty of forming end laps.

Figure 8.54 shows one method of fixing interlocking profiled sheets to the roof sub-structure by special clips attached by mechanical fasteners to the purlins. The liner panels are positioned first with all joints lapped and fixed. All joints are sealed with PVC tape or mastic, and any swarf or debris is removed from the panels before covering the whole installation with vapour control sheets made continuous by lapping all joints and sealing with PVC tape. The mini-zed purlins are then fixed through nylon spacers and liner panel to the roof purlins. A mineral wool blanket is placed between the mini-zed purlins and pushed tightly around each of them. The breather paper is laid over the insulation and the mini-zed purlins so that any moisture vapour getting into the insulation from below can ventilate to the outside. Finally, secret fixing clips are attached by self-tapping screws into the mini-zed purlins and the top sheet is secured by pushing down firmly to engage the clips into the corrugations.

4. *Composite panels*, as described for profiled aluminium, are made as hollow boxes filled with foamed insulation, which provides strength in shear to the panels. The insulation is either stopped short of the edges so conventional lap joints can be formed or, alternatively, adjacent panels interlock or thicken at the edges to make upstands that are sealed with a cover strip (fig. 8.55).

Steel profiled sheets may be designed to be walked on for maintenance and installation purposes. The code stipulation for concentrated loads (see under aluminium sheeting above) is not helpful, as thicker gauge metal than otherwise required may be necessary to withstand the weight of a person bearing at the crown only. Under the area of a shoe, this is equivalent to a concentrated load of approximately 1.3 kN; this is often the determinant of the thickness of sheet to be used.

Unprotected steel would have a very short life, but zinc coating (galvanizing) affords substantial protection at relatively low cost (MBS: *Materials* chapter 9). This can be by hot-dipped galvanizing to a spangled or matt finish. For greater corrosion resistance hot-dipped zinc/aluminium alloy coated steel is used (which is composed of 55 per cent aluminium, 43.4 per cent zinc and 1.6 per cent silicone). To receive a *plastisol* finish this is then rinsed and cleaned to remove impurities and painted with corrosion-resistant primer. The prime coat paint layer is cured in an oven at 220°C and allowed to cool. The weathering coat of 200 μm

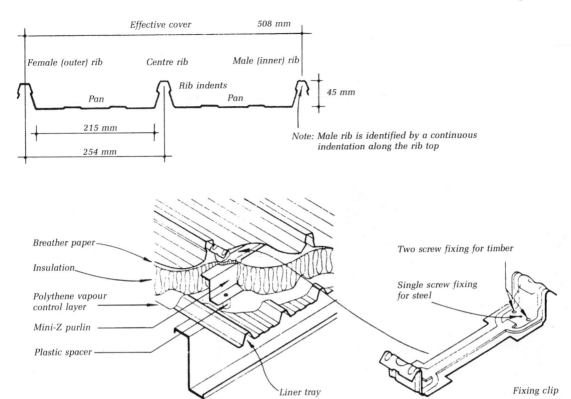

Note: Male rib is identified by a continuous indentation along the rib top

Fig. 8.54 Profiled steel sheeting details

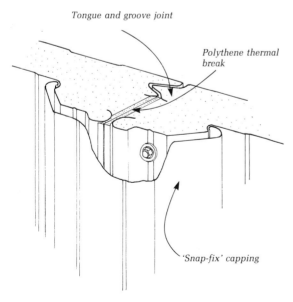

Fig. 8.55 Cover strip detail for composite panels

thick PVC plastisol is factory applied as a liquid and cured in an oven at 265°C to form a textured surface four to eight times thicker than other single-layer paint finishes. This gives plastisol coatings a greater resistance to damage and abrasion than alternative finishes. British Steel publish a performance graph for calculation of the anticipated length of time to first repainting for different coloured sheets in different environments (depending on geographic region, whether inland or marine, orientation, etc). In general, plastisol-coated steel should have a life, anticipated by the manufacturers, to be in excess of 40 years.

Alternatives are silicone polyester stoving, polyvinylidene fluoride (PVF2) coating, acrylic modified polyester coating or enamel finish to form the internal face. Silicone polyester and PVF2 are thinner and less durable than plastisol and have a smooth finish; PVF2 normally needs repainting after 15 years in inland locations. Acrylic modified polyester resin is applied thickly (around 500 μm) on an epoxy base coat. It is more expensive than other finishes, it weathers well and has good resistance to abrasion and colour stability.

Fixing methods for profiled steel roofing, single sheets and sandwich construction, are similar to aluminium profiled roofing, and also have the same fire ratings.

Photos 8.7 and 8.8 Fabric roof at La Grande Arche, La Defense, Paris. Architect: Johann Otto von Spreckelsen; fabric roof structure: Peter Rice, Ove Arup and Partners, 1983–88.

It was originally intended that the roof over the podium beneath the arch would be made from glass. In the event the canopies have been made from teflon-coated fibreglass. They are attached to the lower chord of a series of cable trusses which are hung from the intersections of the concrete office structures at either side, and tied down at the podium. The glass discs are fixed into steel rings that also clamp the edges of the fabric.

8.5 Suspended and air-supported roofing membranes

Increasing use in being made of thin fabric membranes to form roofs either suspended by external tension cables or supported by an internal structure or by air pressure. Since the 1950s, this form of building has developed in use from temporary pavilions for festivals and exhibitions to more permanent structures for many uses, including factories and laboratories (see photos 8.7, 8.8 and 8.9). During this period, the design of membrane structures has been facilitated by progress made in the development of synthetic fabrics. Early versions consisted of cotton, cotton polyester, and simple PVC-coated nylons and polyesters as well as neoprene or Hypalon coated nylons. These were either short lasting or expensive.

The most commonly employed fabrics used for this purpose today are *PTFE (Teflon)-coated glass fibre* and *PVC-coated polyester*. These woven fabrics are strong, stiff and durable as a result of the development of both synthetic fibres and polymer coatings (that resist UV degradation and make the material waterproof).

PVC is modified by the inclusion of additives to produce fabric that is cheap, stable, flexible and has some fire resistance. It does, however, collect dirt and degrade over the course of time. Different manufacturers use varying combinations of basic chemical ingredients so performance differs between the products available. Like other PVC materials used for roofing (see section 8.3.3), embrittlement can result from migration of plasticizer to the surface of the fabric. This can also result from the use of cleaning solvents so careful maintenance instructions have to be given.

Photo 8.9 Detail of fabric connections at Bedfont Lakes Office Development, Feltham, Middlesex. Architect: Michael Hopkins; Engineers: Buro Happold, 1990.
Attachments to fabric roofs are made using details that spread the stress into as large a perimeter of material as possible; a simple point connection would cause the fabric to tear. The cable junctions are made using yacht-rigging components.

PTFE is a fluorocarbon resin; only glass-based cloth can be used with PTFE because of the high temperatures involved in the coating process. The resulting fabric is more durable than PVC-coated polyester with a similar or somewhat lower level of translucence (15−7 per cent daylight transmission); but is more costly. PTFE-coated glass fibre is weldable, the surface coating of the fabric fuses at around 370°C. Fewer ingredients are required than for PVC-coated polyester; plasticizers and UV-absorbing additives are not necessary, but being more brittle, it is less forgiving of errors in construction and cannot withstand the degree of 'flutter' that PVC-coated polyester might.

An alternative waterproof coating with similar properties to PTFE is *silicone resin*; it differs, however, in being highly transparent (40−50 per cent daylight transmission), which makes its use of interest in the restricted daylight conditions of northern Europe. It too is applied to glass-based cloth, and has good flame resistance although to achieve comparable performance to PTFE fabric, fire-retardant fillers have to be added that reduce the material's translucency. Silicone-coated glass fibre roof panels can be glued together using silicone resin.

PVC/polyester and Teflon or silicone-coated glass fibre fabric membranes in themselves, provide virtually no thermal control and their application is limited as a result, although they are finding an increasing number of uses. It is technically possible to add insulation by using two layers of fabric containing a quilt of glass fibre, but daylight transmission is much reduced (a compromise being a double skin membrane with an air space between).

Building Regulations require wall and ceiling materials to be either class 1 for rooms or class 0 (class 1 plus a fire propagation index of less than 12) for escape routes when tested to BS 476: Part 7. Only fabrics with glass-based cloth can achieve class 0 as the base material cannot be a thermoplastic. Teflon coatings have been of restricted use in this context due to fears about the toxicity of gases produced under fire conditions.

Methods of jointing include *sewing* which is still used for special details or *heat welding* by heated irons or heated steel strips. The fabric is heated sufficiently for it to melt or an interlayer joining the thicknesses is placed between them. *Radio frequency* or *microwave welding* is also used; aluminium or brass discharge electrodes are placed below and above the material which is pressed together with pneumatic rams to pressure-seal the layers together. *Hot air welding* is carried out with a jet of hot air, whilst the fabric seams are being sealed together by a roller, a method suitable for lightweight cloth. There are also various types of *mechanical attachment* employed, mostly for site joints, such as bolted plates forming fabric clamps.

There is no British Standard regulating the design of fabric membranes and, when compared with conventional fully supported roof coverings, their use is still fairly unusual. Consequently, insurance warranties may be difficult to obtain. Nevertheless, there are an ever-increasing number of applications resulting in visually interesting roofs. Membrane structures allow relatively large areas to be covered without internal supports and a wide variety of structural solutions are possible. The development of this form of construction has largely been the result of the introduction of sophisticated computer programs. They take into account the physical properties of the fabric and the amount of prestress to be put into it so as to develop a suitable structural shape and cutting schedules for the material.

8.6 Green roofs

Green or *landscaped* roofs have been built in this country for many years (Derry and Toms roof garden in Kensington, dating from the 1930s, was the largest in the world at that time). Recently, many examples have been constructed on the continent; complete systems are being

Photo 8.10   Roof Garden at Derry and Toms Department Store, Kensington, London, 1938.

This roof garden, now over half a century old, was the outcome of a decision not to construct the building to the originally anticipated height because of fire escape difficulties. The excess capacity of the structure was, as a result, able to accommodate the weight of the metre depth of soil beneath which is three-coat asphalt waterproofing. The palm trees thrive because of heat in winter rising from the floors below.

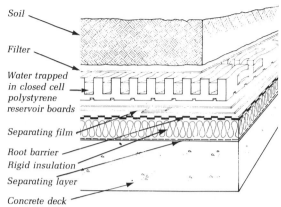

Fig. 8.56   Proprietary roof garden system including reservoirs (Erisco Bauder)

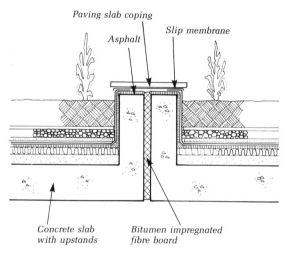

Fig. 8.57   Expansion joint detail. (Courtesy of *The Architects' Journal*)

marketed for their construction, mostly by manufacturers of high performance built-up roofing.

Green roofs offer several *ecological advantages*:

- In urban environments, an increased area of planting improves air quality by reducing *carbon dioxide*, increasing the supply of *oxygen* and filtering *dust* and *pollution*. A roof garden can also form a controlled habitat for *wildlife*.

- Overall, *land utilization* is improved, an extra *amenity* is provided for a building's users and the *appearance* of a flat roof can be enhanced by landscaping.

Whilst green roofs do present technical complications, some of the familiar problems associated with flat roof construction can be ameliorated by landscaping. Flat roofs leak for a number of reasons: wind suction across the roof can tear the waterproofing as can thermal movement between the various layers within the roof; the expansion of any moisture trapped within the construction can expand and cause the waterproofing to burst. A thickness of soil produces a more stable environment within the roof that is as a result relatively buffered from diurnal thermal fluctuations, and the additional weight obviates problems of uplift. When, however, a layer of soil is added to the roof, leaks become much more difficult to find.

Roof landscaping presents the additional problem of *weight*. At Derry and Toms roof garden a soil thickness of a metre or more was used; a lesser thickness of soil will dry out particularly given the relatively windy conditions at the top of buildings and the net shortfall of rainwater

between May and September. If, however, provision can be made for regular watering or if a system of irrigation is installed, many plants (but not trees) can be grown in only 300 mm of soil. Even so the roof loading will be in the region of 6.5 kN/m^2.

The roof will need to *drain* to outlets as in a conventional flat roof; a drainage layer of granular material (through which water can flow freely) is always used beneath the soil to stop it from becoming waterlogged during times of heavy rainfall. Sand or porous board made from large polystyrene beads are sometimes used, also leca (expanded clay granules) which has the advantage of being lightweight whilst absorbing some water to maintain a constant level of humidity within the growing medium. Some continental systems include a drainage layer made from corrugated polystyrene board, the resulting troughs forming waterways for drainage whilst the bottom of the corrugations acts as a sump to store some water for future irrigation (fig. 8.56). Fine material within the soil has to be stopped from leaching away into the drainage layer and eventually clogging the rainwater outlets; a geotextile mat (that is resistant to rot) is usual for this purpose. As a consequence, the waterproof membrane is subject to conditions that are more continuously wet than is usual in flat roof construction. Only a well-established and high-performance material is appropriate. Most large-scale roof gardens in this country as a consequence employ asphalt on a concrete structural deck, and this should preferably be a full tanking three-coat specification. Landscape roofs are constructed as both inverted and warm roofs, also uninsulated where constructed over multi-storey car parks, etc.

Because the several layers of a green roof will form a *thermal blanket* covering the building the necessity for expansion joints in the roof structure may be eliminated (helping offset the additional costs of the landscaping). If at all possible, movement joints (if required), should not be buried within the roof, but back-to-back waterproofed upstands should project above the level of the soil with a coping bridging between them and designed to accommodate the anticipated movement (fig. 8.57). The thickness of soil will contribute to the U-value of the roof, although less so when the soil is wet (so it cannot form a calculable part of the construction). It also buffers the roof membrane from extremes of temperature which can greatly increase its life.

The plants themselves also generate a further set of requirements:

- *Protection.* Roots can penetrate cracks between layers of construction and the waterproofing has to be protected from them. Some built-up roofing systems intended for green roofs incorporate a cap-sheet containing polymers that have been shown to be root repellent. Alternatives such as metal foil laminated felt present the possibility of roots penetrating at the laps. At present there are no BS or DIN standards for root-resistant roofing materials. Asphalt has the advantage of being seamless.

  Usually, above the waterproofing, a protective sheet or board will be required to withstand possible damage by following trades and the garden tools used in the future maintenance of the planting. Proprietary rubber or bitumen-bonded aggregate boards are available for this purpose; in the past brick paving has been bedded in mortar above the asphalt to provide a protective layer below the planting.

- *Irrigation.* The garden has to provide the right environment for plants to grow. Roof gardens are windy and subject to evaporation so a method of making up the summer shortfall of water has to be found either by

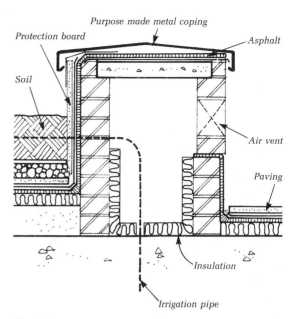

Fig. 8.58 Sprinkler pipe enclosure. (Courtesy of *The Architects' Journal*)

hand or automatic sprinklers linked by pipes beneath the soil (fig. 8.58). 'Pop-up' sprinklers work by water pressure, on turning the water off, the heads retract so as not to form an obstruction to lawn mowers.

- *Plants.* Any plant that grows on the ground can grow on a roof because topsoil is rarely more than 500 mm thick. Shallow soil is, however, likely to become frozen in winter which may preclude planting some varieties. On the other hand, roof gardens that are extensively heated by the building below (as opposed to landscaped roofs over multi-storey car parks, etc.) may make species flourish that otherwise would not in the UK. Initial allowances for the loading imposed on the structure by the planting must anticipate the eventual size of shrubs and trees. Trees grow to be very heavy and are easily uprooted in shallow soil, so particular locations, with deeper soil, have to be allocated to them in relation to the overall structural layout. Perhaps surprisingly, grass is difficult to use on landscaped roofs, it can be awkward to maintain requiring access between levels for lawn mowers; it does not like shade or drought and is subject to many diseases (see photo 8.11).

The typical build-up of layers for a roof garden is shown in fig. 8.59.

Because the most vulnerable parts of a flat roof installation, in terms of the potential for future leaks, is at intersections and changes in direction, the waterproofing of a roof garden should be as simple and consistent as possible. This has particular implications for the way in

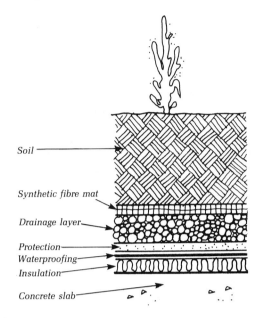

Soil

Synthetic fibre mat

Drainage layer

Protection
Waterproofing
Insulation

Concrete slab

Fig. 8.59 Typical green roof build-up — warm roof. (Courtesy of *The Architects' Journal*)

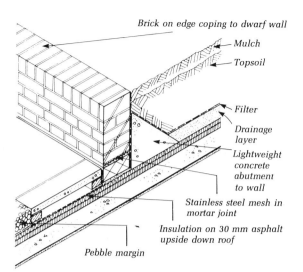

Brick on edge coping to dwarf wall

Mulch

Topsoil

Filter

Drainage layer

Lightweight concrete abutment to wall

Stainless steel mesh in mortar joint

Insulation on 30 mm asphalt upside down roof

Pebble margin

Fig. 8.60 Retaining wall planter detail — inverted warm roof (RMC offices, Runnymede. Architect: Ted Cullinan. Courtesy of *The Architects' Journal*)

which separations between the different areas of the roof are detailed. To keep the waterproofing level, concrete kerbs and brick walls around landscaped areas should be built up off the waterproofing rather than off the structural deck. Consequently additional concrete footings may have to be cast against the kerbs (and hidden beneath the planting) to resist overturning of the kerbs where they are acting as

retaining walls (fig. 8.60). The protection layer, waterproofing membrane and the insulation must have sufficient strength to withstand these localized loads. Similarly, *handrails* around the perimeter of the roof that are required by code to resist sideways thrust will require their own foundation formed above the waterproofing rather than bolting them to the structure in the conventional way (fig. 8.61). Penetrations through the waterproofing should generally be avoided (including those for pipes and conduit)

Photo 8.11 Phase 4 IBM Headquarters, Cosham, Hampshire. Architects, engineers & quantity surveyors: Arup Associates, 1982. The soil here is 300 mm deep. The grass is watered two or three times a week to prevent drying out in summer. The photograph was taken soon after construction and the shrubs at the perimeter are now sizeable.

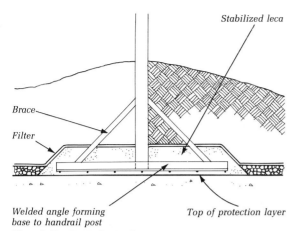

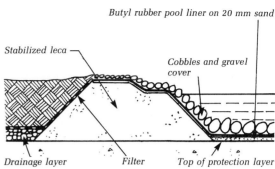

Fig. 8.64 Pool details

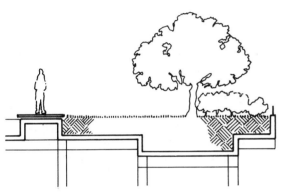

Fig. 8.61 Handrail detail (Wiggins Teape, Basingstoke. Architects, engineers, quantity surveyors: Arup Associates. Courtesy of *The Architects' Journal*)

Fig. 8.62 Tree pit detail. The alternative is to keep the ground surface level and form tree pits within the structural deck. The result is a number of right-angled junctions within the waterproofing, a major disadvantage of this approach. (Courtesy of *The Architects' Journal*)

since they are also potential locations at which leaks may occur.

The greater depth of soil required by *trees* can be accommodated either by forming a recessed area (tree pit) within the roof slab (fig. 8.62) or by mounding the soil (fig. 8.63). Both of these entail locally increased loads which probably require them to be associated with the location of columns and/or principal beams below. Since 'tree pits' have to be accommodated within the section of the building and require potentially hazardous changes in direction for the waterproofing, 'tree mounds' are a preferable solution. *Ponds* may be formed most simply by constructing an enclosure (from *in situ* lightweight concrete for example) containing a bentonite, GRP or welded butyl rubber waterproof liner (fig. 8.64). The pond water has to circulate to stop it going stagnant and if it is to be passed through pumps, it will need to be filtered.

For small roof gardens the *rainwater outlets* within the waterproof layer may have an increased depth of drainage material local to the outlet (fig. 8.65). It is preferable that the outlets should be accessible at a later date by constructing an inspection trap with a removable top. A

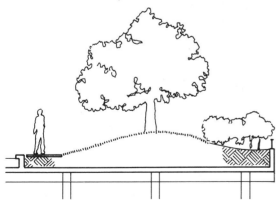

Fig. 8.63 Tree mound detail. One method of giving trees the extra depth of soil they need is to locally form mounds. Trees when full grown are very heavy which, with the additional weight of soil, requires them to be located above structural columns or walls. (Courtesy of *The Architects' Journal*)

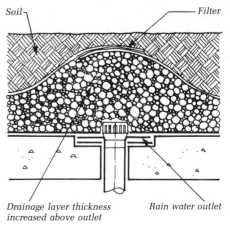

Fig. 8.65 Small roof outlet details (uninsulated)

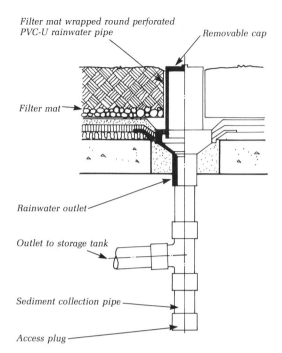

Filter mat wrapped round perforated
PVC-U rainwater pipe

Removable cap

Filter mat

Rainwater outlet

Outlet to storage tank

Sediment collection pipe

Access plug

Fig. 8.66 Drainage outlet details. (Courtesy of *The Architects' Journal*)

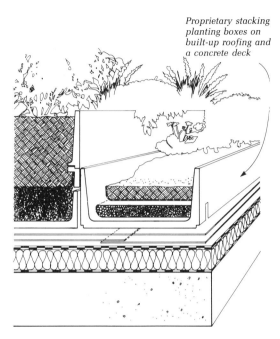

Proprietary stacking
planting boxes on
built-up roofing and
a concrete deck

Fig. 8.67 Proprietary planting box details (Erisco Bauder)

length of perforated pipe to the depth of the drainage layer and topsoil with a grating at ground level is an alternative to the construction of a masonry enclosure (fig. 8.66). During the average British summer the irrigation requirements of a roof garden commonly amount to the equivalent of 300 mm depth of water over the roof area. Reliance upon mains supply, which is increasingly subject to restriction, is likely to be problematic. The roof drainage system may be linked via a sediment collection sump to a storage tank from which the water can be recycled.

Systems now available for the construction of roof gardens not only incorporate the insulant, waterproofing, drainage and filter layers but also pre-seeded soil. One proprietary system includes modular polystyrene planting boxes that can be stacked to accommodate varying soil depths; these are built up off the separating layer above the waterproofing (fig. 8.67). A variety of panels are available to face the vertical sides of the planter boxes where they are visible. Also available are hydroculture systems; the drainage layer is kept permanently flooded so the soil thickness can be reduced to as little as 150 mm. Packaged roof gardens of this sort usually employ built-up bitumen felt waterproofing; some single ply roofing manufacturers also regard their materials as root-resistant and suitable for landscaped roof applications. In all cases, however, the membrane has to be adequately protected from possible damage by gardening implements. The logic of using any jointed waterproofing material has to be questioned in a situation where the roof may be permanently wet although these systems have been extensively used on the continent.

8.7 Pitched roof coverings

Pitched roof coverings may consist of the membranes, large sheets and methods already considered under section 8.3, or more usually, of small dry-jointed units such as slates and tiles. Slates and tiles provide a *semi-permeable* barrier to wind, rain and snow and need to be used in conjunction with an internal layer of impervious felt or plastics sheet to become completely weather-proof (see section 8.2.1). Another system of pitched roof covering is the traditional method of thatching.

Roofing *slates* are made from stone, epoxy resin reinforced aggregate and fibre cement; *tiles* from clay, concrete, fibre cement, reinforced bitumen strip, metal and wood (shingles). There is an important distinction between components which overlap, such as plain tile, pantiles and slates, and those which interlock (see fig. 8.68). Of those which overlap, plain tiles and slates are laid to a *double lap*, whereas pantiles are laid to a *single lap*. Tiles and synthetic slates which interlock may do so on two sides or on all four.

BS 5534 *Code of practice for slating and tiling*: Part 1: 1990 *Design* deals with the design and application of both

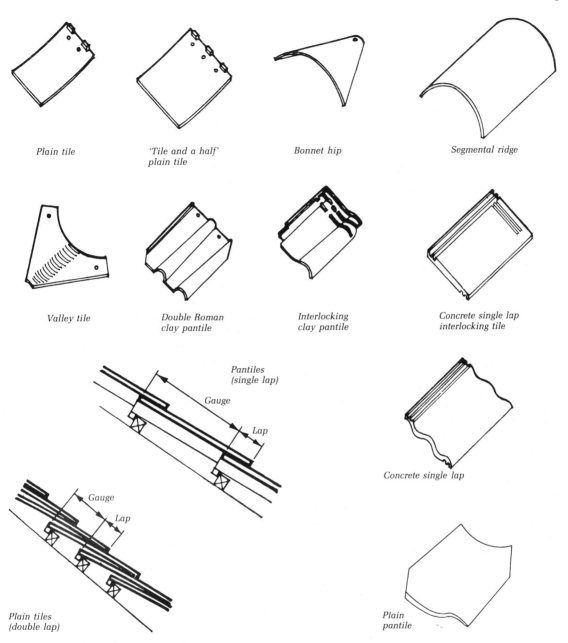

Plain tile

'Tile and a half' plain tile

Bonnet hip

Segmental ridge

Valley tile

Double Roman clay pantile

Interlocking clay pantile

Concrete single lap interlocking tile

Pantiles (single lap)

Gauge

Lap

Gauge

Lap

Plain tiles (double lap)

Concrete single lap

Plain pantile

Fig. 8.68   Types of roofing tiles

slating and tiling, including the provision of underlay, boarding, counterbattens and their fixing. There have been an increasing number of building failures as a result of wind damage to pitched roofs. In the recent revision of the Code of Practice calculation procedures have been included to ensure that pitched roof coverings are resistant to wind uplift. The variables to be taken into account with regard

to slates and tiles include anticipated wind loading (as determined from CP 3: Chapter 5 Part 2: 1972, see section 3.2.2), the height and width of the building, the slope of the roof and the type and method of fixing of the roof covering. See also BS 5534: Part 2: 1986 *Design charts for fixing roof slating against wind uplift in roofs of ridge height not exceeding 30 m above ground*. Manufacturers of roofing components commonly prove their systems by wind tunnel testing.

Procedures on site are described in BS 8000: 1990 *Workmanship on building sites*, Part 5: *Code of practice for slating and tiling of roofs and claddings*. Another useful reference source is the *Redland roofing manual: a guide to good practice and Redland roofing systems* published by Redland Roof Tiles Limited.

Pitched roof construction is required to meet the thermal insulation standards of the *Building Regulations 1991*: Approved Document L1. This is discussed in some detail under section 8.2.3. For dwellings the required thermal transmittance (U) of 0.25 W/m²K is most simply achieved through the use of a 125 mm minimum thickness mineral wool (or similar material) insulation quilt. An alternative is to follow the calculation procedures outlined in Approved Document L1 (as explained under section 8.2.3, if the effect of the rest of the construction is taken into account, the insulation thickness may be reduced).

The provision of thermal insulation must go hand in hand with effective methods of vapour control, as explained under section 8.2.4. Condensation in roof spaces, particularly in domestic buildings, has become problematic because of increasing standards of insulation and the growing use of appliances that cause humidity. The *Building Regulations 1991*: Approved Document F2 and BS 5250:

Table 8.10 Installed weights of pitched roof coverings

Material	Approx. installed weight (kg/m²)
Natural slates	28.0–38.0
Fibre cement slates	20.3–23.1
Plain tiles (double lap)	
clay	67.1–76.2
concrete	80.0–104.0
Interlocking tiles	
clay	39.0–50.0
concrete	43.9–50.6
Bitumen strip	10.0
Metal multi-tiles	7.7
Wood shingles	7.5

1989 *Control of condensation in buildings* require ventilation openings at the eaves of pitched roofs (as shown in figs 8.69 and 8.70). Mono-pitch roofs and rooms within roofs also need to be ventilated at the apex of the roof. BS 5250 further recommends that lofts insulated at ceiling level, if the roof pitch is over 35 degrees and the building is more than 10 m wide, should be ventilated at the ridge as well as the eaves.

When correctly installed on timber rafters, with or without an underlay or boarding, slates and tiles have an

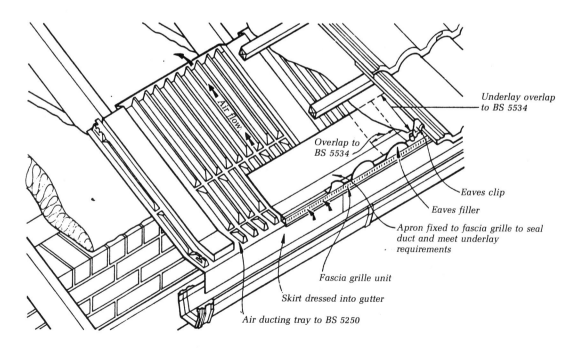

Fig. 8.69 Proprietary plastics airduct tray to BS 5250

AA external fire designation in the *Building Regulations 1991*: Approved Document B (see section 8.2.5). Slates and tiles with a backing of battens, counterbattens, underlay and boarding (or plywood) provide a defence against burglars as well as better wind resistance when compared with a backing of battens, underfelt and open rafters.

The approximate installed weights of pitched roof coverings are as indicated in table 8.10. It is important to check with manufacturers and/or suppliers to obtain precise weights according to specification or materials used.

8.7.1 Slates

This term is used to describe both slates that are natural stone as well as fibre cement slates made from a composition of factory-produced materials. Both are double lap units. In addition, reconstituted slates are made from crushed slate with a resin binder, some have interlocking edges and are single lap units.

Natural slates Natural roofing slate, a dense fine-grained metamorphic rock, is obtained from the rock beds of Wales, north Lancashire, Westmorland and Cornwall and imported, particularly from Spain, France, Portugal and South America. Slates should comply with BS 680 *Roofing slates* Part 2: 1971 *Metric units* which describes those from the Cambrian, Ordovician, Silurian, Devonian, and Dalradian formations, and describes characteristics, standard designations, thickness gradings, marketing descriptions, testing procedure for atmospheric conditions,

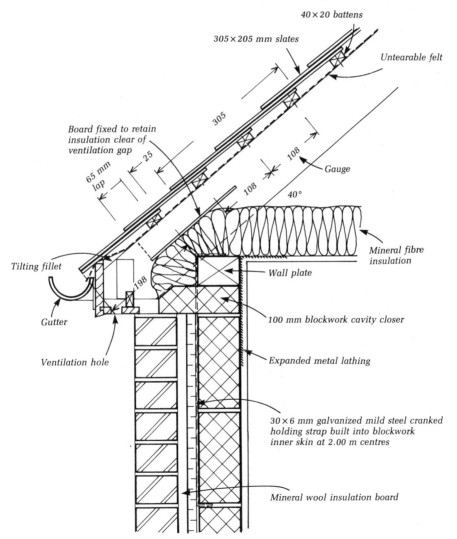

Fig. 8.70 Slating details: head-nailed slates showing counterbattens

lengths and widths. MBS: *Materials* chapter 4 gives specific information regarding metamorphic stone under which slate is categorized; it is formed by immense earth pressures acting upon clay which results in planes of cleavage. Roofing slates are manufactured by splitting along these planes; the resulting surface may not be perfectly flat. Good slate, such as that complying with BS 680, provides a most durable roofing material; poor slate may begin to decay in a few months, especially in damp conditions in industrial areas. Imported slates have proved to be of excellent quality but, before fixing, all slates should be inspected for consistency in size and shape and colour (particularly since imported slates tend to come from more than one quarry). Care should be taken if using a material which is not native to the UK climate and BS 680 tests should be carried out, although they do not establish the ultimate quality of the material they do indicate a minimum standard. Included are two tests that all slates should pass — a water absorption test, and a wetting and drying test. Slates to be used in polluted areas should pass a further test involving immersion in sulphuric acid (this was introduced in the 1940s and is less generally applicable now given the intervening improvements in air quality). British slate has a fine well-compressed grain that enables slates to be of consistent thickness and have an expected life of up to 400 years in the UK. Although continental slates have considerable durability they may not last as well in the climate of this country.

Natural slates normally have fine grain riven texture which may contain ingrained stripes, and vary in overall colour, for example: Welsh, blue/black; Westmorland, blue/grey to grey green; and French, black.

There are more than 20 'standard' sizes of roofing slate, the largest is 610 mm long × 355 mm wide and the smallest 225 mm long × 150 mm wide. When rainwater falls on to a pitched roof it will fan out and run over the surface at a given angle. This angle will depend upon the pitch of the roof and is commonly referred to as the *angle of creep*, see fig. 8.71. The steeper the pitch, the narrower the angle will be and this can be used as a guide to the minimum width of slate to be used. The relationship between size of slate, head lap and angle of creep is defined within BS 5534. It follows that the shallower the pitch the wider the slates will have to be and, as a general principle, the more exposed the position of the roof the smaller will be the slates and the steeper will be the pitch. In order to collect any wind-blown water which passes the rain check provided by the slates, an underlay of untearable felt or plastic sheet must be fixed beneath the roofing battens (see p. 190 and fig. 8.72). Underlay when draped over the rafters should be type 1F reinforced bitumen felt to BS 747 or sheet PVC or polyethylene of sufficient strength and durability.

Table 8.11 gives a range of the metric equivalent sizes of slate most commonly used and the minimum recommended rafter pitch.

The thickness of slates varies according to the source,

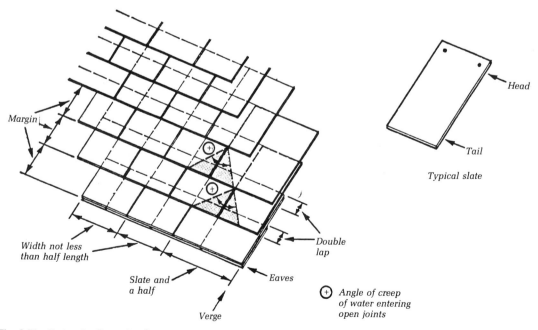

Margin

Width not less than half length

Slate and a half

Verge

Eaves

Head

Tail

Typical slate

Double lap

⊕ *Angle of creep of water entering open joints*

Fig. 8.71 Slating details: angle of creep

Table 8.11 Metric equivalent sizes and minimum pitch

Sizes of slates: length × width (mm)	Minimum pitch in degrees
305 × 205	45
330 × 180	40
405 × 205	35
510 × 255	30
610 × 305	25
610 × 355	$22\frac{1}{2}$

those from Westmorland and north Lancashire being relatively thick and coarser in surface texture. The thickest slates from any quarry are called Bests or Firsts, and the thinner slates Seconds and Thirds. This description does not then refer to quality but indicates thickness. Where slates supplied vary in thickness, the thinner slates should be used at the ridge and the thicker slates at lower courses.

Each row of slates is laid starting from the eaves and is butt jointed at the side and overlapped at the head (see fig. 8.72). Slates are laid double lap with special slates at the eaves and verge. This means that there are two thicknesses of slate over each nail hole as protection, making in all, three thicknesses of material at the overlaps. The side joint should be left very slightly open so that water will drain quickly. Each slate is nailed twice. The slate should be holed so that the 'spoiling' will form a countersinking for the nail heads. The slates are best holed by machine on the site so that the holes can be correctly positioned by the fixers. The nails should be of aluminium alloy, copper or silicon bronze. They are 30 mm long for the lighter and smaller slates and up to 65 mm long for the

heavier slates. The slates may be either *centre-nailed* or *head-nailed*. For centre-nailing, the nail holes are positioned by reference to the gauge and lap so that the nails just clear the head of the slates in the course below. Centre-nailed slates on battens and counterbattens on felt are illustrated in fig. 8.72. For head-nailed slates the holes will be positioned about 25 mm from the upper edge of the slate. Head-nailed slates on battens and counterbattens on felt are illustrated in 8.70. The holes should not be nearer than 20–25 mm to the side of the slate. Centre-nailing gives more protection against lifting in the wind or chattering. Head-nailing should therefore only be used on smaller sizes of slate. Because of the angle of creep the width of slate is chosen bearing in mind the pitch of the roof; the shallower the pitch the larger the unit required. The head lap is chosen according to the degree of exposure, and in relation to the pitch, since the steeper the pitch the quicker the runoff. The following minimum laps can be taken as a guide for moderate exposure:

Rafter (degrees)	Head laps (mm)
20	115
25	90
30	75
40	65

For severe exposure, that is to say on sites which are on high ground, near the coast or where heavy snowfall is common, the lap should be further increased as follows:

Rafter pitches (degrees)	Head laps (mm)
25	100
30	75

(see BS 5534: Part 1: 1990).

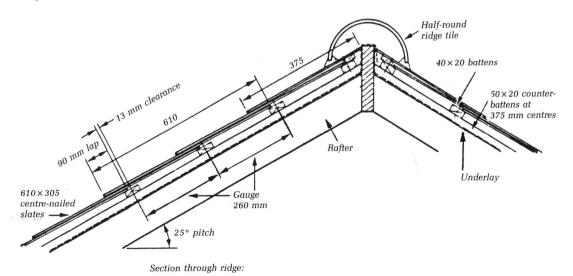

Section through ridge:

Fig. 8.72 Slating details: centre-nailed slates showing counterbattens

Before setting out the slating, the distance from the centre to centre of the battens must be determined. This distance is known as the gauge and is equal to the amount of slate which is exposed measuring up the slope of the roof. The gauge may be worked out as follows: first decide on the head lap required with regard to the degree of exposure, say for example 90 mm at 25 degree pitch, using 610 × 305 mm slates. Then for centre-nailed slates:

$$\text{gauge} = \frac{\text{length of slate lap}}{2}$$

$$= \frac{610 - 90}{2}$$

$$= 260 \text{ mm}$$

(see fig. 8.72). If the slates are head-nailed, allowance must be made for the fact that the nail holes are positioned 25 mm from the top of the slates. For example, 65 mm lap for 305 × 205 mm slates at 40 degree pitch:

$$\text{gauge} = \frac{\text{length of slate} - (\text{lap} + 25 \text{ mm})}{2}$$

$$= \frac{305 - (65 + 25)}{2}$$

$$= 108 \text{ mm}$$

(see fig. 8.70). The preserved timber battens upon which the slates are fixed should not be less than 38 mm wide and of sufficient thickness to prevent undue springing back as the slates are being nailed through them. The thickness of the battens will depend upon the spacing of the rafters; for rafters at say 450 mm centres the battens should be 19 mm thick (note that in Scotland the *Regulations* require thicker battens).

Eaves courses of slates must always be head nailed, the length of the eaves slate is worked out as follows:

length of slate at eaves = gauge + lap + 25 mm

therefore for previous example:

length of slate at eaves = 108 + 65 + 25 = 198 mm

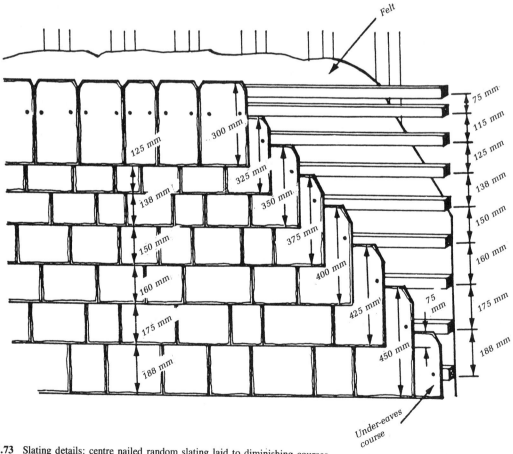

Fig. 8.73 Slating details: centre nailed random slating laid to diminishing courses

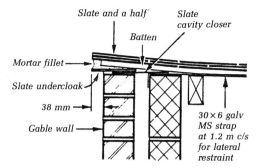

Labels on figure:
- Slate and a half
- Slate cavity closer
- Batten
- Mortar fillet
- Slate undercloak
- 38 mm
- Gable wall
- 30 × 6 galv MS strap at 1.2 m c/s for lateral restraint

Fig. 8.74 Slating details: verge

In order that the maximum width of lateral cover is maintained, the slates are laid half-bond so that the joints occur as near as possible over the centre of slates in the course below. This means that in each alternate course the slate at the verge will be 'slate and a half' in width. Slating can be laid so that the gauge diminishes towards the ridge and this is known as laying in *diminishing courses*. This traditional technique (shown in fig. 8.73) requires skilled craftsmanship to ensure correct bonding. It imparts an attractive appearance, particularly where slates of random width are used. A minimum size of lap should be specified to be increased as required to maintain the diminishing courses.

Slating should overhang slightly at the *verge* in order to protect the structure below. The overhang of the slate should be not more than 50 mm and the edge of the slate is supported by using an undercloak of slate or fibre cement sheeting bedded on the walling. The verge should have an inward tilt and the bedding mortar is usually 1:3 cement/ sand by volume, as shown in fig. 8.74. Alternatively, the roof structure, supported on gable ladders built into the brickwork, may overhang the wall and be finished off with a timber barge board. The verge slating will then project slightly beyond the barge board.

Hips can be finished with lead rolls or with tiles, but for the steeper pitches the neatest solution is to cut the slates and mitre them along the head using lead soakers lapped and bonded with each course and nailed at the top edges. Specially wide slates should be used so that the side bond is maintained when the slate is cut. *Valleys* are usually formed by having a dressed metal valley gutter and raking cut slates. As at hips, specially wide slates are required so that they are sufficiently wide at their tails when cut. The slates are not bedded and do not have an undercloak. The traditional techniques for the swept valley formed by cutting slates to special shapes, and the laced valley require skilled craftsmanship. Details are shown in fig. 8.75.

Slating has always tended to be comparatively expensive because of the skill required to punch nail holes on site and the waste due to breakage. The continental practice of fixing slates with hooks, which obviates the need for punched holes, has been introduced into the UK but has made only a limited impact on this traditional craft.

Fibre cement slates As already mentioned under section 8.4.1, the manufacture of slate substitutes using asbestos fibre cement has been replaced by less hazardous materials. Now, Portland cement reinforced with a blend of natural and synthetic fibres (cellulose or glass fibre) and fillers is pressed into sheets with the addition of pigments to form slates in a restricted range of colours (blue/black, grey and brown), and a variety of shapes and sizes — see fig. 8.76 and table 8.12. They are also available with smooth or riven texture, matt finished or with a semi-gloss acrylic coating. Under the fire test specified in BS 476 Part 3: 1975, fibre cement slates achieve the highest AA rating for performance in external fire exposure of roofs and class 0 surface spread of flame designation under BS 476 Part 7: 1987.

Fibre cement slates are supplied pre-holed for centre-nailing with two nails to each slate or plain for hook fixing. As described in BS 5534 *Slating and tiling*: Part 1: 1990 slates of the fully compressed type are additionally fixed by inverted disc-headed copper rivets projecting between the edges of the two under-slates and through a hole in the tail of the top slate and bent down flat to the roof. Typical details at eaves, verge, ridge and valley positions are shown in figs 8.77 and 8.78. There is no British Standard at present specifically for fibre cement slates, but the recommendations contained in BS 690: *Asbestos cement slates and sheets* Part 4: 1971 *Slates* are useful.

The runoff from fibre cement slates is alkaline and

Table 8.12 Typical range of sizes for fibre cement slates (courtesy of Eternit TAC Ltd)

Slate size (mm)	Lap (mm)	Pitch	
		Severe exposure	Moderate exposure
600 × 350	100	20 and over	20
600 × 300	106	20 to 25	—
	100	25 and over	20 and over
	90	30 and over	25 and over
	80	35 and over	30 and over
	70	40 and over	35 and over
500 × 250	106	25 to 30	—
	100	25 and over	22.5 and over
	90	30 and over	25 and over
	80	30 and over	27.5 and over
	70	45 and over	40 and over
400 × 240	80	45	30
	70	45	40

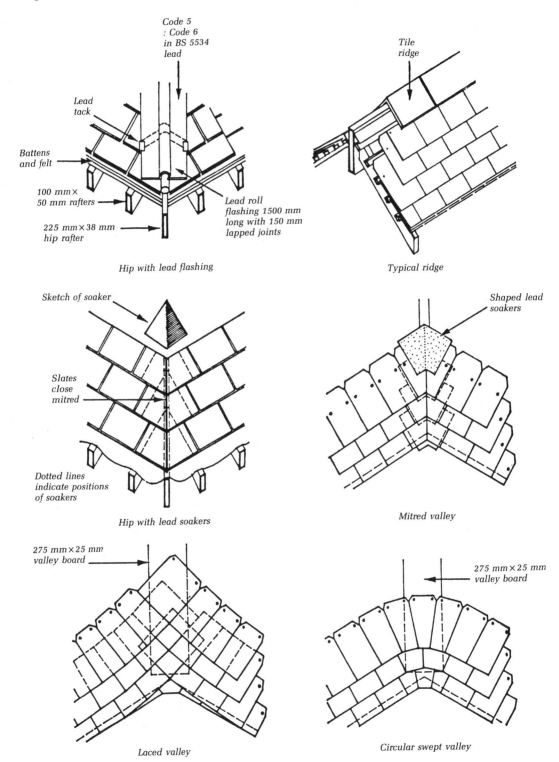

Code 5
: Code 6
in BS 5534
lead

Tile
ridge

Lead
tack

Battens
and felt

100 mm ×
50 mm rafters

225 mm × 38 mm
hip rafter

Lead roll
flashing 1500 mm
long with 150 mm
lapped joints

Hip with lead flashing

Typical ridge

Sketch of soaker

Slates
close
mitred

Dotted lines
indicate positions
of soakers

Shaped lead
soakers

Hip with lead soakers

Mitred valley

275 mm × 25 mm
valley board

275 mm × 25 mm
valley board

Laced valley

Circular swept valley

Fig. 8.75 Slating details: hips, ridge and valleys

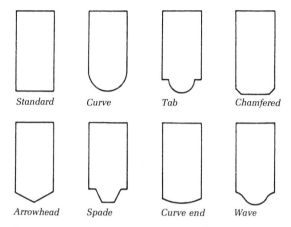

Standard   Curve   Tab   Chamfered

Arrowhead   Spade   Curve end   Wave

Fig. 8.76 Typical range of shapes for fibre cement slates (courtesy of Eternit TAC Ltd)

detrimental to some metals such as aluminium which needs to be protected and maintained with bituminous paint if used for flashings and gutters. Being pressed to shape, special slates or jointing pieces are produced by manufacturers to make a whole roofing system not requiring any other materials (i.e. such as the lead sheet used in traditional slate roofing to form hips, valleys, etc.). These special components include eaves and ridge ventilators, verge closers, valley gutters and ridge tiles.

The use of *liner trays* is increasing as an alternative to traditional carpentry roof construction. Liner trays reduce the number of purlins required or eliminate them entirely; their spanning capability results in an economic structure that speedily enables the building to be made watertight. Fibre cement slates are well suited to this lightweight form of construction as shown in fig. 8.79. The liner tray should preferably run from eaves to ridge in one length; the overlapping upstands of the liner tray support counterbattens fixed with self-tapping screws at centres determined by the expected levels of wind suction. These in turn support a conventional build-up of sarking felt, slating battens and the fibre cement slate covering.

Reconstituted slates   These resemble natural slate but are made from crushed slate with the addition of stone aggregate, resin and glass fibre reinforcement. Natural exposure and accelerated weathering tests incline manufacturers to anticipate a life of between 30 and 60 years for these components. They are highly resistant to pollution and satisfy the sulphuric acid test described in BS 680. Some are made with profiled edges to be interlocking rather than plain tiles, they are consequently lightweight and relatively fast to fix.

Being manufactured with organic binders, their performance in fire, particularly in terms of flame spread,

may restrict their application. Since the *Building Regulations* classify a roof pitch of 70 degrees or more as a wall, considerations of fire propagation may preclude the use of reconstituted slates for vertical hanging or the lower slopes of mansard roofs.

8.7.2 Tiles

Plain tiles   Plain tiles which are available in clay and concrete are classified as double lap units. Like slates, they are laid in bond with double laps and have no interlocking joints. Nearly all plain tiles are cambered from head to tail so that they do not lie flat on each other, which prevents capillary movement of water between the tiles when they are fixed. Some also have a camber in the width, but usually only on the upper surface. It should be noted that the camber in the length of the tile reduces the effective pitch, normally by about 9 degrees at 65 mm lap. Special tiles are available for use as ridges and to form hips and valleys; also 'tile and a half' for verges. Each plain tile has two holes for nailing and most are provided with nibs so that they hang on to the battens, see fig. 8.68.

Clay plain tiles and fittings should comply with BS 402: Part 1: 1990 *Specification for plain tiles and fittings*. For the first time, this recent revision of the standard includes a rigorous frost test to suit the relatively severe climatic conditions of the UK. It is anticipated that a future European standard will specify alternative levels of performance depending on the varieties of climate found throughout the EC. Well-burnt clay tiles are resistant to frost and are not affected by atmospheric pollution.

Standard sizes in mm are: plain tile 265 long × 165 wide: 'tile and a half' 265 long × 248 wide; thickness 10–15 mm. There is a limited production in certain districts of hand-made tiles which are now only used for special work. These are slightly thicker than the machine-made tiles, varying between 13 and 16 mm. Special length tiles are required at the eaves and for the top course. The minimum rafter pitch recommended for clay tiles in BS 5534: Part 1: 1990 is 40 degrees.

Concrete plain tiles should comply with BS 473/550 (Combined) 1990 *Specification for concrete roofing tiles and fittings*. The fittings include half-round, segmental, hogsback and angular ridge and hip tiles, bonnet hip tiles and valley and angle tiles. The metric equivalent standard size for concrete plain tiles is 265 × 165 mm × approximately 10 mm thick. Concrete plain tiles usually cost less than clay tiles; generally they are faced with coloured granules which give a textured finish in a wide range of colours.

As a result of their density and absence of any laminar structure, when manufactured in accordance with the British

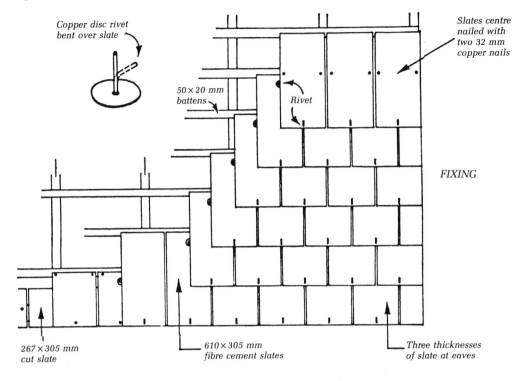

Copper disc rivet
bent over slate

Slates centre
nailed with
two 32 mm
copper nails

50 × 20 mm
battens

Rivet

FIXING

267 × 305 mm
cut slate

610 × 305 mm
fibre cement slates

Three thicknesses
of slate at eaves

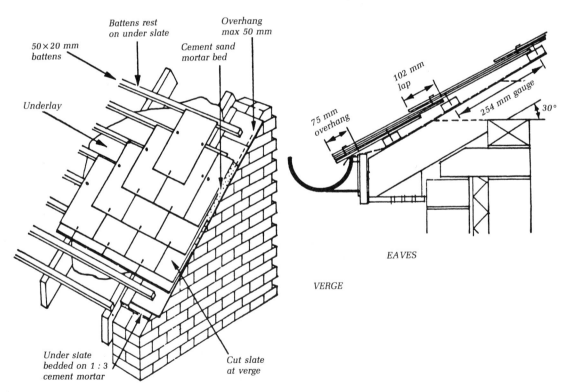

50 × 20 mm
battens

Battens rest
on under slate

Overhang
max 50 mm

Cement sand
mortar bed

Underlay

102 mm
lap

254 mm gauge

75 mm
overhang

30°

EAVES

Under slate
bedded on 1 : 3
cement mortar

Cut slate
at verge

VERGE

Fig. 8.77 Fibre cement slating details: fixings and verge

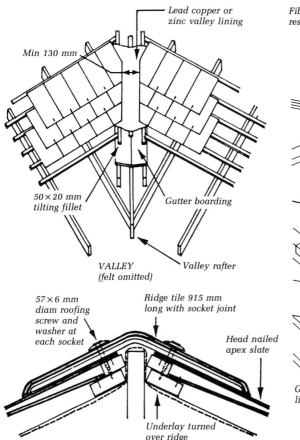

RIDGE OR HIP DETAIL

Fig. 8.78 Fibre cement slating details: ridge, hip and valley

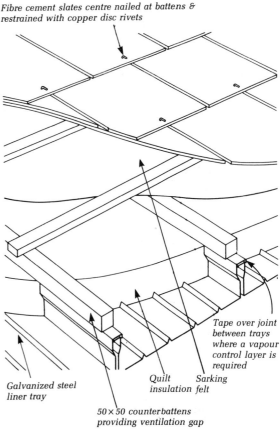

Fig. 8.79 Fibre cement tile roofing on liner trays

$$\text{gauge} = \frac{265 - 65 \text{mm}}{2} = 100 \text{ mm}$$

Standard, concrete tiles are not affected by frost. The minimum rafter pitch for these tiles recommended in BS 5534 in order to prevent rain and snow penetrating the joints is 35 degrees, 5 degrees less than the less frost-resistant clay tiles.

The lap for both clay and concrete plain tiling must not be less than 65 mm for moderate exposure. When exposure is severe the lap of clay tiles should be increased to 75 mm or more. It should be noted that increasing the lap decreases the pitch of individual tiles and for this reason the lap must never exceed one third of the length of the tile. The gauge — or normal spacing of the battens — on the roof slope is worked out as follows:

$$\text{gauge} = \frac{\text{length of tile lap}}{2}$$

For standard 265 × 165 mm tiles:

Nibbed plain clay and concrete tiles for rafter pitches below 60 degrees require nailing as follows:

1. Two nails in each tile every fifth course.
2. At each side of valleys and hips, the end tile in every course should be nailed or otherwise mechanically fixed.
3. Similarly at the ends of each course adjacent to abutments and verges.
4. At eaves and top edges, two courses of tiles should be nailed or otherwise mechanically fixed.
5. See BS 5534 for special recommendations for extra nailing in exposed locations and at steep pitches of 60 degrees and over.

Nails should be made from the following materials:

1. Aluminium alloy complying with BS 1202 Part 3. These are extensively used and have excellent resistance to corrosion.

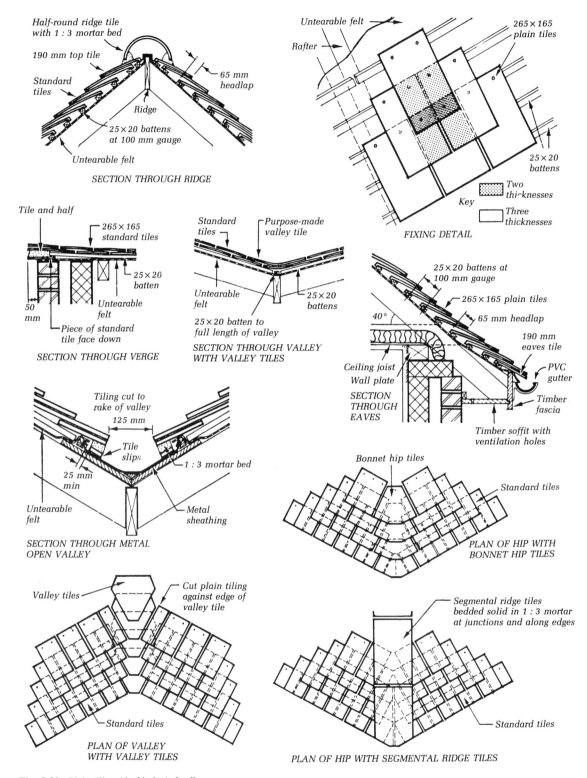

Half-round ridge tile
with 1 : 3 mortar bed

190 mm top tile

Standard
tiles

65 mm
headlap

Ridge

25 × 20 battens
at 100 mm gauge

Untearable felt

SECTION THROUGH RIDGE

Untearable felt

Rafter

265 × 165
plain tiles

25 × 20
battens

Two
thicknesses

Key

Three
thicknesses

FIXING DETAIL

Tile and half

265 × 165
standard tiles

25 × 20
batten

50
mm

Untearable
felt

Piece of standard
tile face down

SECTION THROUGH VERGE

Standard
tiles

Purpose-made
valley tile

Untearable
felt

25 × 20
battens

25 × 20 batten to
full length of valley

**SECTION THROUGH VALLEY
WITH VALLEY TILES**

25 × 20 battens at
100 mm gauge

265 × 165 plain tiles

65 mm headlap

40°

190 mm
eaves tile

Ceiling joist

Wall plate

PVC
gutter

**SECTION
THROUGH
EAVES**

Timber
fascia

Timber soffit with
ventilation holes

Tiling cut to
rake of valley

125 mm

Tile
slips

1 : 3 mortar bed

25 mm
min

Untearable
felt

Metal
sheathing

**SECTION THROUGH METAL
OPEN VALLEY**

Bonnet hip tiles

Standard tiles

**PLAN OF HIP WITH
BONNET HIP TILES**

Valley tiles

Cut plain tiling
against edge of
valley tile

Standard tiles

**PLAN OF VALLEY
WITH VALLEY TILES**

Segmental ridge tiles
bedded solid in 1 : 3 mortar
at junctions and along edges

Standard tiles

PLAN OF HIP WITH SEGMENTAL RIDGE TILES

Fig. 8.80   Plain tiling (double lap) details

2. Copper complying with BS 1202 Part 2. These have a high resistance to corrosion, but tend to be soft.
3. Silicon—bronze alloy of 96 per cent copper, 3 per cent silicon and 1 per cent manganese. These have also a high resistance to corrosion, and are much harder than copper.

Figure 8.80 is a typical plain tiling detail sheet showing the construction at ridge, verge, valley and eaves. Note that an underlay of untearable felt or, more commonly now, plastic sheet must be provided under the battens (of the same type used with slates, see above). The underlay is laid parallel to the ridge and each tier should be overlapped 150 mm at horizontal joints (in low sloping roofs) diminishing to 75 mm overlaps for angles of slope above 35 degrees. The underlay will sag slightly between the rafters, which, providing it is not allowed to be too pronounced, will allow any moisture to find its way to the eaves where there should be ample turndown of the felt into the gutter.

The *Building Regulations 1991*, Approved Document L1 requires that the thermal insulation (U) of roofs should be not more than 0.25 W/m^{2}K in houses, flats and maisonettes. The U-value of tiles over felt including the ceiling is only 2.22 W/m^{2}K, and so insulation will be required. In a cold roof construction, where the insulation is laid horizontally above the ceiling, if gaps in the ceiling are blocked and holes for services are sealed, there is little likelihood of interstitial condensation provided that the loft is adequately ventilated. Consequently a vapour control layer is not usually necessary but care must be taken to see that water tanks, etc. in the roof space are insulated.

If, however, the roof space is used as a room and the ceiling and insulation follow the slope of the rafters, interstitial condensation is likely to be a problem because of the restricted volume of the ceiling void. A vapour control layer is consequently required above the ceiling, on the warm side of the insulation, made from 500 gauge polythene, or its equivalent, with sealed gaps.

A roof constructed using steel liner trays will not require a vapour control layer as the metal surface of the trays will act as a vapour barrier (though the upstand joints between them should be taped together to make the barrier continuous, see fig. 8.79). Increasing use is also being made of rigid insulation boards with rebated edges that are designed to lay over the rafters, the battens being nailed through joints between the boards. This forms a truly 'warm' roof construction (see fig. 8.81) where all the main structure of the roof is contained within the thermal envelope. Cold-bridging is eliminated (the U-value of timber not being very favourable by comparison with the levels of insulation that are now required). There is no necessity for either a vapour control membrane or ventilation of voids within the construction because of the high vapour resistance of the insulation and the overlap of the boards at joints.

Counterbattens are required whenever boarding or rigid sheeting is used over rafters. They should be laid on the line of each rafter over the sheeting and the underlay. In this way the tiling battens are raised clear of the underlay by the thickness of the counterbattens (see fig. 8.72) to allow any wind-blown water penetrating the outer roof covering to drain away on the felt into the eaves gutter. Tiling battens, fixed to the correct gauge for the tiles concerned, should be a minimum of 32 × 19 mm when the supporting rafters are spaced at maximum 450 mm centres.

Single lap tiles   The profiled side lap in single lap tiling takes the place of the bond in plain tiling and slating so that protection at the head lap can be reduced to two thicknesses of material. So this type of tiling has a single overlap (double thickness of one tile upon the other). Included within this category is pantiling, a method that is of ancient origin. Many types of single lap tiles are available, examples of which are shown in figs 8.83 and 8.84. Nearly all single lap tiles are interlocking and some have anti-capillary grooves at the head lap of the tile. This makes it possible to lay them on roofs of comparatively shallow pitch. The amount of side lap is determined by the shape of the tile, the head lap may also be predetermined by the profile but should never be less than 75 mm. Certain patterns of single lap tiles can be laid at variable gauge. This can be used to avoid cutting tiles at top courses.

Figure 8.84 shows typical details of concrete tiles with an interlocking side lap. Tiles should be nailed to comply with the wind uplift requirements of BS 5534 but at verges, each side of valleys and hips and at abutments, the end tiles in every course should be nailed; for tiling on steep pitches reference should be made to the British Standard. At the eaves and top edges one course of tiles should be nailed. The eaves course of tiles projects about 50 mm over the edge of the fascia board and the underlay is drawn taut and fixed in this case by a proprietary eaves clip nailed to the top edge of the fascia board. Purpose-made valley tiles on the underlay form the valley detail. The verge can be formed with specially profiled tiles that turn down and mask the top of the wall below and are fixed with concealed chips. The ridge is covered in the example by a segmental ridge tile bedded and pointed in 1:3 cement/sand mortar.

As an alternative to traditional jointing with mortar, which is highly dependent for its success on a good standard of workmanship and site supervision, manufacturers of concrete tiles provide accessories that are dry jointed. Figure 8.82 shows examples of ridge tiles that are mechanically fixed without mortar and which are also designed to incorporate, where necessary, the ventilation requirements of BS 5250.

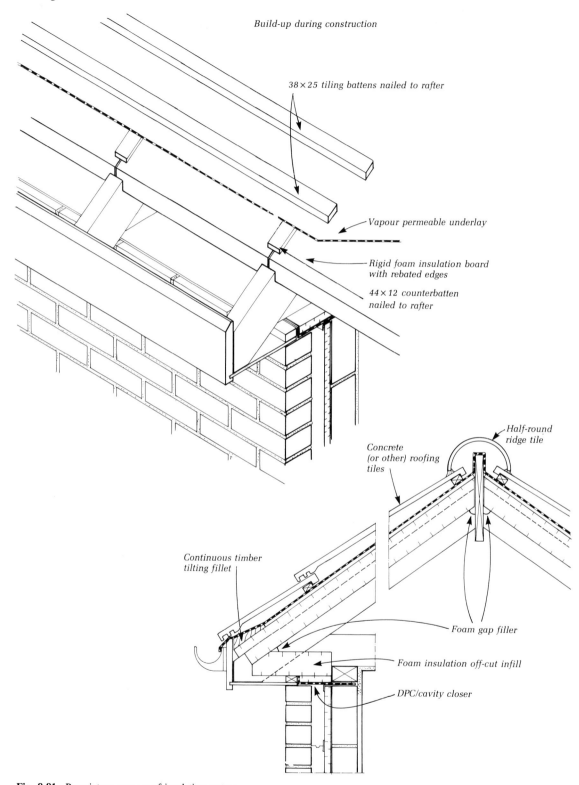

Build-up during construction

38 × 25 tiling battens nailed to rafter

Vapour permeable underlay

Rigid foam insulation board with rebated edges

44 × 12 counterbatten nailed to rafter

Half-round ridge tile

Concrete (or other) roofing tiles

Continuous timber tilting fillet

Foam gap filler

Foam insulation off-cut infill

DPC/cavity closer

Fig. 8.81 Proprietary warm roof insulation system

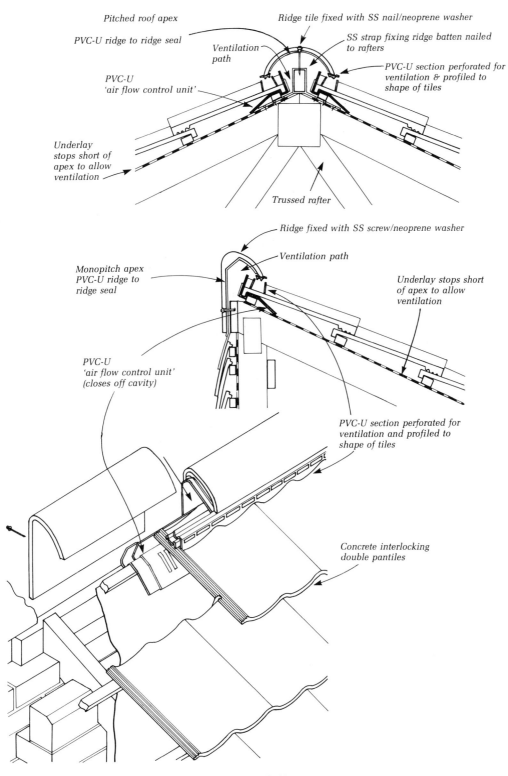

Pitched roof apex

PVC-U ridge to ridge seal

Ventilation path

Ridge tile fixed with SS nail/neoprene washer

SS strap fixing ridge batten nailed to rafters

PVC-U 'air flow control unit'

PVC-U section perforated for ventilation & profiled to shape of tiles

Underlay stops short of apex to allow ventilation

Trussed rafter

Ridge fixed with SS screw/neoprene washer

Ventilation path

Monopitch apex PVC-U ridge to ridge seal

Underlay stops short of apex to allow ventilation

PVC-U 'air flow control unit' (closes off cavity)

PVC-U section perforated for ventilation and profiled to shape of tiles

Concrete interlocking double pantiles

Fig. 8.82   Ventilated ridge dry assembly components (Redland Roof Tiles Ltd)

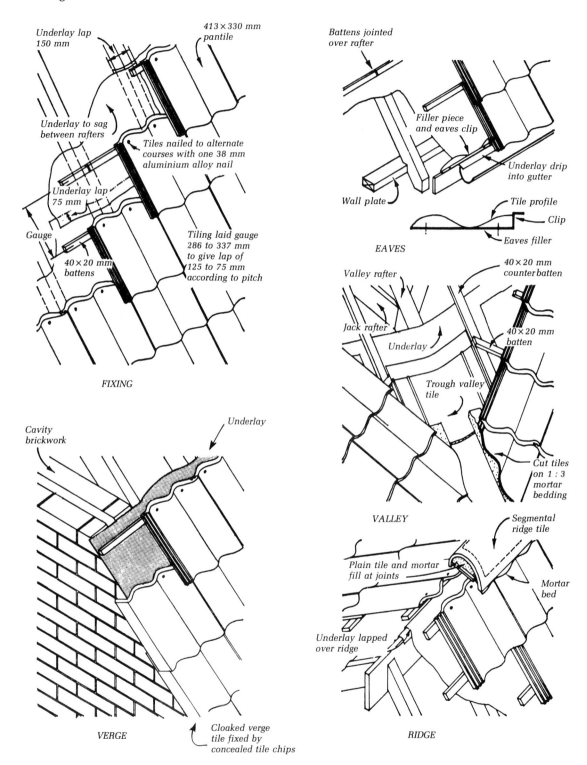

Underlay lap
150 mm

413 × 330 mm
pantile

Underlay to sag
between rafters

Tiles nailed to alternate
courses with one 38 mm
aluminium alloy nail

Underlay lap
75 mm

Gauge

40 × 20 mm
battens

Tiling laid gauge
286 to 337 mm
to give lap of
125 to 75 mm
according to pitch

FIXING

Battens jointed
over rafter

Filler piece
and eaves clip

Underlay drip
into gutter

Wall plate

Tile profile

Clip

Eaves filler

EAVES

Valley rafter

40 × 20 mm
counter batten

Jack rafter

40 × 20 mm
batten

Underlay

Trough valley
tile

Cut tiles
on 1 : 3
mortar
bedding

VALLEY

Cavity
brickwork

Underlay

Segmental
ridge tile

Plain tile and mortar
fill at joints

Mortar
bed

Underlay lapped
over ridge

Cloaked verge
tile fixed by
concealed tile chips

VERGE

RIDGE

Fig. 8.83 Pantile roofing (single lap) details

Figure 8.84 shows details using a single lap concrete tile of simple profile. The neat interlocking detail at the side of the tile allows the adjacent units to lie in the same plane, giving the appearance of slating. Tiles of this and similar pattern on sites of moderate exposure can be laid on pitches down to between $22\frac{1}{2}$ degrees and $17\frac{1}{2}$ degrees. Some single lap clay tiles can be laid to a shallower pitch than the 35 degree minimum recommended in BS 5534 but the BS advises that designers should be satisfied that a pattern of tile has been satisfactorily in use for 15 years or more before specifying its use at a reduced pitch.

The tiles shown can be laid to a variable gauge, so that the head lap can be increased to avoid cut tiles at the ridge. The minimum headlap is 75 mm. The tiles are shown laid with broken joints and this is advisable on roofs of less than 30 degrees pitch. Where tiles are laid on roofs of lower pitch, the wind uplift increases, consequently the extent of fixings becomes more important. On exposed sites, at lower pitches, each tile is secured by a special clip nailed to the back of the batten carrying the course below; see CP 3 chapter V *Loading* Part 2 1972 *Wind loads* and BS 5534: Part 2: 1986. Under these conditions special verge clips are also used as shown on the detail. A special valley tile designed for use at low pitches is used in this example to form the valley gutter as an alternative to a metal open valley. Ridge tiles can either be bedded solid at joints in 1:3 sand and cement mortar and edge bedded along both sides or a dry-jointed ridge can be used as described above.

Figure 8.85 shows the junction of a roof covered with pantiles and a chimney, the joint is made watertight by the use of lead flashings. The flashing at each side of the chimney is in one piece, the upper edge being stepped to follow the coursing of the brickwork. The horizontal edges of the steps are turned about 25 mm into joints of the brickwork and secured by lead wedges and pointed in. The free edge is dressed over the nearest tile roll and down into the pan of the tile beyond the roll. The front of the chimney stack is flashed with a lead apron which is carried down on to the tiles at least 150 mm and dressed to a close fit. The top edge of the apron is turned into a brickwork joint, wedged and pointed in. The flashing to the back of the chimney is prefabricated as short valley gutter with a separate lead cover flashing. A horizontal DPC within the chimney will also be necessary; this can be made of lead, its projecting front edge being dressed down over the apron. For further details of lead work in conjunction with tiles and slated roofs reference should by made to *The Lead Sheet Manual*, Vol. 1: *Lead Sheet Flashings* by the Lead Sheet Association.

For details of dormer windows and skylights in pitched roofs see sections 4.3.2, 5.7 and 5.8.

Figure 8.86 shows a typical detail of a proprietary ridge ventilator which is connected to a flexible pipe within the roof space. The half-round ridge ventilation terminal provides a fairly unobtrusive method for terminating the ductwork from a mechanical extract system or for venting a WC located within a building without the need for an external soil and vent pipe.

8.7.3 Reinforced bitumen felt slates

This method of laying roofing felt has the advantage that free movement at the edges prevents the material from becoming overstressed, blisters cannot form and the life of the felt may be significantly longer than in the conventional application of roofing felt in sheets.

The slates are produced in 1000×336 mm strips of roofing felt to BS 747, reinforced with glass fibre and surfaced with mineral granules, in shades of red, green, grey and brown (fig. 8.87). The leading edge is cut either at centres into rectangular tiles or to decorative tile shapes. They are laid double lapped so only the tile divisions are visible, staggered as in conventional tile roofing. One manufacturer produces copper-faced felt tiles; the potential problem of differential thermal expansion of the copper and felt is alleviated by the small size of the units. Felt slates should not be used below 12 degrees pitch, for they have no inherent strength and must be fully supported and nailed to a preserved, tongued and grooved, boarded or WBP plywood roof deck. They are classified as either BB to BS 476: Part 3 *External fire exposure roof test* (requiring, to comply with the *Building Regulations*, that they are located at least 6 m from a boundary) or CC according to the type of felt, so it is important to obtain a fire test certificate. There is no British Standard which describes their use.

8.7.4 Metal multi-tiles

Metal multi-tiles are pressed from galvanized steel or aluminium sheet into 1330×415 mm units profiled to resemble multiples of conventional clay roofing tiles (fig. 8.87). They can be finished with acrylic resin incorporating a mineral granule surface (grey, terracotta and green), or with a smooth PVF2 or polyester painted surface (blue/black, black and red). They are marketed as complete systems including eaves and gable trim and gutter sections and in one case pressed metal battens for use instead of conventional timber battens.

Tile strips are laid with a single side lap to make staggered joints in adjacent courses. They are fixed through the upper and lower interlocking flanges by self-tapping plastic-headed screws with EPDM washers of the sort used for fixing profiled metal cladding (see fig. 8.51). Coursing is achieved by using half-tile lengths at the verge or by a double side lap every alternate course. The sheets can be

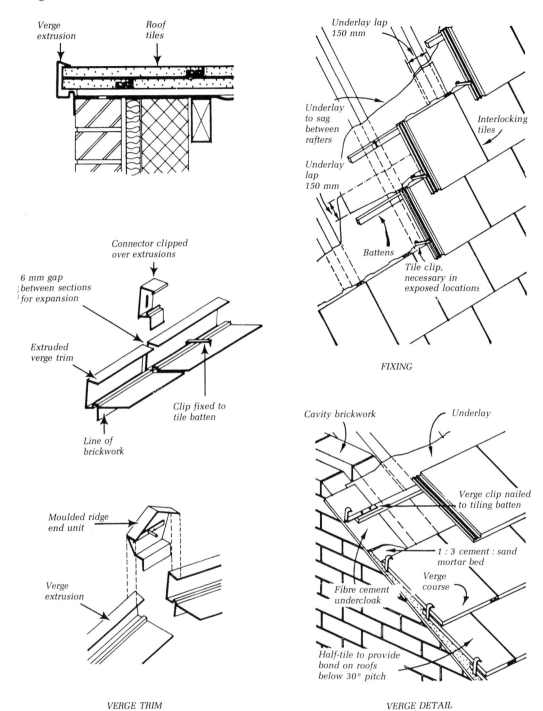

Verge extrusion

Roof tiles

Underlay lap 150 mm

Underlay to sag between rafters

Underlay lap 150 mm

Interlocking tiles

Connector clipped over extrusions

6 mm gap between sections for expansion

Extruded verge trim

Clip fixed to tile batten

Line of brickwork

Battens

Tile clip, necessary in exposed locations

FIXING

Moulded ridge end unit

Verge extrusion

Cavity brickwork

Underlay

Verge clip nailed to tiling batten

1 : 3 cement : sand mortar bed

Verge course

Fibre cement undercloak

Half-tile to provide bond on roofs below 30° pitch

VERGE TRIM

VERGE DETAIL

Fig. 8.84 Interlocking concrete tiling details

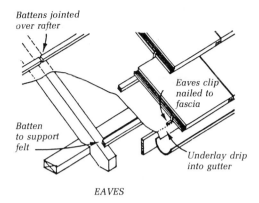

Battens jointed over rafter

Eaves clip nailed to fascia

Batten to support felt

Underlay drip into gutter

EAVES

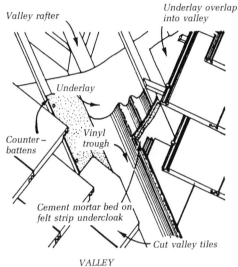

Valley rafter

Underlay overlap into valley

Underlay

Counter-battens

Vinyl trough

Cement mortar bed on felt strip undercloak

Cut valley tiles

VALLEY

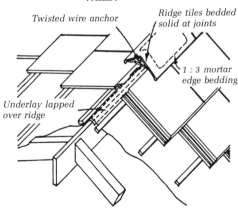

Twisted wire anchor

Ridge tiles bedded solid at joints

1 : 3 mortar edge bedding

Underlay lapped over ridge

RIDGE DETAIL

cut on site with metal shears or a jigsaw. They are lightweight having only one-seventh the weight of concrete tiles and can be laid to a minimum pitch of 12 degrees.

There is no British Standard for metal multi-tiles but they have adequate strength when tested to BS 5534, are classified as AA for fire control purposes, are unaffected by frost, and according to their manufacturers, have an anticipated life of 30 years.

8.7.5 Wood shingles

Neither wood shingles or shakes (which are formed by splitting rather sawing), have attained widespread use in the UK. They are more usual in colder countries; the moderate climate of the UK is ideal for the growth of fungi and insects so even the western red cedar used to make shingles, although it has a natural resistance to decay, needs to be pressure impregnated with copper chrome arsenate preservative to BS 4072. They have no classification for fire control and may not in most cases be used on roofs less than 22 m from a boundary; however, they can be fire-proofed to achieve class 1 spread of flame to BS 476: Part 7: 1971. Being very lightweight, only a tenth of the weight of a tiled or slated roof, rafters and battens can be reduced in size or spaced further apart. A variety of factory-shaped shingles are made but they can readily be cut on site to fit at valleys and hips. The preservative treatment requires that they are fixed with phosphor or silicon—bronze nails.

Properly preserved and installed, shingles have a life expectancy of more than 50 years. There is no British Standard, they are normally specified in the UK to Canadian catagories, the only suitable type, other than for undercoursing, being *Canadian Standards Association no. 1 grade*. Shingles are cut from the blocks of western red cedar in a tapered cross-section, with edges trimmed square and parallel to uniform length of 405 mm. Widths are random from 100 to 350 mm. Hip and ridge cappings are supplied made from two shingles fastened together at an angle. The natural brown colour of the timber weathers to silver grey after a period of months. Western red cedar is the best insulator of all commonly used softwoods and in themselves, shingles are the most insulative of all weather-coat roofing materials.

Shingles should be installed at 14 degrees pitch, according to lap, using conventional pitched roof construction techniques or by nailing through an underfelt into a continuous deck of WBP plywood. All shingle roofing commences from the eaves with a double course, the shingles are then laid as they come from the bundle so each course contains a random variety of widths. Nails are

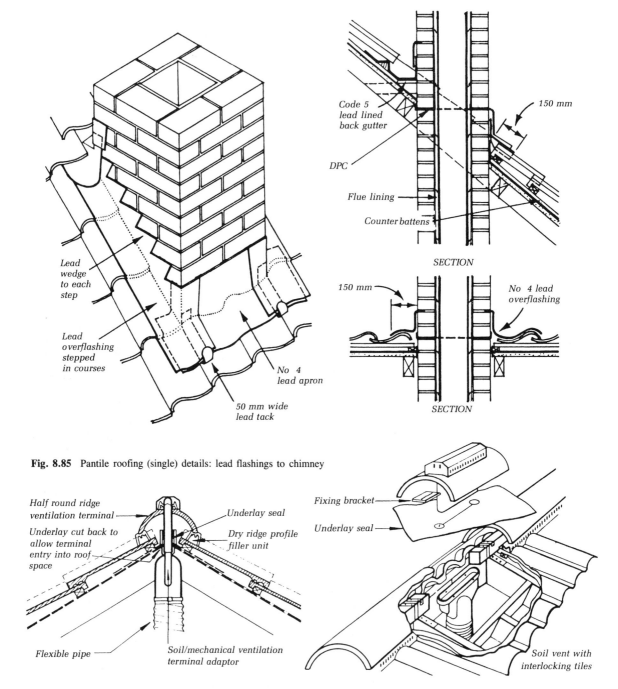

Fig. 8.85 Pantile roofing (single) details: lead flashings to chimney

Fig. 8.86 Interlocking concrete tiling details: proprietary ridge ventilator for WC

positioned at least 20 mm in from the edge of the shingles, pre-drilling is not required as they can easily be nailed without splitting, and cutting at valleys and hips is carried out on site using conventional power tools. A minimum side lap is allowed between the edge of each shingle and the joint between the shingles in the course below. The most common pattern is to run the bottom edge of the shingles in straight horizontal lines, but other patterns can be formed

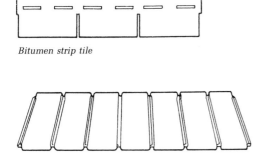

Bitumen strip tile

Metal multi-tile

Fig. 8.87 Bitumen strip tile and metal multi-tile

also using shaped decorative shingles. Correctly laid shingles lie flat and provide good protection from the weather perhaps making underlay unnecessary. However, on sites subject to severe winds and driving rain or if a change of roof pitch is required, underlay should be used.

8.7.6 Thatch

Thatch is a traditional form of roof covering which has gained in popularity in recent years due to the revival of interest in craft-based construction skills. There is no British Standard for thatching.

There are six basic types of thatch:

1. Norfolk reed, which is the common water reed *Phragmites australis* and the most durable natural

Photo 8.12 Hatch Warren Junior School, Basingstoke, Hampshire. Architects: Hampshire County Architects Department; job architect: John Collins.

This roof utilizes a variety of materials according to the different slopes across the section: zinc sheeting to form the 'eyebrow', clay tiled roofing above and patent glazing at the apex.

Table 8.13 Durability of roofing materials

Roofing material	Average life (years)	Overhaul
3 layer bitumen felt	20–30	Additional felt or renew
3 layer bitumen felt with 75 mm shingle	30–40	Additional felt or renew
2 layer mastic asphalt and solar paint	60+	10-year cycle after 20 years
32 g copper	30–40	Renewal
24 g copper	100 or more	Renewal
Lead Roofing		
Code 4	40–60	10-year cycle of overhaul after 40 years
Code 8	100 or more	10-year cycle of overhaul after 80 years
Cast lead	200 or more	St. Pauls' dome lead work survived 250 years
Slating		
New Welsh	80–90	Many slates renewable, copper nails essential
Westmoreland slate	200 or more	Patching and renewal
Cement fibre	30–40	Renewal
Cedar shingles (with preservative treatment reapplied frequently)	40–60	Renewal
Tiling		
Well-burnt fully vitrified clay tiles	Indefinite	Patch and renew where damage occurs
Gas-fired clay tiles	Not known, probably 60	Patch and renew where damage occurs
Concrete tiles	Oldest examples now 70 yrs old	Patch and renew where damage occurs
Corrugated and formed sheet		
Stainless steel	Probably indefinite	Oldest roof; Chrysler Building, New York (1930) cleaned after 50 yrs, found perfect
Galvanized iron	30	Overpaint after 20 years to prolong life to 30 yrs
Coated sheet steel	20 for paint film	Overpaint to prolong life
Coated aluminium sheet	20 for paint film	Overpaint to prolong life

thatching material.

2. Long straw, which is threshed winter straw in common use through the country, but generally has the shortest life expectancy.

3. Combined wheat reed (Devon reed), which is winter wheat straw that is 'combed' rather than threshed but is less readily available owing to modern farming methods.

4. Sedge, which is similar to, but more pliable and less durable than reed, and is used to form ridges on reed thatching.

5. Heather, which comes in much shorter lengths than straw or reed.

6. Bark, which comes from chestnut fencing stakes and can be used as thatching.

7. Synthetic, which is made from polypropylene oil-based yarn, tufted into a woven polypropylene base with a rubber latex backing.

Thatch of conventional materials is laid in bundles, termed bunches or shoves for Norfolk reed, or nitches for combed wheat reed. Roofing commences at the eaves and the first course is tied in place to preserved timber battens with tarred hemp twine. Subsequent courses are held in position with hazel or elm rods (sways) laid across the thatch and held to the rafters or boarded deck by wrought iron hooks. The eaves should tilt upwards, and the pitch of the roof should be a minimum of 45 degrees. In order to protect the thatch from birds, roofs are covered with a netting of galvanized wire. A felt underlay to the thatch reduces the accumulation of debris in the roof spaces when rafters are not boarded.

Thatching has the advantage of being adaptable to different plan forms, lightweight, and providing good thermal insulation and sound control. In addition, gutters and downpipes are not required, and the materials make use of a renewable resource (unless synthetic). Its main disadvantage is in relation to fire control; thatch has no classification within the *Building Regulations 1991* which consequently require that a thatched roof must be more than 22 m from a boundary. For this reason, insurance companies are reluctant to provide a fire policy for a thatch roof covering. The Thatching Advisory Service Ltd recommend the use of a fire-retardant treatment consisting of a chemical water-based solution which is injected by means of a multi-headed spray lance into the thatch. When the chemical dries, the individual straws or reeds of the thatch remain coated with a deposit of the fire-resistant material. However, weathering causes this deposit to wear off at the top of the ridge and the outer surface of the roof; a better method is to dip the bundles in the retardant prior to thatching. It is also possible to install external sparge pipes at the ridge position which will cause the roof to be sprayed with water in the event of a fire. Other precautions can also be taken, such as providing an underlining of a material having a class 0 spread of flame rating, running electric wiring in conduit to BS 31/4568 or MICC to BS 6207 within roof spaces, locating television aerials and cables on masonry, and ensuring properly constructed and maintained chimneys.

SI units

Quantities in this volume are given in SI units which have been adopted by the construction industry in the United Kingdom. Twenty-five other countries (not including the USA or Canada) have also adopted the SI system although several of them retain the old metric system as an alternative. There are six SI basic units. Other units derived from these basic units are rationally related to them and to each other. The international adoption of the SI will remove the necessity for conversions between national systems. The introduction of metric units gives an opportunity for the adoption of modular sizes.

Multiples and sub-multiples of SI units likely to be used in the construction industry are as follows:

Multiplication factor	Prefix	Symbol
1 000 000	10^6 mega	M
1 000	10^3 kilo	k
100	10^2 hecto	h
10	10^1 deca	da
0.1	10^{-1} deci	d
0.01	10^{-2} centi	c
0.001	10^{-3} milli	m
0.000 001	10^{-6} micro	μ

Note: further information concerning metrication is contained in BS PD 6031 *A Guide for the use of the Metric System in the Construction Industry*, and BS 5555: 1976 *SI units and recommendations for the use of their multiples and of certain other units.*

Quantity	Unit	Symbol	Imperial unit × Conversion factor = SI value		
LENGTH	kilometre	km	1 mile	=	1.609 km
	metre	m	1 yard	=	0.914 m
			1 foot	=	0.305 m
	millimetre	mm	1 inch	=	25.4 mm
AREA	square kilometre	km^2	$1\ mile^2$	=	$2.590\ km^2$
	hectare	ha	1 acre	=	0.405 ha
	square metre	m^2	$1\ yard^2$	=	$0.836\ m^2$
			$1\ foot^2$	=	$0.093\ m^2$
	square millimetre	mm^2	$1\ inch^2$	=	$645.16\ mm^2$
VOLUME	cubic metre	m^3	$1\ yard^3$	=	$0.765\ m^3$
			$1\ foot^3$	=	$0.028\ m^3$
	cubic millimetre	mm^3	$1\ inch^3$	=	$1\ 638.7\ mm^3$
CAPACITY	litre	l	1 UK gallon	=	4.546 litres

Quantity	Unit	Symbol	Imperial unit × Conversion factor = SI value		
MASS	kilogramme gramme	kg g	1 lb 1 oz 1 lb/ft(run) 1 lb/ft^2	= = = =	0.454 kg 28.350 g 1.488 kg/m 4.882 kg/m^2
DENSITY	kilogramme per cubic metre	kg/m^3	1 lb/ft^3	=	16.019 kg/m^3
FORCE	newton	N	1 lbf 1 tonf	= = =	4.448 N 9 964.02 N 9.964 kN
PRESSURE, STRESS	newton per square metre meganewton per square metre	N/m^2 MN/m^2† or N/mm^2	1 lbf/in^2 1 tonf/ft^2 1 tonf/in^2 1 lb/ft run 1 lb/ft^2 1 ton/ft run	= = = = = =	6 894.8 N/m^2 107.3 kN/m^2 15.444 MN/m^2 14.593 N/m 47.880 N/m^2 32 682 kN/m
	*bar (0.1 MN/m^2) *hectobar (10 MN/m^2) *millibar (100 MN/m^2)	bar h bar m bar			
VELOCITY	metre per second	m/s	1 mile/h	=	0.447 m/s
FREQUENCY	cycle per second	Hz	1 cycle/sec	=	1Hz
ENERGY, HEAT	joule	J	1 Btu	=	1 055.06 J
POWER, HEAT FLOW RATE	watts newtons metres per second joules per second	W Nm/s J/s	1 Btu/h 1 hp 1 ft/1bf	= = =	0.293 W 746 W 1.356 J

THERMAL
    CONDUCTIVITY (k)
	watts per metre
	degree Celsius
		W/m deg C	1 Btu in/ft^{2}h
			    deg F
				=	0.144 W/m
					deg C

THERMAL
    TRANSMITTANCE (U)
	watts per square metre
	degree Celsius
		W/m^2 deg C	1 Btu/ft^{2}h
			    deg F
				=	5.678
					W/m^2 deg C

TEMPERATURE	degree Celsius (difference)	° C	1° F	=	$\frac{5}{9}$ ° C
	degree Celsius (level)	° C	° F	=	$\frac{2}{5}$ ° C + 32

* Alternative units, allied to the SI, which will be encountered in certain industries
† BSI preferred symbol

CI/SfB

The following information from the *Construction Indexing Manual 1976* is reproduced by courtesy of RIBA Publications Ltd.

Used sensibly and in appropriate detail, as explained in the manual, the CI/SfB system of classification facilitates filing and retrieval of information. It is useful in technical libraries, in specifications and on working drawings. *The National Building Specification* is based on the system, and BRE Digest 172 describes its use for working drawings.

The CI/SfB system comprises tables 0 to 4, tables 1 and 2/3 being the codes in most common use. For libraries, classifications are built up from:

Table 0	Table 1	Table 2/3	Table 4
-a number code	-a number code in brackets	-upper and lower case letter codes	-upper case letter code in brackets
eg 6	eg (6)	eg Fg	eg (F)

An example for clay brickwork in walls is: (21) Fg2, which for trade literature, would be shown in a reference box as:

```
CI/SfB 1976 reference by SfB Agency
    (21)   Fg2

```

The lower space is intended for UDC (Universal decimal classification) codes — see BS 1000A 1961. Advice in classification can be obtained from the SfB Agency UK Ltd at 39 Moreland Street, London EC1V 8BB.

In the following summaries of the five tables, chapter references are made to the six related volumes and chapters of *Mitchell's Building Series* in which aspects of the classifications are dealt with. The following abbreviations are used:

Environment and Services	ES
Materials	M
Structure and Fabric, Part 1	SF (1)
Structure and Fabric, Part 2	SF (2)
External Components	EC
Internal Components	IC
Finishes	F

Table 0 Physical Environment
(main headings only)

Scope: End results of the construction process

0	Planning areas
1	Utilities, civil engineering facilities
2	Industrial facilities
3	Administrative, commercial, protective service facilities
4	Health, welfare facilities
5	Recreational facilities
6	Religious facilities
7	Educational, scientific, information facilities
8	Residential facilities
9	Common facilities, other facilities

Table 1 Elements

Scope: Parts with particular functions which combine to make the facilities in table 0

(0-)	**Sites, projects**
	Building plus external works
	Building systems *IC* 2

(1-)	**Ground, substructure**
(11)	Ground *SF(1)* 4, 8, 11; *SF(2)* 2, 3, 11
(12)	Vacant
(13)	Floor beds *SF(1)* 4, 8; *SF(2)* 3
(14), (15)	Vacant
(16)	Retaining walls, foundations *SF(1)* 4; *SF (2)* 3, 4
(17)	Pile foundations *SF(1)* 4; *SF(2) 3, 11*
(18)	Other substructure elements
(19)	Parts, accessories, cost summary, etc

(2-)	**Structure, primary elements, carcass**
(21)	Walls, external walls *SF(1)* 1, 5; *SF (2)* 4, 5, 10
(22)	Internal walls, partitions *SF(1)* 5; *SF(2)* 4, 10; *IC* 7
(23)	Floors, galleries *SF(1)* 8; *SF(2)* 6, 10
(24)	Stairs, ramps *SF(1)* 10; *SF(2)* 8, 10
(25), (26)	Vacant
(27)	Roofs *SF(1) 1, 7; SF(2)* 9, 10
(28)	Building frames, other primary elements *SF(1)* 1, 6; *SF(2)* 5, 10
	Chimneys *SF(1)* 9
(29)	Parts, accessories, cost summary, etc

(3-)	**Secondary elements, completion of structure**
(31)	Secondary elements to external walls, including windows, doors *SF(1)* 5; *SF(2)* 10; *EC* 2, 4, 5, 7
(32)	Secondary elements to internal walls, partitions including borrowed lights and doors *SF(2)* 10; *EC* 2, 3
(33)	Secondary elements to floors *SF(2)* 10
(34)	Secondary elements to stairs including balustrades *EC* 5
(35)	Suspended ceilings *IC* 8
(36)	Vacant
(37)	Secondary elements to roofs, including roof lights, dormers *SF(1)* 7; *SF(2)* 10; *EC* 4
(38)	Other secondary elements
(39)	Parts, accessories, cost summary, etc.

(4-)	**Finishes to structure**
(41)	Wall finishes, external *SF(2)* 4, 10; *F* 3, 4, 5
(42)	Wall finishes, internal *F* 2, 4, 5
(43)	Floor finishes *F* 1
(44)	Stair finishes *F* 1
(45)	Ceiling finishes *F* 2
(46)	Vacant
(47)	Roof finishes *SF(2)* 10; *F* 7
(48)	Other finishes
(49)	Parts, accessories, cost summary, etc

(5-)	**Services** (mainly piped and ducted)
(51)	Vacant
(52)	Waste disposal, drainage *ES* 13/*ES* 11, 12
(53)	Liquids supply *ES* 9, 10; *SF(1)* 9; *SF(2)* 6, 10
(54)	Gases supply
(55)	Space cooling
(56)	Space heating *ES* 7; *SF(1)* 9; *SF(2)* 6
(57)	Air conditioning, ventilation *ES* 7; *SF(2)* 10
(58)	Other piped, ducted services
(59)	Parts, accessories, cost summary, etc Chimney, shafts, flues, ducts independent *SF(2)* 7

(6-)	**Services** (mainly electrical)
(61)	Electrical supply
(62)	Power *ES* 14
(63)	Lighting *ES* 8
(64)	Communications *ES* 14
(65)	Vacant
(66)	Transport *ES* 15
(67)	Vacant
(68)	Security, control, other services
(69)	Parts, accessories, cost summary, etc

(7-)	**Fittings** with subdivisions (71) to (79)
(74)	Sanitary, hygiene fittings *ES* 10

(8-)	**Loose furniture, equipment** with subdivisions (81) to (89)
	Used where the distinction between loose and fixed fittings, furniture and equipment is important.

(9-)	**External elements, other elements**
(90)	External works, with subdivisions (90.1) to (90.8)
(98)	Other elements
(99)	Parts, accories etc. common to two or more main element divisions (1-) to (7-)
	Cost summary

Note: The SfB Agency UK do not use table 1 in classifying manufacturers' literature

Table 2 Constructions, Forms

Scope: Parts of particular forms which combine to make the elements in table 1. Each is characterised by the main product of which it is made.

A Constructions, forms — used in specification applications for Preliminaries and General conditions

B Vacant — used in specification applications for demolition, underpinning and shoring work

C Excavation and loose fill work

D Vacant

E Cast *in situ* work *M* 8; *SF(1)* 4, 7, 8; *SF(2)* 3, 4, 5, 6, 8, 9

Blocks

F Blockwork, brickwork
Blocks, bricks *M* 6, 12; *SF(1)* 5, 9; *SF(2)* 4, 6, 7

G Large block, panel work
Large blocks, panels *SF(2)* 4

Sections

H Section work
Sections *M* 9, *SF(1)* 5, 6, 7, 8; *SF(2)* 5, 6

I Pipework
Pipes *SF(1)* 9; *SF(2)* 7

J Wire work, mesh work
Wires, meshes

K Quilt work
Quilts

L Flexible sheet work (proofing)
Flexible sheet work (proofing) *M* 9, 11

M Malleable sheet work
Malleable sheets *M* 9

N Rigid sheet overlap work
Rigid sheets for overlappings *SF(2)* 4; *F* 7

P Thick coating work *M* 10, 11; *SF(2)* 4; *F* 1, 2, 3, 7

Q Vacant

R Rigid sheet work
Rigid sheets *M* 3, 12, 13; *SF(2)* 4; *EC* 7

S Rigid tile work
Rigid tiles *M* 4, 12, 13; *F* 1, 4

T Flexible sheet and tile work
Flexible sheets eg carpets, veneers, papers, tiles cut from them *M* 3, 9; *F* 1, 6

U Vacant

V Film coating and impregnation work *F* 6; *M* 2

W Planting work
Plants

X Work with components
Components *SF(1)* 5, 6, 7, 8, 10; *SF(2)* 4; *IC* 5, 6; *EC* 2, 3, 4, 5, 6, 7

Y Formless work
Products

Z Joints, where described separately

Table 3 Materials

Scope: Materials which combine to form the products in table 2

---

a   **Materials**

b,c,d Vacant

Formed materials e to o

e   **Natural stone** *M* 4; *SF(1)* 5, 10; *SF(2)* 4

e1   Granite, basalt, other igneous

e2   Marble

e3   Limestone (other than marble)

e4   Sandstone, gritstone

e5   Slate

e9   Other natural stone

f   **Precast with binder** *M* 8; *SF(1)* 5, 7, 8, 9, 10; *SF(2)* 4 to 9; *F* 1

f1   Sandlime concrete (precast)
Glass fibre reinforced calcium silicate (gres)

f2   All-in aggregate concrete (precast) *M* 8
Heavy concrete (precast) *M* 8
Glass fibre reinforced cement (gre) *M* 10

f3   Terrazzo (precast) *F* 1
Granolithic (precast)
Cast/artificial/reconstructed stone

f4   Lightweight cellular concrete (precast) *M* 8

f5   Lightweight aggregate concrete (precast) *M8*

f6   Asbestos based materials (preformed) *M* 10

f7   Gypsum (preformed) *EC* 2
Glass fibre reinforced gypsum *M* 10

f8   Magnesia materials (preformed)

f9   Other materials precast with binder

g   **Clay (Dried, Fired)** *M* 5; *SF(1)* 5, 9, 10; *SF(2)* 4, 6, 7

g1   Dried clay eg pisé de terre

g2   Fired clay, vitrified clay, ceramics
Unglazed fired clay eg terra cotta

g3   Glazed fired clay eg vitreous china

g6   Refractory materials eg fireclay

g9   Other dried or fired clays

h   **Metal** *M 9; SF(1) 6, 7, SF(2) 4, 5, 7*
h1   Cast iron
      Wrought iron, malleable iron
h2   Steel, mild steel
h3   Steel alloys eg stainless steel
h4   Aluminium, aluminium alloys
h5   Copper
h6   Copper alloys
h7   Zinc
h8   Lead, white metal
h9   Chromium, nickel, gold, other metals, metal alloys

i   **Wood** including wood laminates *M 2, 3; SF(1) 5 to 8, 10; SF(2) 4, 9; EC 2*
i1   timber (unwrot)
i2   Softwood (in general, and wrot)
i3   Hardwood (in general, and wrot)
i4   Wood laminates eg plywood
i5   Wood veneers
i9   Other wood materials, except wood fibre boards, chipboards and wood-wool cement

j   **Vegetable and animal materials** — including fibres and particles and materials made from these
j1   Wood fibres eg building board *M 3*
j2   Paper *M 9, 13*
j3   Vegetable fibres other than wood eg flaxboard *M 3*
j5   Bark, cork
j6   Animal fibres eg hair
j7   Wood particles eg chipboard *M 3*
j8   Wood-wool cement *M 3*
j9   Other vegetable and animal materials

k, l   Vacant

m   **Inorganic fibres**
m1   Mineral wool fibres *M 10; SF(2) 4, 7*
      Glass wool fibres *M 10, 12*
      Ceramic wool fibres
m2   Asbestos wool fibres *M 10*
m9   Other inorganic fibrous materials eg carbon fibres *M 10*

n   **Rubber, plastics, etc**
n1   Asphalt (preformed) *M 11, F 1*
n2   Impregnated fibre and felt eg bituminous felt *M 11; F 7*
n4   Linoleum *F 1*

Synthetic resins n5, n6
n5   Rubbers (elastomers) *M 13*

n6   Plastics, including synthetic fibres *M 13*
      Thermoplastics
      Thermosets
n7   Cellular plastics
n8   Reinforced plastics eg grp, plastics laminates

o   **Glass** *M 12; SF(2) 4; EC 3*
o1   Clear, transparent, plain glass
o2   Translucent glass
o3   Opaque, opal glass
o4   Wired glass
o5   Multiple glazing
o6   Heat absorbing/rejecting glass
      X-ray absorbing/rejecting glass
      Solar control glass
o7   Mirrored glass, 'one-way' glass
      Anti-glare glass
o8   Safety glass, toughened glass
      Laminated glass, security glass, alarm glass
09   Other glass, including, cellular glass

Formless materials p to s
p   **Aggregates, loose fills** *M 8*
p1   Natural fills, aggregates
p2   Artificial aggregates in general
p3   Artificial granular aggregates (light) eg foamed blast furnace slag
p4   Ash eg pulverised fuel ash
p5   Shavings
p6   Powder
p7   Fibres
p9   Other aggregates, loose fills

q   **Lime and cement binders, mortars, concretes**
q1   Lime (calcined limestones), hydrated lime, lime putty, *M 7*
      Lime-sand mix (coarse stuff)
q2   Cement, hydraulic cement eg Portland cement *M 7*
q3   Lime-cement binders *M 15*
q4   Lime-cement-aggregate mixes
      Mortars (ie with fine aggregates) *M 15; SF(2) 4*
      Concretes (ie with fine and/or coarse aggregates) *M 8*
q5   Terrazzo mixes and in general *F 1*
      Granolithic mixes and in general *F 1*
q6   Lightweight, cellular, concrete mixes and in general *M 8*
q9   Other lime-cement-aggregate mixes eg asbestos cement mixes *M 10*

r **Clay, gypsum, magnesia and plastics binders, mortars**
r1 Clay mortar mixes, refractory mortar
r2 Gypsum, gypsum plaster mixes
r3 Magnesia, magnesia mixes *F* 1
r4 Plastics binders
Plastics mortar mixes
r9 Other binders and mortar mixes

s **Bituminous materials** M 11; *SF(2)* 4
s1 Bitumen including natural and petroleum bitumens, tar, pitch, asphalt, lake asphalt
s4 Mastic asphalt (fine or no aggregate), pitch mastic
s5 Clay-bitumen mixes, stone bitumen mixes (coarse aggregate)
Rolled asphalt, macadams
s9 Other bituminous materials

Functional materials t to w
t **Fixing and jointing materials**
t1 Welding materials *M* 9; *SF(2)* 5
t2 Soldering materials *M* 9
t3 Adhesives, bonding materials *M* 14
t4 Joint fillers eg mastics, gaskets *M* 16 *SF(1)* 2
t6 Fasteners, 'builders ironmongery'
Anchoring devices eg plugs
Attachment devices eg connectors *SF(1)* 6, 7
Fixing devices eg bolts, *SF(1)* 5
t7 'Architectural ironmongery' *IC* 7
t9 Other fixing and jointing agents

u **Protective and Process/property modifying materials**
u1 Anti-corrosive materials, treatments *F* 6
Metallic coatings applied by eg electroplating *M* 9
Non-metallic coatings applied by eg chemical conversion
u2 Modifying agents, admixtures eg curing agents *M* 8
Workability aids *M* 8
u3 Materials resisting specials forms of attack such as fungus, insects, condensation *M* 2
u4 Flame retardants if described separately *M* 1
u5 Polishes, seals, surface hardeners *F* 1; *M* 8
u6 Water repellants, if described separately
u9 Other protective and process/property modifying agents eg ultra-violet absorbers

v **Paints** *F* 6
v1 Stopping, fillers, knotting, paint preparation materials including primers
v2 Pigments, dyes, stains
v3 Binders, media eg drying oils
v4 Varnishes, lacquers eg resins
Enamels, glazes
v5 Oil paints, oil-resin paints
Synthetic resin paints
Complete systems including primers
v6 Emulsion paints, where described separately
Synthetic resin-based emulsions
Complete systems including primers
v8 Water paints eg cement paints
v9 Other paints eg metallic paints, paints with aggregates

w **Ancillary materials**
w1 Rust removing agents
w3 Fuels
w4 Water
w5 Acids, alkalis
w6 Fertilisers
w7 Cleaning materials F 1
Abrasives
w8 Explosives
w9 Other ancillary materials eg fungicides

x **Vacant**

y **Composite materials**
Composite materials generally *M* 11
See p. 63 *Construction Indexing Manual*

z **Substances**
z1 By state eg fluids
z2 By chemical composition eg organic
z3 By origin eg naturally occurring or manufactured materials
z9 Other substances

Table 4 Activities, Requirements
(main headings only)

Scope: Table 4 identifies objects which assist or affect constuction but are not incorporated in it, and factors such as activities, requirements, properties, and processes.

Activities, aids
(A) Administration and management activities, aids, *IC* 2; *M* Introduction, *SF(1)* 2; *SF(2)* 1, 2, 3
(B) Construction plant, tools *SF(1)* 2; *SF(2)* 2, 11
(C) Vacant
(D) Construction operations *SF(1)* 2, 11; *SF(2)* 2, 11

Requirements, properties, building science, construction technology

Factors describing buildings, elements, materials, etc

(E) Composition, etc *SF(1)* 1, 2; *SF(2)* 1, 2

(F) Shape, size, etc *SF(1)* 2

(G) Appearance, etc *M* 1; *F* 6

Factors relating to surroundings, occupancy

(H) Context, environment

Performance factors

(J) Mechanics *M* 9; *SF(1)* 3, 4; *SF(2) 3, 4*

(K) Fire, explosion *M* 1; *SF(2)* 10

(L) Matter

(M) Heat, cold *ES* 1

(N) Light, dark *ES* 1

(O) Sound, quiet *ES* 1

(Q) Electricity, magnetism, radiation *ES* 14

(R) Energy, other physical factor *ES* 7

(T) Application

Other factors

(U) Users, resources

(V) Working factors

(W) Operation, maintenance factors

(X) Change, movement, stability factors

(Y) Economic, commercial factors *M* Introduction; *SF(1)* 2; *SF(2)* 3, 4, 5, 6, 9

(Z) Peripheral subjects, form of presentation, time, place — may be used for subjects taken from the UDC (*Universal decimal classification*), see BS 1000A 1961

Subdivision: All table 4 codes are subdivided mainly by numbers

Index

Figures in italics refer to illustrations